EMERGING CARBON MATERIALS FOR CATALYSIS

EMERGING CARBON MATERIALS FOR CATALYSIS

Edited by

SAMAHE SADJADI

ELSEVIER

Elsevier
Radarweg 29, PO Box 211, 1000 AE Amsterdam, Netherlands
The Boulevard, Langford Lane, Kidlington, Oxford OX5 1GB, United Kingdom
50 Hampshire Street, 5th Floor, Cambridge, MA 02139, United States

Library of Congress Cataloging-in-Publication Data
A catalog record for this book is available from the Library of Congress

British Library Cataloguing-in-Publication Data
A catalogue record for this book is available from the British Library

ISBN: 978-0-12-817561-3

For information on all Elsevier publications
visit our website at https://www.elsevier.com/books-and-journals

Publisher: Susan Dennis
Acquisitions Editor: Emily M. McCloskey
Editorial Project Manager: Billie Jean Fernandez
Production Project Manager: Joy Christel Neumarin Honest Thangiah
Cover Designer: Matthew Limbert

Typeset by SPi Global, India

Dedication

To my father, who loved me unconditionally; thanks to him for supporting me as he could read my words in my silence and guided me by all the means at his disposal.

Contents

Contributors

K.S. Adarsh
Electroplating Metal Finishing and Technology Division, CSIR-Central Electrochemical Research Institute, Karaikudi, Tamil Nadu, India

Francisco Alcaide Monterrubio
CIDETEC, Donostia-San Sebastián, Spain

C. Alegre
Instituto de Carboquímica, CSIC, Zaragoza, Spain

M. Bernardo
LAQV/REQUIMTE, Department of Chemistry, Faculty of Science and Technology, NOVA University of Lisbon, Caparica, Portugal

Naveen Chandrasekaran
Electroplating Metal Finishing and Technology Division, CSIR-Central Electrochemical Research Institute, Karaikudi, Tamil Nadu, India

Catherine Collett
Department of Chemical & Biological Engineering, University of Sheffield, Sheffield, United Kingdom

Gareth Davies
Department of Chemical & Biological Engineering, University of Sheffield, Sheffield, United Kingdom

C. Durán-Valle
Department of Organic and Inorganic Chemistry, University of Extremadura, Badajoz, Spain

Ahmed El Sheikh
Department of Chemical & Biological Engineering, University of Sheffield, Sheffield, United Kingdom

I. Fonseca
LAQV/REQUIMTE, Department of Chemistry, Faculty of Science and Technology, NOVA University of Lisbon, Caparica, Portugal

José M. Fraile
Faculty of Sciences, Institute of Chemical Synthesis and Homogeneous Catalysis (ISQCH), CSIC-University of Zaragoza, Zaragoza, Spain

Junkuo Gao
Institute of Fiber Based New Energy Materials, School of Materials Science and Engineering, Zhejiang Sci-Tech University, Hangzhou, China

Gonzalo García
Department of Chemistry, Institute of Materials and Nanotechnology, University of La Laguna, La Laguna, Tenerife, Spain

Enrique García-Bordejé
Institute of Carbochemistry (ICB-CSIC), Zaragoza, Spain

Gunniya Hariyanandam Gunasekar
Clean Energy Research Centre, Korea Institute of Science and Technology, Cheongryang, Seoul, Republic of Korea

Babak Karimi
Department of Chemistry, Institute for Advanced Studies in Basic Sciences (IASBS), Zanjan, Iran

M.J. Lázaro
Instituto de Carboquímica, CSIC, Zaragoza, Spain

I. Matos
LAQV/REQUIMTE, Department of Chemistry, Faculty of Science and Technology, NOVA University of Lisbon, Caparica, Portugal

E. Pérez Mayoral
Department of Inorganic Chemistry and Technical Chemistry, Faculty of Science, National University of Distance Education, UNED, Madrid, Spain

James McGregor
Department of Chemical & Biological Engineering, University of Sheffield, Sheffield, United Kingdom

Elena Pastor
Department of Chemistry, Institute of Materials and Nanotechnology, University of La Laguna, La Laguna, Tenerife, Spain

S. Pérez-Rodríguez
Instituto de Carboquímica, CSIC, Zaragoza, Spain

Elisabet Pires
Faculty of Sciences, Institute of Chemical Synthesis and Homogeneous Catalysis (ISQCH), CSIC-University of Zaragoza, Zaragoza, Spain

Samahe Sadjadi
Gas Conversion Department, Faculty of Petrochemicals, Iran Polymer and Petrochemical Institute, Tehran, Iran

Sodeh Sadjadi
Nuclear Science and Technology Research Institute, Tehran, Iran

D. Sebastián
Instituto de Carboquímica, CSIC, Zaragoza, Spain

Yuhang Wu
Institute of Optoelectronic Materials and Devices, China Jiliang University, Hangzhou, People's Republic of China; Institute of Fiber Based New Energy Materials, School of Materials Science and Engineering, Zhejiang Sci-Tech University, Hangzhou, China

Hui Xu
Institute of Optoelectronic Materials and Devices, China Jiliang University, Hangzhou, People's Republic of China

Ibrahim Yakub
Department of Chemical & Biological Engineering, University of Sheffield, Sheffield, United Kingdom

Sungho Yoon
Department of Chemistry, Chung-Ang University, Dongjak-gu, Seoul, Republic of Korea

Acknowledgment

"The support of Iran Polymer and Petrochemical Institute is appreciated."

CHAPTER 1

New aspects of covalent triazine frameworks in heterogeneous catalysis

Gunniya Hariyanandam Gunasekar[a] and Sungho Yoon[b]
[a]Clean Energy Research Centre, Korea Institute of Science and Technology, Cheongryang, Seoul, Republic of Korea
[b]Department of Chemistry, Chung-Ang University, Dongjak-gu, Seoul, Republic of Korea

1 Introduction

The continuous environmental and economic challenges in the world strongly impulse the chemical industries to develop simple and more efficient chemical processes that utilize environmentally benign catalysts, reactants, solvents, and minimum energy inputs to produce selective products with almost no or minimal wastes. To date, most industrial chemical conversions (>90%) use catalysts at least in a single step to speed up the reaction rate [1,2], and hence one of the most promising strategy would be developing economically simple and environmentally friendly active catalytic systems for various transformations with utmost (100%) selectivity and durability at minimum energy inputs. This, on the other side, currently drives the research on catalysis across chemistry and chemical engineering.

To date, industries mainly use "classic" heterogeneous catalysts for the chemical conversions [3–6], owing to their robust nature, easy catalyst separation, recovery, regeneration and reuse, and their facile practical applicability in continuous operating equipment systems. However, these catalysts usually show lower catalytic efficiency and selectivity and usually require harsh reaction conditions including high temperature and pressure, etc. In addition, these catalysts often have multiple active sites in the catalytic entity, and thus, developing catalyst design strategies for introducing specific active sties with greater uniformity is generally difficult. Hence, numerous trial-and-error experiments are historically required to produce highly active and selective catalytic systems. Such experiments have been mainly limited to altering the particle size of active metals, catalyst support and its acidity/basicity, the use of promoters and alloy formation, etc. [7–9].

Emerging Carbon Materials for Catalysis
https://doi.org/10.1016/B978-0-12-817561-3.00001-9

Therefore, the design and development of single-site well-defined catalysts that enable rapid and selective transformation with easy separation of catalyst/product is still a paramount challenge in the field of catalysis.

In this regard, heterogenized or immobilized catalysts are gaining increasing attention across the scientific and technological society owing to their conceptual viability of having high catalytic activity, selectivity, finely distributed and well-defined active single-sites, and facile catalyst handling and separation [9–15]. With this in mind, substantial effort has been focused to immobilize homogeneous catalysts onto suitable solid supports for procuring maximal activity and stability. There are four common methods that are classified based on the interaction between the catalyst and the solid scaffold for the heterogenization of homogeneous catalysts onto solid support materials: (1) covalent binding [16–20]; (2) electrostatic interaction [21–23]; (3) adsorption [24,25]; and (4) encapsulation [26–29]. Among them, covalent bonding is the most frequently used method for the immobilization of the homogeneous catalysts. For a long time, conventional solid supports such as silica, zeolite, alumina, polyethylene glycol and polystyrene, etc. were applied to anchor homogeneous complexes [30–34]. However, the interest on grafting the complexes on the conventional solid support is gradually fading owing to their low stability and activity and high cost [35]. The main reason for their low stability is the undesirable interaction between the support scaffold and the catalyst active sites, caused frequently by the use of linkers. Therefore, the viability of immobilized catalysts in industrial catalytic transformations has been questioned [35]. Nevertheless, research on realizing this conceptually ideal catalyst is still dynamic, especially owing to the recent emergence of thermally and chemically robust high-surface-area porous materials and novel methods for the immobilization.

For the past two decades, high-surface-area porous solid polymers have been gaining significant interest across diverse research fields including catalysis, gas capture and separation technology, semiconductors, photochemistry, and biology [36,37]. These polymers are broadly classified into metal organic frameworks (MOFs) and porous organic frameworks (POFs). MOFs are generally composed of inorganic metal ions or clusters as building units and organic functional groups as linkers, and they are connected via coordination bonds [38–40]. POFs, on the other hand, are solely constructed from organic units connected via covalent bonds [41–44]. These materials usually possess surface area in the range from a few hundred to several thousands $m^2 g^{-1}$, with uniform and tunable pore sizes from micro- to mesopores. In addition, a wide range of chemical functionalities, including organic functional ligands, can be introduced in the skeleton of these frameworks. Generally, MOFs exhibit poor chemical stability under harsh reaction condition, such as under a highly basic and acidic solution, compared to POFs because of their intrinsic coordination chemical bonds [45–48]. On the other hand, POFs show greater chemical stability because of the strong covalent bonds between their lightweight elements and have, thus, emerged as attractive and effective porous materials, especially in the field of catalysis.

Different types of POFs that are classified based on the structure of the molecular building block have been developed in recent years, including covalent triazine frameworks (CTFs) [49], porous aromatic frameworks (PAFs) [50], covalent organic frameworks (COFs) [51,52], benzimidazole-linked polymers (BILPs) [53,54], polymers of intrinsic porosity (PIMs) [55], hyper-cross-linked polymers (HCPs) [56,57], conjugated microporous polymers (CMPs) [58,59], and porous imine polymers (CIFs) [60].

CTF is one of the most interesting classes of POFs (Fig. 1), receiving intensive limelight in the field of catalysis. They are nitrogen-rich porous polymers constructed using triazine building blocks. They often lack long-range order, but have excellent robust and rigid structures, immense thermal and chemical stability, high acid-base resistivity, large surface area, and tunable pore sizes and structures [61–64]. Contrary to other POFs, the porous properties of CTFs can be easily tuned by varying the CTF synthesis conditions, such as temperature, time, and catalyst (zinc chloride) ratio. Most interestingly, coordinating functional groups incorporated in the skeleton of CTFs can enable anchoring transition metal complexes on the robust and high-surface-area solid supports and generate well-defined porous immobilized metal complexes. Consequently, diffusion of reactants, solvent(s), and product molecules, which plays a key role in heterogeneous catalysis, would be facile and could lead to the activities similar to or better than homogeneous complexes. In addition, the numerous coordinating sites available in the skeleton of CTFs allow the immobilization of a large number of molecular complexes on the support, i.e., number of active site per gram of the support can be higher, which is also important from an industrial viewpoint [65–69]. Finally, the undesirable interactions caused by the use of linkers in conventional immobilization method can be prevented. Hence, CTF-based heterogenized complexes can offer both enhanced activity and stability.

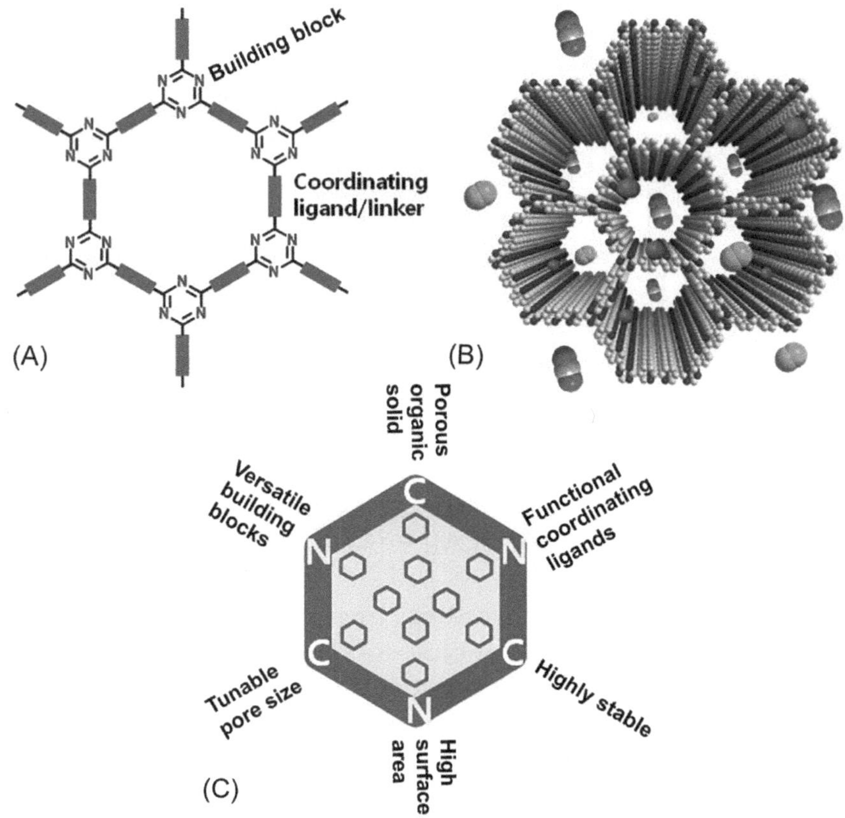

Fig. 1 (A) Basic structure of CTF; (B) Ideal pore networks of CTF; (C) Salient features of CTF.

The formation of cross-linked triazine-based polymer via transition metal-catalyzed trimerization of dinitriles was first reported in 1973 [70]. However, this material gained significant scientific attention in 2008 by Kuhn, Antonietti, and Thomas, who were interested in the synthesis of microporous organic polymers with intrinsic porosity and tailor-made functionalities [61,62]. These researchers discovered CTFs as new class of high performance polymer frameworks with regular and irregular porosity. A variety of aromatic dinitrile compounds were trimerized in their report using $ZnCl_2$ at high temperatures, particularly above the molten temperature of $ZnCl_2$ [61–65]. Inspired by the excellent characters and performances of CTFs, several methods have been developed for the preparation of CTFs; however, the properties of the final products have been strongly influenced by the synthetic process.

To date, CTFs can be prepared through: (1) ionothermal trimerization of carbonitrile groups at temperatures ranging from 300 to 600°C using $ZnCl_2$ as a catalyst and salt melt [61–68]; (2) the Schiff base reaction between melamine with different aldehydes [71–77]; (3) nucleophilic substitution of cyanuric chloride with different nucleophiles [78–83]; (4) the Sonogashira coupling between substituted bromo derivatives of triazine rings with various derivatives of terminal alkynes [84]; (5) the Yamamoto self-coupling reaction of substituted bromo derivatives of triazine rings [85,86]; and (6) the Friedel-Crafts reaction between cyanuric chloride with a variety of electron-rich aromatic compounds [87–89]. The recent reviews published independently by Puthiyaraj et al. [69] and Artz [90] provide detailed information on the synthesis of CTFs.

As stated, the need for highly active, selective, and durable catalysts that withstand a harsh reaction atmosphere is driving scientists to develop thermally and chemically stable solid support materials for the heterogenization of molecular complexes. In this platform, we introduce the aspects of CTFs used for developing well-defined heterogenized catalysts for various catalytic transformations. Here, we limit our discussion to CTFs prepared by ionothermal synthesis because they have varied characters, including robustness and pore rigidity, compared to those prepared by other methods, and most of the heterogenized catalysts employ this synthetic-based CTFs. To date, three classes of coordinating ligands embedded into the CTF skeleton via ionothermal synthesis have been employed as solid chelating ligands for the preparation of heterogenized catalysts: Pyridine, Acetyl acetone, and *N*-heterocyclic carbenes. Therefore, we segmented this chapter according to the coordinating ligands incorporated within CTF.

2 CTF incorporated with pyridinic ligands

The most widely employed CTFs for the construction of CTF-based heterogenized catalysts are pyridinic-based CTFs. There are two kinds of pyridinic ligands-based CTFs that have been constructed and used in the heterogenized catalyst preparation: (1) a CTF constructed using 2,6-dicyanopyridine monomer (Fig. 2), where the metals are expected to coordinate via one pyridinic nitrogen and one triazinic nitrogen; (2) a CTF constructed using 5,5-dicyano-2,2′-bipyridine monomer, where the metals are expected to coordinate via 2,2′-bipyridinic nitrogen. Although both are similar at first glance, their characters including porosity, surface area, electron density, and/or electron-donating abilities are supposedly different.

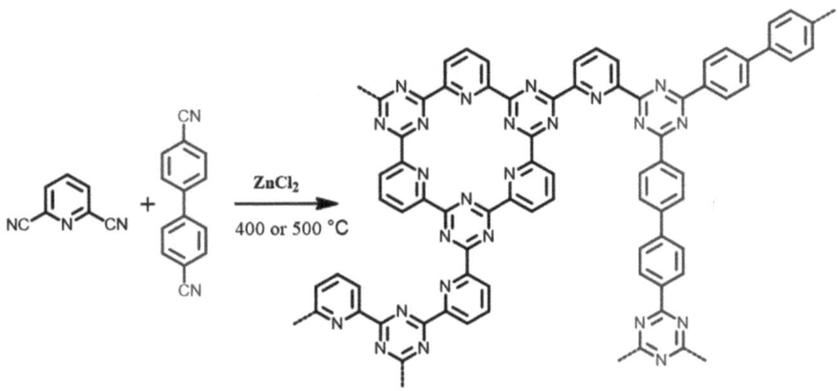

Fig. 2 Route of synthesis of CTF derived from 2,6-dicyanopyridine building block. *(Adapted from A.V. Bavykina, M.G. Goesten, F. Kapteijn, M. Makkee, J. Gascon, Efficient production of hydrogen from formic acid using a covalent triazine framework supported molecular catalyst, ChemSusChem 8 (2015) 809–812, with permission of John Wiley and Sons. R. Palkovits, M. Antonietti, P. Kuhn, A. Thomas, F. Schüth, Solid catalysts for the selective low-temperature oxidation of methane to methanol, Angew. Chem. Int. Ed. 48 (2009) 6909–6912, with permission of John Wiley and Sons.)*

2.1 Pyridinic-CTF derived from 2,6-dicyanopyridine building block

The potential viability of CTFs for the immobilization of molecular catalysts was first demonstrated by Palkovits et al. using 2,6-dicyanopyridine-based CTF for the oxidation of methane to methanol (Fig. 3) [91]. The 2,6-dicyanopyridine-based CTF was prepared in molten $ZnCl_2$ through a stepwise increase of temperature (at 400°C for 40h and then 600°C for 40h). The details of porous properties after Pt immobilization were not provided in that study. A nitrogen binding site of the pyridinic unit and a nitrogen binding site of the triazine unit cooperatively enabled the coordination of Pt via N^N fashion. The resulting complex was structurally similar to the molecular Pt-bipyrimidine complex reported by Periana et al., the commercial application of which was restricted by difficulties in the separation and recycling of this precious metal complex [92]. The immobilized Pt catalyst efficiently oxidized methane into methanol with almost similar activity and selectivity to the molecular catalyst at 200°C in the presence of SO_3 in concentrated sulfuric acid. The exact nature and chemical environment of the Pt sites prior to and after the catalysis were studied using a combination of several sophisticated analytical methods including solid-state ^{195}Pt NMR spectroscopy and aberration-corrected scanning transmission electron microscopy (AC-STEM) [93]. Although the catalytic reaction was performed under harsh reaction

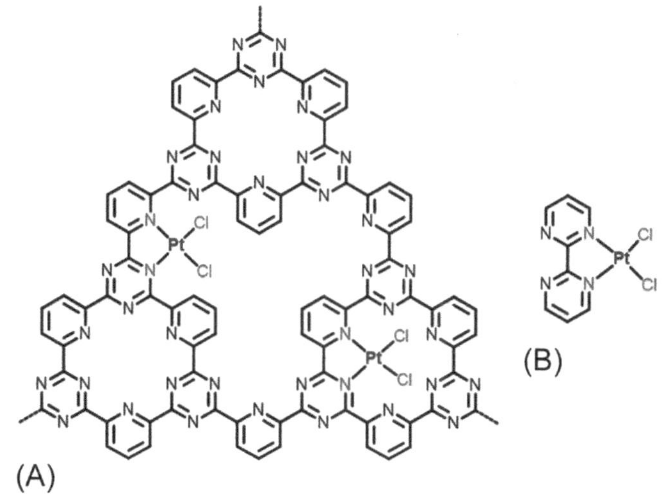

Fig. 3 Representative structure of Pt-CTF (A) and its homogeneous counterpart (Periana Catalyst) (B). *(Reproduced from R. Palkovits, M. Antonietti, P. Kuhn, A. Thomas, F. Schüth, Solid catalysts for the selective low-temperature oxidation of methane to methanol, Angew. Chem. Int. Ed. 48 (2009) 6909–6912, with permission of John Wiley and Sons.)*

conditions, the efficiency of the immobilized Pt catalyst was well-maintained upon successive runs. This indicates that the CTF-based catalyst is thermally and chemically stable, and most importantly, the coordinating ability of the nitrogen species with the metal (Pt) cation in the CTF is remarkably strong.

Inspired by this interesting approach, Bavykina et al. employed a CTF constructed by mixing 2,6-dicyanopyridine and 4,4′-biphenyldicarbonitrile (1:2 ratio) building blocks for the immobilization of IrCp* unit (Cp* = 1,2,3,4,5-pentamethylcyclopentadient) via N^N coordination (Fig. 4A) [94], similar to Pt coordination strategy reported by Palkovits et al. Mixing 4,4′-biphenyldicarbonitrile with 2,6-dicyanopyridine building block may facilitate the diffusion of reactant and product molecules. The immobilized Ir complex, IrCp*@CTF, was employed for the catalytic dehydrogenation of formic acid into CO_2/H_2 under base-free conditions. The catalyst produced initial turnover frequencies (TOFs) of up to 27,000 h^{-1} and turnover numbers (TONs) of up to 1,060,000 during continuous operations at 80°C; this initial TOF was the highest at the time of publication. The authors linked the working capability of IrCp*@CTF in a base-free medium with the inherent basicity of the pyridinic sites present in the CTF matrix. The catalyst was recycled for at least four runs without any significant Ir leaching and changes in the oxidation state of the Ir sites.

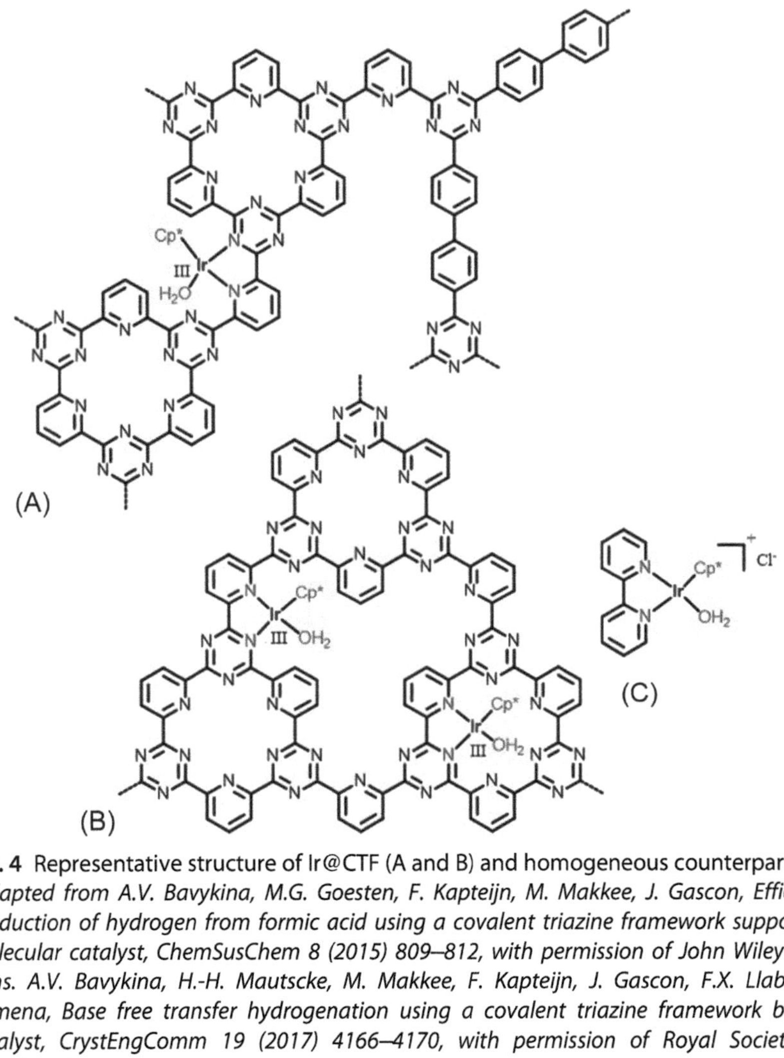

Fig. 4 Representative structure of Ir@CTF (A and B) and homogeneous counterpart (C). *(Adapted from A.V. Bavykina, M.G. Goesten, F. Kapteijn, M. Makkee, J. Gascon, Efficient production of hydrogen from formic acid using a covalent triazine framework supported molecular catalyst, ChemSusChem 8 (2015) 809–812, with permission of John Wiley and Sons. A.V. Bavykina, H.-H. Mautscke, M. Makkee, F. Kapteijn, J. Gascon, F.X. Llabres i Xamena, Base free transfer hydrogenation using a covalent triazine framework based catalyst, CrystEngComm 19 (2017) 4166–4170, with permission of Royal Society of Chemistry.)*

Bavykina et al. further demonstrated the dual role of the CTF as solid support and an intrinsic base in the self-transfer hydrogenation of allylic alcohols to saturated ketones under a base-free medium [95]. At this time, the authors solely employed the 2,6-dicyanopyridine-based CTF (instead of mixed CTF) for the immobilization of IrCp* unit via the aforementioned N^N coordination fashion (Fig. 4B). The immobilized catalyst Ir@CTF exhibited the TOF of 24 min^{-1} at 120°C, which was the best compared to other Ir(III)-based systems at the time. The efficiency of Ir@CTF was maintained over the first three runs, but declined significantly from the fourth cycle onward. The oxidation state of Ir was well-maintained as +3 in the recovered catalyst, and a small amount of Ir leaching was observed in the filtrate. The author attributed this reduced activity mainly to the catalyst deactivation caused by the build-up of adsorbed products on the catalyst surface, which progressively blocks the active sites.

Following this, Rozhko et al. used various $NiBr_2$ catalysts supported on CTFs and CIFs for the oligomerization of ethylene [96]. CTFs prepared using only the 2,6-dicyanopyridine monomer or mixed 4,4′-biphenyldicarbonitrile and 2,6-dicyanopyridine monomers were used as supports (Fig. 5). It has

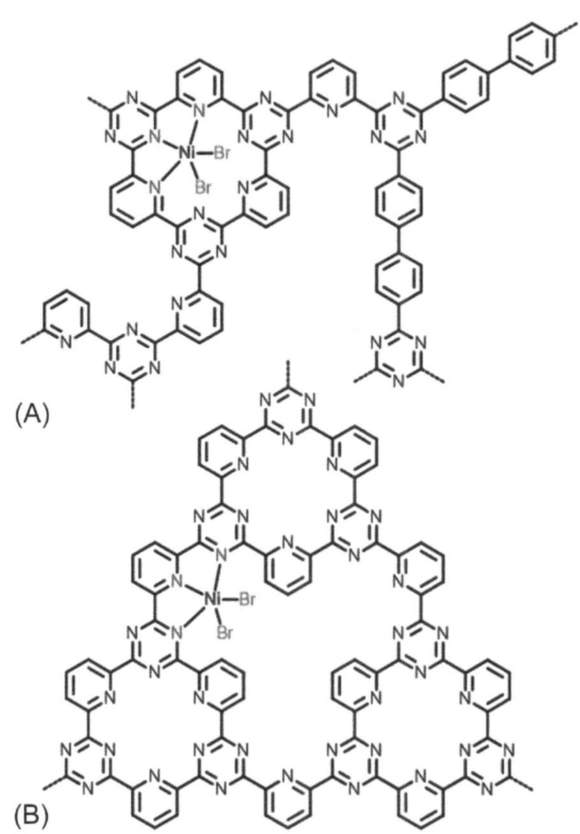

Fig. 5 Representative structure of $NiBr_2$@meso-CTF (A) and $NiBr_2$@micro-CTF (B). *(Reproduced from E. Rozhko, A. Bavykina, D. Osadchii, M. Makkee, J. Gascon, Covalent organic frameworks as supports for a molecular Ni based ethylene oligomerization catalyst for the synthesis of long chain olefins, J. Catal. 345 (2017) 270–280, with permission of Elsevier.)*

been reported that the Ni^{2+} cation was immobilized onto solid supports via tridentate N^N^N coordination fashion. The catalytic reactions were performed in the presence of triethylaluminum as a cocatalyst in a heptane solution. The efficiency of the supported Ni^{2+} catalyst was about one order of magnitude smaller compared to the homogeneous catalyst. It was shown that the activity and selectivity depended on the porosity of the respective support. Upon recycling, the activity and selectivity of the catalysts were dramatically reduced owing to the accumulation of long-chain olefins, which resulted in the blocking of pores and active sites of the catalysts. Nevertheless, the activity was retained for at least five catalytic runs by washing out these long-chain olefins with dichlorobenzene. The authors are currently focusing on the inclusion of a cocatalyst function within these solid scaffolds, as they do not rule out the pore or active site blockage by the adsorption of the cocatalyst.

Recently, Xu et al. anchored a Re carbonyl complex [Re(CO)$_3$Cl] on a 2,6-dicyanopyridine-based CTF (*Re*-CTF-py) via N^N coordination for the photocatalytic CO_2 reduction to CO in a solid-gas system (Fig. 6), owing to the dual characters of nitrogen in CTF as a coordinating ligand and CO_2 absorber [97]. The characteristic peaks in IR spectroscopy confirmed the presence of carbonyl groups in the catalyst. Through photoluminescence (PL) technique and electrochemical impedance spectroscopy (EIS)

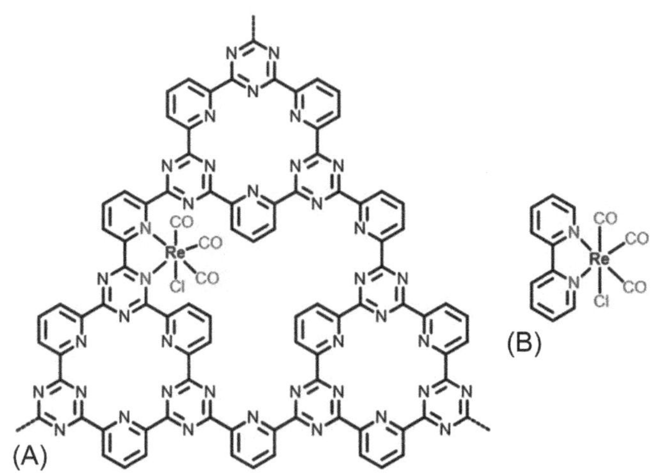

Fig. 6 Representative structure of *Re*-CTF (A) and its homogeneous counterpart (B). *(Adapted from R. Xu, X.S. Wang, H. Zhao, H. Lin, Y.B. Huang, R. Cao, Rhenium-modified porous covalent triazine framework for highly efficient photocatalytic carbon dioxide reduction in a solid–gas system, Catal. Sci. Technol. 8 (2018) 2224–2230. Royal Society of Chemistry.)*

measurements, the authors indicated that the photo-generated electrons on the CTF support could easily be transferred to Re sites through coordination bond between Re and CTF. The Re-CTF-py catalyst efficiently catalyzed the photoreduction with a CO evolution rate of upto 353.05 $\mu mol\,g^{-1}\,h^{-1}$ in 10 h and TON of 4.8 under full-light irradiation in a solid-gas system. Notably, the CO evolution rate was significantly higher than the CTF support itself and the $Re(CO)_5Cl$ precursor. Although a very small fraction of eight-electron-reduction product CH_4 was observed, the catalytic system almost majorly produced two-electron-reduction product CO. The catalyst also exhibited higher stability with good recycling ability compared to the homogeneous *Re*-bipyridine systems. The authors attributed this enhanced stability of the Re-CTF-py to the inaccessibility of the Re species to generate dimerized species during photoreduction, which is the main deactivation pathway for the homogeneous Re-bipyridine catalyst.

In a step closer to practical application of CTFs-based molecular catalysts, Bavykina et al. reported a facile one-step method for producing porous and mechanically rigid CTF-based spheres prepared using polyimide Matrimid as a binder (CTF constructed with 4,4′-biphenyldicarbonitrile and 2,6-dicyanopyridine building blocks was employed in that study) [98]. Although the pores of the CTF were substantially blocked due to polymer chain penetration, mesoporosity was still partially preserved. The fabricated CTF sphere (few millimeters in diameter) enabled the immobilization of the IrCp* complex in N^N coordination. Scanning electron microscopy (SEM) and energy-dispersive spectroscopy (EDS) mapping indicated that most of the Ir species were present within the outer shell of the catalyst spheres. The yielded Ir@CTF spheres were used to catalyze both the hydrogenation of CO_2 to formate and the dehydrogenation of formic acid into CO_2/H_2. Although the Ir@CTF sphere showed relatively diminished efficiency (40%–90%) compared to powdered catalyst, the deviation during the recycling experiments was significantly lower (5%), indicating facile catalyst recycling and the improved reproducibility of the Ir@CTF sphere.

In the continued efforts to overcome the diffusion limitations and handling difficulties of CTF-based catalysts in large-scale production, Bavykina et al. deposited CTF-based stable films on the surface of cordierite monoliths by applying a novel quasi-chemical vapor deposition synthetic protocol [99]. In this method, cordierite monoliths with specified size, channel diameter and wall thickness, etc. were dipped and stirred overnight with a solution of monomer and $ZnCl_2$ in acetone, then dried at 60°C to remove the acetone, and subjected to ionothermal polymerization. Depending on the monomer, either micro- or mesoporous structures of typical of CTFs were used to coat the monolith. (CTF constructed with 4,4′-biphenyldicarbonitrile and 2,6-dicyanopyridine building blocks provides mesoporous structures, and CTF constructed with 2,6-dicyanopyridine building block provides microporous structures). The use of γ-Al_2O_3 as monoliths was unsuccessful, owing to its decomposition during the washing procedure with HCl to remove residual $ZnCl_2$ in the CTF preparation. Similarly, coating on metal plates did not provide good results, as the coating was easily removed from the surface when handling the material due to its poor mechanical strength. Two different homogeneous catalysts were immobilized on the cordierite monoliths: (1) an IrCp*-based catalyst for the production of H_2 by formic acid dehydrogenation, and (2) a Pt(II)-based catalyst for the oxidation of methane to methanol. The metal-modified monolith catalysts presented enhanced catalytic efficiency in both conversions compared to the unsupported CTF powder catalysts. The authors ascribed this enhancement to the improved mass transport of coated catalysts. Although the recycling ability of coated Pt@CTF was not tested in methane oxidation, the efficiency of the coated Ir@CTF remained excellent over five successive runs.

Although CTF-based heterogeneous catalysts were gaining considerable interests since 2009, the first report on CTF-based electrocatalyst was published in 2014 by the groups of Hashimoto and Nakanishi [100]. The poor electrical conductivity of CTFs was the main bottleneck in achieving this goal. These Japanese researchers induced conductivity to the materials by hybridizing CTFs with conductive carbon nanoparticles (CPs) via introducing CPs during the trimerization process of 2,6-dicyanopyridine building block (at a 1:1 weight ratio) (Fig. 7A). Using an approach similar to Palkovits et al., Pt(II) was immobilized on this hybridized CTF via N^N coordination bond. The Pt-modified CTF (Pt-CTF/CP) exhibited clear and selective electrocatalytic activity for oxygen reduction reactions (ORR) in acidic solutions. Unlike conventional carbon-supported Pt catalysts, the Pt-CTF/CP showed almost no activity for methanol oxidation. The high selectivity of this catalyst for ORR with a high methanol tolerance makes it very attractive in direct methanol fuel cell applications. The authors correlated this extraordinary selectivity and methanol tolerance to the presence of solely single-site molecular Pt catalytic centers within the CTF support. Nevertheless, the ORR activity of Pt-CTF/CP was inferior to that of bulk Pt catalysts; this decreased activity of Pt-CTF/CP was attributed to the outcome of excessively strong interaction between the isolated Pt atoms and the intermediate oxygen species.

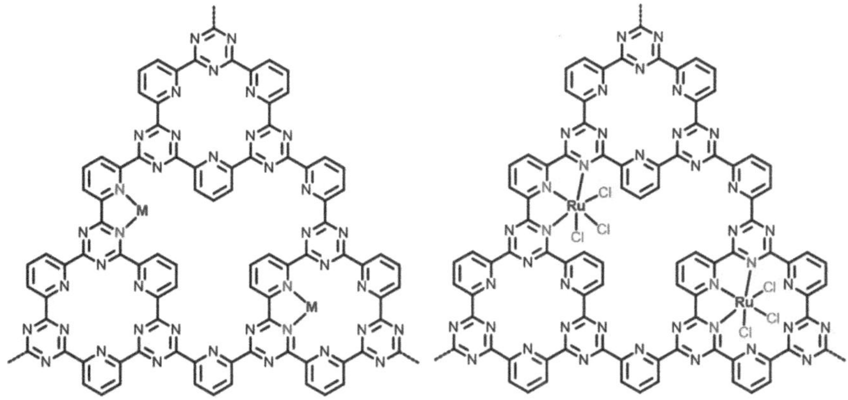

Fig. 7 Representative structures of Pt and/or Cu-based (A) and Ru-based (B) electrocatalysts hybridized with carbon nanoparticles. *(Adapted from K. Kamiya, T. Tatebe, S. Yamamura, K. Iwase, T. Harada, S. Nakanishi, Selective reduction of nitrate by a local cell catalyst composed of metal-doped covalent triazine frameworks, ACS Catal. 8 (2018) 2693–2698, with permission of American Chemical Society. S. Yamaguchi, K. Kamiya, K. Hashimoto, S. Nakanishi, Ru atom-modified covalent triazine framework as a robust electrocatalyst for selective alcohol oxidation in aqueous electrolytes, Chem. Commun. 53 (2017) 10437–10440. Royal Society of Chemistry.)*

The authors also successfully applied the same catalyst for the hydrogen oxidation reaction (HOR) [101]. The catalyst (0.29 wt% Pt) showed superior electrocatalytic HOR activity without requirement of overpotential and exhibited high oxygen tolerance. The authors correlated these behaviors to the dispersion of single Pt atoms throughout the CTF matrix. The HOR activity of Pt-CTF/CP was comparable to the conventional Pt/C (20 wt%) when the Pt loading increased to 2.8 wt%. Most importantly, the selectivity toward ORR was drastically low even in the presence of dissolved oxygen, enhancing PEFC cathode stability during start-up/shut-down cycles of the fuel cells.

In the continued work, Hashimoto and coworkers developed a Cu-modified CTF (Cu-CTF/CP) electrocatalyst for ORR in neutral solutions (Fig. 7A) [102]. To overcome the strong interaction between the isolated Pt atoms in Pt-CTF/CP and the intermediate oxygen species, which decreased the activity of Pt-CTF/CP compared to the bulk Pt catalysts, they anticipated from the study of Cu-N organometallic complexes and bulk Cu metal that atomically dispersed Cu(II) species may have more optimal adsorption energy with oxygen species than Pt. The coordinating strategy and the process followed for synthesis of Pt-CTF/CP was used to prepare Cu-CTF/CP. The Cu-CTF/CP showed the highest onset potential (810 mV vs. RHE at pH7.0) for the ORR among the synthetic Cu-based ORR catalysts at the time of publication. The Cu-CTF/CP electrocatalyst also displayed higher stability than that of a Cu-based molecular complex at neutral pH solutions.

In the later work, the group of Hashimoto and Kamiya applied the Cu-CTF/CP electrocatalyst for the reduction of NO_3^- to N_2O [103]. The catalyst exhibited an onset potential of −50 mV vs. RHE for the reduction, and the faradaic efficiency for N_2O formation was reached 18% at −200 mV vs. RHE. Shortly after understanding the HOR activity of Pt-CTF/CP and the NO_3^- reduction activity of Cu-CTF/CP, Kamiya et al. demonstrated a local cell resulting from the coupling of HOR and NO_3^- reduction on the same conductive substrate using metal-doped covalent triazine frameworks as catalytic units [104]. A conductive carbon substrate loaded with both Pt-CTF/CP and Cu-CTF/CP promotes HOR and NO_3^- reduction, respectively. Nakanishi and coworkers also reported that a Ru-modified CTF had a higher selectivity for electrooxidation of benzyl alcohol over O_2 evolution reaction in water medium (Fig. 7B) [105]. The authors correlated this behavior to the singly isolated Ru atoms in Ru-modified CTF, characterized by TEM and EXAFS analyses. Interestingly, the Ru-modified CTF showed enhanced stability than its structurally related homogeneous Ru complex Ru (tptz)Cl_3, which resulted from the rigid cross-linked network of covalent bonds in CTF.

2.2 Pyridinic-CTFs derived from 5,5-dicyano-2,2′-bipyrdine building block

Yoon and coworkers developed a new strategy for immobilizing the bipyridine complexes, which exactly resembles the structure of homogeneous bipyridine complexes [106–110]. The CTF embedded with 2,2′-bipyridine functional ligand (BPY-CTF), prepared upon ionothermal trimerization of 5,5-dicyano-2,2′-bipyridine succeeded by Hug et al. (Fig. 8) [66], was used as catalytic support for the immobilization of molecular complexes. The ready availability of numerous bipyridine moieties in the porous network and the steric demand created by the monomer unit enabled the facile coordination of metal complexes in a tailored N^N fashion.

Initially, they immobilized the half-sandwich Ir complex (IrCp*(bpy)Cl)Cl onto BPY-CTF (Fig. 9) by treating the Ir precursor ($[IrCp^*Cl_2]_2$) with BPY-CTF in methanol-chloroform solution, for the hydrogenation of CO_2

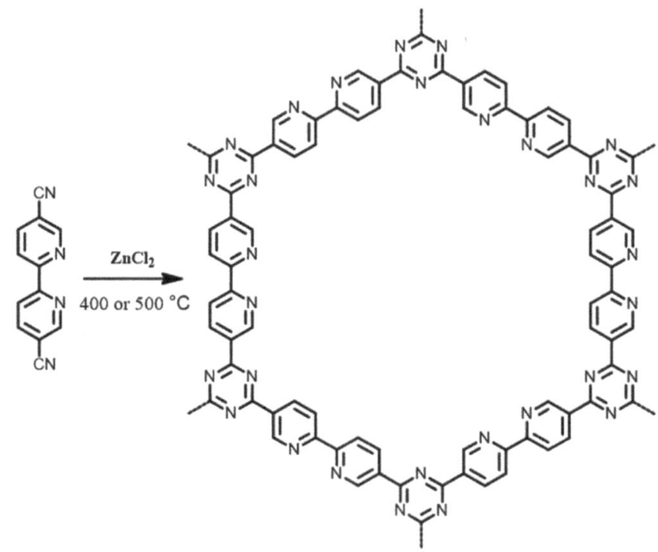

Fig. 8 Synthesis route of CTF derived from 5,5-dicyano-2,2′-bipyridine building block. *(Adapted from S. Hug, M.E. Tauchert, S. Li, U.E. Pachmayr, B.V. Lotsch, A functional triazine framework based on N-heterocyclic building blocks, J. Mater. Chem. 22 (2012) 13956–13964. Royal Society of Chemistry.)*

to formate [106]. The coordination environment of Ir in the heterogenized catalyst, [IrCp*(BPY-CTF)Cl]Cl, was exactly similar to that of its homogeneous analogue, as the authors expected. The catalyst demonstrated the best initial TOF of 5300 h^{-1} at that time and a maximum TON of 5000 in the presence of triethylamine (Et_3N). The efficiency of heterogenized Ir catalyst was slightly reduced during recycling; approximately 92% of the activity was retained after each cycle.

A deep study on the influence of various parameters, including the CTF architecture, the central metal cation, and the metal loadings, was then reported by considering the following CTF characters [107]: (1) CTFs synthesized at 300 or 400°C have the structure of two-dimensional (2D) sheets stacked together via van der Waals forces, which commonly exhibit a laminar architecture with periodic pore structures; (2) Recently, the structural evolution of the 2D structure of CTFs into a three-dimensional (3D) architecture via cross-linking reactions between the stacked sheets was explored using m-CTF by synthesizing the CTF at temperatures exceeding 400°C (The temperature at which the structural evolution occurs may vary according to the aromatic nitrile chosen); (3) The numerous coordinating sites

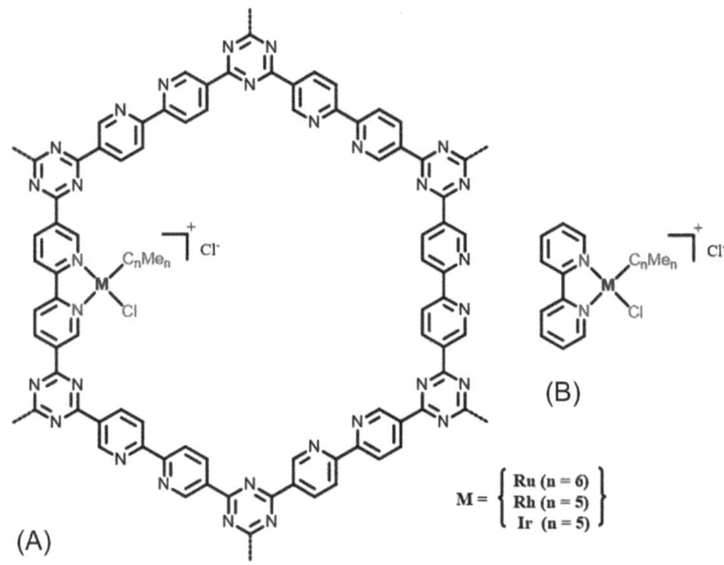

Fig. 9 Representative structures of $[M(C_nMe_n)(BPY\text{-}CTF)Cl]Cl$ (A) and their homogeneous counterparts (B). *(Reproduced from G.H. Gunasekar, K. Park, H. Jeong, K.D. Jung, S. Yoon, Molecular Rh(III) and Ir(III) catalysts immobilized on bipyridine-based covalent triazine frameworks for the hydrogenation of CO_2 to formate, Catalysts 8 (2018) 295. G.H. Gunasekar, J. Shin, K.D. Jung, S. Yoon, Design strategy toward recyclable and highly efficient heterogeneous catalysts for the hydrogenation of CO_2 to formate, ACS Catal. 8 (2018) 4346–4353, with permission of ACS.)*

available in the skeleton of CTF enable an increase in the number of active sites per gram of the support, which is one of the important considerations from the industry viewpoint.

Before examining these CTF-based characters, they studied the effect of a central metal cation on the catalytic activity of CTF-based half-sandwich heterogenized catalysts [107]. Following similar synthetic procedures, Ru and Rh counterparts were prepared (Fig. 9) and characterized to have coordination environment identical to that of their homogeneous analogues. The heterogenized Ru and Rh complexes showed lower TOFs (2640 and 960 h^{-1}, respectively) compared to the heterogenized Ir catalyst. Since CTFs can be prepared with different dimensional and physical properties by altering the trimerization temperature, as stated, the catalytic performance of BPY-CTFs prepared at 400 and 500°C for CO_2 hydrogenation was examined. The efficiency of the Ir catalyst immobilized on BPY-CTF-prepared at 500°C was significantly lower (TOF = 1360 h^{-1}) compared to BPY-CTF-prepared at 400°C, although the surface area/pore volume was remarkably higher for the former catalyst. Metal loading studies on 2D BPY-CTF revealed that a critical balance between the CTF porosity and metal loadings must be reached to obtain the desired structure of complexes; if the metal loadings cross the balance, the generation of metal (Ir) nanoparticles with zero-valent state might be possible due to the unavailability of CTF pores.

Finally, the reduced catalytic efficiency of heterogenized catalysts was reexamined. ICP-AES analysis of the recovered catalyst and filtrate of [IrCp*(BPY-CTF)Cl]Cl revealed that about 6% of Ir was leached into the solution. Nevertheless, the Ir that remained intact in BPY-CTF maintained its coordination environment after the catalysis. Similar to the heterogenized Ir catalyst, the efficiency of Ru and Rh catalysts decreased (maintained only 88 and 65% of the activity in each cycles for Ru and Rh, respectively) upon successive runs. These results suggested that the leaching pathway might be the same for all three catalysts, albeit with varied range. The authors studied the plausible leaching pathway of Ir in [IrCp*(BPY-CTF)Cl]Cl by density functional theory (DFT) calculations and suggested that the undesirable interaction of H_2 with bpy N-sites during H_2 heterolysis causes Ir leaching.

Considering this leaching pathway, Yoon and coworkers proposed a new strategy that introduces oxyanionic ligand(s) in the coordination sphere to overcome this drawback (Fig. 10) [108]. This oxyanionic ligand(s) deterred the undesirable interactions between H_2 and BPY N-site and activated the heterolysis of H_2 simultaneously. An unprecedented initial TOF of 22,700 h^{-1} with the highest formate concentration of 1.8 M in 3 h was obtained at 120°C and 8 MPa total pressure using the tailored Ru catalyst [BPY-CTF-Ru(III)(acac)$_2$]Cl. Most importantly, the efficiency of the catalyst was excellently maintained in both the 1 and 3 M Et_3N solutions upon consecutive runs. XPS analysis of the recovered catalyst revealed that the coordination environment of Ru was slightly altered. DFT calculations suggested that the coordinated carbonate anion, generated by the reaction of CO_2 with H_2O in the presence Et_3N and Ru center, actually formed a six-membered transition state with H_2 during heterolysis and prevented the deleterious interaction between H_2 and the BPY N-site.

In the continuous efforts on developing CTF-based heterogenized catalysts, Yoon and coworkers parallely applied the immobilized half-sandwich Rh and Ir catalysts for the transfer hydrogenation of carbonyl compounds to alcohols [109]. Both catalysts efficiently reduced the carbonyl compounds with the help of cheap and environmentally benign formic acid/formate hydrogen source in water medium at pH 3.5. In contrast to CO_2

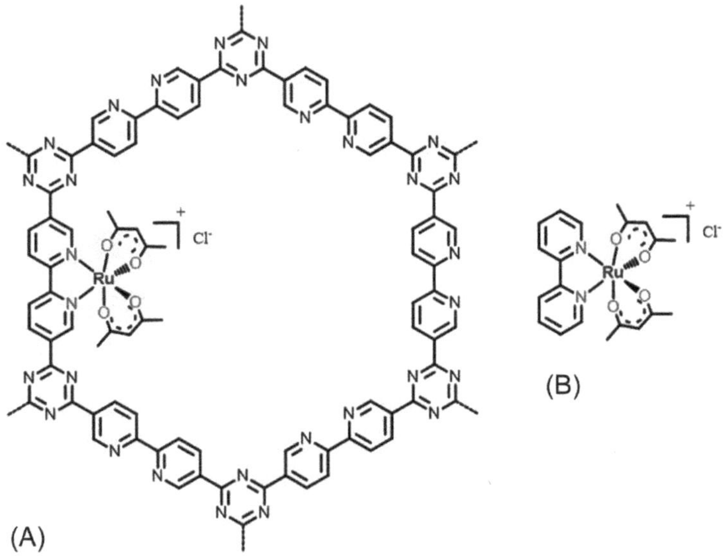

Fig. 10 Representative structure of [BPY-CTF-Ru(III)(acac)$_2$]Cl (A) and its homogeneous counterpart (B). *(Reproduced from G.H. Gunasekar, J. Shin, K.D. Jung, S. Yoon, Design strategy toward recyclable and highly efficient heterogeneous catalysts for the hydrogenation of CO_2 to formate, ACS Catal. 8 (2018) 4346–4353, with permission of American Chemical Society.)*

hydrogenation chemistry, the immobilized Rh catalyst outperformed with the Ir catalyst. Nevertheless, the catalytic efficiency of the Ir catalyst was greatly maintained over four consecutive runs than the Rh catalyst.

Yoon and coworkers also developed CTF-based heterogenized catalysts for the ring-expansion carbonylation of propylene oxide (PO) to butyrolactone [110]. The homogeneous N_2O_2 coordination type [(salph)Al(THF)$_2$][Co(CO)$_4$] complex is one of the most famous catalyst in ring-expansive carbonylation chemistry and it was reported by Coates et al. in 2002 [111]. Although conversions above 95% were achieved, the frustrating suitability of this bimetallic catalyst in commercial applications drove the authors to develop heterogenized counterparts for this conversion. The coordination of Al(OTf)$_3$ onto N^N sites of BPY in BPY-CTF provided a N_2O_2 coordination environment for Al, and exchange of OTf^- with KCo(CO)$_4$ generated the structurally similar heterogeneous [BPY-CTF-Al(OTf)$_2$][Co(CO)$_4$] catalyst (Fig. 11), which was confirmed through XPS analyses. The [BPY-CTF-Al(OTf)$_2$][Co(CO)$_4$] catalyst effectively carbonylated PO with >99% conversion and 90% selectivity toward

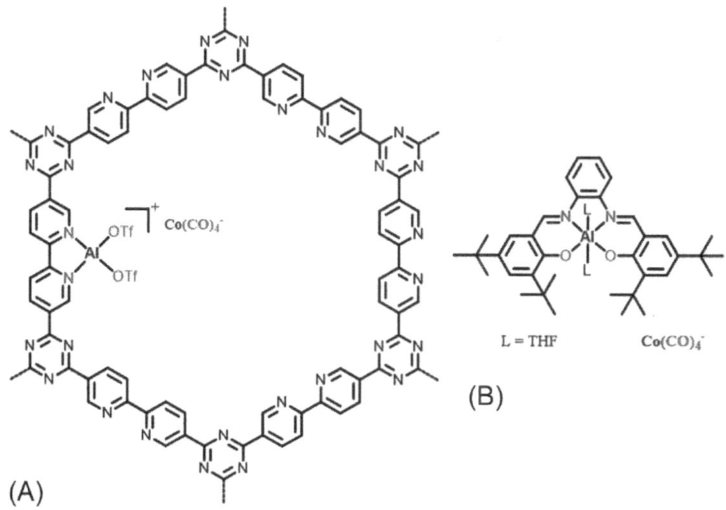

Fig. 11 Representative structure of [bpy-CTF-Al(OTf)$_2$][Co(CO)$_4$] (A) and its structurally similar analogue (B). *(Adapted from S. Rajendiran, P. Natarajan, S. Yoon, A covalent triazine framework-based heterogenized Al–Co bimetallic catalyst for the ring-expansion carbonylation of epoxide to β-lactone, RSC Adv. 7 (2017) 4635–4638. Royal Society of Chemistry.)*

butyrolactone. Nevertheless, this catalytic efficiency was lower compared to that achieved by the homogeneous [(salph)Al(THF)$_2$][Co(CO)$_4$]. Recycling experiments showed that the activity and selectivity decreased during recycling (about 10% decrement after each run); however, upon regeneration with K[Co(CO)$_4$] precursor, the efficiency and the selectivity were restored.

3 CTF incorporated with *N*-heterocyclic imidazolium (carbene) ligands

Although pyridine-functionalized CTFs have been widely applied for the construction of heterogenized pyridine-ligated molecular complexes, to date, a vast number of complexes apart from pyridine-ligated complexes have been realized in the field of coordination chemistry. The electron sharing characters of those ligands are entirely different from the pyridine-based ligands [112,113]. *N*-Heterocyclic Carbene (NHC) is one of such ligands and very famous for its strong σ-donating and poor π-accepting characteristics [114,115]. Unfortunately, incorporating such organic functional ligands in the network of CTFs is scarcely designed and synthesized.

3.1 NHC-CTF constructed with mono-imidazolium (carbene) ligand

With the above considerations, Yoon and coworkers, in 2017, prepared a new CTF with imidazolium ligands (bpim-CTF) via trimerization of 1,3-bis(pyridyl) imidazolium cation using $ZnCl_2$ as a salt melt and a catalyst (Fig. 12) [116]. Similar to other CTFs, the physiochemical properties of bpim-CTF were tuned by changing the trimerization temperature. The prepared CTF showed excellent thermal stability and CO_2 capture ability.

In the continued efforts on developing highly efficient CTF-based CO_2 hydrogenation catalyst, Yoon and coworkers demonstrated that the numerous NHC structural units available in the above-prepared bpim-CTF enabled the N^C coordination of the {IrCp*} unit in the presence of a base, to form the half-sandwich Ir(III) NHC complex (Fig. 13) [117]. The N^C coordination of Ir was confirmed through XPS studies in comparison to its homogeneous IrCp*NHC counterpart. Excellent values for the initial TOF (up to 16,000 h^{-1}) and TONs (up to 24,300 in 15 h) were obtained at 120°C and 8 MPa equimolar CO_2/H_2 pressure. The high activity was attributed to enhanced electron density at the Ir(III) cation due to strong σ-donating and weak π-accepting characters of the NHC ligands within the framework.

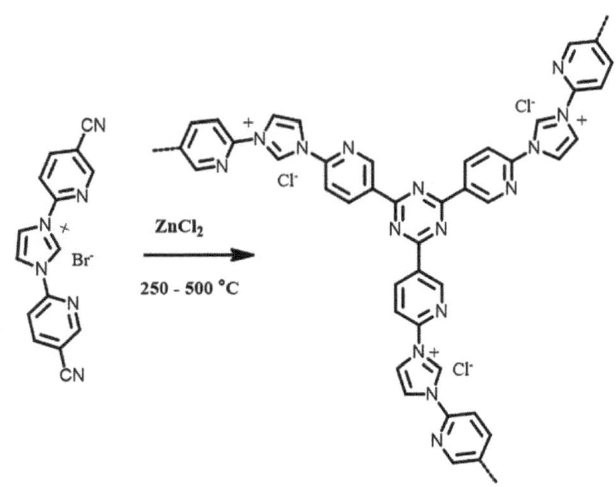

Fig. 12 Synthetic route of NHC-CTF constructed with mono-carbene (imidazolium) ligand. *(Adapted from K. Park, K. Lee, H. Kim, V. Ganesan, K. Cho, S.K. Jeong, S. Yoon, Preparation of covalent triazine frameworks with imidazolium cations embedded in basic sites and their application for CO_2 capture, J. Mater. Chem. A 5 (2017) 8576–8582. Royal Society of Chemistry.)*

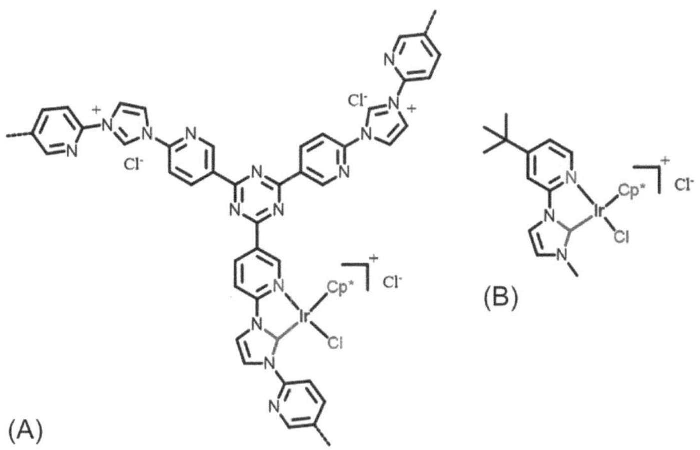

Fig. 13 Representative structure of [Ir-NHC-CTF] (A) and its homogeneous analogue (B). *(Reproduced from G.H. Gunasekar, K. Park, V. Ganesan, K. Lee, N.-K. Kim, K.-D. Jung, S. Yoon, A covalent triazine framework, functionalized with Ir-N-Heterocyclic carbene sites for the efficient hydrogenation of CO_2 to formate, Chem. Mater. 29 (2017) 6740–6748, with permission of American Chemical Society.)*

Similar to the observation of half-sandwich BPY complexes, the activity upon recycling decreased slightly and Ir leaching was observed in the filtrate.

They also explored the potential of the embedded cationic imidazolium unit in bpim-CTF for the generation of ionic liquids within CTFs, through exchange of halide anions with cobaltate $[Co(CO)_4]^-$ anion (Fig. 14) [118]. The generation of ionic liquid within the CTF and the structural similarity to the homogeneous imidazolium cobaltate [Bmim]$[Co(CO)_4]$ were confirmed by XPS analysis. This CTF-based cobaltate ionic liquid was employed as a catalyst for the direct ring-opening carbonylation of PO to methyl 3-hydroxybutyrate (MHB). The highest selectivity (86%) toward MHB was achieved, and comparable conversions (>99%) to the homogeneous counterpart were observed, albeit with the use of high temperature and pressure for a long time. Recycling studies revealed that the catalyst could be used for at least five runs with slight decrease in conversion and selectivity. Nevertheless, the activity and selectivity can be restored upon regeneration with $K[Co(CO)_4]$ precursor.

Very recently, Rh(I)-based CTF catalysts were developed for methanol carbonylation. Both BPY-CTF and bpim-CTF were used as supports for the carbonylation (Fig. 15) [119]. The Rh catalysts were prepared by treating the Rh(I) precursor ($Rh_2(CO)_4I_2$) with CTF supports in methanol solution under a CO atmosphere at room temperature. The Rh catalyst

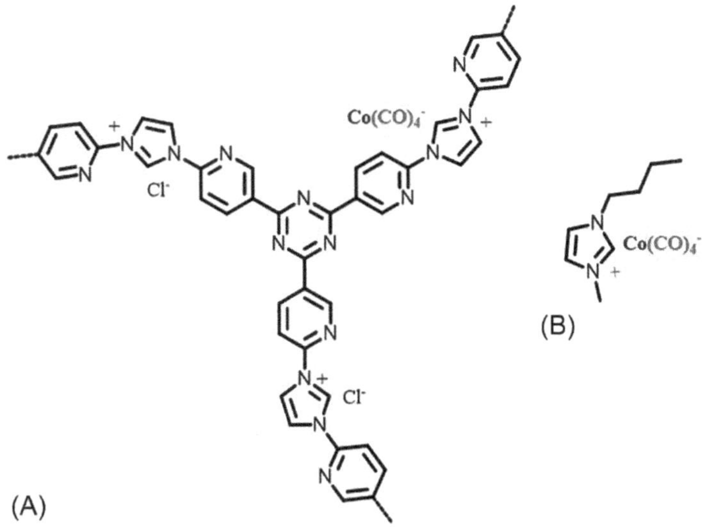

Fig. 14 Structure of [imidazolium-CTF][$Co(CO)_4$] (A) and its homogeneous counterpart (B). *(Reproduced from S. Rajendiran, K. Park, K. Lee, S. Yoon, Ionic-liquid-based heterogeneous covalent triazine framework cobalt catalyst for the direct synthesis of methyl 3-hydroxybutyrate from propylene oxide, Inorg. Chem. 56 (2017) 7270–7277, with permission of American Chemical Society.)*

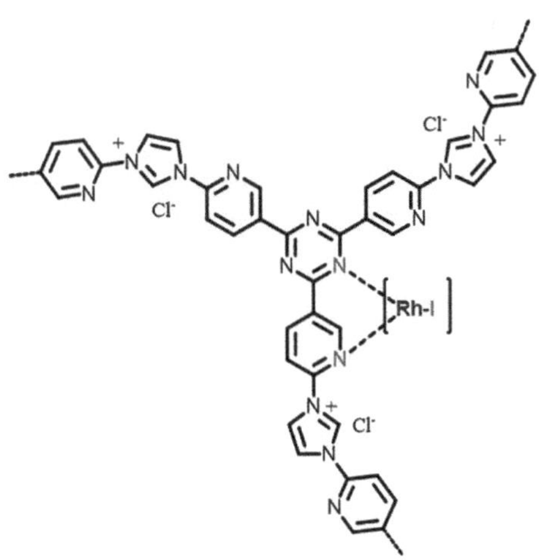

Fig. 15 Structural representation of Rh-bpim-CTF. *(Adapted from K. Park, S. Lim, J.H. Baik, K.D. Jung, S. Yoon, Exceptionally stable Rh-based molecular catalyst heterogenized on a cationically charged covalent triazine framework support for efficient methanol carbonylation, Catal. Sci. Technol. 8 (2018) 2894–2900. Royal Society of Chemistry.)*

immobilized on bpim-CTF demonstrated excellent catalytic conversion of methanol at 93%, corresponding to a TOF of 2100 h^{-1} and a productivity of 124.4 mol kg^{-1} h^{-1} at 240°C under 1 MPa pressure, and a W/fm ratio of 7.6 g h mol^{-1}. The selectivity, productivity, and TOF increased while performing the carbonylation with reduced contact time (W/fm ratio of 1.9 g h mol^{-1}). In addition, bpim-CTF exhibited an outstanding stability over 40 h of carbonylation; conversely, the BPY-CTF-based catalyst showed slight deactivation over time. The main reason for the long-term stability of imidazolium-based CTF may be attributed to the strong ion-pair interaction between the in situ-generated Rh center and support. Both TEM and HAADF-STEM analyses confirmed that there was no generation of Rh nanoparticles either prior to or after catalysis.

3.2 NHC-CTF constructed with bis-imidazolium ligand

In their very recent publication, Yoon and coworkers incorporated bis-imidazolium cations in CTF for enhancing the stability of cobaltate $[Co(CO)_4]^-$ anion during PO carbonylation (Fig. 16) [120]. The bis-imidazolium CTF was prepared through ionothermal trimerization of bis-imidazolium monomer with $ZnCl_2$ at 400°C. Structural intactness of triazine and imidazolium moieties was confirmed by various techniques including XPS and IR analyses. Immobilization of the $[Co(CO)_4]^-$ anion onto CTF was performed by following the previously reported standard procedure, and the prepared catalyst was characterized using SEM-EDS, CHN, and XPS analyses. The catalyst efficiently converted a variety of epoxides into

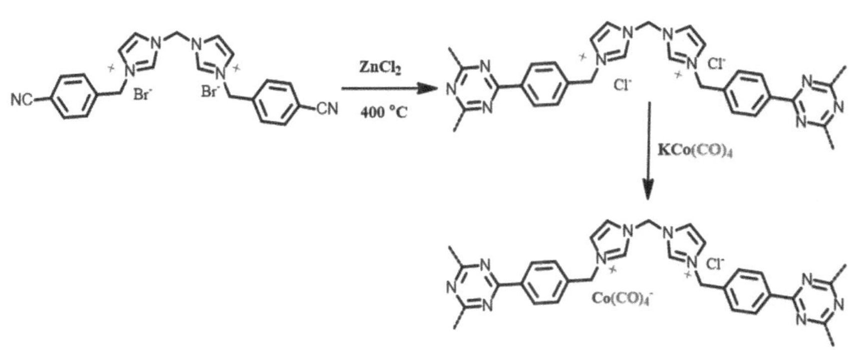

Fig. 16 Synthetic route of [Bis-imidazolium-CTF] $[Co(CO)_4]$. *(Adapted from S. Rajendiran, G.H. Gunasekar, S. Yoon, A heterogenized cobaltate catalyst on a bis-imidazolium-based covalent triazine framework for hydroesterification of epoxides, New J. Chem. 42 (2018) 12256–12262. Royal Society of Chemistry.)*

β-hydroxyesters with good conversions and selectivities. The efficiency of the catalyst during recycling was well-maintained up to five cycles, compared to mono-imidazolium-based catalyst. The improved activity and stability may originate from the efficient activation of epoxides and prevention of cobaltate leaching through strong intramolecular anion stabilization, respectively.

4 CTF incorporated with acetyl acetone ligand

In designing new CTFs with different active binding sites, Jena et al. prepared a CTF with acetylacetonate group (acac) (Fig. 17) [121]. 4,4′-

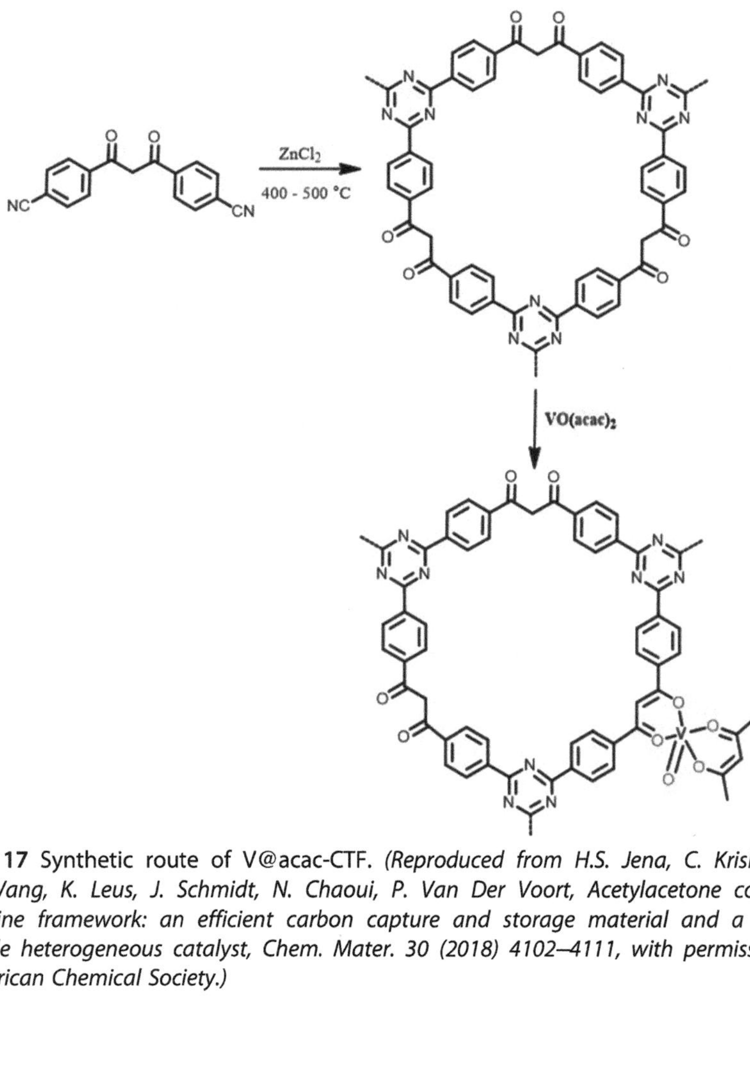

Fig. 17 Synthetic route of V@acac-CTF. *(Reproduced from H.S. Jena, C. Krishnaraj, G. Wang, K. Leus, J. Schmidt, N. Chaoui, P. Van Der Voort, Acetylacetone covalent triazine framework: an efficient carbon capture and storage material and a highly stable heterogeneous catalyst, Chem. Mater. 30 (2018) 4102–4111, with permission of American Chemical Society.)*

malonyldibenzonitrile was subjected to ionothermal conditions to prepare acac-CTF. CTFs synthesized at different temperature and $ZnCl_2$ ratios showed varied physicochemical properties, as reported previously. The existence of the acac group in the CTF was confirmed by IR analysis. The synthesized CTFs were stable up to 500°C, as confirmed by TGA analyses. The acac-CTF synthesized at 500°C using 5 equivalents of $ZnCl_2$ showed both the best CO_2 uptake at 1 bar (at 273 K = 3.30 mmol/g and 298 K = 1.97 mmol/g) and the highest H_2 storage uptake (1.53 wt%) at 77 K and 1 bar. To explore the viability of acac-CTF as support, $VO(acac)_2$ was anchored through an O^O coordination bond with acac-CTF. The prepared catalyst was thoroughly characterized by XPS, IR, ^{13}C-CP-MAS-ss-NMR analysis to confirm that V(IV) was coordinated solely via the O^O binding sites. The V(IV)-based catalyst was highly active compared to homogeneous $VO(acac)_2$ catalyst in the Mannich reaction, and in fact, it effectively catalyzed numerous substrates, even where homogeneous catalyst failed. The authors also mentioned that the catalyst retained similar reactivity over at least four reaction cycles.

5 Conclusions

After a decade of research, CTF has become one of the most important catalyst support materials in the field of catalysis owing to their rigid pore structures, good stability over a wide range of pH and temperature, resistance toward solvent swelling, and robust characters with various chemical functionalities including H_2 gas, etc. The solid chelating ligands embedded in CTFs enable formation of strong coordination bond(s) with metal precursors and allow construction of structurally identical catalytic entities with homogeneous catalysts. The efficiency of CTF-based catalysts is more or less similar to the homogeneous catalysts in many transformations, and the isolation of catalysts from the reaction medium is facile by simple filtration. The durability of CTF-based heterogenized catalysts is highly encouraging and remarkable for many conversions, and most specifically, it is close to practical realization in the case of CTF-based CO_2 hydrogenation catalysts.

To date, CTFs were constructed mainly with three classes of coordinating ligands: pyridinic, *N*-heterocyclic, and acetyl acetone. Among them, extensive studies have been carried out on pyridinic-based ligands including the preparation of various N^N metal complexes and applications in catalysis. Although a myriad of catalytic transformation have been documented in the literature, the CTF-based heterogenized catalysts have been applied to a very few catalytic transformation such as methane oxidation to methanol, formic acid dehydrogenation, CO_2 hydrogenation to formate, transfer hydrogenation of carbonyl, alkene isomerization, and epoxides and methanol carbonylation.

Although the field has progressed significantly, more studies must be carried out to enlarge the scope of CTF-based heterogenized catalysts for various catalytic transformations. Most importantly, CTFs with a diverse class of ligands, including phosphine (P^P, P^N and P^C), porphyrin, salen, salphene, and pincer motifs (P^N^P, P^C^P, P^N^N, P^C^C, C^N^C and N^C^C), and their numerous combinations with various metal precursors need to be constructed. Moreover, it is important to prepare the CTFs with great economic benefit, including low temperatures and scalable methods, while preserving the attractive and practically imperative properties of CTFs.

References

[1] I. Fechete, Y. Wang, J.C. Védrine, The past, present and future of heterogeneous catalysis, Catal. Today 189 (2012) 2–27.
[2] J.N. Armor, A history of industrial catalysis, Catal. Today 163 (2011) 3–9.
[3] C.H. Bartholomew, R.J. Farrauto, Fundamentals of Industrial Catalytic Processes, second ed., John Wiley and Sons, Hoboken, NJ, 2006. https://doi.org/10.1002/cjce.5450850114.
[4] J. Smith, History of catalysis, in: I. Horvath (Ed.), Encyclopedia of Catalysis, vol. 3, John Wiley & Sons, 2003, pp. 447–479.
[5] H. Heinemann, A brief history of industrial catalysis, in: J. Anderson, M. Boudart (Eds.), Catalysis Science & Technology, vol. 1, Springer-Verlag, Berlin, 1981, p. 1.
[6] A. Kieboom, J. Moulijn, P. van Leeuwen, R. van Santen, History of catalysis, Stud. Surf. Sci. 123 (2000) 3–28.
[7] C. Copéret, F. Allouche, K. Wing Chan, M.P. Conley, M.F. Delley, A. Fedorov, I. B. Moroz, V. Mougel, M. Pucino, K. Searles, K. Yamamoto, P.A. Zhizhko, Bridging the gap between industrial and well-defined supported catalysts, Angew. Chem. Int. Ed. 57 (2018) 6398–6440.
[8] S. Chaturvedi, P.N. Dave, N.K. Shah, J. Saudi, Applications of nano-catalyst in new era, Chem. Soc. 16 (2012) 307–325.
[9] X. Cui, W. Li, P. Ryabchuk, K. Junge, M. Beller, Bridging homogeneous and heterogeneous catalysis by heterogeneous single-metal-site catalysts, Nat. Catal. 1 (2018) 385–397.
[10] X.S. Zhao, X.Y. Bao, W. Guo, F.Y. Lee, Immobilizing catalysts on porous materials, Mater. Today 9 (2006) 32–39.
[11] Z. Wang, G. Chen, K. Ding, Self-supported catalysts, Chem. Rev. 109 (2009) 322–359.
[12] D.E. De Vos, M. Dams, B.F. Sels, P.A. Jacobs, Ordered mesoporous and microporous molecular sieves functionalized with transition metal complexes as catalysts for selective organic transformations, Chem. Rev. 102 (2002) 3615–3640.

[13] F. Liguiro, in: P. Barbaro (Ed.), Heterogenized Homogeneous Catalysts for Fine Chemicals Production, Springer, New York, 2010, p. 1.

[14] C.W. Jones, On the stability and recyclability of supported metal–ligand complex catalysts: myths, misconceptions and critical research needs, Top. Catal. 53 (2010) 942–952.

[15] N. Madhavan, C.W. Jones, M. Weck, Rational approach to polymer-supported catalysts: synergy between catalytic reaction mechanism and polymer design, Acc. Chem. Res. 41 (2008) 1153–1165.

[16] X.S. Zhao, G.Q. Lu, X. Hu, A novel method for tailoring the pore-opening size of MCM-41 materials, Chem. Commun. (1999) 1391–1392.

[17] J.M. Thomas, B.F.G. Johnson, R. Raja, G. Sankar, P.A. Midgley, High-performance nanocatalysts for single-step hydrogenations, Acc. Chem. Res. 36 (2003) 20–30.

[18] X.S. Zhao, M.G.Q. Lu, C. Song, Immobilization of aluminum chloride on MCM-41 as a new catalyst system for liquid-phase isopropylation of naphthalene, J. Mol. Catal. A Chem. 191 (2003) 67–74.

[19] B.F.G. Johnson, S.A. Raynor, D.S. Shephard, T. Mashmeyer, J.M. Thomas, G. Sankar, S. Bromley, R. Oldroyd, L. Gladden, M.D. Mantle, Superior performance of a chiral catalyst confined within mesoporous silica, Chem. Commun. (1999) 1167–1168.

[20] J.M. Thomas, R. Raja, D.W. Lewis, Single-site heterogeneous catalysts, Angew. Chem. Int. Ed. 44 (2005) 6456–6482.

[21] H.H. Wagner, H. Hausmann, W.F. Holderich, Immobilization of rhodium diphosphine complexes on mesoporous Al-MCM-41 materials: catalysts for enantioselective hydrogenation, J. Catal. 203 (2001) 150–156.

[22] P. Piaggio, P. McMorn, C. Langham, D. Bethell, P.C. Bulman-Page, F.E. Hancock, G.J. Hutchings, Asymmetric epoxidation of stilbene by manganese(III) chiral salen complex immobilized in Al-MCM-41, New J. Chem. 22 (1998) 1167–1169.

[23] P. Piaggio, C. Langham, P. McMorn, D. Bethell, P.C. Bulman-Page, F.E. Hancock, C. Sly, G.J. Hutchings, Catalytic asymmetric epoxidation of stilbene using a chiral salen complex immobilized in Mn-exchanged Al-MCM-41, J. Chem. Soc. Perkin Trans. 2 (2000) 143–148.

[24] J. Jamis, J.R. Anderson, R.S. Dickson, E.M. Campi, W.R. Jackson, Aqueous enantioselective hydrogenations involving silica-heterogenised catalysts, J. Organomet. Chem. 603 (2000) 80–85.

[25] J. Jamis, J.R. Anderson, R.S. Dickson, E.M. Campi, W.R. Jackson, Modified silica-heterogenised catalysts for use in aqueous enantioselective hydrogenations, J. Organomet. Chem. 627 (2001) 37–43.

[26] C. Li, Chiral synthesis on catalysts immobilized in microporous and mesoporous materials, Catal. Rev. 46 (2004) 419–492.

[27] P. McMorn, G.J. Hutchings, Heterogeneous enantioselective catalysts: strategies for the immobilisation of homogeneous catalysts, Chem. Soc. Rev. 33 (2004) 108–122.

[28] S.B. Ogunwami, T. Bein, Intrazeolite assembly of a chiral manganese salen epoxidation catalyst, Chem. Commun. (1997) 901–902.

[29] A. Zsigmond, K. Bogar, F. Notheisz, Comparative study of "ship-in-a-bottle" and anchored heterogenized Rh complexes, J. Catal. 213 (2003) 103–108.

[30] C. Coperet, A. Comas-Vives, M.P. Conley, D.P. Estes, A. Fedorov, V. Mougel, H. Nagae, F. Nunez-Zarur, P.A. Zhizhko, Surface organometallic and coordination chemistry toward single-site heterogeneous catalysts: Strategies, methods, structures, and activities, Chem. Rev. 116 (2016) 323–421.

[31] J. Guzman, B.C. Gates, Supported molecular catalysts: metal complexes and clusters on oxides and zeolites, Dalton Trans. (2003) 3303–3318.

[32] D.E. De Vos, M. Dams, B.F. Sels, P.A. Jacobs, Chem. Rev. 102 (2002) 3615–3640.

[33] D.R. Godhani, H.D. Nakum, D.K. Parmar, N.C. Desai, Zeolite-Y immobilized metallo-ligand complexes: a novel heterogenous catalysts for selective oxidation, Inorg. Chem. Commun. 72 (2016) 105–116.
[34] T.K. Dey, K. Ghosh, P. Basu, R.A. Mollab, S.M. Islam, Chloromethylated polystyrene immobilized ruthenium complex of 2-(2-pyridyl)benzimidazole catalyst for the synthesis of bioactive disubstituted ureas by carbonylation reaction, New J. Chem. 42 (2018) 9168–9176.
[35] S. Hubner, J.G. de Vries, V. Farina, Why does industry not use immobilized transition metal complexes as catalysts? Adv. Synth. Catal. 358 (2016) 3–25.
[36] M. Rose, Nanoporous polymers: bridging the gap between molecular and solid catalysts? ChemCatChem 6 (2014) 1166–1182.
[37] J. Wu, F. Xu, S. Li, P. Ma, X. Zhang, Q. Liu, R. Fu, D. Wu, Porous polymers as multifunctional material platforms toward task-specific applications, Adv. Mater. 30 (2018) 1802922.
[38] G. Algara-Siller, N. Severin, S.Y. Chong, T. Bjorkman, R.G. Palgrave, A. Laybourn, M. Antonietti, Y.Z. Khimyak, A.V. Krasheninnikov, J.P. Rabe, U. Kaiser, A.I. Cooper, A. Thomas, M.J. Bojdys, Triazine-based graphitic carbon nitride: a two-dimensional semiconductor, Angew. Chem. Int. Ed. 53 (2014) 7450–7455.
[39] A. Schoedel, Z. Ji, O.M. Yaghi, The role of metal–organic frameworks in a carbon-neutral energy cycle, Nat. Energy 1 (2016) 16034.
[40] A. Dhakshinamoorthy, Z. Li, H. Garcia, Catalysis and photocatalysis by metal organic frameworks, Chem. Soc. Rev. 47 (2018) 8134–8172.
[41] P. Kaur, J.T. Hupp, S.T. Nguyen, Porous organic polymers in catalysis: opportunities and challenges, ACS Catal. 1 (2011) 819–835.
[42] P.J. Waller, F. Gandara, O.M. Yaghi, Chemistry of covalent organic frameworks, Acc. Chem. Res. 48 (2015) 3053–3063.
[43] K. Huang, J.-Y. Zhang, F. Liu, S. Dai, Synthesis of porous polymeric catalysts for the conversion of carbon dioxide, ACS Catal. 8 (2018) 9079–9102.
[44] D.S. Ahmed, G.A. El-Hiti, E. Yousif, A.A. Ali, A.S. Hameed, Design and synthesis of porous polymeric materials and their applications in gas capture and storage: a review, J. Polym. Res. 25 (2018) 75.
[45] M. Opanasenko, A. Dhakshinamoorthy, J. Cejka, H. Garcia, Deactivation pathways of the catalytic activity of metal–organic frameworks in condensation reactions, ChemCatChem 5 (2013) 1553–1561.
[46] Q. Pang, B. Tu, E. Ning, Q. Li, D. Zhao, Distinct packings of supramolecular building blocks in metal–organic frameworks based on imidazoledicarboxylic acid, Inorg. Chem. 54 (2015) 9678–9680.
[47] T. Ishida, M. Nagaoka, T. Akita, M. Haruta, Deposition of gold clusters on porous coordination polymers by solid grinding and their catalytic activity in aerobic oxidation of alcohols, Chem. Eur. J. 14 (2008) 8456–8460.
[48] A. Dhakshinamoorthy, M. Alvaro, P. Concepcion, H. Garcia, Chemical instability of $Cu_3(BTC)_2$ by reaction with thiols, Catal. Commun. 12 (2011) 1018–1021.
[49] A. Thomas, Functional materials: From hard to soft porous frameworks, Angew. Chem. Int. Ed. 49 (2010) 8328–8344.
[50] T. Ben, S. Qiu, Porous aromatic frameworks: synthesis, structure and functions, CrystEngComm 15 (2013) 17–26.
[51] W. Zhao, L. Xia, X. Liu, Covalent organic frameworks (COFs): perspectives of industrialization, CrystEngComm 20 (2018) 1613–1634.
[52] S.-Y. Ding, W. Wang, Covalent organic frameworks (COFs): from design to applications, Chem. Soc. Rev. 42 (2013) 548–568.
[53] M.G. Rabbani, H.M. El-Kaderi, Synthesis and characterization of porous benzimidazole-linked polymers and their performance in small gas storage and selective uptake, Chem. Mater. 24 (2012) 1511–1517.

[54] M.G. Rabbani, H.M. El-Kaderi, Chem. Mater. 23 (2011) 1650–1653.
[55] D. Ramimoghadam, E. MacAGray, C.J. Webb, Review of polymers of intrinsic microporosity for hydrogen storage applications, Int. J. Hydrog. Energy 41 (2016) 16944–16965.
[56] S. Xu, Y. Luo, B. Tan, Recent development of hypercrosslinked microporous organic polymers, Macromol. Rapid Commun. 34 (2013) 471–484.
[57] L. Tan, B. Tan, Hypercrosslinked porous polymer materials: design, synthesis, and applications, Chem. Soc. Rev. 46 (2017) 3322–3356.
[58] Y. Xu, S. Jin, H. Xu, A. Nagai, D. Jiang, Conjugated microporous polymers: design, synthesis and application, Chem. Soc. Rev. 42 (2013) 8012–8031.
[59] Y.-B. Zhou, Z.-P. Zhan, Conjugated microporous polymers for heterogeneous catalysis, Chem. Asian J. 13 (2018) 9–19.
[60] N. Popp, T. Homburg, N. Stock, J. Senker, Porous imine-based networks with protonated imine linkages for carbon dioxide separation from mixtures with nitrogen and methane, J. Mater. Chem. A 3 (2015) 18492–18504.
[61] P. Kuhn, A. Forget, D. Su, A. Thomas, M. Antonietti, From microporous regular frameworks to mesoporous materials with ultrahigh surface area: dynamic reorganization of porous polymer networks, J. Am. Chem. Soc. 130 (2008) 13333–13337.
[62] P. Kuhn, M. Antonietti, A. Thomas, Porous, covalent triazine-based frameworks prepared by ionothermal synthesis, Angew. Chem. Int. Ed. 47 (2008) 3450–3453.
[63] P. Kuhn, A. Thomas, M. Antonietti, Toward tailorable porous organic polymer networks: a high-temperature dynamic polymerization scheme based on aromatic nitriles, Macromolecules 42 (2009) 319–326.
[64] R. Palkovits, C. von Malotki, M. Baumgarten, K. Müllen, C. Baltes, M. Antonietti, P. Kuhn, J. Weber, A. Thomas, F. Schüth, Development of molecular and solid catalysts for the direct low-temperature oxidation of methane to methanol, ChemSusChem 3 (2010) 277–282.
[65] P. Kuhn, A. Forget, J. Hartmann, A. Thomas, M. Antonietti, Template-free tuning of nanopores in carbonaceous polymers through ionothermal synthesis, Adv. Mater. 21 (2009) 897–901.
[66] S. Hug, M.E. Tauchert, S. Li, U.E. Pachmayr, B.V. Lotsch, A functional triazine framework based on N-heterocyclic building blocks, J. Mater. Chem. 22 (2012) 13956–13964.
[67] C.E. Chan-Thaw, A. Villa, P. Katekomol, D. Su, A. Thomas, L. Prati, Covalent triazine framework as catalytic support for liquid phase reaction, Nano Lett. 10 (2010) 537–541.
[68] L. Hao, J. Ning, B. Luo, B. Wang, Y. Zhang, Z. Tang, J. Yang, A. Thomas, L. Zhi, Structural evolution of 2D microporous covalent triazine-based framework toward the study of high-performance supercapacitors, J. Am. Chem. Soc. 137 (2015) 219–225.
[69] P. Puthiaraj, Y.-R. Lee, S. Zhang, W.-S. Ahn, Triazine-based covalent organic polymers: design, synthesis and applications in heterogeneous catalysis, J. Mater. Chem. A 4 (2016) 16288–16311.
[70] G. Miller, US3775380, 1973.
[71] M.X. Tan, Y.N. Sum, J.Y. Ying, Y. Zhang, A mesoporous poly-melamine-formaldehyde polymer as a solid sorbent for toxic metal removal, Energy Environ. Sci. 6 (2013) 3254–3259.
[72] W.-C. Song, X.-K. Xu, Q. Chen, Z.-Z. Zhuang, X.-H. Bu, Nitrogen-rich diaminotriazine-based porous organic polymers for small gas storage and selective uptake, Polym. Chem. 4 (2013) 4690–4696.
[73] M.K. Bhunia, S.K. Das, P. Pachfule, R. Banerjee, A. Bhaumik, Nitrogen-rich porous covalent imine network (CIN) material as an efficient catalytic support for C–C coupling reactions, Dalton Trans. 41 (2012) 1304–1311.

[74] Q. Liu, Z. Tang, M. Wu, B. Liao, H. Zhou, B. Ou, G. Yu, Z. Zhou, X. Li, Novel ferrocene-based nanoporous organic polymers for clean energy application, RSC Adv. 5 (2015) 8933–8937.
[75] H. Zhou, S. Xu, H. Su, M. Wang, W. Qiao, L. Ling, D. Long, Facile preparation and ultra-microporous structure of melamine–resorcinol–formaldehyde polymeric microspheres, Chem. Commun. 49 (2013) 3763–3765.
[76] P. Puthiaraj, K. Pitchumani, Triazine-based mesoporous covalent imine polymers as solid supports for copper-mediated Chan–Lam cross-coupling N-arylation reactions, Chem. Eur. J. 20 (2014) 8761–8770.
[77] L. Stegbauer, K. Schwinghammer, B.V. Lotsch, A hydrazone-based covalent organic framework for photocatalytic hydrogen production, Chem. Sci. 5 (2014) 2789–2793.
[78] H. Zhao, Z. Jin, H. Su, X. Jing, F. Sun, G. Zhu, Targeted synthesis of a 2D ordered porous organic framework for drug release, Chem. Commun. 47 (2011) 6389–6391.
[79] H.A. Patel, F. Karadas, A. Canlier, J. Park, E. Deniz, Y. Jung, M. Atilhan, C.T. Yavuz, High capacity carbon dioxide adsorption by inexpensive covalent organic polymers, J. Mater. Chem. 22 (2012) 8431–8437.
[80] H.A. Patel, F. Karadas, J. Byun, J. Park, E. Deniz, A. Canlier, Y. Jung, M. Atilhan, C.T. Yavuz, Highly stable nanoporous sulfur-bridged covalent organic polymers for carbon dioxide removal, Adv. Funct. Mater. 23 (2013) 2270–2276.
[81] P. Puthiaraj, K. Pitchumani, Palladium nanoparticles supported on triazine functionalised mesoporous covalent organic polymers as efficient catalysts for Mizoroki–Heck cross coupling reaction, Green Chem. 16 (2014) 4223–4233.
[82] S.K. Kundu, A. Bhaumik, A triazine-based porous organic polymer: a novel heterogeneous basic organocatalyst for facile one-pot synthesis of 2-amino-4H-chromenes, RSC Adv. 5 (2015) 32730–32739.
[83] S. Zulfiqar, M.I. Sarwar, C.T. Yavuz, Melamine based porous organic amide polymers for CO_2 capture, RSC Adv. 4 (2014) 52263–52269.
[84] S. Ren, R. Dawson, A. Laybourn, J.-X. Jiang, Y. Khimyak, D.J. Adams, A.I. Cooper, Functional conjugated microporous polymers: from 1,3,5-benzene to 1,3,5-triazine, Polym. Chem. 3 (2012) 928–934.
[85] Z. Xiang, D. Cao, Synthesis of luminescent covalent–organic polymers for detecting nitroaromatic explosives and small organic molecules, Macromol. Rapid Commun. 33 (2012) 1184–1190.
[86] Z. Xiang, D. Cao, L. Huang, J. Shui, M. Wang, L. Dai, Nitrogen-doped holey graphitic carbon from 2D covalent organic polymers for oxygen reduction, Adv. Mater. 26 (2014) 3315–3320.
[87] S. Xiong, X. Fu, L. Xiang, G. Yu, J. Guan, Z. Wang, Y. Du, X. Xiong, C. Pan, Liquid acid-catalysed fabrication of nanoporous 1,3,5-triazine frameworks with efficient and selective CO_2 uptake, Polym. Chem. 5 (2014) 3424–3431.
[88] H. Lim, M.C. Cha, J.Y. Chang, Preparation of microporous polymers based on 1,3,5-triazine units showing high CO_2 adsorption capacity, Macromol. Chem. Phys. 213 (2012) 1385–1390.
[89] P. Puthiaraj, S.-S. Kim, W.-S. Ahn, Covalent triazine polymers using a cyanuric chloride precursor via Friedel–Crafts reaction for CO_2 adsorption/separation, Chem. Eng. J. 283 (2016) 184–192.
[90] J. Artz, Covalent Triazine-based frameworks—tailor-made catalysts and catalyst supports for molecular and nanoparticulate species, ChemCatChem 10 (2018) 1753–1771.
[91] R. Palkovits, M. Antonietti, P. Kuhn, A. Thomas, F. Schüth, Solid catalysts for the selective low-temperature oxidation of methane to methanol, Angew. Chem. Int. Ed. 48 (2009) 6909–6912.
[92] R.A. Periana, D.J. Taube, S. Gamble, H. Taube, T. Satoh, H. Fujii, Platinum catalysts for the high-yield oxidation of methane to a methanol derivative, Science 280 (1998) 560–564.

[93] M. Soorholtz, L.C. Jones, D. Samuelis, C. Weidenthaler, R.J. White, M.-M. Titirici, D.A. Cullen, T. Zimmermann, M. Antonietti, J. Maier, R. Palkovits, B.F. Chmelka, F. Schuth, Local platinum environments in a solid analogue of the molecular periana catalyst, ACS Catal. 6 (2016) 2332–2340.
[94] A.V. Bavykina, M.G. Goesten, F. Kapteijn, M. Makkee, J. Gascon, Efficient production of hydrogen from formic acid using a covalent triazine framework supported molecular catalyst, ChemSusChem 8 (2015) 809–812.
[95] A.V. Bavykina, H.-H. Mautscke, M. Makkee, F. Kapteijn, J. Gascon, F.X. Llabres i Xamena, Base free transfer hydrogenation using a covalent triazine framework based catalyst, CrystEngComm 19 (2017) 4166–4170.
[96] E. Rozhko, A. Bavykina, D. Osadchii, M. Makkee, J. Gascon, Covalent organic frameworks as supports for a molecular Ni based ethylene oligomerization catalyst for the synthesis of long chain olefins, J. Catal. 345 (2017) 270–280.
[97] R. Xu, X.S. Wang, H. Zhao, H. Lin, Y.B. Huang, R. Cao, Rhenium-modified porous covalent triazine framework for highly efficient photocatalytic carbon dioxide reduction in a solid–gas system, Catal. Sci. Technol. 8 (2018) 2224–2230.
[98] A.V. Bavykina, E. Rozhko, M.G. Goesten, T. Wezendonk, B. Seoane, F. Kapteijn, M. Makkee, J. Gascon, Shaping covalent triazine frameworks for the hydrogenation of carbon dioxide to formic acid, ChemCatChem 8 (2016) 2217–2221.
[99] A.V. Bavykina, A.I. Olivos-Suarez, D. Osadchii, R. Valecha, R. Franz, M. Makkee, F. Kapteijn, J. Gascon, Facile method for the preparation of covalent triazine framework coated monoliths as catalyst support: applications in C1 catalysis, ACS Appl. Mater. Interfaces 9 (2017) 26060–26065.
[100] K. Kamiya, R. Kamai, K. Hashimoto, S. Nakanishi, Platinum-modified covalent triazine frameworks hybridized with carbon nanoparticles as methanol-tolerant oxygen reduction electrocatalysts, Nat. Commun. 5 (2014) 5040.
[101] R. Kamai, K. Kamiya, K. Hashimoto, S. Nakanishi, Oxygen-tolerant electrodes with platinum-loaded covalent triazine frameworks for the hydrogen oxidation reaction, Angew. Chem. Int. Ed. 55 (2016) 13184–13188.
[102] K. Iwase, T. Yoshioka, S. Nakanishi, K. Hashimoto, K. Kamiya, Copper-modified covalent triazine frameworks as non-noble-metal electrocatalysts for oxygen reduction, Angew. Chem. Int. Ed. 54 (2015) 11068–11072.
[103] T. Yoshioka, K. Iwase, S. Nakanishi, K. Hashimoto, K. Kamiya, Electrocatalytic reduction of nitrate to nitrous oxide by a copper-modified covalent triazine framework, J. Phys. Chem. C 120 (2016) 15729–15734.
[104] K. Kamiya, T. Tatebe, S. Yamamura, K. Iwase, T. Harada, S. Nakanishi, Selective reduction of nitrate by a local cell catalyst composed of metal-doped covalent triazine frameworks, ACS Catal. 8 (2018) 2693–2698.
[105] S. Yamaguchi, K. Kamiya, K. Hashimoto, S. Nakanishi, Ru atom-modified covalent triazine framework as a robust electrocatalyst for selective alcohol oxidation in aqueous electrolytes, Chem. Commun. 53 (2017) 10437–10440.
[106] K. Park, G.H. Gunasekar, N. Prakash, K.D. Jung, S. Yoon, A highly efficient heterogenized iridium complex for the catalytic hydrogenation of carbon dioxide to formate, ChemSusChem 8 (2015) 3410–3413.
[107] G.H. Gunasekar, K. Park, H. Jeong, K.D. Jung, S. Yoon, Molecular Rh(III) and Ir(III) catalysts immobilized on bipyridine-based covalent triazine frameworks for the hydrogenation of CO_2 to formate, Catalysts 8 (2018) 295.
[108] G.H. Gunasekar, J. Shin, K.D. Jung, S. Yoon, Design strategy toward recyclable and highly efficient heterogeneous catalysts for the hydrogenation of CO_2 to formate, ACS Catal. 8 (2018) 4346–4353.
[109] P. Sudakar, G.H. Gunasekar, I.H. Baek, S. Yoon, Recyclable and efficient heterogenized Rh and Ir catalysts for the transfer hydrogenation of carbonyl compounds in aqueous medium, Green Chem. 18 (2016) 6456–6461.

[110] S. Rajendiran, P. Natarajan, S. Yoon, A covalent triazine framework-based heterogenized Al–Co bimetallic catalyst for the ring-expansion carbonylation of epoxide to β-lactone, RSC Adv. 7 (2017) 4635–4638.
[111] Y.D.Y.L. Getzler, V. Mahadevan, E.B. Lobkovsky, G.W. Coates, Synthesis of β-lactones: a highly active and selective catalyst for epoxide carbonylation, J. Am. Chem. Soc. 124 (2002) 1174–1175.
[112] C. Gunanathan, D. Milstein, Bond activation and catalysis by ruthenium pincer complexes, Chem. Rev. 114 (2014) 12024–12087.
[113] W.A. Herrmann, N-heterocyclic carbenes: a new concept in organometallic catalysis, Angew. Chem. Int. Ed. 41 (2002) 1290–1309.
[114] Z.G. Specht, D.B. Grotjahn, C.E. Moore, A.L. Rheingold, Iridium and ruthenium complexes with chelating N-heterocyclic carbenes: efficient catalysts for transfer hydrogenation, β-alkylation of alcohols, and N-alkylation of amines, Organometallics 32 (2013) 6400–6409.
[115] C. Segarra, E. Mas-Marza, M. Benitez, J.A. Mata, E. Peris, Unconventional reactivity of imidazolylidene pyridylidene ligands in iridium(III) and rhodium(III) complexes, Angew. Chem. Int. Ed. 51 (2012) 10841–10845.
[116] K. Park, K. Lee, H. Kim, V. Ganesan, K. Cho, S.K. Jeong, S. Yoon, Preparation of covalent triazine frameworks with imidazolium cations embedded in basic sites and their application for CO_2 capture, J. Mater. Chem. A 5 (2017) 8576–8582.
[117] G.H. Gunasekar, K. Park, V. Ganesan, K. Lee, N.-K. Kim, K.-D. Jung, S. Yoon, A covalent triazine framework, functionalized with Ir-N-Heterocyclic carbene sites for the efficient hydrogenation of CO_2 to formate, Chem. Mater. 29 (2017) 6740–6748.
[118] S. Rajendiran, K. Park, K. Lee, S. Yoon, Ionic-liquid-based heterogeneous covalent triazine framework cobalt catalyst for the direct synthesis of methyl 3-hydroxybutyrate from propylene oxide, Inorg. Chem. 56 (2017) 7270–7277.
[119] K. Park, S. Lim, J.H. Baik, K.D. Jung, S. Yoon, Exceptionally stable Rh-based molecular catalyst heterogenized on a cationically charged covalent triazine framework support for efficient methanol carbonylation, Catal. Sci. Technol. 8 (2018) 2894–2900.
[120] S. Rajendiran, G.H. Gunasekar, S. Yoon, A heterogenized cobaltate catalyst on a bis-imidazolium-based covalent triazine framework for hydroesterification of epoxides, New J. Chem. 42 (2018) 12256–12262.
[121] H.S. Jena, C. Krishnaraj, G. Wang, K. Leus, J. Schmidt, N. Chaoui, P. Van Der Voort, Acetylacetone covalent triazine framework: an efficient carbon capture and storage material and a highly stable heterogeneous catalyst, Chem. Mater. 30 (2018) 4102–4111.

CHAPTER 2

Heteroatom-doped carbon materials derived from ionic liquids for catalytic applications

Samahe Sadjadi[a] and Babak Karimi[b]
[a]Gas Conversion Department, Faculty of Petrochemicals, Iran Polymer and Petrochemical Institute, Tehran, Iran
[b]Department of Chemistry, Institute for Advanced Studies in Basic Sciences (IASBS), Zanjan, Iran

1 Introduction

The utility of carbon materials for diverse range of applications ranging from (photo/electro) catalysis [1], sensors [2–4], waste water treatment [5], energy storage (batteries, super capacitors) [6–9] to the state-of-the-art technologies related to the development of renewable and green energies such as fuel cells [10, 11] has been established. As the physical and chemical nature of carbons such as specific surface area, morphology, etc. may significantly influence the properties of carbon materials and consequently their function for a specific application, adjusting their properties is of great importance. The studies confirmed that doping heteroatoms such as nitrogen, sulfur, and boron atoms in the structure of carbon materials can improve both the electric and chemical properties of carbon materials. In this line, many attempts have been devoted to the development of varied heteroatom-doped carbons [12–16]. There are two main approaches for the introduction of heteroatoms in the structure of the carbon materials: (1) treating the as-synthesized carbons with appropriate gases such as H_2S and NH_3; and (2) carbonization of heteroatom containing precursors [17]. Each approach has its own pros and cons. In the former, posttreatment is often hard to control, and the functional groups are certainly restricted due to the harsh processing condition, while in the second approach, yield of carbonization is low due to the volatile nature of small molecules. One solution to this problem could be based upon using ionic liquids (ILs) as heteroatom-rich carbon precursors [17]. Although this concept has already been emerged during last decades, the promising results have now motivated increasing research on this field, and to date, various research groups disclosed diverse doped

Emerging Carbon Materials for Catalysis
https://doi.org/10.1016/B978-0-12-817561-3.00002-0

(porous) carbon materials with broad range of properties by using different ILs as well as templates. Along this line, several excellent review articles have been published and covered many aspects of ILs-derived carbon materials, including their synthetic procedures, classification and some of their applications [17–19]. In this chapter, after a brief introduction to ILs, various methods for the preparation of doped carbon materials derived from ILs are thoroughly discussed. Furthermore, the utility of these carbonaceous materials as (electro) catalysts is also elaborated to some extent. In the last part of this chapter, porous doped carbonaceous structures prepared by utilizing ILs as precursors are covered. It is tried to emphasize on the most recent advances in this field in order to provide useful information for the readers. The readers, interested in this topic, are encouraged to study the aforementioned review articles for more detail.

2 Ionic liquids as carbon precursors

Ionic liquids (ILs) are heteroatom-containing organic salts, often formed from various types of organic cations such as pyridinium, imidazolium, or phosphonium and inorganic anion, which can either be a single atom like Cl^- and Br^-, or larger complex like ethyl sulfate, tetrafluoroborate, or hexafluorophosphate [20]. These organic salts benefit from some advantages such as low toxicity, high electric conductivity, tune-ability, and low melting point (below 100° C or even below room temperature). The low vapor pressure of ILs stems from Coulombic interactions between ions. Since their discovery in 1914, ILs have attracted immense attention and their utilities for diverse range of applications such as catalysis, chemical and electrochemical synthesis, gas adsorption, polymers, and magnetic and luminescent materials have been disclosed [19]. Moreover, ILs can be used as the nonvolatile reaction solvent for the synthesis of wide range of either inorganic or functional carbon materials [17]. The appeal for using ILs as carbon precursors is owing, to a great extent, to the unique properties of ILs. ILs possess heteroatoms (such as N, P, S, B) in their backbone, which upon carbonization could be potentially incorporated in the final carbon material. To date, various B, S-doped, and N-doped ILs-derived carbon materials have been reported [21], among which N-doped carbon materials have received significant attention due to their wide range of applications in catalysis [22, 23]. In fact, considering the diversity of ILs' structures, various carbon materials with desired properties such as surface basicity and/or acidity, electrical conductivity, and catalytic activity can be prepared by proper choice of cations and anions.

On the other hand, the anion and cation of ILs can serve as porogen and lead to the formation of porous carbon material [17]. Another advantage of ILs is their low vapor pressure even at high temperature used for the carbonization process which can facilitate processing, shaping, and leads to high yield of carbonization even at high temperatures [21]. Considering high ability of ILs for dissolving both inorganic and organic materials, it is practically possible to obtain various types of structurally diverse carbon materials by simple dissolution of a second compound in IL followed by thermal treatment [19].

Regarding carbonization of ILs, it has to be taken into consideration that the choice of ILs precursor is of a great importance. Despite the fact that ILs exhibit negligible vapor pressure, the first thermal decomposition step can result in the formation of highly volatile intermediates, and in the case of many conventional ILs, thermal treatment can lead to full decomposition. According to the literature [19, 21], one of the most suitable ILs for the carbonization purposes is IL with functional groups, mostly cyano/nitrile functionality. In fact, polymerization of these functionalities at lower temperatures avoids full decomposition and results in the formation of carbon materials at higher temperatures. In Table 1, the structure of common ILs used as carbon precursors is summarized. As shown, cyano/nitrile functionality can be present either on the anion or cation backbone. However, it should be noted that the formation of volatiles intermediates and even the use of trace amounts of volatile solvents during the carbonization stages would be sometimes beneficial for nano-fibrillated carbonaceous structures [24].

2.1 Reaction mechanism of carbonization

Carbonization of nitrile/cyano-containing ILs proceeds through a three-step process including: (1) decomposition at lower temperatures (~300°C) via cleavage of alkyl chains and the elimination of ammonia that furnishes water-soluble oligomer or a viscous liquid; (2) trimerization of cyano groups, and aromatic condensation (~500°C) that leads to the formation of a compound, referred as framework intermediate, with high content of nitrogen (30–40 wt%); and (3) condensation of the entire system at elevated temperatures (~1000°C) with the elimination of hydrogen and nitrogen, and formation of graphitic domain (Fig. 1) [18, 19, 21]. Considering this mechanism, it can be concluded that the presence of cyano/nitrile functionalities is imperative for promoting the first step of the carbonization.

Table 1 The chemical formula and abbreviation of ILs used as carbon precursors.

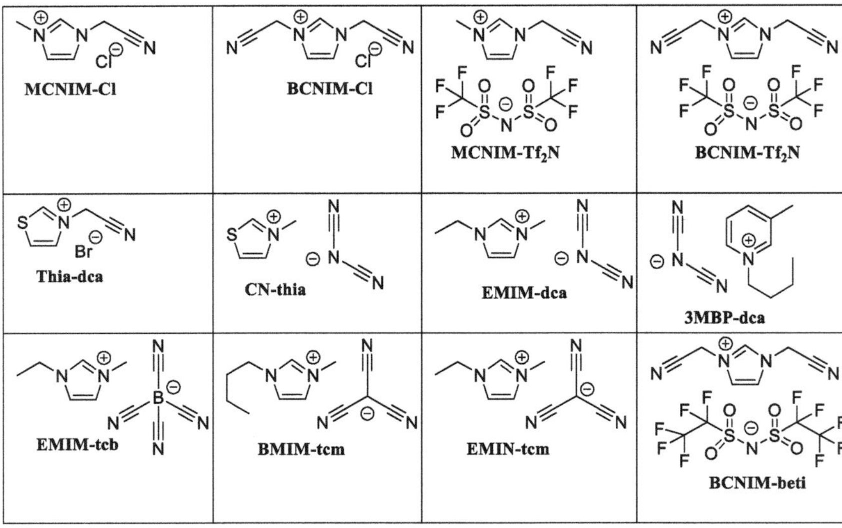

Adapted from J.P. Paraknowitsch, A. Thomas, Functional carbon materials from ionic liquid precursors, Macromol. Chem. Phys. 213(10–11) (2012) 1132–1145; T.-P. Fellinger, et al., 25th anniversary article: "Cooking carbon with salt": carbon materials and carbonaceous frameworks from ionic liquids and poly(ionic liquid)s, Adv. Mater. 25(41) (2013) 5838–5855, with permission from John Wiley and Sons.

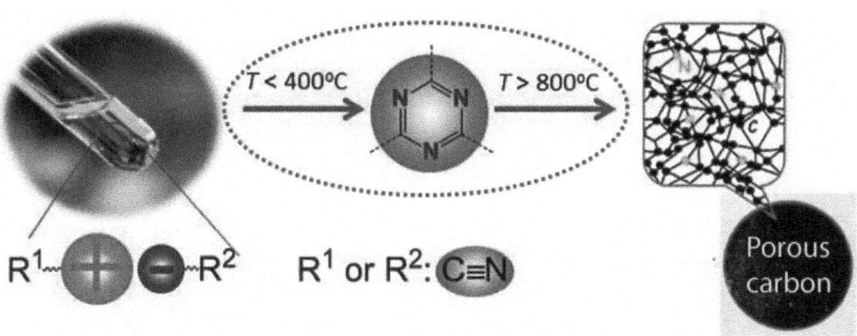

Fig. 1 Direct synthesis of nitrogen-doped carbon from specific nitrile/cyano-containing IL precursors under ambient pressure. *(Adapted from S. Zhang, K. Dokko, M. Watanabe, Carbon materialization of ionic liquids: from solvents to materials, Mater. Horiz. 2(2) (2015) 168–197, with permission from Royal Society of Chemistry.)*

2.2 Effecting parameters on the properties of IL-derived carbon

It is worth noting that the nature of ILs can significantly affect the properties of the final carbon material. Among these features, specific surface area, morphology, heteroatom (nitrogen) content, etc. can be counted.

Nitrogen content as well as the nature of nitrogen in the final carbon material can be often adjusted by the appropriate choice of cation and/or anion. Another important parameter that may remarkably influence the content of nitrogen in the final carbon structure is the carbonization temperature. Higher carbonization temperature generally results in lower nitrogen content. In Table 2, the nitrogen contents and specific surface areas of some of the ILs-derived carbon materials, prepared at 800°C, are collected [19].

Although there is no linear correlation between the structure of ILs precursor and the content of the nitrogen in the final carbon material, the data in Table 2 suggest that the presence of dangling cyano/nitrile functionality in the imidazolium ring is beneficial for attaining high nitrogen content in the resulting carbon material. Moreover, it can be concluded that anions with higher nitrogen content may lead to carbon materials with higher nitrogen content [19].

As can be seen, the choice of ILs may have a considerable impact on the specific surface area and the porosity of the final carbon material. As an example, use of cyano-free anion such as beti and Tf_2N can lead to the formation of porous carbon. In these cases, the anion served as porogen. More precisely, in the course of cyclotrimerization of cyano-containing cation at low temperature, the anion is trapped in the carbon matrix. Upon elevating the temperature, the trapped anion decomposition results in the formation of pores in the backbone of the carbon [25]. Interestingly, use of Cl^- as anion instead of Tf_2N leads to the loss of the porosity of the final carbon [21].

Table 2 The nitrogen content and specific surface area of some of the ILs-derived carbon materials, prepared at 800°C.

ILs precursor	N-content (wt%)	S_{BET} ($m^2\ g^{-1}$)
BCNIM-Cl	–	15.5
BCNIM-Tf_2N	2–3	640.4
MCNIM-Tf_2N	–	780.6
EMIM-tcm	19.9	3.8
BMIM-tcm	15.8	64.8
BCNIM-tcm	13.0	56.6
BCNIM-beti	23.0	662.7
EMIM-dca	26.0	<10
3MBP-dca	16.0	<10

Adapted from J.P. Paraknowitsch, A. Thomas, Functional carbon materials from ionic liquid precursors, Macromol. Chem. Phys. 213 (10–11) (2012) 1132–1145, with permission from John Wiley and Sons.

On the other hand, comparing the specific areas of carbon materials from different ILs precursors (Table 2), it can be concluded that for preparing carbon materials with higher surface area, the use of ILs with cyano/nitrile functionality on their cation is more advantageous than those ILs bearing cyano/nitrile groups on their anion.

Notably, multi-heteroatom-doped carbon materials can be obtained by applying IL precursors with heteroatoms other than nitrogen. For instance, boron and sulfur can be introduced in the backbone of the resulting carbon by applying EMIM-tcb and thiazolium-based ILs, respectively [18].

2.3 Examples of catalysts based on cyano/nitrile containing ILs-derived carbon

ILs-derived nitrogen-doped carbon materials are of great importance and promising in catalytic application, and to date, various examples have been reported regarding the catalytic utility of these materials. It is worth noting that these carbon materials can be used in their free form or applied as catalyst supports for immobilizing other catalytic-active species such as metallic nanoparticles. Moreover, ILs-derived carbons are not only promising catalysts for the conventional organic transformation, but also are very appealing candidates for electrocatalysis. In this part, the utility of ILs-derived carbons for the catalysis is elaborated by means of a few recent examples. Considering the fact that $Ni(OAc)_2$ and $Pd(OAc)_2$ could form metal-*N*-heterocyclic carbene (NHC) complexes through the reaction with C2 of imidazolium ring to afford the corresponding coordination polymers, Li, Xia et al. disclosed a one-pot method for the preparation of mono/bi-metallated nitrogen-doped carbon materials, M@CN, (Pd, Ni or Ni_xPd_y) through direct carbonization of metal-NHC coordination polymer. To prepare the precursor, [TPBAIm][NTf_2]$_3$ was first synthesized (Fig. 2 and the reference cited in Fig. 2 caption) and then metallated and carbonized. It was found that carbonization temperature played an important role in the growth and properties of the final catalyst and the polymer carbonized at 800°C, $M@CN^{800}$, led to a catalyst with strong interaction between metallic nanoparticles and carbon material, uniform dispersion of metallic particles as well as graphene-like layers of nitrogen-doped carbon. Moreover, a sample of $Pd@CN^{800}$ exhibited much superior catalytic activity and recyclability as compared with that of $PdCl_2(PPh_3)_2$ in a typical domino carbonylative synthesis of pyrazole derivatives from the reaction of aryl iodide, terminal alkyne, hydrazine, and CO gas.

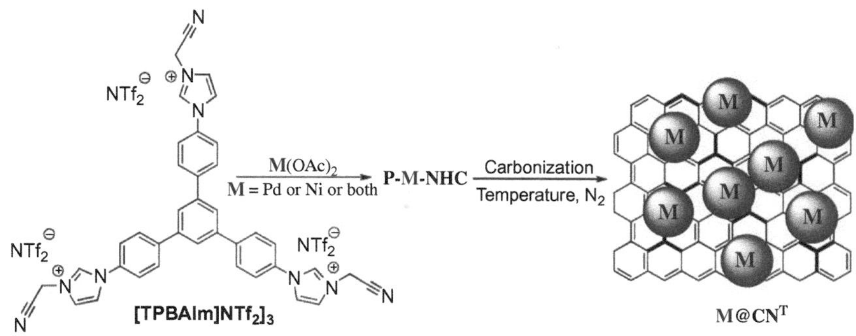

Fig. 2 Synthesis of P-M-NHC and M@CN. *(Adapted from Z. Li, et al., One-pot synthesis of Pd nanoparticle catalysts supported on N-doped carbon and application in the domino carbonylation, ACS Catal. 3(5) (2013) 839–845, with permission from American Chemical Society.)*

As another example, Wang et al. showed that nitrogen-doped carbon materials having different mesoporous structure and nitrogen contents could be successfully prepared by using eight structurally diverse nitrogen containing ILs under essentially the same carbonization conditions. The resulting carbonaceous materials were then employed for the preparation of palladium-supported catalyst system (Pd@CN$_x$, $x = 1$–8) using a simple ultrasonic-assisted protocol [26]. The prepared catalysts were thus used in selective hydrogenation of phenol to cyclohexanone in aqueous media. It was shown that, among all investigated samples, the catalyst prepared using ionic **1** (Fig. 3) exhibited much superior catalytic performance under the described reaction condition.

Although the comparison of the eight samples confirmed that there should be some correlation between the amount of Pd^0 and the catalytic activity of the samples, a direct impact of other physicochemical properties of the carbon materials on the overall catalytic activity of the samples cannot be made. Notably, the catalytic activity of Pd@CN-1 was superior compared to the conventional catalysts such as Pd@MgO, Pd@γ-Al_2O_3, commercial Pd@C, and Pd@TiO_2.

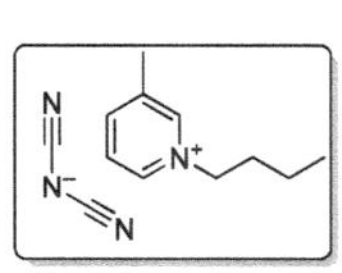

Fig. 3 The structure of IL 1 [26].

3 Carbonization of confined ionic liquids

As mentioned, carbonization of conventional ILs that do not possess cross-linking functionalities is not successful and does not lead to the formation of carbon material. This issue can be attributed to the lack of polymeric intermediates and the formation of volatile components in the course of carbonization [17]. To circumvent this problem, carbonization of ILs entrapped within a nano-confined framework has been suggested to be beneficial. In a pioneer research, Dai et al. carbonized IL-containing monolithic silica gels, prepared through sol-gel polymerization of tetraethyl orthosilicate (TEOS) in the presence of formic acid and an IL, such as BMIM-NTf_2. The comparison of thermal treatment of BMIM-NTf_2 and BMIM-NTf_2-containing silica gel confirmed that the former led to no char residue, while the latter resulted in the formation of silica-carbon composite. It is believed that the noncovalent interactions such as H-bonding and interactions between ILs and silica surface could alter the properties of IL and justify the improved thermal stability of the confined ILs [17]. Notably, the carbon material can be achieved simply by etching silica with HF [27]. The protocol was also successfully applied for the preparation of titania-carbon composite. Karimi and his coworkers reported on the preparation and characterization of novel nano-fibrillated mesoporous carbon (IFMC) through carbonization of the ionic liquid 1-methyl-3-phenethyl-1*H*-imidazolium hydrogen sulfate at 900°C in the presence of SBA-15 and a mold [24, 28]. They also developed other IL-derived carbons such as IL-derived ordered mesoporous carbons functionalized with sulfonic acid groups IOMC-$ArSO_3H$ [29]. These ordered mesoporous carbons have been discussed in more details in Section 7.4. Following these reports, other examples of carbonization of conventional ILs have been disclosed. As an additional example, Yu, Zhang et al. reported on the synthesis of nitrogen-doped hollow carbon sphere (HCSs) with very thin walls (5–12 nm) by using monodisperse silica spheres as the hard template and conventional ILs as carbon precursors. HCSs were synthesized through assembly of ILs on the surface of negatively charged silica followed by carbonization and etching of silica template. Interestingly, the wall thickness could be tuned by altering the IL nature and molar ratio of IL to silica sphere [30].

4 Poly ionic liquids carbonization

Poly(ionic liquid)s (PILs), also known as polymerized ILs, are polyelectrolytes that comprise a polymeric backbone and an IL species in monomer

Table 3 The chemical formula and abbreviations of PLIs used for carbonization.

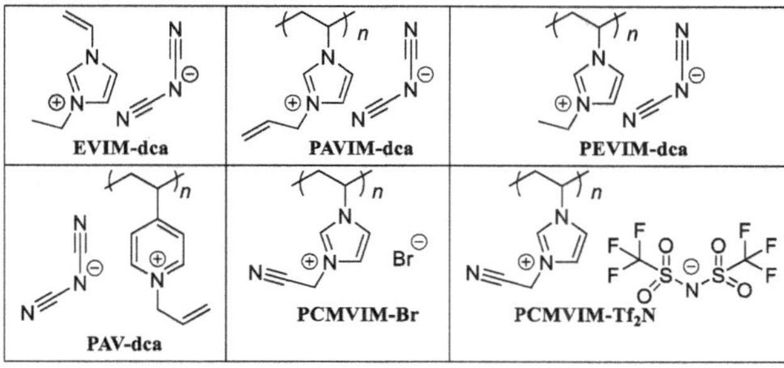

Adapted from T.-P. Fellinger, et al., 25th anniversary article: "Cooking carbon with salt": carbon materials and carbonaceous frameworks from ionic liquids and poly(ionic liquid)s, Adv. Mater. 25(41) (2013) 5838–5855, with permission from John Wiley and Sons.

repeating units [31], benefiting from the advantages of ILs and the polymer nature. PILs can be considered as promising carbon precursors. Compared to ILs, the polymer nature of PILs allowed their facile shaping and processing through some classical polymer methods, such as spin-coating, extrusion, electro spinning, wet spinning, embossing, etc. [21]. Moreover, it is expected that carbonization of PILs leads to higher yield of carbon. In Table 3, some of the common PILs are collected.

4.1 Nitrogen-doped carbon materials with definite morphology derived from PILs and their utilities

Similar to ILs, PILs were also exploited for the formation of nitrogen-doped carbon materials with definite morphology. As an example, EVIM-BF_4, as carbon precursor, was coated on monodisperse SiO_2 particles as the template, through in situ polymerization to afford core/shell composite sphere. Carbonization of the latter followed by the removal of silica template led to the formation of HCS with high surface area, good conductivity, and suitable porosity for mass transport. The resulting HCS was further applied as a support for Pt electrocatalyst (Fig. 4). Fine Pt nanoparticles (average size of 2.8 nm) were homogeneously dispersed onto the HCSs. The obtained Pt/HCSs exhibited a significant catalytic activity for the oxidation of methanol superior to Pt/Vulcan XC-72 catalyst [32].

In another example, Yuan et al. reported synthesis of nitrogen-doped carbon capsules by employing PILs as precursors (Fig. 5 and the reference cited in Fig. 5 caption). As shown in Fig. 5, the procedure included coating of PILs on the SiO_2 template via layer-by-layer assembly approach followed

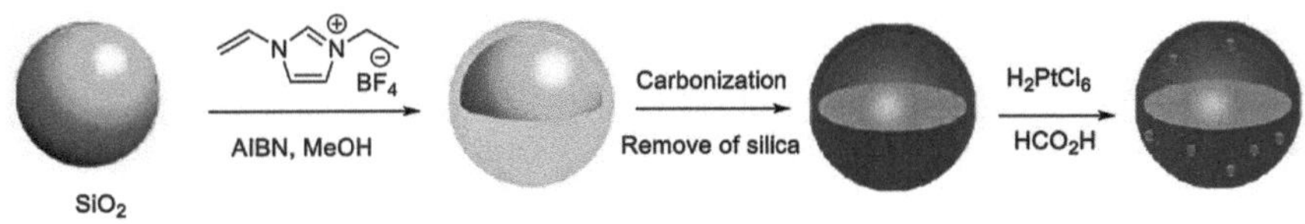

Fig. 4 Preparation pathway of the HCSs and Pt/HCSs nanocomposites. *(Adapted from X. Bo, et al., Highly dispersed Pt nanoparticles supported on poly(ionic liquids) derived hollow carbon spheres for methanol oxidation, J. Power Sources 196(20) (2011) 8360–8365, with permission from Elsevier.)*

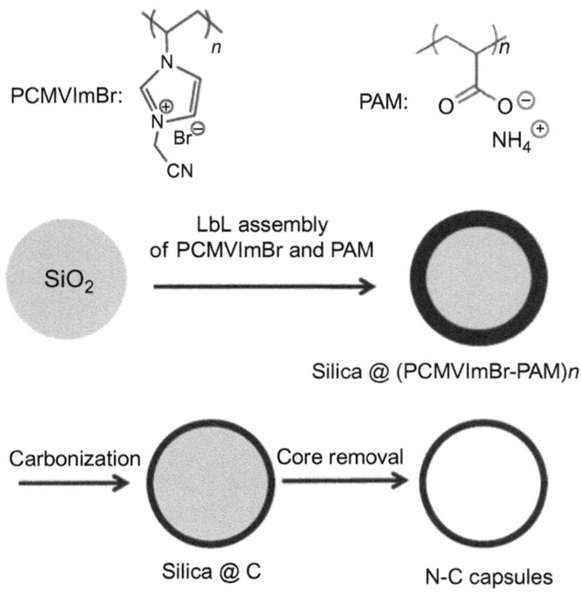

Fig. 5 Preparation of nitrogen-doped carbon capsules via PILs-based layer-by-layer assembly. *(Adapted from Q. Zhao, et al., Nitrogen-doped carbon capsules via poly(ionic liquid)-based layer-by-layer assembly, Macromol. Rapid Commun. 33(13) (2012) 1149–1153, with permission from John Wiley and Sons.)*

by carbonization and etching of silica template. The authors applied PCMVIM-Br as cationic PILs and poly(ammonium metacrylate) (PAM) as anionic polyelectrolyte to prepare carbon capsules with specific surface area of $423\,m^2\,g^{-1}$ and nitrogen content of ≈ 7 wt%.

In another research, Nabid et al. described the synthesis of a yolk/shell Fe_3O_4@PIL-derived nitrogen-doped graphitic porous carbon material (Fig. 6) through a two-stage radical polymerization of 3-cyanomethyl-1-vinylimidazolium chloride (CMVImCl) in the presence of 2,2′-azobisisobutyronitrile (AIBN) followed by carbonization at 900°C. The

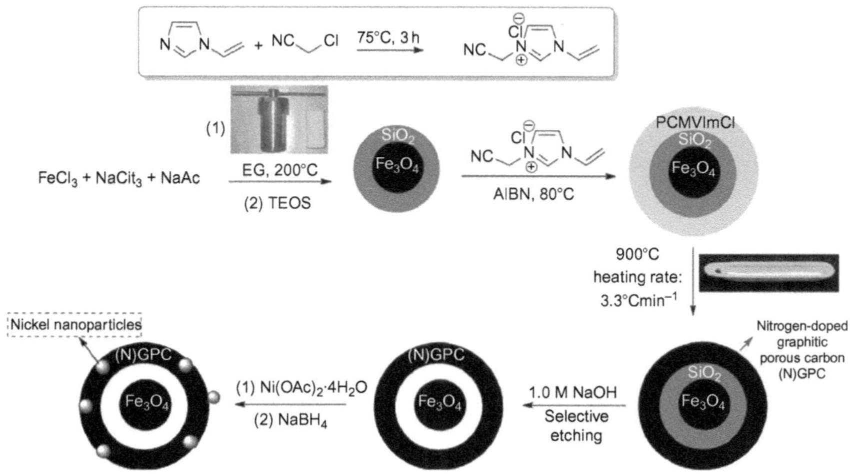

Fig. 6 Schematic diagram for the synthesis of yolk/shell Fe_3O_4@PIL-derived nitrogen-doped graphitic porous carbon material. *(Adapted from M.R. Nabid, Y. Bide, Z. Habibi, Synthesis of a yolk/shell Fe_3O_4@poly(ionic liquid)s-derived nitrogen doped graphitic porous carbon materials and its application as support for nickel catalysts, RSC Adv. 5(3) (2015) 2258–2265, with permission from Royal Society of Chemistry.)*

resulting composite material was then utilized for the immobilization of Ni nanoparticles to furnish a magnetic heterogeneous and recyclable catalyst for promoting the cascade dehydrogenation of $NaBH_4$ and hydrogenation of various nitro/nitrile compounds in aqueous media [33].

In an alternative approach for the preparation of magnetic nitrogen-doped hallow spheres, Wei, Yuan et al. applied silica nanoparticles as template for the low temperature polymerization of CMVImBr to give the corresponding poly(3-cyanomethyl-1-vinylimidazolium bromide) (PCMVIMBr)-derived nitrogen-doped carbon spheres with relatively high surface area and 7–10 wt% nitrogen content. The latter was then used for the immobilization of magnetic nanoparticles through impregnation-thermal decomposition strategy (Fig. 7). The magnetic composite exhibited high performance and good cycling stability as anode material for lithium-ion batteries. The authors suggested that the confinement effect of the magnetic nanoparticles as well as the high electric property of HCS may be the major sources of the observed improvement of the anode performance [34].

Apart from spherical morphology, there are several reports on the synthesis of tubular nanocomposites by using PILs as carbon precursors [35, 36]. As an example, Chen et al. reported on an efficient electrocatalyst for

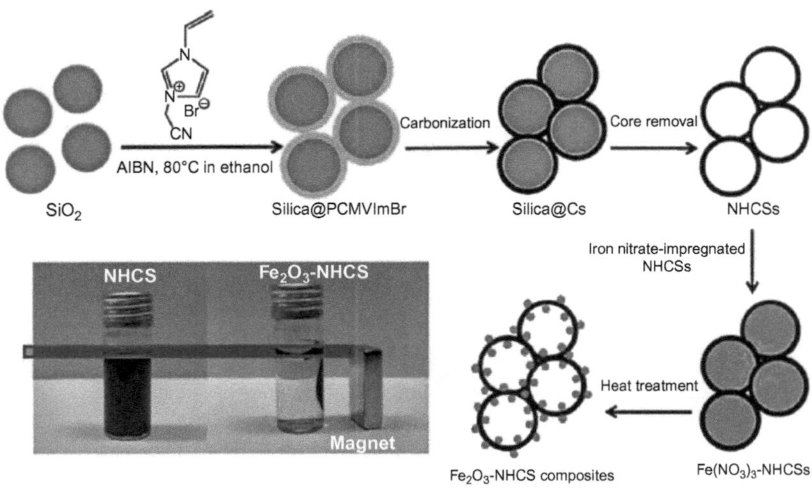

Fig. 7 Schematic illustration of the formation of NHCSs based on the polymerization of 3-cyanomethyl-1-vinylimidazolium bromide (CMVImBr) monomer onto silica nanoparticle templates, and the further preparation of Fe_2O_3-HCS composites via impregnation-thermal decomposition of iron nitrate. *(Adapted from J. Balach, et al., Poly(ionic liquid)-derived nitrogen-doped hollow carbon spheres: synthesis and loading with Fe_2O_3 for high-performance lithium ion batteries, RSC Adv. 3(21) (2013) 7979–7986, with permission from Royal Society of Chemistry.)*

MeOH oxidation by immobilization of bimetallic (Pt-Ru) nanoparticles on a nitrogen-enriched carbon nanotube (SNE-CNTs) prepared by carbonization of IL-covered multiwalled carbon nanotubes (Fig. 8) [36]. The surface functionalization of carbon nanotubes was first started with polymerization of EVIM-Br in the presence of AIBN and then ended with the carbonization of the resulting composite products at 900°C. Notably, the Pt-Ru loading in the catalyst was higher than that of bare carbon nanotube, confirming the efficiency of the support for anchoring the nanoparticles.

It is worth noting that the utility of nitrogen-doped carbon nanotubes and fibers prepared from functionalization with PIL and subsequent carbonization is not limited to the catalytic purposes. As an example, Yuan et al. have recently developed different kinds of nitrogen-doped microporous carbon fibers (NPCN) through initial synthesis of PCMVIMBr onto outer surface of porous carbon fibers (PCF) followed by carbonization at 550 and 750°C (Fig. 9). The developed carbon/carbon improved the interaction between carbon dioxide and PCF for unusually high CO_2 uptake and resistive sensing [35].

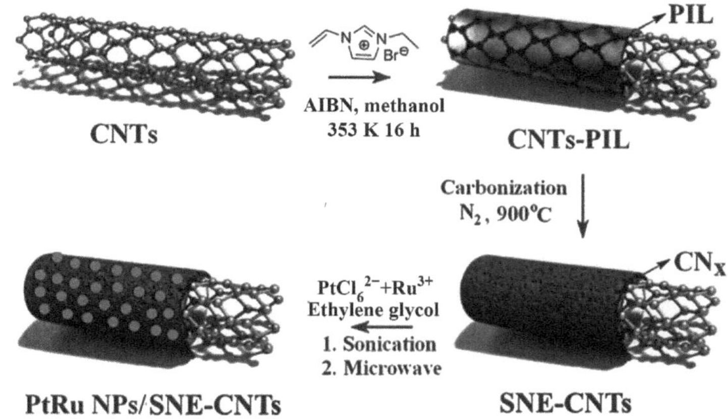

Fig. 8 Schematic diagram for the preparation of SNE-CNTs and Pt-Ru NPs/SNE-CNT nanohybrids. *(Adapted from B. Wu, et al., Carbonization of ionic liquid polymer-functionalized carbon nanotubes for high dispersion of PtRu nanoparticles and their electrocatalytic oxidation of methanol, J. Mater. Chem. 22(26) (2012) 13085–13090, with permission from Royal Society of Chemistry.)*

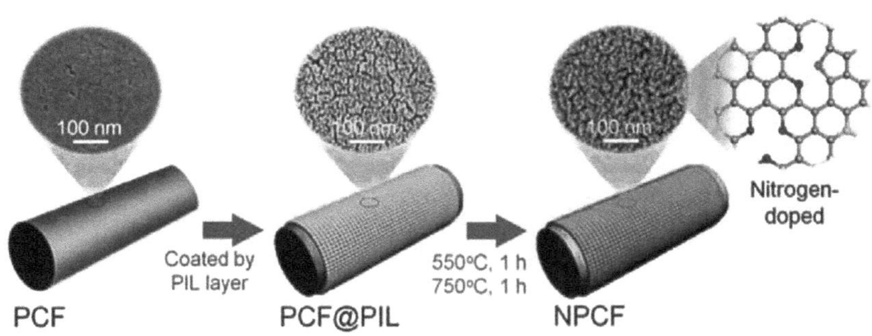

Fig. 9 Schematic illustration of NPCF preparation. *(Adapted from J. Gong, M. Antonietti, J. Yuan, Poly(ionic liquid)-derived carbon with site-specific N-doping and biphasic heterojunction for enhanced CO_2 capture and sensing, Angew. Chem. Int. Ed. 56(26) (2017) 7557–7563, with permission from John Wiley and Sons.)*

4.2 Examples of the utility of PILs-derived carbon for the catalysis

Higher electrical conductivity of nitrogen-doped ILs-derived carbon materials compared to heteroatom-free carbon materials renders them practical candidates for electrocatalysis.

In an example, Yan et al. developed a novel electrocatalyst by copolymerization of acrylonitrile with [BVIM][Br] followed by ion-exchange with

H_2PtCl_6 and carbonization according to the procedure outlined in Fig. 10 [37]. It was found that the content of nitrogen in the final carbon as well as yields of carbonization stage could be tuned by adjusting the molar ratio of IL in the copolymer. Moreover, the capability of IL for ion exchanging with Pt species resulted in high Pt loading and homogeneous dispersion of Pt nanoparticles. Notably, the prepared catalyst showed high electrocatalytic activity and stability for oxidation of methanol.

In another research, using poly(1-butyl-4-vinylpyridinium dicyanamide) and polyacrylonitrile (PAN) precursors, N-doped carbon fibers and thin films were prepared by applying electrospinning (Fig. 11) and dip-coating approaches, followed by carbonization. The nitrogen content of the resulting carbons depended on the carbonization temperature and reached to 11–13 wt% at 800°C. Moreover, the materials displayed high electric conductivity that was attributed to the presence of predominantly graphitic nitrogen atoms incorporated throughout the samples and to the

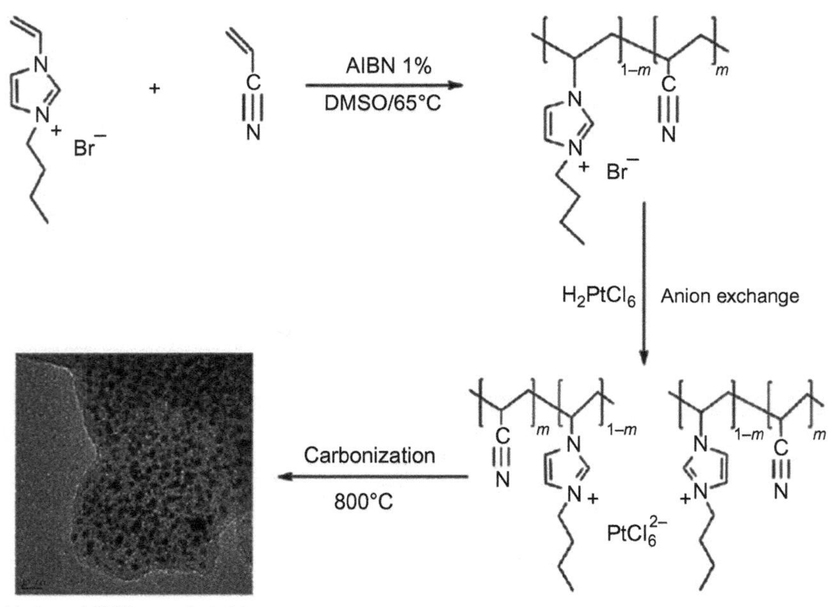

Fig. 10 Schematic diagram of preparing nitrogen-doped carbon-Pt nanohybrids. *(Adapted from C. Pan, et al., Facile synthesis of nitrogen-doped carbon–Pt nanoparticle hybrids via carbonization of poly([Bvim][Br]-co-acrylonitrile) for electrocatalytic oxidation of methanol, J. Mater. Chem. 2(27) (2012) 13578–13584, with permission from Royal Society of Chemistry.)*

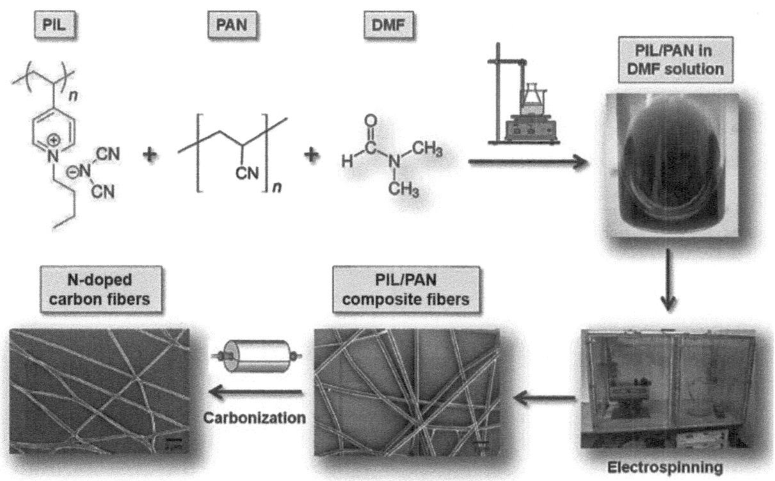

Fig. 11 Schematic illustration of the preparation-path to N-doped carbon nanofibers: the precursors PIL and PAN were dissolved in DMF. *(Adapted from M. Einert, et al., Nitrogen-doped carbon electrodes: influence of microstructure and nitrogen configuration on the electrical conductivity of carbonized polyacrylonitrile and poly(ionic liquid) blends, Macromol. Chem. Phys. 216(19) (2015) 1930–1944, with permission from John Wiley and Sons.)*

formation of a π-conjugated aromatic carbon network composed of graphene stacks with a small number of single sp^2 layers [38]. It was claimed that the obtained N-doped carbons could potentially be used for electrode materials in Li-ion batteries.

In another study [13], a Co-N-carbon catalyst was prepared in which a hydrolyzed vinyl imidazolium dihydrogen phosphate served as polymerizable ionic liquid and N-doping precursor. The polymerized materials were first allowed to react with $Co(NO_3)_2 \cdot 6H_2O$, followed by carbonization at 600°C to furnish the final Co-N-C catalyst. The catalyst could efficiently promote ORR reaction. Similar to the previous example, the electrocatalytic activity of the present catalytic system was superior to the samples that were prepared from either nonpolymerizable IL or non-IL precursor under almost the same conditions. The observed higher catalytic activity was attributed to the presence of more nitrogen with more pyridinic and pyrrolic nitrogen group in the catalyst.

Very recently, the synthesis of P and N co-doped carbon material with significant electrocatalytic activity for oxygen reduction reaction (ORR) was reported [39]. To prepare the catalyst, vinyl imidazolium dihydrogen phosphate possessing both N and P atoms in its structure was employed as a starting ionic liquid monomer. After polymerization of vinylic ionic liquid and subsequent pyrolysis at 560°C for 1 h the catalyst was prepared. It was clearly found that the resulting carbonaceous materials displayed superior catalytic activity when compared to that obtained from its N-doped carbonaceous analogs. Moreover, the catalyst exhibited higher catalytic activity than P,N-doped carbon prepared from nonpolymerizable IL, methyl imidazolium dihydrogen phosphate.

5 Protic ILs and salts for synthesis of (porous) doped carbons

As mentioned, only limited number of expensive ILs containing cyano/nitrile functionalities, which are generally prepared through tedious and time-consuming multistep synthetic procedures, can be successfully applied for the preparation of N-doped carbons in the absence of a template. Very recently, Watanabe et al. disclosed the utility of a range of protic ILs and salts (PILs/PSs) as carbon precursors without the need for either metal catalysts or cross-linkable functionalities. PILs/PSs are cost-effective and can be obtained simply by neutralization of an acid and a base followed by carbonization (Fig. 12). Moreover, diverse range of nitrogen-containing bases, including heterocycles, and amines with low boiling points and high volatilities can be used to directly react with inorganic or organic acids such

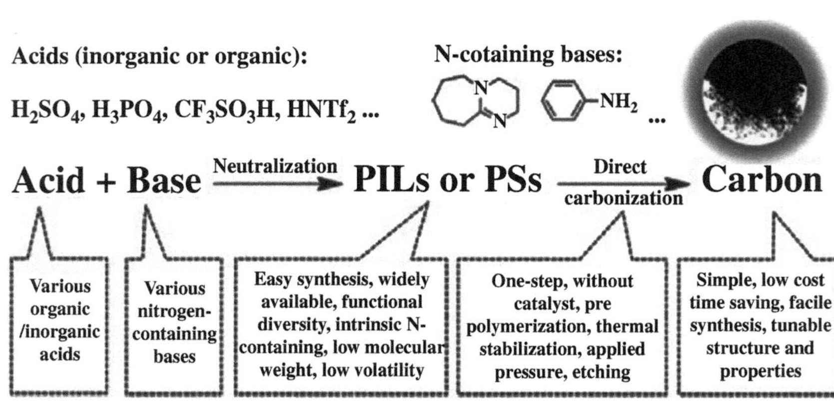

Fig. 12 Schematic representation of the synthesis of carbon materials by direct carbonization of PILs/PSs. *(Adapted from S. Zhang, K. Dokko, M. Watanabe, Direct synthesis of nitrogen-doped carbon materials from protic ionic liquids and protic salts: structural and physicochemical correlations between precursor and carbon, Chem. Mater. 26(9) (2014) 2915–2926, with permission from American Chemical Society.)*

as *p*-toluenesulfonic acid (TsOH), 1,1,1-trifluoromethansulfonic acid (TfOH), HNO_3, H_3PO_4, and H_2SO_4 to afford the corresponding protonated salts. It is believed that protonation of a base by an acid not only depresses the volatility and enhances the thermal stability of the original base, but also mediates the in situ thermal decomposition and carbonization process, thus affecting the structures and properties of the final carbons [40].

Using this approach, carbon materials with tunable nitrogen contents up to 11.1% and high conductivities, superior even to that of graphite, can be furnished [40]. The yield of carbonization as well as the features of the resulting carbon such as porosity, thermal stability, and graphitic structure can be adjusted by smart choice of the acid, bases, and other functional groups in their structure. In Table 4, the structures of some PILs/PSs used for the direct carbonization are depicted.

Following the first report on the utility of PILs/PSs as carbon precursors, other research groups have developed a few carbon materials with more or less the same approach for diverse range of applications, such as catalysis and supercapacitors. Along this line, Dong, Yang et al. reported on a yolk-shell Fe_3O_4 composite nanoparticle with individual Fe_3O_4 nanoparticle as the yolk and N-doped carbon derived from carbonization of a typical ionic liquid as the shell [41]. The materials were prepared by initial coating of magnetic nanoparticles with silica layer followed by two-step carbonization of 3-cyanopyridine/H_2SO_4 at 875°C and etching of silica layer by a 5% NaOH aqueous solution. This unique structure would not only afford adequate void to accommodate the large volume change during charge/

Table 4 The structure of some PILs/PSs used for the direct carbonization.

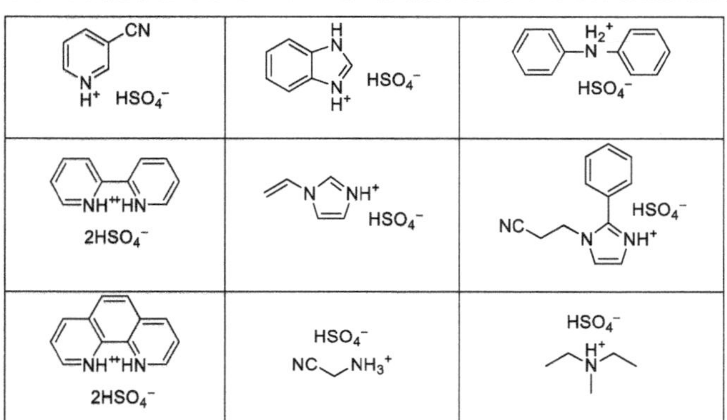

Adapted from S. Zhang, K. Dokko, M. Watanabe, Carbon materialization of ionic liquids: from solvents to materials, Mater. Horiz. 2(2) (2015) 168–197, with permission from Royal Society of Chemistry.

discharge process, but also improved both the structural stability and electrical conductivity of the materials. Electrochemical tests demonstrated that Fe_3O_4@void@N-doped carbon showed a high reversible capacity of 860 and 500 mAh/g were retained after 500 charge/discharge cycles at current rates of 1000 and 2000 mA/g, respectively [41].

In another report, N, S-doped carbons were prepared by using the carbonization of protic salt of 5-carboxybenzene-1,3-diamine sulfate [DABA][$2HSO_4$] which was proposed to serve as the carbon and heteroatom source as well as the pore-forming agent during carbonization. The pore structure parameters and chemical compositions of carbon could be simply tuned by adjusting the carbonization temperature. It was found that carbonization at 900°C resulted in a carbon with large surface area (1543 $m^2 g^{-1}$) and nanorod morphology with moderate content of heteroatom functionalities (5.80 at.% N and 0.44 at.% S) [42].

5.1 Carbons derived from protic ILs and salts as catalysts

In one of the pioneer studies, N,P co-doped carbon material with electrocatalytic activity for ORR was developed by polymerization of vinyl imidazolium dihydrogen phosphate followed by pyrolysis (Fig. 13) [39]. In this approach, the anions provided P atoms, while the cation introduced N atoms in the final carbon material. Notably, the electrocatalytic performance of the N,P-co-doped carbon (an onset potential of 0.92 V in alkaline electrolytes via a four-electron pathway) was superior compared to its N-doped counterpart and the N,P-co-doped carbon obtained from nonpolymerizable IL. This observation was attributed to the good intrinsic catalytic activity derived from the co-doping effect of N and P.

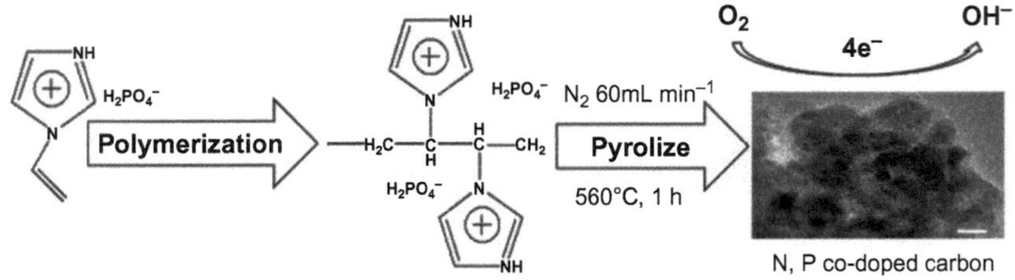

Fig. 13 Illustration of the synthesis procedure of the catalyst. *(Adapted from J. Gao, et al., Polymerizable ionic liquid as a precursor for N, P co-doped carbon toward the oxygen reduction reaction, Catal. Sci. Technol. 8(4) (2018) 1142–1150, with permission of Royal Society of Chemistry.)*

In another example, using PIL-hydrolyzed vinyl imidazolium nitrate, Yin, Lu, and Feng and their coworkers reported on the synthesis of a nitrogen-doped and curved carbon nanosheet in one-pot procedure [43]. The authors investigated the electrocatalytic activity of the metal-free N-doped curved sheet as ORR catalyst. It was found that due to the nitrogen doping as well as three-dimensional cross-linkable porous structure, the catalyst provided an onset potential of 0.92 V RHE for ORR in 0.1 M KOH solutions. Moreover, it exhibited significant CH_3OH tolerance and a catalytic activity toward oxygen evolution reaction (OER).

Recently, Liu, Gan et al. disclosed preparation of N,S-co-doped ultramicroporous carbon nanoparticles (N/S-UCNs) via simple polymerization and carbonization of an ionic liquid (IL) following the steps demonstrated in Fig. 14 [44]. As it can be clearly seen, PIL p[pPD][2HSO$_4$] was first generated by means of chemical oxidative polymerization of *p*-phenylenediamine sulfate ([pPD][2HSO$_4$]). In the next stage, PIL served as both carbon precursor and the source of heteroatoms as well as self-template to create regular ultramicropores throughout the carbonaceous materials during carbonization stage. The resulting carbon material (N/S-UCNs) possessed high surface area (505–1018 $m^2\,g^{-1}$) and high doping content of heteroatoms (N and S) with uniform ultramicropores (0.50–0.59 nm). Moreover, it showed high electrochemical performance as electrode for supercapacitors. The electrode exhibited high specific capacitance (225 $F\,g^{-1}$ at 2.0 $A\,g^{-1}$), long-term cycle stability (90.8% retention after 10,000 cycles), and excellent rate capability (160 $F\,g^{-1}$ at 20 $A\,g^{-1}$) in alkaline electrolyte.

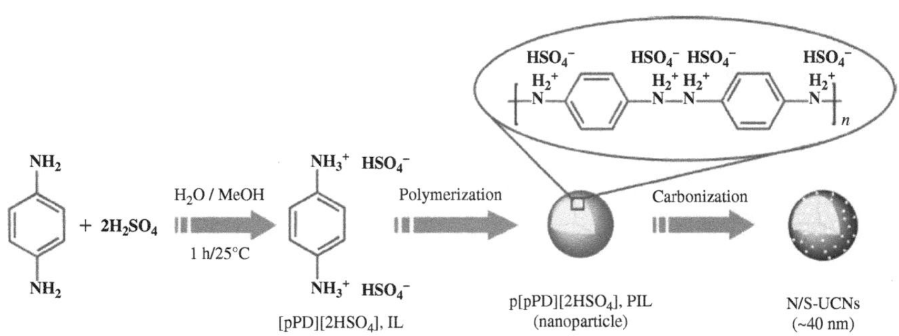

Fig. 14 Schematic synthesis route of N/S-UCNs via polymerization of IL of [pPD][2HSO$_4$], followed by a carbonization process. *(Adapted from L. Miao, et al., Poly(ionic liquid)-derived, N, S-codoped ultramicroporous carbon nanoparticles for supercapacitors, Chem. Eng. J. 317 (2017) 651–659, with permission from Elsevier.)*

The same research group also developed N, S-doped porous carbon nanosheets (N/S-HCSs) by precarbonization and postactivation of a low-cost protic salt of *p*-phenylenediamine toluenesulfate ([pPD] [$2CH_3PhSO_3$]). [pPD][$2CH_3PhSO_3$] which was prepared by a simple neutralization reaction at ambient temperature (Fig. 15) [45]. It was supposed that the employed protic salt acted as a "all-in-one" precursor, while acting as self-templating agent and serving as the source of carbon and heteroatoms in the final carbon materials. N/S-HCSs have high specific surface area and nanosheet geometry which provide enough adsorbing sites for charge accumulation and shorten diffusion distance of electrolyte ions at electrode/electrolyte interface. Besides, N/S-HCSs exhibit unique porous nanoarchitecture with uniform ultramicropores and a well-developed network of supermicropores and mesopores, which altogether make the material highly efficient ion-highways for electrolyte diffusion and transport. Furthermore, incorporated heteroatoms in the carbon framework structure improve the electrical conductivity and surface wettability, thus providing extra pseudocapacitance. As a result, N/S-HCS electrode delivers superior electrochemical performance including high gravimetric capacitance ($280\,F\,g^{-1}$ at $1.0\,A\,g^{-1}$), good rate performance ($134\,F\,g^{-1}$ at $50\,A\,g^{-1}$), and cycling stability (94.4% retention after 10,000 cycles at $2.0\,A\,g^{-1}$) in 6M KOH electrolyte.

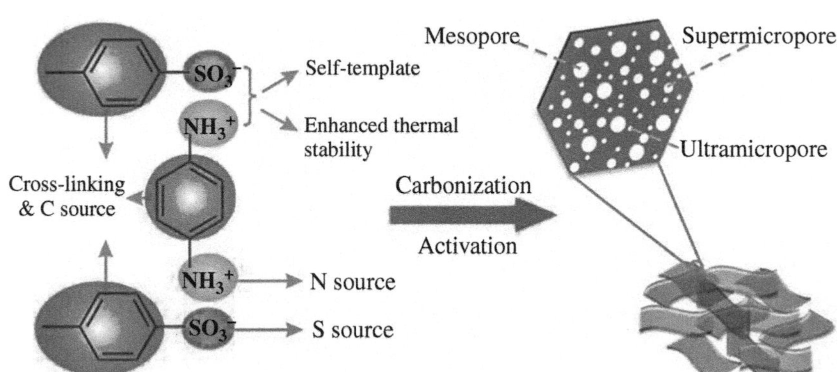

Fig. 15 Schematic synthesis route of N/S-HCSs with ultramicro-, supermicro-, and mesoporous structure via carbonization-activation of [pPD][$2CH_3PhSO_3$] protic salt. *(Adapted from L. Miao, et al., Cooking carbon with protic salt: nitrogen and sulfur self-doped porous carbon nanosheets for supercapacitors, Chem. Eng. J. 347 (2018) 233–242, with permission from Elsevier.)*

6 Coating other materials with ILs-derived carbons

It is worth to mention that ILs can not only be used as precursors for the preparation of N-doped carbons, but are also applicable for creating a thin coating on the conventional carbon materials such as carbon nanotubes, CNT [46]. Because of the fact that in this approach less amount of costly ILs is used, it is cost-effective and economical. Moreover, this method allows benefiting from the properties of both carbon nanotubes and N-doped carbons in a single material. As one of the examples of these carbon composites, Tuaev et al. reported coating of surface-oxidized CNTs with nitrogen-doped carbon derived from 3MBP-dca [46]. The choice of CNT as host was based on the fact that its defined tubular morphology could allow homogeneous coverage. The nitrogen-doped carbon, N-CNTs, was then used as a support for the deposition of Pt nanoparticles. To compare the effect of nitrogen in the resulted carbon, the authors applied Pt-CNT and Pt-N-CNT for the preparation of a liquid catalyst ink suspension, cast into a thin catalyst electrode film to study and compare their oxygen reduction reaction (ORR) performance. In situ surface area measurements and particle size trajectories suggested a significantly reduced particle coarsening and degradation behavior of Pt-N-CNT. It was believed that the surface nitrogen could mediate the enhancement of Pt adsorption on the surface of carbon materials.

In another attempt, Joo et al. prepared carbon nanotube/heteroatom-doped carbon core-sheath nanostructure (CNT/HDC) by adsorption of IL on CNT followed by carbonization (Fig. 16) [47]. The nanocomposite prepared in this way benefited from the reactivity of heteroatom-doped carbon layer as well as electric conductivity of CNT and exhibited high electrocatalytic activity and kinetic for the ORR, comparable to conventional Pt/C catalyst. Interestingly, the metal-free catalyst showed superior tolerance toward methanol and durability compared to Pt/C.

It is worth noting that coating of ILs-derived carbon not only was carried out on carbon materials, but also on inorganic compounds such as porous $Li_4Ti_5O_{12}$ [48].

7 Porous IL-derived carbon

Despite promising properties of heteroatom-doped carbon materials derived from ILs, their textural properties, such as specific surface area and porosity, are not appealing. Hence, modification of ILs-derived carbons to improve

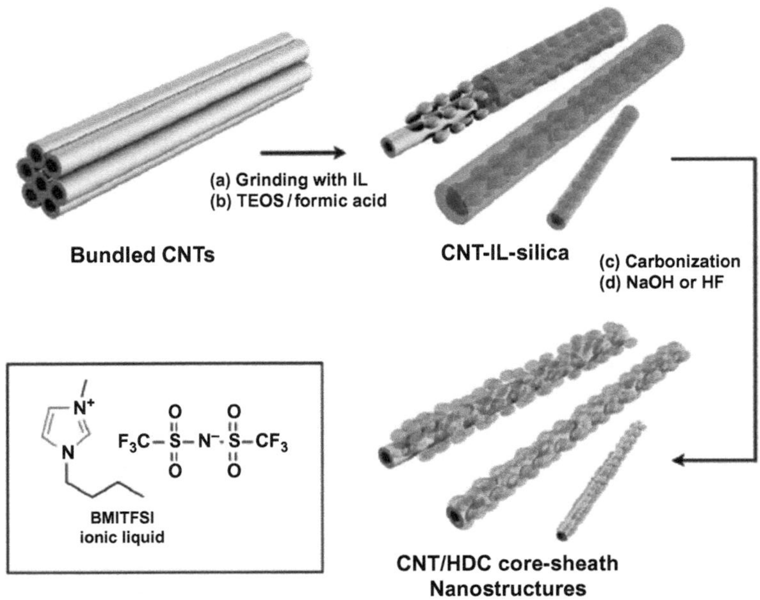

Fig. 16 Synthesis of the CNT/HDC core-sheath nanostructures. *(Adapted from Y.J. Sa, et al., Carbon nanotubes/heteroatom-doped carbon core–sheath nanostructures as highly active, metal-free oxygen reduction electrocatalysts for alkaline fuel cells, Angew. Chem. Int. Ed. 53(16) (2014) 4102–4106, with permission from John Wiley and Sons.)*

their features including the structural and textural porosity is of great interest and can expand their applications. As porous carbon materials are rich in active sites and can provide better mass transport and diffusion, they are considered as potential candidates for (electro) catalytic purposes. The porous ILs-derived carbons can be prepared in three pathways: (1) template-free; (2) the use of salt template; and (3) use of porous templates [18]. In the following, these methods are briefly described.

7.1 Template-free approach

In this protocol, no hard template or structural mold is used and porous carbons are obtained via direct carbonization. Although this method is facile, the wise choice of the carbon precursors is a crucial and determining factor. More precisely, most of the precursors have low thermal stability and evaporate or decompose under thermal treatment. Moreover, in the cases that the precursors yield in carbon materials, the resulting carbon is not porous due to the lack of suitable and stable self-sacrificial porogens [18].

7.2 Use of porous templates

As there are not many choices for the preparation of porous carbon materials through template-free method, an alternative approach has been introduced, in which hard templates such as mesoporous silica or disordered colloidal silica are required. Notably, the interaction of ILs with the inorganic surface of the hard template can trigger homogeneous coverage of ILs on the surface of the template. One of the templates used for this purpose is SBA-15 [24, 28, 49, 50]. Using this structural guide, mesoporous carbon materials with high specific surface area can be prepared. However, other silica materials such as LUDOX® that are cost-effective and commercially available are also well-considered as hard template [51, 52]. Notably, other inorganic porous materials such as magadiite and porous alumina membrane have been used for this purpose [53]. Moreover, metal-organic frameworks (MOFs) that benefit from exceptionally large surface areas and controlled pore textures have also been used successfully to furnish porous carbon materials [54]. As an example in this regard, Jhung et al. revealed the utility of Al-based metal-organic frameworks (MOF) for the synthesis of nitrogen-doped porous carbon material. To synthesize the MOF-derived porous carbon material, CDIL@AlPCP, the MOF (AlPCP) was loaded with IL via ship-in-bottle (SIB) technique and then carbonized under inert flow (Fig. 17) [55]. The results indicated that the nitrogen content in the final catalyst increased upon increasing the content of IL. Notably, the prepared porous carbon exhibited high adsorptive performance for both aqueous and nonaqueous phase purifications despite the decreased porosity with IL loading in AlPCP.

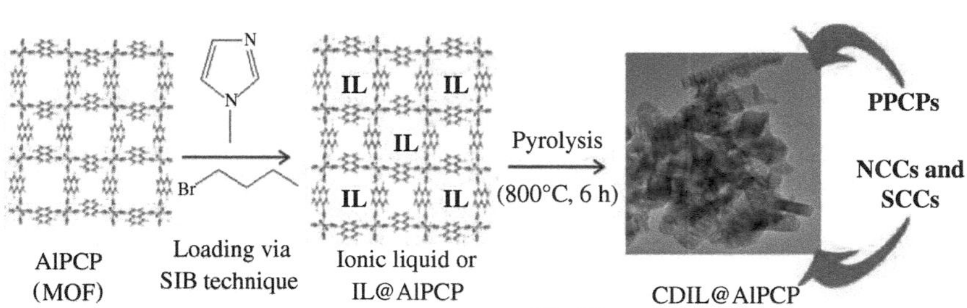

Fig. 17 The synthetic procedure for CDIL@AlPCP. *(Adapted from M. Sarker, et al., Nitrogen-doped porous carbon from ionic liquid@Al-metal-organic framework: a prominent adsorbent for purification of both aqueous and non-aqueous solutions, Chem. Eng. J. 338 (2018) 107–116, with permission from Elsevier.)*

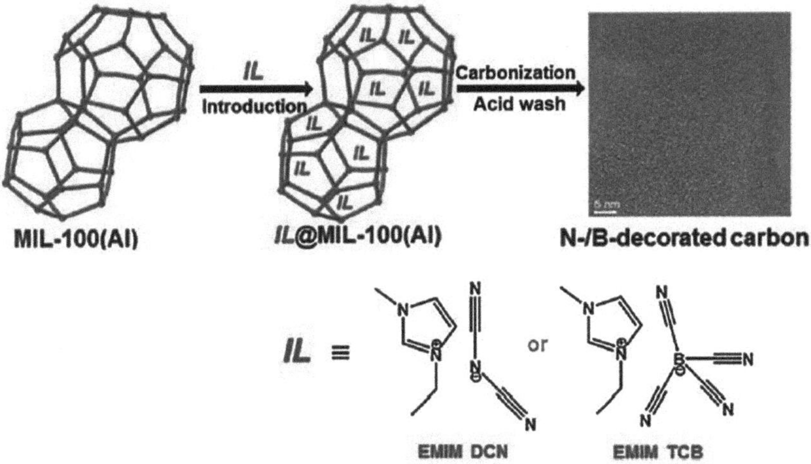

Fig. 18 Schematic illustration of the synthetic procedure for N- and BN-decorated nanoporous carbons with molecular structures of the used ionic liquids (ILs). *(Adapted from A. Aijaz, et al., From ionic-liquid@metal–organic framework composites to heteroatom-decorated large-surface area carbons: superior CO_2 and H_2 uptake, Chem. Commun. 50(49) (2014) 6498–6501, with permission from Royal Society of Chemistry.)*

In another instance, an aluminum-based zeolite-type MOF, MIL-100 (Al), that possessed high surface area and 3D structure with two large cavities (free diameters of 2.5 and 2.9 nm) was employed for the synthesis of porous boron and nitrogen-doped carbon. In this line, two types of ILs, EMIM DCN and EMIM TCB (Fig. 18 and the reference cited in Fig. 18 caption), were applied as carbon precursors and impregnated with MIL-100(Al) to afford IL@MOF. Subsequently, the composite was thermally treated and carbonized. To remove the hard template, HF was applied to wash MOF and afford doped carbon materials with narrow pore size and high specific area. Notably, the resulting carbon showed high CO_2 and H_2 uptake capacity, rendering them an appealing candidate for flue gas separation.

7.3 Salt templates

Another important approach for the preparation of porous doped carbon materials was first introduced by Fechler et al. as "Salt templating," in which carbon precursor and a noncarbonizable inorganic eutectic salt that acted as a template were mixed to furnish a glassy solid that condensed and scaffolded in the presence of the molten salt at elevated temperatures [56]. In this process, the miscibility between the carbonizing precursors and the salt melt

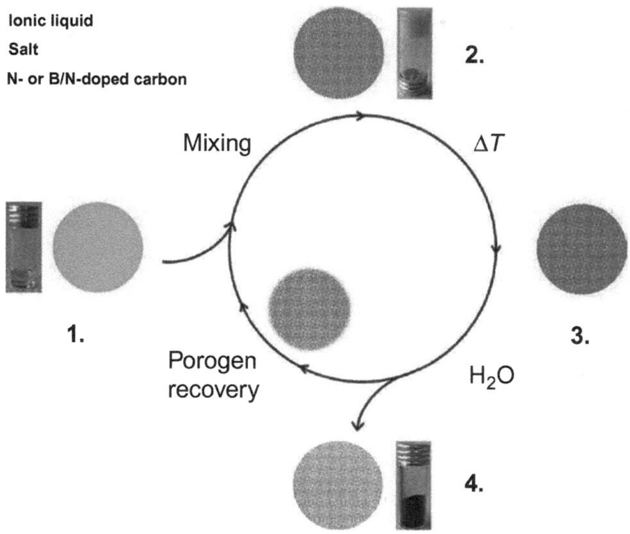

Fig. 19 Description of product and surface area formation using the salt templating approach. *(Adapted from N. Fechler, T.-P. Fellinger, M. Antonietti, "Salt templating": a simple and sustainable pathway toward highly porous functional carbons from ionic liquids, Adv. Mater. 25(1) (2013) 75–79, with permission from John Wiley and Sons.)*

generally leads to high specific surface area materials. Moreover, the size of the pores in the resulting porous carbons corresponded to the salt clusters and their percolation structures. After carbonization, the removal of salt can be easily achieved by simply washing with water. The schematic representation of the salt templating method is illustrated in Fig. 19.

7.4 Porous ILs-derived carbon for catalysis derived from nanocasting

In a pioneer work, Karimi et al. reported synthesis of a novel nano-fibrillated mesoporous carbon, IFMC, using1-methyl-3-phenethyl-1*H*-imidazolium hydrogen sulfate as carbon precursor and SBA-15 as structural guide. IFMC was then applied as an efficient support for the immobilization of Pd catalytic species to furnish Pd@IFMC catalyst (Fig. 20) with the utility for promoting organic transformations such as aerobic oxidation of aliphatic and allylic alcohols and Ullmann homocoupling reactions of aryl halides in aqueous media [24, 28].

In the case of alcohol oxidation reaction, the catalyst showed high catalytic activity and could promote the aerobic oxidation of benzylic, heterocyclic, and heteroatom containing alcohols with either molecular oxygen or

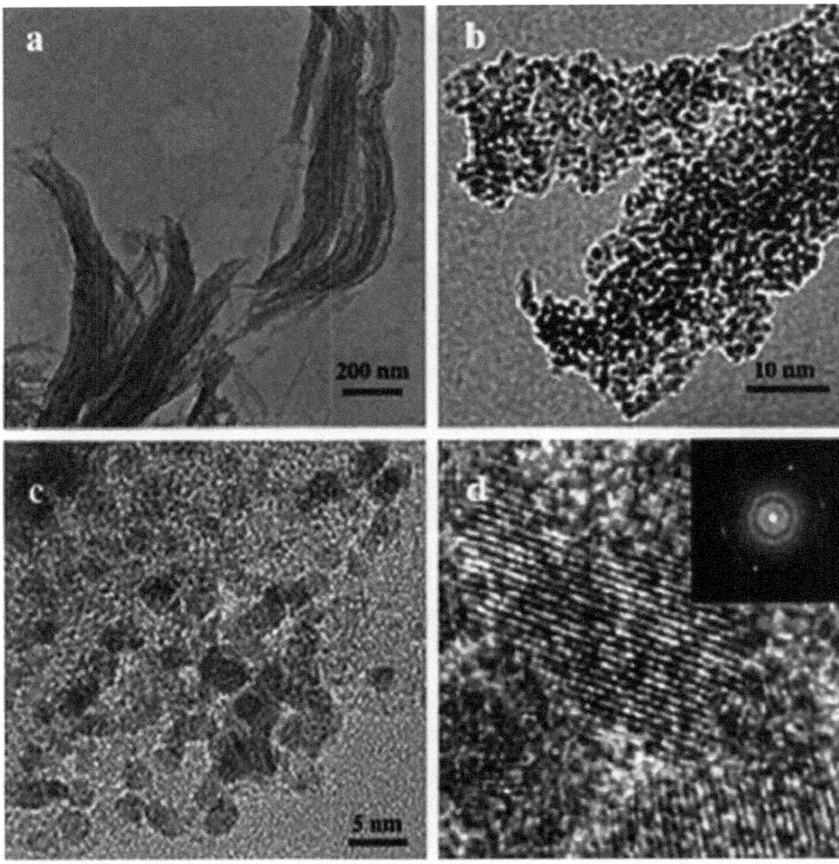

Fig. 20 TEM images of Pd@IFMC. (A) The fibrilar structure is clearly visible at low magnification. (B) PdNPs and nanopores of the carbon matrix are easily detectable at higher magnification at under-focus values. (C) Individual PdNPs with diameter between 2 and 4 nm are visible at high magnification. (D) HRTEM image shows the crystalline structure of PdNPs having a d-spacing of 0.28 nm with the corresponding selected area electron diffraction pattern (inset). *(Reprinted from B. Karimi, et al., A nano-fibrillated mesoporous carbon as an effective support for palladium nanoparticles in the aerobic oxidation of alcohols "on pure water", Chem. A Eur. J. 18(28) (2012) 8634–8640, with permission from John Wiley and Sons.)*

air in aqueous media at ambient temperature and 40°C. Additionally, the developed catalyst was highly recyclable and could be recovered and reused for six reaction runs. It is worth mentioning that the catalytic activity Pd@IFMC was superior to that of Pd/C and Pd@CMK-3. The better performance of Pd@IFMC was assigned to the fibrous nature and nitrogen functionalities in the structure of carbon support which could induce uniform and nearly atomic scale dispersion of Pd nanoparticles and prevent their aggregation in the course of reaction [24]. Regarding Ullmann homocoupling reaction, the catalyst showed high catalytic activity for promoting the reaction of various substrates ranging from aryl chlorides, iodides, bromides, and particularly, heteroaryl bromides in the presence of Cs_2CO_3 as base in aqueous media. Notably, the catalyst could promote the reaction without any need for any other chemicals and co-reducing agent. Moreover, the catalyst showed high recyclability and could be reused for several reaction runs. The analysis showed that, after five consecutive reaction runs, the catalyst preserved almost 80% of its initial Pd loading. It was suggested that IFMC not only served as support, but also as a reducing agent. Furthermore, the nitrogen heteroatoms in the structure of carbon could contribute to the catalysis [28].

Regarding the utility of mesoporous carbons for coupling reaction, Wang et al. reported on the utility of mesoporous N-doped carbon for Cu-mediated Ullmann C-O/N/-S cross-coupling reaction [57]. To prepare the mesoporous carbon that possessed high specific surface area and nitrogen content, the authors utilized EMIM-dca, 3-cyanomethyl-1-vinylimidazolium bromide, and silica nanoparticles as structural guide. The catalytic experiments confirmed that use of mesoporous carbon along with CuI and KOH could furnish the desired products in moderate to good yield. Notably, it was found that the reaction did not proceed in the presence of copper that was essential for promoting the reaction and the type of the applied copper species affected the yield of the reaction. The authors also used the synthesized mesoporous carbon as a support for CuO by deposition-precipitation method. The resulting catalyst also exhibited high catalytic activity and recyclability for Ullmann-type cross-coupling of aryl halides with phenols. The observed catalytic activity was assigned to the heterojunction between special nitrogen-containing structure in the carbon materials and copper species.

Nitrogen-doped carbon materials were also successfully used for upgrading the biofuel. In an interesting instance, Wang et al. prepared nitrogen-doped carbon via hard template approach by using 3MBP-dcp as carbon precursor and silica nanoparticles with size of 12 nm as hard template [58]. Upon carbonization and removal of template, nitrogen-doped carbon with 12% nitrogen was obtained and used for the immobilization of Pd nanoparticles by using ultrasonic irradiation. Characterization of the catalyst showed well-dispersed fine Pd nanoparticles (4.1 nm). The as-prepared compound exhibited high catalytic activity and recyclability (up to six reaction runs) for the hydrodeoxygenation of vanillin, a common component in lignin-derived bio-oil, under mild reaction condition. Notably, the catalytic activity of the catalyst (100% conversion of vanillin and 100% selectivity of 2-methoxy-4-methyl phenol) was superior compared to commercially available Pd@C as well as metal oxide-supported Pd nanoparticles, including Pd@TiO_2, Pd@ γ-Al_2O_3, Pd@MgO, and Pd@CeO_2, indicating the importance of the nature of the support. It was assumed that nitrogen atom doping increased the hydrophilic property of the catalyst and improved its dispersion in aqueous media and consequently improved the exposure of the catalyst toward the substrates that led to the higher catalytic activity.

Karimi et al. reported on a new IL-derived ordered mesoporous carbons functionalized with sulfonic acid functionalities, IOMC-ArSO_3H and GIOMC-ArSO_3H, for selective dehydration of fructose into 5-hydroxymethylfurfural in aqueous and nonaqueous media [29]. The schematic synthesis procedures for those two catalysts are depicted in Fig. 21. As shown, the carbon materials were prepared by using 1-methyl-3-phenethyl-1*H*-imidazolium hydrogen sulfate IL and SBA-15 as structural guide in the absence of any cosolvent. In the case of GIOMC-ArSO_3H, guanine was also used along with IL. Sulfur functionalization was achieved by using 4-benzenediazoniumsulfonate. Using CHN analysis, it was found that the content of nitrogen in GIOMC-ArSO_3H was 8%, while this value for IOMC-ArSO_3H was 5%.

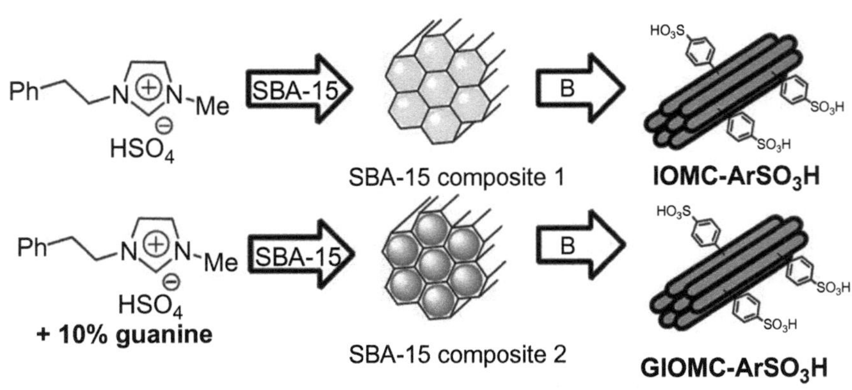

Fig. 21 Schematic representation of ordered mesoporous carbon solid acid preparation (B is three steps; (1) carbonization; (2) hard template removal; and (3) carbon sulfonation). *(Reprinted from B. Karimi, et al., Novel ordered mesoporous carbon based sulfonic acid as an efficient catalyst in the selective dehydration of fructose into 5-HMF: the role of solvent and surface chemistry, ACS Appl. Mater. Interfaces 7(34) (2015) 19050–19059, with permission from American Chemical Society.)*

To study and correlate the surface properties of these carbocatalysts and some other SBA-15 typed solid acids with 5-HMF yield, hydrophilicity index (H-indices) was employed in the fructose dehydration. The H-index values for IOMC-$ArSO_3H$ and GIOMC-$ArSO_3H$ were 0.83 and 1.28, respectively.

It was found that by increasing H-index value up to 0.8, the HMF yield enhances. Thus, the catalysts with H-index around 0.8 seem to be more preferable in fructose dehydration from the standpoint of catalyst durability. However, additional increasing in H-index up to 1.3 did not change the HMF yield profoundly. Notably, IOMC-$ArSO_3H$ showed high recyclability for at least four reaction runs in terms of fructose conversion, HMF yield, and selectivity; GIOMC-$ArSO_3H$, however, was slightly deactivated. This issue can be attributed to the more water uptake in this sample. Notably, this research group disclosed the utility of GIOMC for ORR [59].

Karimi et al. also developed the application of their developed IFMC for electrochemistry [49]. Considering the beneficial effects of IL binders in carbon paste electrodes such as higher rate of electron transfer and lower overvoltage, the authors for the first time prepared a disc-type electrode, via impregnating graphite electrodes with 1-octylpyridinum hexafluorophosphate ionic liquid as binder and fine powder of fibrous mesoporous carbon and investigated its electrochemical performance for electrochemical oxidation of catechol derivatives. Interestingly, the novel electrode exhibited improved performance due to very high surface area as well as edge plane-like defective sites of the ordered structure of carbon material.

Notably, MnO_2-incorporated IFMC was also prepared by Karimi, Kazemi et al. and applied as a novel electrode material with utility for supercapacitors [60]. Interestingly, the unique features of IFMC such as ordered mesoporous structure with inner connections improved faradaic reactions for the MnO_2 nanoparticles and the resulting electrode showed excellent performance with preserving its initial capacitance after 2000 charge-discharge cycles.

IFMC was also employed for modification of glassy carbon electrode (GCE). More precisely, 4-phenylurazole was grafted via electrochemical reduction of 4-(4-nitrophenyl)urazole (4-NPU) in the presence of sodium nitrite on IFMC/GCE (Fig. 22).

The resulted electrode showed the prominent electrochemical performance. The efficiency of the electrode was evaluated over the entire range of pHs. It was found that the oxidized form of ungrafted 4-phenyluazole, 4-phenyl-1,2,4-triazoline-3,5-dione, was highly unstable due to fast ring

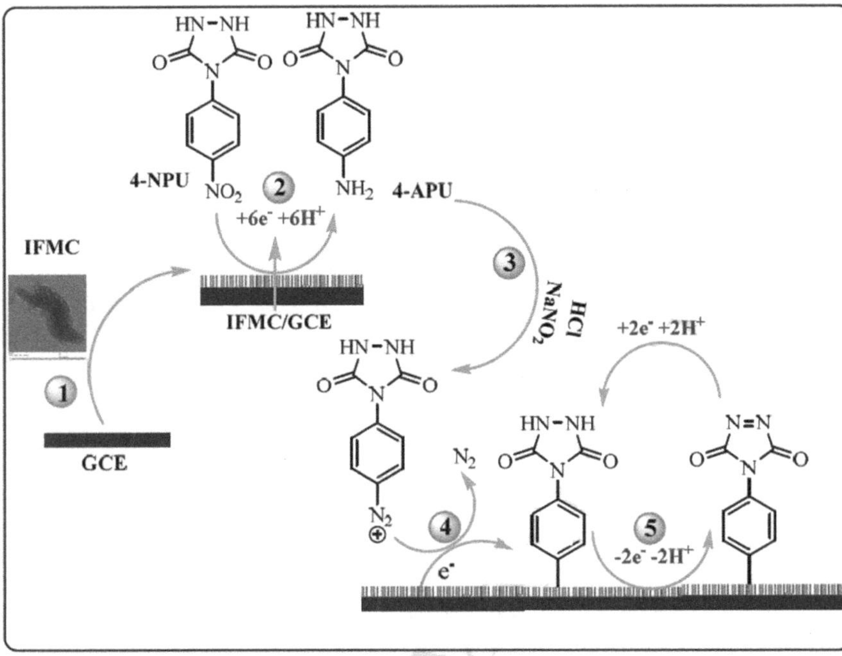

Fig. 22 Modification of IFMC/GCE by electroreduction of 4-NPU in the presence of sodium nitrite in 0.5 M HCl. *(Adapted from F. Varmaghani, B. Karimi, S. Mallakpour, Stabilization of 4-phenylurazole by electrografting on a nano-fibrillated mesoporous carbon modified electrode. Reactivity of anchored triazolinedione groups against Michael-type addition at electrode/electrolyte interface, Electrochim. Acta 269 (2018) 312–320, with permission from Elsevier.)*

cleavage. In contrast to oxidation of the ungrafted 4-phenylurazole, the electrografted 4-phenyl-1,2,4-triazoline-3,5-dione exhibited a high stability in a wide range of pHs. The anchored 4-phenyl-1,2,4-triazoline-3,5-dione groups were reactive against Michael-type addition. Further modification of the electrode could be achieved by the Michael-type addition reaction of the triazolinedione groups with nucleophiles at the solid/liquid interface [61].

It is worth noting that IFMC that benefits from high surface area and porous structure not only proved to be an excellent candidate for catalytic purposes, but also could be successfully applied as highly efficient material for the preparation of microextraction sorbents in the extraction of some volatile organic compounds from aqueous media [62].

The utility of a IL-nitrogen-doped carbon for the electrocatalysis in ORR was disclosed by Kim et al. [63]. The catalyst was prepared through pyrolysis of EMIM-dca as carbon precursor as well as transition metal and silica as hard template. Investigation of the electrocatalytic activities of various catalysts with different types (Ni and Co) and contents of transition metals via electrochemical measurement and XPS analysis confirmed the significant role of these two parameters in the observed catalytic activity. In detail, it was found that the use of optimum amount of proper transition metal (Co) in the course of carbonization of IL could create a large amount of graphitic-N and high degree of sp^2-carbon network, which is favorable for improving the ORR catalytic activity by providing active sites for the ORR and an effective electron path during the ORR. In comparison to Ni, Co-derived nitrogen-doped carbon catalysts exhibited a much higher ORR activity, which was primarily due to the formation of a higher degree of sp^2-carbon network [63].

Li, Sun, Xia, et al. reported preparation of Fe-N-doped ordered mesoporous carbon materials, Fe^X@NOMC (*X* stands for the mass percentage of Fe). The synthesis of the catalyst was carried out by using Fe-IL ([FcN][NTf_2]) as metal-containing precursor and N-IL ([MCNIm][$N(CN)_2$]) as a compatible nitrogen content modulator as well as SBA-15 as hard template (Fig. 23). Investigation of the catalytic activities of Fe^X@NOMC materials in ORR under alkaline condition indicated excellent performance of Fe^{10}@NOMC with nitrogen content of 11.9 wt% (comparable reaction current density and onset potential with the commercial 20 wt% Pt/C, but much better fuel crossover resistance and long-term durability in alkaline condition). High performance of the catalyst was attributed to the high surface area and ordered mesoporous structure of carbon material as well as the populated Fe (Fe_3O_4) and N (N_2) active sites and their improved interactions [64].

In another example, Karimi and Kazemi et al. reported the utility of a novel nanocomposite, PANi@IOMC, for energy storage purposes. To fabricate the nanocomposite, ordered mesoporous carbon (IOMC) was first prepared through carbonization of 1-methyl-3-phenethyl-1*H*-imidazolium hydrogen sulfate in the presence of SBA-15. To incorporate PANi, chemical oxidative polymerization of aniline in the presence of IOMC was accomplished. The TEM images of IOMC and PANi@IOMC nanocomposite showed that PANi rods grew at the surface of IOMC substrate (Fig. 24) [65]. PANi@IOMC exhibited a high specific capacitance ($542\,F\,g^{-1}$ at $5\,mV\,s^{-1}$) and long cycle life, suggesting a highly promising prospective for high performance supercapacitors. Interestingly, the electrochemical performance of the nanocomposite was superior to that of the individual components. It was believed that this improved performance was due to the covalent bonding between PANi and IOMC.

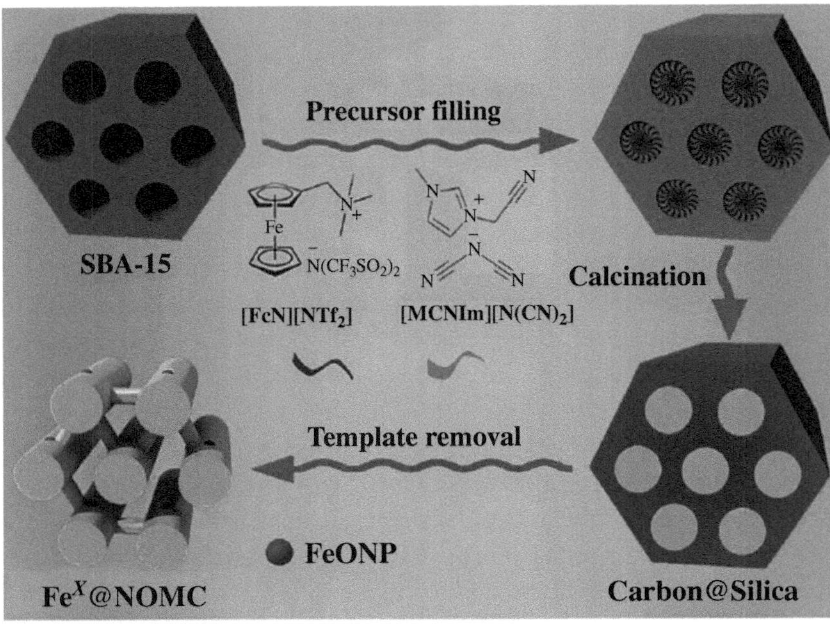

Fig. 23 Illustration of the synthesis of FeX@NOMC. *(Adapted from Z. Li, et al., Ionic liquids as precursors for efficient mesoporous iron-nitrogen-doped oxygen reduction electrocatalysts, Angew. Chem. 127(5) (2015) 1514–1518, with permission from John Wiley and Sons.)*

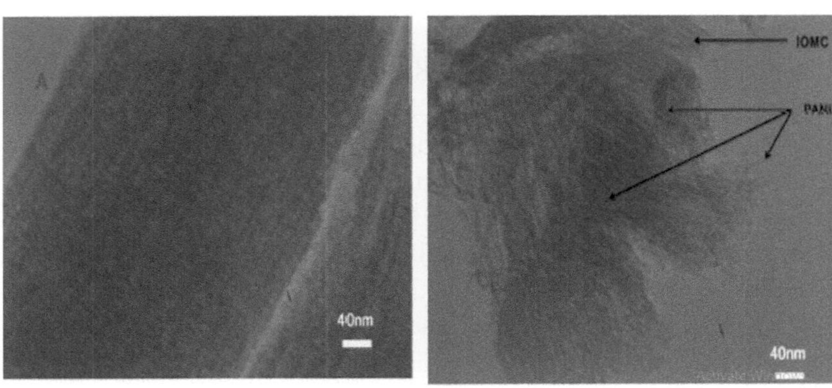

Fig. 24 TEM images of (A) pure IOMC, and (B) PANi@IOMC nanocomposite. *(Reprinted from S.H. Kazemi, et al., Polyaniline–ionic liquid derived ordered mesoporous carbon nanocomposite: synthesis and supercapacitive behavior, RSC Adv. 5(84) (2015) 69032–69041, with permission from Royal Society of Chemistry.)*

Recently, Salimi, Karimi et al. employed Au-incorporated GIOMC as electrocatalyst for the oxidation of NADH and a platform for NADH biosensor construction and immobilization of glucose dehydrogenase GDH enzyme [66]. It was believed that Au nanoparticles could enhance the biocompatibility of GIOMC and improve the surface features of the electrode. Consequently, GDH enzyme could be conjugated on the surface of the electrode in a covalent manner. Moreover, use of IL along with guanine improved the conductivity of mesoporous carbon. The covalently attached enzyme preserved its bioelectrocatalytic activity for glucose oxidation. Noteworthy, using SO_3^--modified carbon, a biocathode was designed via stabilization of bilirubin oxidase enzyme (BOD) for oxygen reduction. The improvement of the bioelectrocatalytic activity of the stabilized enzyme was achieved through electrodeposition of Au nanoparticles on IOMC-$PhSO_3H$ carbon.

Assembly of the fabricated bioanode and biocathode resulted in a membrane-less glucose/O_2 biofuel cell (Fig. 25) with power density of 33 ($mW\,cm^{-2}$) at 257 mV. The open circuit voltage of this biofuel cell and maximum produced current density were 508 mV and 0.252 ($mA\,cm^{-2}$), respectively.

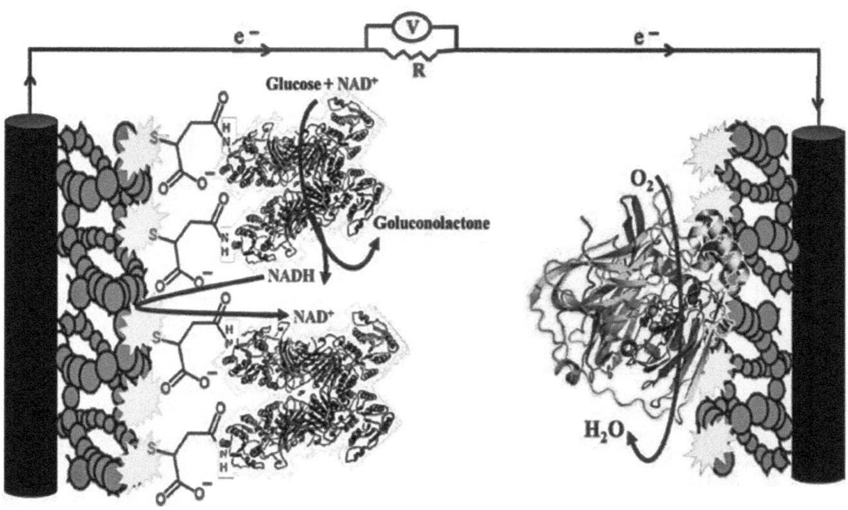

Fig. 25 Schematic representation of the assembled membrane-less glucose/oxygen biofuel cell. *(Reprinted from A. Korani, A. Salimi, B. Karimi, Guanine/ionic liquid derived ordered mesoporous carbon decorated with AuNPs as efficient NADH biosensor and suitable platform for enzymes immobilization and biofuel cell design, Electroanalysis 29(11) (2017) 2646–2655, with permission of John Wiley and Sons.)*

In another example, Thomas et al. reported on the synthesis of ordered mesoporous nitrogen-doped carbon electrocatalyst through self-assembly method. In detail, EMIM-DCA co-assembled with triblock copolymer poly (ethylene oxide)-*b*-poly (propylene oxide)-*b*-poly (ethylene oxide) (PEO-PPO-PEO, Pluronic F127) micelles. In this protocol (Fig. 26), IL was used for nitrogen doping, while resol was employed for the formation of bulk mesoporous carbon. The study of electrocatalytic activity of the resulting nitrogen-doped ordered carbon material confirmed its excellent performance (catalytic activity and stability) for hydrogen evolution reaction. The observed catalytic activity was assigned to the accessible nitrogen atoms on ordered carbon surface as well as open mesoporous channel structure [67].

Very recently, Karimi research group reported immobilizing Au-Pd bimetallic nanoparticles on IFMC via facile impregnation procedure to afford a catalyst for Ullmann coupling reaction of chloroarenes in aqueous media at ambient temperature. The results showed that the catalyst could furnish the desired product in 46% conversion [68]. Although the observed catalytic activity was not high, it was superior to Au-Pd immobilized on nitrogen-free supports such as Au-Pd@CMK-3, confirming the role of nitrogen content in IFMC in the improvement of the catalytic activity. Armed with this result, the authors prepared another N-rich ordered mesoporous carbon, NMCI, with different N-content through carbonization of 1-methyl-3-phenethyl-1*H*-imidazolium hydrogen sulfate containing 10% and 15% dissolved guanine, respectively, using SBA-15 as a structural guide. The novel carbons possessed higher nitrogen contents (9.2–12.6 wt% for NMCI-1 and NMCI-2) compared to IFMC (about 6 wt%). Notably, use of higher amount of guanine led to highly disordered carbon. The authors then immobilized Au-Pd bimetallic nanoparticles on the support and characterized the resulting catalyst (Fig. 27). The catalytic tests showed that the developed catalyst could promote the reaction of various aryl bromides and chlorides with different electronic features at room temperature in the presence of K_2CO_3 in high yields. It is worth noting that challenging substrates such as thiophenes could also tolerate this reaction to afford the corresponding product in good yield. Compared to some previously reported catalysts such as Au-Pd@PVP and Au-Pd@CMK-3, Au-Pd@NMCI exhibited superior catalytic activity. Moreover, the catalyst was recyclable at least for five reaction runs.

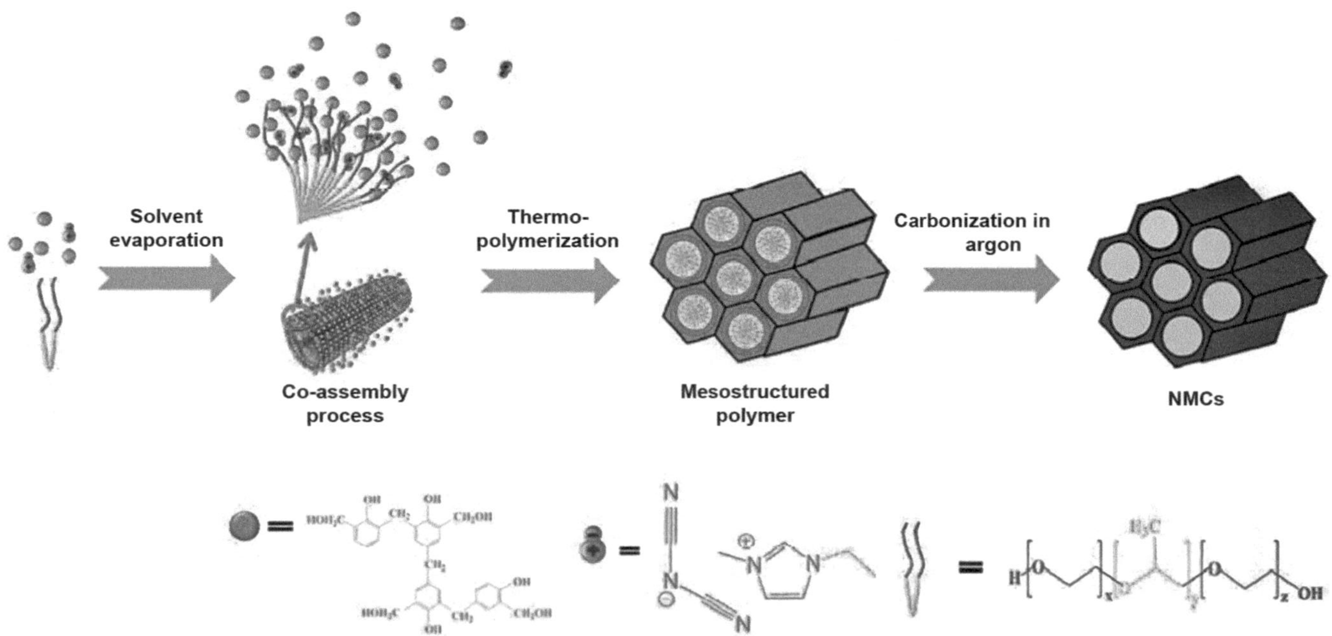

Fig. 26 Schematic illustration of the ionic liquid-assisted co-assembly process for the synthesis of mesoporous carbon with surface-enriched nitrogen. *(Adapted from X. Zhao, et al., Ionic liquid-assisted synthesis of mesoporous carbons with surface-enriched nitrogen for the hydrogen evolution reaction, ACS Appl. Mater. Interfaces 10(4) (2018) 3912–3920, with permission from American Chemical Society.)*

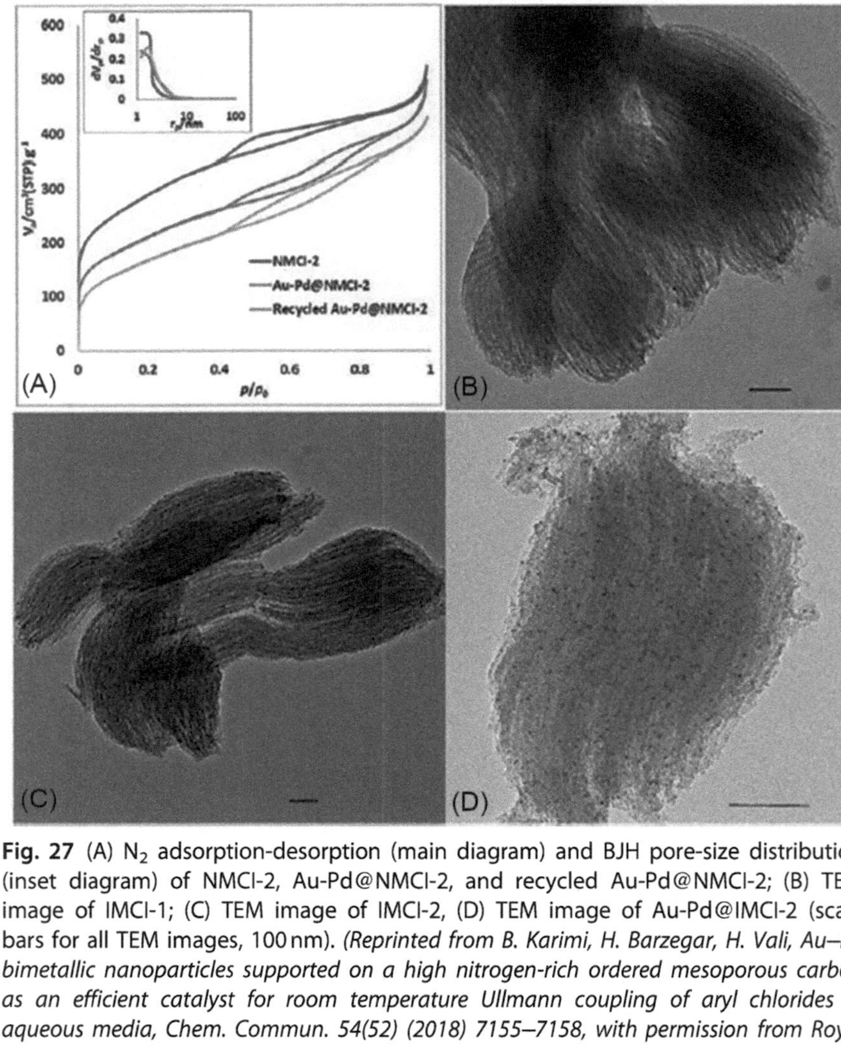

Fig. 27 (A) N_2 adsorption-desorption (main diagram) and BJH pore-size distribution (inset diagram) of NMCl-2, Au-Pd@NMCl-2, and recycled Au-Pd@NMCl-2; (B) TEM image of IMCl-1; (C) TEM image of IMCl-2, (D) TEM image of Au-Pd@IMCl-2 (scale bars for all TEM images, 100 nm). *(Reprinted from B. Karimi, H. Barzegar, H. Vali, Au–Pd bimetallic nanoparticles supported on a high nitrogen-rich ordered mesoporous carbon as an efficient catalyst for room temperature Ullmann coupling of aryl chlorides in aqueous media, Chem. Commun. 54(52) (2018) 7155–7158, with permission from Royal Society of Chemistry.)*

8 Conclusion

In this chapter, the use of ILs as potential precursors for the synthesis of heteroatom-doped carbon materials has been presented. As discussed, there are several approaches for achieving ILs-derived carbon materials. The first method is carbonization of ILs with specific functionalities (cyano/nitrile). As the ILs with these functionalities are costly and their preparation is time-consuming and tedious, another approach has been developed, in which conventional ILs were entrapped within a confined framework. Another method for the synthesis of ILs-derived carbon is carbonization of PILs and PILs/PSs. Notably, the textural properties of ILs-derived carbons can also be modified. The methods for this purpose included template-free, nanocasting, and salt templates. As exemplified, ILs-derived heteroatom-doped carbon materials in their free form can be considered as potential (electro) catalysts. Moreover, they can be applied as catalyst support to stabilize various catalytic species to furnish efficient catalysts with utility for promoting diverse range of (electro) chemical transformations. The comparative studies in some reports indicated that ILs-derived carbons showed superior catalytic activity compared to the conventional carbon-based catalysts.

Considering the promising results that have been achieved to date, it is expected that the use of ILs-derived carbon materials will grow rapidly, and in the near future, open a new window toward functional carbon materials with tunable properties.

Acknowledgment

The authors appreciate the support of Iran Polymer and Petrochemical Institute and Institute for Advanced Studies in Basic Sciences.

References

[1] D. Salinas-Torres, et al., Nitrogen-doped carbon materials as a promising platform toward the efficient catalysis for hydrogen generation, Appl. Catal. A. Gen. 571 (2019) 25–41.
[2] B. Gao, et al., Wearable eye health monitoring sensors based on peacock tail-inspired inverse opal carbon, Sens. Actuators B 288 (2019) 734–741.
[3] S.W. Lee, et al., Recent advances in carbon material-based NO_2 gas sensors, Sens. Actuators B 255 (2018) 1788–1804.
[4] V. Schroeder, et al., Carbon nanotube chemical sensors, Chem. Rev. 119 (1) (2019) 599–663.
[5] S. Jose, et al., Improvement of water quality of remnant from chemical retting of coconut fibre through electrocoagulation and activated carbon treatment, J. Clean. Prod. 210 (2019) 630–637.
[6] J.-S. Wei, et al., Robust negative electrode materials derived from carbon dots and porous hydrogels for high-performance hybrid supercapacitors, Adv. Mater. 31 (5) (2019) 1806197.
[7] V.C. Hoang, V.G. Gomes, High performance hybrid supercapacitor based on doped zucchini-derived carbon dots and graphene, Mater. Today Energy 12 (2019) 198–207.
[8] X. Fu, et al., Carbon nanotube@N-doped mesoporous carbon composite material for supercapacitor electrodes, Chem. Asian J. 14 (5) (2019) 634–639.
[9] Y. Zhang, et al., Honeycomb hard carbon derived from carbon quantum dots as anode material for K-ion batteries, Mater. Chem. Phys. 229 (2019) 303–309.

[10] B. Liang, et al., Nitrogen and phosphorus dual-doped carbon derived from chitosan: an excellent cathode catalyst in microbial fuel cell, Chem. Eng. J. 358 (2019) 1002–1011.
[11] L. Yang, et al., Carbon-based metal-free ORR electrocatalysts for fuel cells: past, present, and future, Adv. Mater. 31 (13) (2019) 1804799.
[12] M. Inagaki, et al., Nitrogen-doped carbon materials, Carbon 132 (2018) 104–140.
[13] J. Gao, et al., Polymerizable ionic liquid as nitrogen-doping precursor for Co–N–C catalyst with enhanced oxygen reduction activity, Ind. Eng. Chem. Res. 54 (32) (2015) 7984–7989.
[14] R. Paul, et al., 3D heteroatom-doped carbon nanomaterials as multifunctional metal-free catalysts for integrated energy devices, Adv. Mater. 31 (13) (2019) 1805598.
[15] A. Wütscher, et al., Nitrogen-doped metal-free carbon materials derived from cellulose as electrocatalysts for the oxygen reduction reaction, ChemElectroChem 6 (2) (2019) 514–521.
[16] J. Luo, et al., Pyridinic-N protected synthesis of 3D nitrogen-doped porous carbon with increased mesoporous defects for oxygen reduction, Small 15 (11) (2019) 1805325.
[17] Z.-L. Xie, D.S. Su, Ionic liquid based approaches to carbon materials synthesis, Eur. J. Inorg. Chem. 2015 (7) (2015) 1137–1147.
[18] S. Zhang, K. Dokko, M. Watanabe, Carbon materialization of ionic liquids: from solvents to materials, Mater. Horiz. 2 (2) (2015) 168–197.
[19] J.P. Paraknowitsch, A. Thomas, Functional carbon materials from ionic liquid precursors, Macromol. Chem. Phys. 213 (10–11) (2012) 1132–1145.
[20] A. Abo-Hamad, et al., Ionic liquid-carbon nanomaterial hybrids for electrochemical sensor applications: a review, Electrochim. Acta 193 (2016) 321–343.
[21] T.-P. Fellinger, et al., 25th anniversary article: "Cooking carbon with salt": carbon materials and carbonaceous frameworks from ionic liquids and poly(ionic liquid)s, Adv. Mater. 25 (41) (2013) 5838–5855.
[22] N. Zdolšek, et al., Electrochemical investigation of ionic liquid-derived porous carbon materials for supercapacitors: pseudocapacitance versus electrical double layer, Electrochim. Acta 298 (2019) 541–551.
[23] Y. Su, et al., Nitrile chain reactions for cyano-based ionic liquid derived mesoporous carbon as efficient bifunctional electrocatalyst, Electrochim. Acta 280 (2018) 258–265.
[24] B. Karimi, et al., A nano-fibrillated mesoporous carbon as an effective support for palladium nanoparticles in the aerobic oxidation of alcohols "on pure water" Chem. A Eur. J. 18 (28) (2012) 8634–8640.
[25] J.H. Jeong, et al., Multimodal porous carbon derived from ionic liquids: correlation between pore sizes and ionic clusters, Nanoscale 9 (38) (2017) 14672–14681.
[26] X. Xu, H. Li, Y. Wang, Selective hydrogenation of phenol to cyclohexanone in water over Pd@N-doped carbon derived from ionic-liquid precursors, ChemCatChem 6 (12) (2014) 3328–3332.
[27] X. Wang, S. Dai, Ionic liquids as versatile precursors for functionalized porous carbon and carbon–oxide composite materials by confined carbonization, Angew. Chem. Int. Ed. 49 (37) (2010) 6664–6668.
[28] B. Karimi, H. Behzadnia, H. Vali, Palladium on ionic liquid derived nanofibrillated mesoporous carbon: a recyclable catalyst for the Ullmann homocoupling reactions of aryl halides in water, ChemCatChem 6 (3) (2014) 745–748.
[29] B. Karimi, et al., Novel ordered mesoporous carbon based sulfonic acid as an efficient catalyst in the selective dehydration of fructose into 5-HMF: the role of solvent and surface chemistry, ACS Appl. Mater. Interfaces 7 (34) (2015) 19050–19059.
[30] A. Chen, et al., Thin-walled, mesoporous and nitrogen-doped hollow carbon spheres using ionic liquids as precursors, J. Mater. Chem. A 1 (4) (2013) 1045–1047.
[31] W. Qian, J. Texter, F. Yan, Frontiers in poly(ionic liquid)s: syntheses and applications, Chem. Soc. Rev. 46 (4) (2017) 1124–1159.

[32] X. Bo, et al., Highly dispersed Pt nanoparticles supported on poly(ionic liquids) derived hollow carbon spheres for methanol oxidation, J. Power Sources 196 (20) (2011) 8360–8365.
[33] M.R. Nabid, Y. Bide, Z. Habibi, Synthesis of a yolk/shell Fe_3O_4@poly(ionic liquid) s-derived nitrogen doped graphitic porous carbon materials and its application as support for nickel catalysts, RSC Adv. 5 (3) (2015) 2258–2265.
[34] J. Balach, et al., Poly(ionic liquid)-derived nitrogen-doped hollow carbon spheres: synthesis and loading with Fe_2O_3 for high-performance lithium ion batteries, RSC Adv. 3 (21) (2013) 7979–7986.
[35] J. Gong, M. Antonietti, J. Yuan, Poly(ionic liquid)-derived carbon with site-specific N-doping and biphasic heterojunction for enhanced CO_2 capture and sensing, Angew. Chem. Int. Ed. 56 (26) (2017) 7557–7563.
[36] B. Wu, et al., Carbonization of ionic liquid polymer-functionalized carbon nanotubes for high dispersion of PtRu nanoparticles and their electrocatalytic oxidation of methanol, J. Mater. Chem. 22 (26) (2012) 13085–13090.
[37] C. Pan, et al., Facile synthesis of nitrogen-doped carbon–Pt nanoparticle hybrids via carbonization of poly([Bvim][Br]-co-acrylonitrile) for electrocatalytic oxidation of methanol, J. Mater. Chem. 22 (27) (2012) 13578–13584.
[38] M. Einert, et al., Nitrogen-doped carbon electrodes: influence of microstructure and nitrogen configuration on the electrical conductivity of carbonized polyacrylonitrile and poly(ionic liquid) blends, Macromol. Chem. Phys. 216 (19) (2015) 1930–1944.
[39] J. Gao, et al., Polymerizable ionic liquid as a precursor for N, P co-doped carbon toward the oxygen reduction reaction, Cat. Sci. Technol. 8 (4) (2018) 1142–1150.
[40] S. Zhang, K. Dokko, M. Watanabe, Direct synthesis of nitrogen-doped carbon materials from protic ionic liquids and protic salts: structural and physicochemical correlations between precursor and carbon, Chem. Mater. 26 (9) (2014) 2915–2926.
[41] L. Yang, et al., Ionic liquid as the C and N sources to prepare yolk-shell Fe_3O_4@N-doped carbon nanoparticles and its high performance in lithium-ion battery, Electrochim. Acta 190 (2016) 797–803.
[42] L. Miao, et al., N, S Co-doped hierarchical porous carbon rods derived from protic salt: facile synthesis for high energy density supercapacitors, Electrochim. Acta 274 (2018) 378–388.
[43] J. Gao, et al., Polymerizable ionic liquid-derived carbon for oxygen reduction and evolution, J. Appl. Electrochem. 47 (2017) 351–359.
[44] L. Miao, et al., Poly(ionic liquid)-derived, N, S-codoped ultramicroporous carbon nanoparticles for supercapacitors, Chem. Eng. J. 317 (2017) 651–659.
[45] L. Miao, et al., Cooking carbon with protic salt: nitrogen and sulfur self-doped porous carbon nanosheets for supercapacitors, Chem. Eng. J. 347 (2018) 233–242.
[46] X. Tuaev, et al., Nitrogen-doped coatings on carbon nanotubes and their stabilizing effect on Pt nanoparticles, Phys. Chem. Chem. Phys. 14 (18) (2012) 6444–6447.
[47] Y.J. Sa, et al., Carbon nanotubes/heteroatom-doped carbon core–sheath nanostructures as highly active, metal-free oxygen reduction electrocatalysts for alkaline fuel cells, Angew. Chem. Int. Ed. 53 (16) (2014) 4102–4106.
[48] L. Zhao, et al., Porous $Li_4Ti_5O_{12}$ coated with N-doped carbon from ionic liquids for Li-ion batteries, Adv. Mater. 23 (11) (2011) 1385–1388.
[49] B. Karimi, et al., Electrochemical performance of a novel ionic liquid derived mesoporous carbon, Chem. Commun. 48 (22) (2012) 2776–2778.
[50] S.H. Kazemi, et al., High-performance supercapacitors based on an ionic liquid-derived nanofibrillated mesoporous carbon, J. Solid State Electrochem. 18 (2014) 2419–2424.
[51] J.P. Paraknowitsch, A. Thomas, M. Antonietti, A detailed view on the polycondensation of ionic liquid monomers towards nitrogen doped carbon materials, J. Mater. Chem. 20 (32) (2010) 6746–6758.

[52] T.-P. Fellinger, et al., Mesoporous nitrogen-doped carbon for the electrocatalytic synthesis of hydrogen peroxide, J. Am. Chem. Soc. 134 (9) (2012) 4072–4075.
[53] P.F. Fulvio, et al., Magadiite templated high surface area graphene-type carbons from metal-halide based ionic liquids, J. Mater. Chem. A 1 (1) (2013) 59–62.
[54] A. Aijaz, et al., From ionic-liquid@metal–organic framework composites to heteroatom-decorated large-surface area carbons: superior CO_2 and H_2 uptake, Chem. Commun. 50 (49) (2014) 6498–6501.
[55] M. Sarker, et al., Nitrogen-doped porous carbon from ionic liquid@Al-metal-organic framework: a prominent adsorbent for purification of both aqueous and non-aqueous solutions, Chem. Eng. J. 338 (2018) 107–116.
[56] N. Fechler, T.-P. Fellinger, M. Antonietti, "Salt templating": a simple and sustainable pathway toward highly porous functional carbons from ionic liquids, Adv. Mater. 25 (1) (2013) 75–79.
[57] P. Zhang, et al., Mesoporous nitrogen-doped carbon for copper-mediated Ullmann-type C–O/–N/–S cross-coupling reactions, RSC Adv. 3 (6) (2013) 1890–1895.
[58] X. Xu, et al., Synthesis of palladium nanoparticles supported on mesoporous N-doped carbon and their catalytic ability for biofuel upgrade, J. Am. Chem. Soc. 134 (41) (2012) 16987–16990.
[59] M. Rafiee, B. Karimi, H. Shirmohammadi, Graphitized nitrogen-doped ordered mesoporous carbon derived from ionic liquid; catalytic performance toward ORR, Electrocatalysis 9 (5) (2018) 632–639.
[60] S.H. Kazemi, et al., Manganese dioxide nanoparticles incorporated within ionic liquid derived fibrillated mesoporous carbon: electrode material for high-performance supercapacitors, RSC Adv. 5 (103) (2015) 84840–84848.
[61] F. Varmaghani, B. Karimi, S. Mallakpour, Stabilization of 4-phenylurazole by electrografting on a nano-fibrillated mesoporous carbon modified electrode. Reactivity of anchored triazolinedione groups against Michael-type addition at electrode/electrolyte interface, Electrochim. Acta 269 (2018) 312–320.
[62] M.M. Abolghasemi, et al., Ionic liquid-derived nano-fibrillated mesoporous carbon based on solid-phase microextraction fiber for the analysis of volatile organic compounds from aqueous solutions, New J. Chem. 39 (8) (2015) 6085–6091.
[63] K.H. Lim, H. Kim, Nitrogen-doped carbon catalysts derived from ionic liquids in the presence of transition metals for the oxygen reduction reaction, Appl. Catal. Environ. 158–159 (2014) 355–360.
[64] Z. Li, et al., Ionic liquids as precursors for efficient mesoporous iron-nitrogen-doped oxygen reduction electrocatalysts, Angew. Chem. 127 (5) (2015) 1514–1518.
[65] S.H. Kazemi, et al., Polyaniline–ionic liquid derived ordered mesoporous carbon nanocomposite: synthesis and supercapacitive behavior, RSC Adv. 5 (84) (2015) 69032–69041.
[66] A. Korani, A. Salimi, B. Karimi, Guanine/ionic liquid derived ordered mesoporous carbon decorated with AuNPs as efficient NADH biosensor and suitable platform for enzymes immobilization and biofuel cell design, Electroanalysis 29 (11) (2017) 2646–2655.
[67] X. Zhao, et al., Ionic liquid-assisted synthesis of mesoporous carbons with surface-enriched nitrogen for the hydrogen evolution reaction, ACS Appl. Mater. Interfaces 10 (4) (2018) 3912–3920.
[68] B. Karimi, H. Barzegar, H. Vali, Au–Pd bimetallic nanoparticles supported on a high nitrogen-rich ordered mesoporous carbon as an efficient catalyst for room temperature Ullmann coupling of aryl chlorides in aqueous media, Chem. Commun. 54 (52) (2018) 7155–7158.

CHAPTER 3

Metal-organic framework-derived porous carbon templates for catalysis

Yuhang Wu[a,b], Hui Xu[a], and Junkuo Gao[b]
[a]Institute of Optoelectronic Materials and Devices, China Jiliang University, Hangzhou, People's Republic of China
[b]Institute of Fiber Based New Energy Materials, School of Materials Science and Engineering, Zhejiang Sci-Tech University, Hangzhou, China

1 Introduction

With the deterioration of the issues from continuing population growth, consumption of fossil fuels, and environmental pollution, it is crucial to search and develop clean renewable energy to sustainable development in modern society [1–5]. In the past few decades, both academia and industry have spent tremendous efforts on seeking advanced technologies for renewable energy (such as biofuels, wind, tidal energy, solar energy, etc.) [6, 7]. A large number of various new-generation energy technologies (including metal-air batteries, fuel cells, electrochemical water splitting, and electrochemical CO_2 reduction) show the advantages such as high energy conversion, lower energy consumption, and environmental friendliness, suggesting the great potential to replace the traditional energy technologies [8, 9]. At the same time, in the process, technological innovation is urgently needed to be developed to reduce consumption of energy, improving the efficiency of energy conversion and large-scale exploitation and energy storage [10–12]. In fact, the energy conversion reactions involved in the electrochemical systems or devices (e.g., oxygen evolution reaction (OER), oxygen reduction reaction (ORR), hydrogen evolution reaction (HER), and CO_2 reduction reaction (CO_2RR)) convert the chemical energy into the electrical energy directly [13, 14]. In order to reduce the inherent energy barrier of these reactions to accelerate the reaction kinetics, the development of advanced electrocatalysts and electrode materials is the key to accelerate the actual industrialization process [15].

Although precious metal materials (such as Pt, RuO_2, etc.) are recognized as the most efficient catalysts applied in various energy devices [16–18],

Emerging Carbon Materials for Catalysis
https://doi.org/10.1016/B978-0-12-817561-3.00003-2

the high cost and scarcity severely hinder their development in large-scale industrialization. Thus, efficient, stable, and easily synthesizable alternative catalysts need to be studied extensively [19, 20]. Among various alternative catalysts, nanoporous materials have been extensively investigated due to their excellent properties such as high specific surface and functionalized pore structure. As a new generation of porous materials with unique structures and physicochemical properties, metal-organic frameworks (MOFs) have become a hot research topic in various fields (e.g., drug release, gas sorption/separation, sensors, catalysis, energy conversion and storage, etc.) [21–27]. MOFs possess the features of ultrahigh surface areas, adjustable pore structure, and easy modification, which show the great potential in the field of catalysis [28]. Nevertheless, the weak coordination bond between the metal cluster and the ligand is particularly unstable in moisture, hindering their applications in the actual catalysis environment [29]. It is worth noting that MOFs-derived materials exhibit more stable characteristics compared with original MOFs and greatly inherit the features of the original MOFs (e.g., large surface areas, pore structure, composition diversity, etc.) [30–33]. Furthermore, the transitional metals (e.g., Co, Fe, Ni, Cu, etc.) and the elements (e.g., C, H, O, N, S, etc.) involved in organic ligands, which are composed of most MOFs, are necessary elements for the catalysis [34–37]. In addition, MOF-derived porous materials possess high porosity, large specific surface, and adjustable morphology compared to the traditional catalysts, which afford better performance in the catalysis.

Actually, MOF as precursors/templates for the fabrication of various unique porous materials, compared to the traditional nanoporous materials, possess the following advantages: (i) facile and convenient preparation method, (ii) adjustable porous structure and customizable aperture, which ensure the mass transfer and active sites' access, (iii) large surface area and high porosity, which are beneficial to expose more active sites, (iv) tunability, i.e., the structure is easy to modify and doping with heteroatoms is also easy, which allow to adjust the electronic structure of the catalysts. Thus, MOFs have been considered as promising sacrificial precursors/templates for the fabrication of porous carbon compounds [22, 34, 35, 37, 38].

In this chapter, we summarize the following: first, the synthesis strategies of various MOF-derived porous carbon materials are introduced systematically. Then, we summarize the recent catalytic applications of MOF-derived porous carbon materials. Reasonable loading of active sites and doping of heteroatoms play a vital role in the reactions of oxidation, reduction, and CO_2 fixation, which show the enormous potential of

MOF-derived porous materials in heterogeneous catalysis. Meanwhile, for photocatalysis, metal-based composites (e.g., metal oxides, sulfides, and so on) derived from MOF exhibit excellent performance, which is also attributed to the porosity of MOF-derived porous materials. For electrocatalysis, the conductivity and specific surface area of MOF-derived porous materials possess a huge impact on various electrocatalytic performances. In the end, particular challenges and opportunities of MOF-derived porous materials for the future development are critically discussed.

2 Fabrication strategies for MOF-derived porous carbons

In view of the diversity of composition and the tunability of structure, MOFs have been considered as the most potential alternative for the preparation of porous carbons [32]. Thus, various fabrication strategies are proposed to prepare different types of porous carbons. Generally, the conversion of MOFs precursors to MOF-derived porous carbon is mainly affected by two key factors: precursor design and pyrolysis conditions (such as temperature, atmosphere, and pyrolysis times). On the one hand, MOF-derived porous carbon can inherit part of the morphology, pore size, and properties of the MOF precursors. On the other hand, the morphology and composition of MOF-derived porous carbons can be adjusted by controlling the pyrolysis conditions. MOFs-derived porous carbons are mainly divided into: metal-free porous carbons, carbon-supported single-atom catalysts, metal/carbons, and metal compounds/carbons.

2.1 MOF-derived metal-free porous carbons (including pure carbon and carbon with heteroatom doping, carbon composites)

Metal-free porous carbon catalysts are widely used in the fields of electrocatalysis [39]. Among them, the defect structure and the introduced heteroatoms (e.g., N, B, S, P) play an important role in the study of the electrocatalytic performance of porous carbon nanocatalysts. Doping of heteroatoms not only provides active sites to improve conductivity, but also adjusts optoelectronic properties and chemical activity [40]. It is worth noting that MOF-derived carbon materials can be used to design the structure of porous, heteroatom doping, and carbon composites, which can limit the active sites to permeable carbon and effectively improve the catalytic performance [41–43]. During the pyrolysis process, the MOF precursors are directly converted to carbon while maintaining the inherent framework structure to form highly porous catalysts and metal species are removed by evaporating or acid etching after carbonization. In addition, the heteroatom-doped MOF-derived porous carbons possess many advantages such as excellent electrical conductivity, high specific surface area, and uniform distribution of active sites, which can facilitate electron transport and proton transfer.

At present, ZIF-8 as a typical N-rich precursor is widely used to prepare pure carbon porous materials. Jiang and coworkers [44] first prepared MOF-derived porous carbon with extremely high surface area (3405 $m^2\ g^{-1}$) and controlled the surface area of the porous carbon by adjusting the annealing temperature. In Fig. 1, the porous carbon is obtained by carbonization of porous ZIF-8 precursor, after introduction of furfuryl alcohol.

In addition, heteroatom doping not only increases the active site, but also improves the conductivity and hydrophilicity of porous carbon. Recently, Zhang et al. [45] successfully prepared N, P, S *co*-doped carbon hollow polyhedron based on ZIF-8 precursor. As shown in Fig. 2, poly(cyclotriphosphazene-*co*-4,4′-sulfonyldiphenol) is coated as a dopant on the surface of ZIF-8

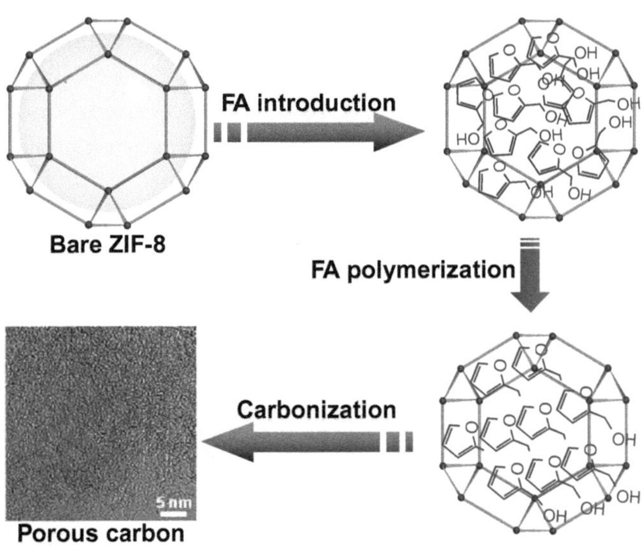

Fig. 1 The scheme of the preparation of nanoporous carbon. *(Adapted with the permission from H.-L. Jiang, B. Liu, Y.-Q. Lan, K. Kuratani, T. Akita, H. Shioyama, F. Zong, Q. Xu, From metal–organic framework to nanoporous carbon: toward a very high surface area and hydrogen uptake, J. Am. Chem. Soc. 133 (2011) 11854–11857. Copyright 2011, American Chemical Society.)*

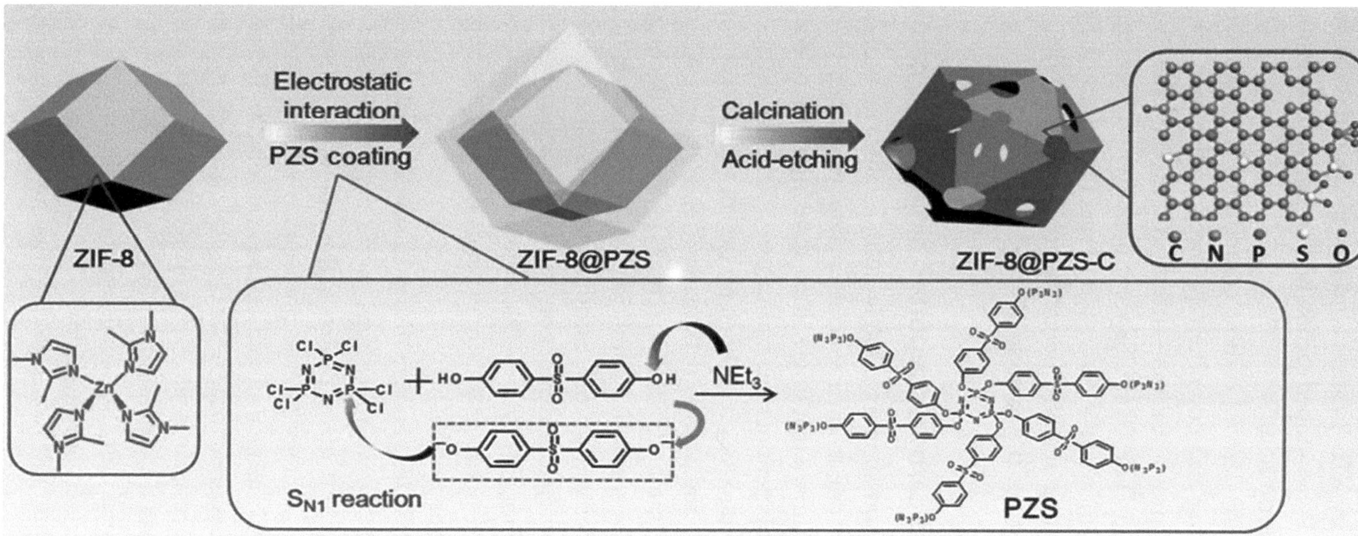

Fig. 2 Schematic illustration of the synthetic process of ZIF-8@PZS-C. *(Adapted with the permission from J. Zhang, J. Fang, J. Han, T. Yan, L. Shi, D. Zhang, N, P, S Co-doped hollow carbon polyhedra derived from MOF-based core–shell nanocomposites for capacitive deionization, J. Mater. Chem. A 6 (2018) 15245–15252. Copyright 2018, Royal Society of Chemistry.)*

to provide N, P, S atoms, and the composite is annealed in N_2 atmosphere. The vapor released by ZIF-8 causes a hollow mesoporous structure during calcination. The N, P, S *co*-doped carbon hollow polyhedron is obtained after etching in hydrochloric acid.

Meanwhile, MOF-derived carbon compounds have also been extensively studied. Zhong and coworkers [46] reported novel 2D graphene-based N-doped porous carbon sheets (GNPCSs) derived from a sandwich-like ZIF-8. The GNPCSs were obtained via growing of ZIF-8 on GO with the addition of PVP after annealing. The synergistic effect, high conductivity, and high stability between graphene and NPC effectively enhanced the activity of the catalysts.

2.2 MOF-derived carbon-supported single-atom catalysts

In recent years, carbon-supported single-atom catalysts (SACs) have caused extensive researches due to their special catalytic properties [47–49]. Nevertheless, it is well-known that single atoms have high surface energy, which limits further development of their fabrication. Therefore, a large number of strategies for preparing SACs have been proposed, such as wet impregnation [50], pyrolysis [51], and atomic deposition [52]. Pyrolysis is currently the most widely studied method, which can prepare SACs by decomposing precursors under various controlled conditions (i.e., atmosphere, time, or temperature). MOFs-derived porous carbons have become an ideal support material because of their ordered pore structure, adjustable morphology and chemical stability.[53] Meanwhile, the metal-nitrogen coordination bond formed during pyrolysis greatly expands the range of application for MOF-derived porous carbon.

Metal nodes in MOFs generate dispersed single atoms during pyrolysis, while bimetallic MOFs with sacrificial metals are mainly used to prepare SACs in order to avoid metal center agglomeration. Yin and coworkers [54] propose a new strategy for the preparation of high metal loading of Co single atomic in nitrogen-doped porous carbon. As shown in Fig. 3, bimetallic MOF is designed to fabricate Co single-atom porous carbon; during the pyrolysis process, Zn is selectively evaporated, while Co is reduced to form single atom and dispersed uniformly.

Recently, Chen group [55] achieved the anchoring of W atoms on N-doped MOF-derived porous carbon. The strategy is to encapsulate the tungsten precursor in UiO-66-NH_2 for the pyrolysis, and the W-SAC was obtained after removing precipitated zirconia by soaking in hydrofluoric

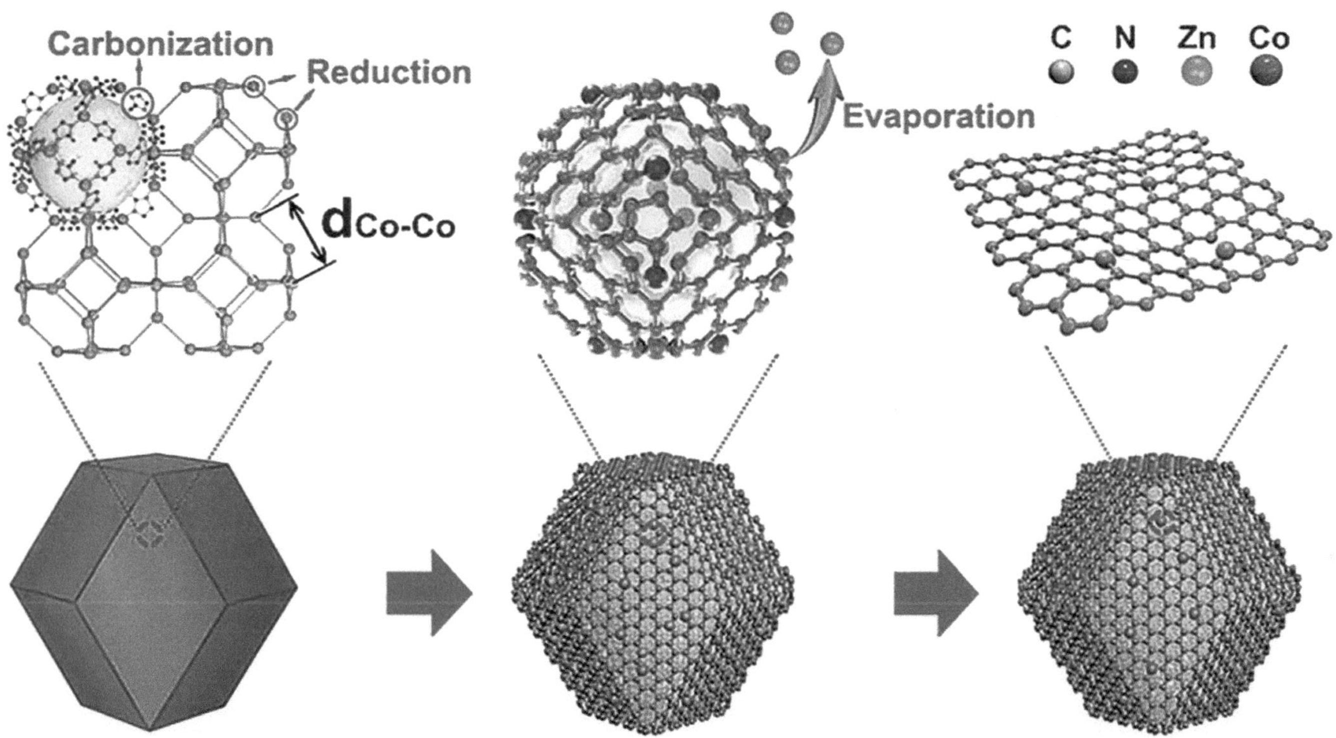

Fig. 3 The scheme of the pyrolysis process of Co-SACs. *(Adapted with the permission from P. Yin, T. Yao, Y. Wu, L. Zheng, Y. Lin, W. Liu, H. Ju, J. Zhu, X. Hong, Z. Deng, G. Zhou, S. Wei, Y. Li, Single cobalt atoms with precise N-coordination as superior oxygen reduction reaction catalysts, Angew. Chem. Int. Ed. 55 (2016) 10800–10805. Copyright 2016, John Wiley and Sons.)*

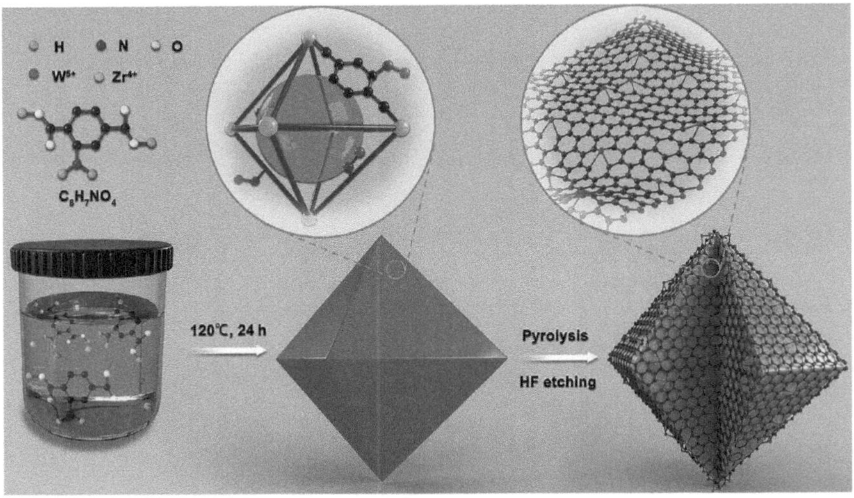

Fig. 4 The preparation of W-SAC. *(Adapted with the permission from W. Chen, J. Pei, C.-T. He, J. Wan, H. Ren, Y. Wang, J. Dong, K. Wu, W.-C. Cheong, J. Mao, X. Zheng, W. Yan, Z. Zhuang, C. Chen, Q. Peng, D. Wang, Y. Li, Single tungsten atoms supported on MOF-derived N-doped carbon for robust electrochemical hydrogen evolution, Adv. Mater. 30 (2018) 1800396. Copyright 2018, John Wiley and Sons.)*

acid solution (Fig. 4). In addition, the uncoordinated amino group effectively prevents the agglomeration of W species.

In another example, Chen et al.[56] employed cage-encapsulated-precursor pyrolysis strategy and successfully prepared highly active and stably separated Fe/N single-atom doped porous carbon with high loading of Fe. As shown in Fig. 5, ZIF-8 was used as a molecular cage to encapsulate the precursor $Fe(acac)_3$, which was pyrolyzed and converted into nitrogen-doped porous carbon anchored with isolated Fe single atoms.

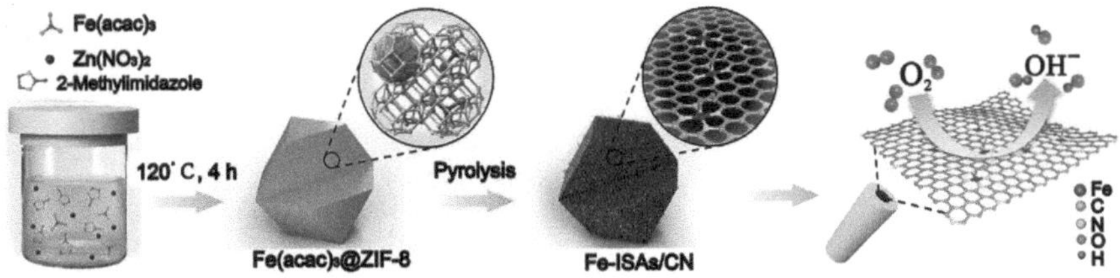

Fig. 5 The scheme of the formation of Fe-ISAs/Cn by cage-encapsulated-precursor pyrolysis strategy. *(Adapted with the permission from Y. Chen, S. Ji, Y. Wang, J. Dong, W. Chen, Z. Li, R. Shen, L. Zheng, Z. Zhuang, D. Wang, Y. Li, Isolated single iron atoms anchored on N-doped porous carbon as an efficient electrocatalyst for the oxygen reduction reaction, Angew. Chem. Int. Ed. 56 (2017) 6937–6941. Copyright 2017, John Wiley and Sons.)*

2.3 MOF-derived metal/carbons

The introduction of metals is a very popular strategy for the preparation of MOF-derived porous materials. Actually, MOFs can be applied as precursors/templates to fabricate their derived metal-doped porous carbons, which can well-maintain and evenly distribute metal sites to optimize the density of active sites [57]. This makes MOF-derived metal-doped porous carbon catalysts widely applied in the field of electrocatalysis. Meanwhile, the metal catalytic sites contained in the carbon layer can work in various harsh environments, providing a promising strategy for improving the consumption of precious metal catalysts. In fact, the introduction of different metals species can expose more active sites. Recently, Zhang and coworkers [58] reported a pyrolysis approach to fabricate Co—Fe alloy@N-doped carbon hollow spheres derived from dual-MOF. *Co*-based MOF self-assembles on the surface of polystyrene microspheres, which contain discrete Fe-based MOF nanocrystals, and then the resulting composite were heat-treated to obtain a hollow structure material with uniform distribution of Co—Fe alloy in the porous carbon (Fig. 6A). Li et al. proposed an in suit growth method to inlaid a three-dimensional ordered macroporous structure with ultrathin NiFe MOF nanosheets [59]. The fabrication is shown in Fig. 6B. As a semiconductor template, ordered macroporous NiFe hydroxide (OM-NFH) was used to grow and implant ultrathin NiFe MOF nanosheets to obtain NiFe MOF/OM-NFH.

2.4 MOF-derived metal compounds/carbons

Metal compounds (e.g., metal carbides, nitrides, oxides, sulfides, phosphides, and selenides) witness rapid development in the field of electrochemical catalysts [60–62]. Therefore, the functional composite materials using MOF as templates/precursors to derive these metal compounds have been extensively studied. However, the metal ions are more easily reduced to metal elements during pyrolysis process. It is still possible to convert metal ions into the corresponding metal compounds directly by controlling pyrolysis conditions [63–67]. Metal carbides or metal nitrides can be directly obtained during pyrolysis process due to the presence of carbon, nitrogen, and corresponding heteroatoms in the organic ligands of MOFs. Shi and coworkers [68] developed an in situ carburization approach based on Mo-based MOF to synthesize a graphite carbon shell catalyst encapsulating highly active and stable MoC (Fig. 7).

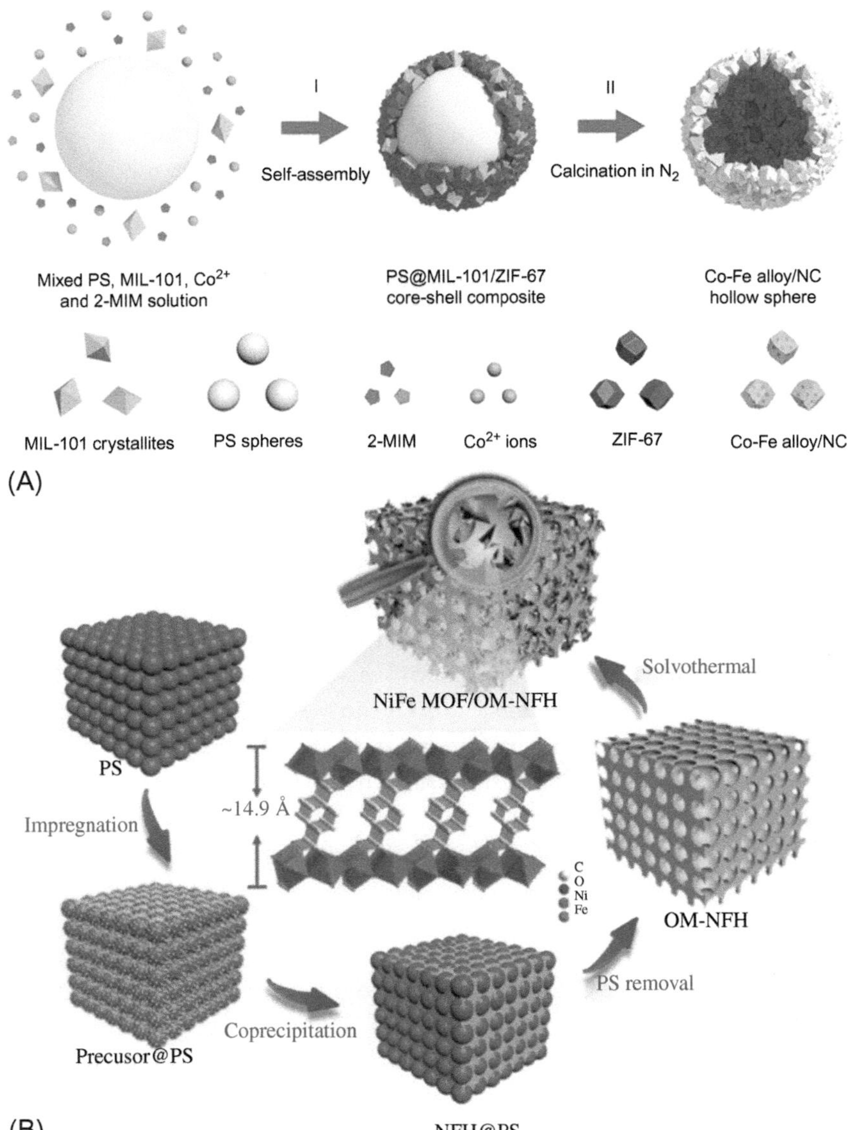

Fig. 6 (A) The scheme of the formation of Co—Fe alloy/N-doped carbon hollow spheres. (B) Schematic representation for the preparation of NiFe MOF/OM-NFH. *((A) Adapted with the permission from S.L. Zhang, B.Y. Guan, X.W. Lou, Co–Fe alloy/N-doped carbon hollow spheres derived from dual metal–organic frameworks for enhanced electrocatalytic oxygen reduction, Small 15 (2019) 1805324. Copyright 2019, John Wiley and Sons. (B) Adapted with the permission from X. Li, D.D. Ma, C. Cao, R. Zou, Q. Xu, X.T. Wu, Q.L. Zhu, Inlaying ultrathin bimetallic MOF nanosheets into 3d ordered macroporous hydroxide for superior electrocatalytic oxygen evolution, Small (2019) 1902218. Copyright 2018, John Wiley and Sons.)*

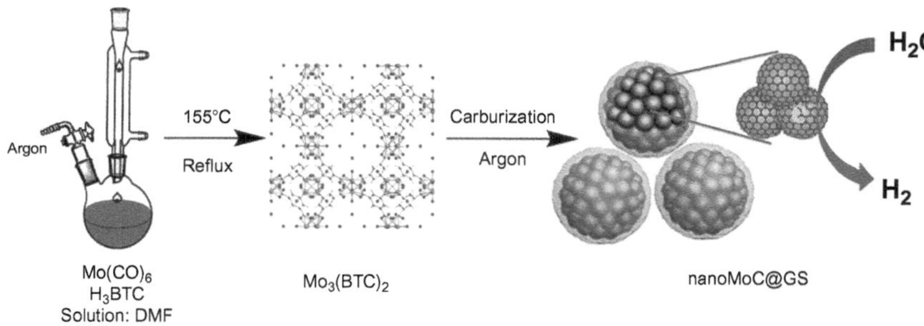

Fig. 7 The scheme of procedure for synthesis of nanoMoC@GS. *(Adapted with the permission from Z. Shi, Y. Wang, H. Lin, H. Zhang, M. Shen, S. Xie, Y. Zhang, Q. Gao, Y. Tang, Porous nanoMoC@graphite shell derived from a MOFs-directed strategy: an efficient electrocatalyst for the hydrogen evolution reaction, J. Mater. Chem. A 4 (2016) 6006–6013. Copyright 2016, Royal Society of Chemistry.)*

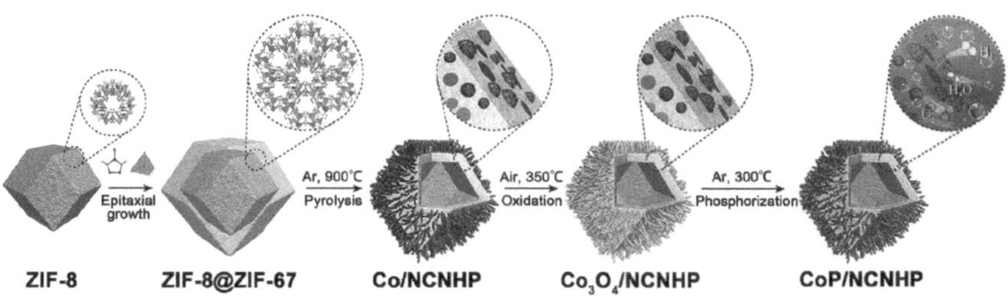

Fig. 8 The scheme of procedure for synthesis of CoP/NCNHP. *(Adapted with the permission from Y. Pan, K. Sun, S. Liu, X. Cao, K. Wu, W.-C. Cheong, Z. Chen, Y. Wang, Y. Li, Y. Liu, D. Wang, Q. Peng, C. Chen, Y. Li, Core–shell ZIF-8@ZIF-67-derived CoP nanoparticle-embedded N-doped carbon nanotube hollow polyhedron for efficient overall water splitting, J. Am. Chem. Soc. 140 (2018) 2610–2618. Copyright 2018, American Chemical Society.)*

Other metal compounds can be converted by simple chemical modification after coating the transition metal into MOF-derived porous carbon [69, 70]. In Fig. 8, Pan et al. [71] reported a pyrolysis-oxidation-phosphating strategy for the synthesis of hollow polyhedron with N-doped carbon nanotube embedded with CoP nanoparticles. The novel hybrid nanostructure was derived from core-shell structure ZIF-8@ZIF-67.

Certainly, the facile strategy can be expanded to fabricate various other MOF-derived porous carbons embedded with the corresponding metal compounds. For example, Xu and coworkers [72] rationally developed a facile in situ selenidation strategy, which is to synthesize MOF-derived carbon-coated selenide. Fig. 9 shows the fabrication route of MOF-derived selenide encapsulated in carbon. A series of various MOFs are synthesized via facile solution method, followed by annealing and

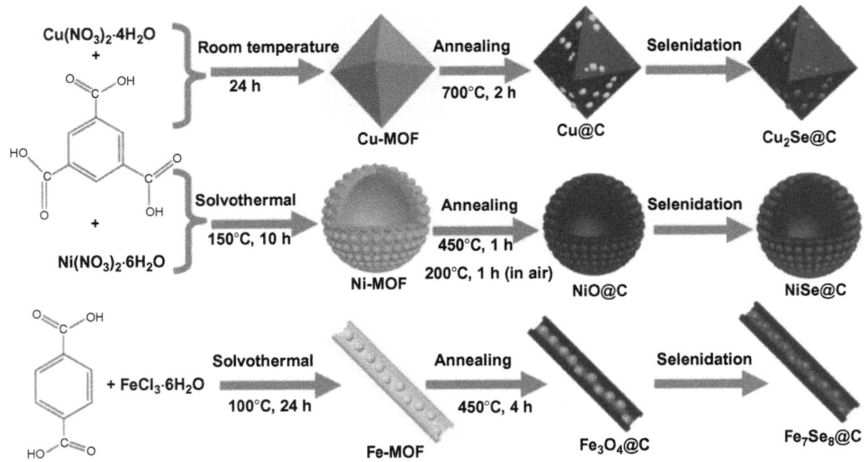

Fig. 9 Schematic illustration for the fabrication of uniform MOF-derived Cu_2Se@C porous octahedra, NiSe@C hollow microspheres, and peapod-like Fe_7Se_8@C nanorods. *(Adapted with the permission from X. Xu, J. Liu, J. Liu, L. Ouyang, R. Hu, H. Wang, L. Yang, M. Zhu, A general metal-organic framework (MOF)-derived selenidation strategy for in situ carbon-encapsulated metal selenides as high-rate anodes for Na-ion batteries, Adv. Funct. Mater. 28 (2018) 1707573. Copyright 2018, John Wiley and Sons.)*

selenidation to obtain the MOF-derived materials with morphology-maintained embedded metal-selenide.

3 MOF-derived porous materials for heterogeneous catalysis

MOF-derived porous materials are ideal candidates for the construction of heterogeneous catalysts [34, 73–75]. They have shown several advantages in these organic reactions [76, 77]. Firstly, the active elements/sites involved in the MOF precursors can be inherited to their derived materials, while the stability is greatly enhanced in the MOF-derived material. Secondly, the ligands between metal ions/clusters might prevent the active metal/metal oxide agglomeration after pyrolysis. Thirdly, the MOF-derived materials might inherit the morphology of the MOF precursors, exhibiting high surface area and high porosity which will facilitate the diffusion of the reactants and products. In previous reports, the ZIF, MIL series, Cu-BTC, and other transition metal-based MOFs and their composites have been frequently used as precursors for the preparation of porous materials for promoting various chemical transformations, including oxidation reaction, reduction reaction, Fischer-Tropsch synthesis, and so on [34, 74, 76].

3.1 Oxidation reaction

In recent years, MOF-derived nanocomposites have been widely explored for various oxidation reactions, such as oxidation of alcohols, oxidation coupling of amines and aerobic epoxidation of styrene, and so on [34, 35, 78–81]. A series of Fe, Co, and Ni-involved porous carbon catalysts have been synthesized from transition metal-based MOFs and utilized for the oxidation reaction catalysis. Li's group showed that Co nanoparticles embedded in nitrogen-doped graphite derived from Co-ZIF-67 could catalyze the aerobic oxidation of alcohols to esters at room temperature under base-free and atmospheric conditions [82]. The Co@C-N catalytic system could promote the reaction of a broad scope of substrates ranging from aromatic, aliphatic alcohols, and diols with good to excellent yields. Jiang's group also utilized Co-ZIF-67 as a self-template to afford N-doped porous carbon incorporating Co@CoO catalytic materials [83]. The optimized sample, NC-700-3h, exhibited excellent catalytic activity, selectivity, and magnetic recyclability for the direct oxidation of alcohols to esters with O_2 as a benign oxidant under mild conditions (80°C, 1 bar O_2). The pyrolysis of $Cu_3(BTC)_2$ afforded a robust and chemically inert copper-carbon nanocomposite Cu@C, which exhibited outstanding performance for the aerobic oxidation of alcohols with a broad scope for the substrate and excellent reusability [84]. In addition, the oxidation of hydrocarbons and CO has been catalyzed by pyrolysis of MOF-derived materials, such as Co-MOF [85], Fe_3O_4@C [86], Fe-Co/C [87], MIL-125(Ti) [88], MIL-45b [87], HKUST-1 [89], and so on.

The emission of poisonous CO gas from motor vehicles and industrial smokestacks causes serious environmental pollution. It is important to convert and utilize CO. Different transition metal-based catalysts obtained by MOF precursors have also shown high activity toward CO oxidation reaction. A number of porous Cu-based nanocomposites, such as CeO_2: Cu^{2+} [90], CuO/TiO_2 [91], CuO/Cu_2O [92], Cu/CuO_x/C [93], and CuO/CeO_2 [94], with different morphologies were synthesized from Cu-MOFs and Cu-MOF composites under different pyrolysis conditions. For instance, Zhang and coworkers have employed $Cu_3(BTC)_2$ loaded with Ce^{3+} as precursor to afford a series of catalysts with well-dispersed copper species and applied them for preferential CO oxidation [94]. Li, Tian, and coworkers reported the preparation of CuO/CeO_2 catalysts by utilizing Cu-containing Ce-MOF [95]. The CuO/CeO_2 catalysts showed excellent catalytic performance for CO conversion reaction.

3.2 Reduction reaction

MOF-derived porous materials have been widely applied as heterogeneous catalysts for reduction reactions, such as the hydrogenation reactions of many compounds, including nitrocompounds, aldehydes, imines, ketones, and so on [96–100]. Jiang's group reported the stabilization of palladium nanoparticles in the metal-organic framework-derived porous carbon for biofuel upgrading [101]. The obtained Pd/NPC-ZIF-8 exhibited remarkable catalytic performance, which was attributed to the well-dispersed and electron-rich Pd sites, high surface area, hierarchical pores, and favorable hydrophilicity. Recently, Jiang's group fabricated novel hollow porous carbons (HPC) with well-dispersed dopants of nitrogen and single Zn atoms by utilizing hollow ZIF-8 as precursor through pyrolysis (Fig. 10) [100]. The optimized sample, HPC-800, exhibited high catalytic performance for CO_2 cycloaddition with epoxides under light irradiation. High catalytic activity was ascribed to ultrahigh loading (11.3 wt%) of single-atom Zn and uniform N active sites, efficient photothermal conversion, and the hierarchical pores in the carbon shell. Li and Wu' group explored a precise control of isolated single ruthenium site supported on nitrogen-doped porous carbon (Ru SAs/N-C) [102].

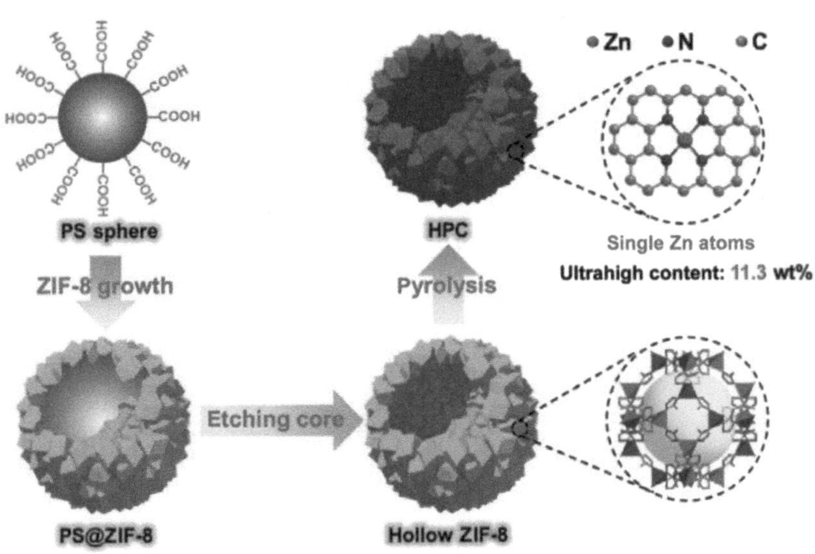

Fig. 10 The scheme of the fabrication process of the HPC. *(Adapted with the permission from Q. Yang, C.C. Yang, C.H. Lin, H.L. Jiang, Metal–organic-framework-derived hollow N-doped porous carbon with ultrahigh concentrations of single Zn atoms for efficient carbon dioxide conversion, Angew. Chem. Int. Ed. 58 (2019) 3511–3515. Copyright 2019, John Wiley and Sons.)*

The atomic dispersion of Ru sites was realized through the coordination between Ru^{3+} and the free amine groups ($-NH_2$) at the skeleton of a metal-organic framework precursor. The single Ru sites with single-mind N coordination catalyzed effectively chemoselective hydrogenation of functionalized quinolones.

Considering the high cost of noble metals (e.g., Ru, Pt, and Pd), non-noble transition metal catalysts, such as Co-based [103–108], Ni-based materials [109, 110], Fe-based materials [111], and transition metal alloy-based materials [112, 113], have been greatly developed for the hydrogenation reactions.

Li and coworkers found that nonnoble Co@C-N catalytic system obtained by Co-MOF presented high activity and selectivity in the hydrogenation of a variety of unsaturated bonds, including C=C, C≡C, N=O, and C=O bonds, and this reaction was carried out in the absence of base additives [103]. Zhao and coworkers reported a porous carbon-supported copper composite ($Cu/Cu_2O/C$), synthesized by utilizing $Cu_3(BTC)_2$ as template and copper precursor [114]. The obtained $Cu/Cu_2O/C$ exhibited comparable catalytic activity with noble metal nanoparticles, which could completely reduce nitrophenol within 160 s with only tiny amount of the catalyst. Jagadeesh, Murugesan, and coworkers have reported that cobalt nanoparticles encapsulated by a graphitic shell exhibited to be broadly effective reductive amination catalysts [105]. The cobalt nanoparticle-based catalysts were obtained by pyrolysis of cobalt diamine-dicarboxylic acid MOFs, which were active for the synthesis of primary, secondary, tertiary, and *N*-methylamines under industrially viable and scalable conditions. Han and coworkers created a new strategy to immobilize ultrafine (0.9 nm) nonnoble metal particles on supports using MOFs as metal precursors [110]. The ultrafine Ni/SiO_2 and Co/SiO_2 particles exhibited very high catalytic activity for liquid-phase hydrogenation of benzene to cyclohexane below 100°C.

Aside from transition metal-based nanoparticles embedded in the carbon matrix, the MOF-derived metal-free catalysts have been developed and the crucial structural and component roles in determining the catalytic performance have been explored for the catalytic hydrogenation reactions [115, 116]. Jiang and coworkers have utilized different types of MOFs as templates/precursors to synthesize porous carbon materials with various nitrogen dopant forms and contents, degrees of graphitization, porosities, and surface areas [116]. The theoretical studies showed that the content and type of the pyrrolic nitrogen dopant played crucial roles in determining the catalytic performance, which led to the superior activity of the sample PCN-224-700 in the catalytic reduction of 4-nitrophenol to 4-aminophenol.

3.3 Fischer-Tropsch synthesis

The Fischer-Tropsch synthesis (FT synthesis) is the process of chemical reactions that converts syngas ($CO + H_2$) into liquid hydrocarbons. Gascon and coworkers synthesized highly dispersed iron carbides embedded in a matrix of porous carbon with very high iron loadings (>40 wt%) via pyrolysis of Fe-based MOF Basolite F300 Fe-BTC [117, 118]. The high dispersion of the metal phase, the absence of large iron particles, and its encapsulation in a highly porous carbon matrix in the obtained solids minimized catalyst deactivation, resulting in high catalytic activity and stability in the FT synthesis. Lin and coworkers prepared unique $Fe_3O_4@Fe_5C_2$ core-shell nanoparticles on carbon catalysts through the pyrolysis of Fe-MIL-88 and Fe-MIL-88-HN_2 [119]. The Fe-time yields (the amount of CO converted to hydrocarbons per gram of Fe per second) of the nitrogen-doped catalyst reached 720 $\mu mol_{CO}\ g_{Fe}^{-1}\ s^{-1}$ under the conditions of 300°C, 2 MPa and $H_2/CO = 1$. The unique $Fe_3O_4@Fe_5C_2$ core-shell structure dispersed on carbon supports structure was favorable for FT synthesis. The authors also found the coverage of carbonates on the surfaces of $Fe_3O_4@Fe_5C_2$ nanoparticles, which may result from decarboxylation of the MOFs during pyrolysis, leading to an active surface rich with dangling bonds for catalytic turnover. Zou and coworkers explored the influence of pore texture and nitrogen species of the carbon support for the F-T synthesis using well-defined catalysts derived from two typical MOFs, nitrogen-rich ZIF-67 and nitrogen-free Co-MOF-74 [120]. The Co-MOF-74-derived nanocomposite (Co@C) showed 30% conversion for CO, 65% selectivity for long-chain hydrocarbons (C_5^+), and 10% selectivity for short-chain hydrocarbons (C_2–C_4) after 100 h on stream, whereas the ZIF-67-derived nanocomposite (Co@NC) exhibited 10% conversion for CO and 31% selectivity for C_5^+ products and 37% selectivity for C_2–C_4 after 100 h on stream. The excellent CO conversion was attributed to the large pore size of the carbon support and the high C_2–C_4 selectivity originated from the influence of nitrogen species in the carbon support.

4 MOF-derived porous materials for photocatalysis

Photocatalysis is a type of catalytic reaction that converts solar energy to chemical energy in a convenient and eco-friendly approach. Since 1972 that

Honda and Fujishima discovered the photocatalytic splitting of water on TiO_2 electrodes, semiconductor-promoted photocatalysis has become a promising and attractive approach to produce clean energy [121]. In recent years, MOF-derived porous composites have been explored as promising photocatalysts toward various reactions, including photocatalytic H_2 production, photocatalytic CO_2 conversion, degradation of organic contaminants, and so on.

4.1 Photocatalytic H_2 production

Photocatalytic water splitting is a photosynthesis process in a photoelectrochemical cell used for the dissociation of water into hydrogen and oxygen, with absorption of light. The photocatalytic activity is strongly influenced by photo-absorption capability, bandgap energy, active sites and the charge separation properties, etc. Numerous researches have been devoted to develop high activity photocatalysts by rational design and controllable synthesis of photocatalytic materials. TiO_2 is an excellent candidate for high efficient photocatalysis for various photochemical reactions. However, TiO_2 only absorbs UV light, which occupies about 5% of sunlight. To fully utilize solar energy and improve activity, great efforts have been dedicated to develop highly efficient photocatalysts exhibiting strong visible-light absorption capability and low cost.

Compared with TiO_2, Fe_2O_3 is able to absorb visible light. However, the conduction of Fe_2O_3 is too low to drive photocatalytic H_2 production. Lin's group has developed a promising Fe-MOF coated with TiO_2-templated approach to combine the merits of both TiO_2 and Fe_2O_3 [122]. The crystalline Fe_2O_3/TiO_2 composite nanoparticles were synthesized by pyrolysis of Fe-MIL-101 coated with amorphous titania. The material enabled better photocatalytic hydrogen production from water using visible light, than individual Fe_2O_3 or TiO_2 or a mixture of the two, demonstrating the advantages of the synergy between Fe_2O_3 and TiO_2.

The construction of heterojunctions can also enhance the photocatalytic activity [123, 124]. Recently, Wang and coworkers reported the simultaneous construction of heterostructures and cocatalysts, g-C_3N_4/Fe_2O_3@FeP composites, with a tight contact favoring charge transfer for improved photocatalytic H_2 production of g-C_3N_4 [124]. The Fe_2O_3@FeP-loaded g-C_3N_4 was synthesized by annealing and phosphidation of a g-C_3N_4/Fe-MOF and served as both heterojunction and cocatalyst material to improve the photocatalytic H_2 evolution performance of g-C_3N_4, with a

H_2 evolution rate as high as 12.03 μmol g^{-1} h^{-1} and the apparent quantum efficiency (AQE) at 420 nm was 38.8%. The improved photocatalytic activity of the composite can be ascribed to its enhanced visible-light absorption intensity, the efficient electron-hole separation by the formation of the heterojunction with Fe_2O_3 and FeP, as well as the accelerated H^+ reductive reaction resulting from the FeP cocatalyst.

Besides metal oxides, MOF-derived metal sulfides can also be used as efficient catalysts. Huang et al. synthesized hollow Co-based bimetallic sulfide ($M_xCo_{3-x}S_4$, M=Zn, Ni, and Cu) polyhedra through sulfidation and annealing of bimetallic MCo-MOFs with superior HER activity and stability [67]. The hollow structure and the homogeneous doping in Co_3S_4 lattice significantly improved the HER activity of Co_3S_4, leading to an excellent rate of H_2 evolution rate of 155.2 μmol g^{-1} h^{-1}. Similarly, Kim and coworkers developed noble metal-free MOF-derived onion slice-type hollow-structured Co_4S_3/CdS from Co-MOF [125]. The Co_4S_3 with the CdS particles effectively accelerated charge separation due to the low density, hollow interior, and shell permeability of the onion-morphology. The optimized Co_4S_3/CdS exhibited high rate of H_2 production of 12,360 μmol $g^{-1}h^{-1}$, could be reused for five runs, and was highly stable for 65 h.

Jiang and coworkers also utilized thermally stable MOFs, MIL53(Al), as hard templates to afford porous metal oxides and subsequently metal sulfides by a nanocasting method [126]. The resultant metal oxides/sulfides show considerable Brunauer-Emmett-Teller (BET) surface areas and exhibited excellent performance in photocatalytic water splitting for H_2 production with a rate of 634 μmol g^{-1} h^{-1}, much higher than both corresponding bulk and nanosized CdS counterparts, which could be attributed to the porous structure and nanosized effect.

4.2 Photocatalytic CO_2 conversion

Besides the photocatalytic production of hydrogen, the photocatalytic transformation of CO_2 into fuels such as CH_4, CO, and CH_3OH is also highly significant in order to compensate excessive CO_2 emission and address energy crisis. Ye's group has realized the solar-driven CO_2 conversion into CO using plasmonic Fe@C nanostructures with iron NPs ($<$10 nm) as core and ultrathin carbon layers as shell via pyrolysis of Fe-MIL-101 precursor [127]. The strong adsorption of visible light and infrared radiation induced a significant photo thermal effect and provided adequate energy for the reaction. Moreover, UV-light-induced iron local surface-plasmon resonance was utilized to activate the nonpolar CO_2 molecules at high temperature and facilitated the final reaction activity. As a result, the core-shell Fe@C catalyst exhibited 2196.17 μmol CO after photoirradiation for 120 min.

Lou's group fabricated the hierarchical In_2S_3-$CdIn_2S_4$ heterostructured nanotubes through a liquid-phase sulfidation process and cation-exchange reaction utilizing In-based MOF (MIL-68) hexagonal prism as the precursor, for efficient and stable photocatalysis of visible light CO_2 reduction [128]. The novel self-templated strategy, including sequential anion- and cation-exchange reactions, led to the successful fabrication of hierarchical tubular hybrids with two distinct sulfide semiconductors. The hierarchical heterostructured nanotubes manifested remarkable performance for photocatalytic conversion of CO_2 with high CO generation rate of 825 μmol h^{-1} g^{-1} and outstanding stability under visible light irradiation.

Recently, Deng and Lin's group has synthesized the ultrathin two-dimensional porous Co_3O_4 nanosheet catalysts by a facile MOF-engaged strategy (Fig. 11) [129]. The ultrathin holey Co_3O_4 nanosheets with a thickness of ~1.8 nm were fabricated by air calcining of ultrathin ZIF-67 nanosheets and then applied as catalysts with Ru-based photosensitizers for photocatalytic CO_2 reduction. The Co_3O_4 nanosheets exhibited enhanced activity and selectivity, faster charge migration ability, and greater number of exposed active sites and reached high CO generation rate of 4.52 μmol h^{-1}, which was considerably higher than that of Co_3O_4-BK (2.48 μmol h^{-1}). Su and coworkers fabricated an efficient and economic

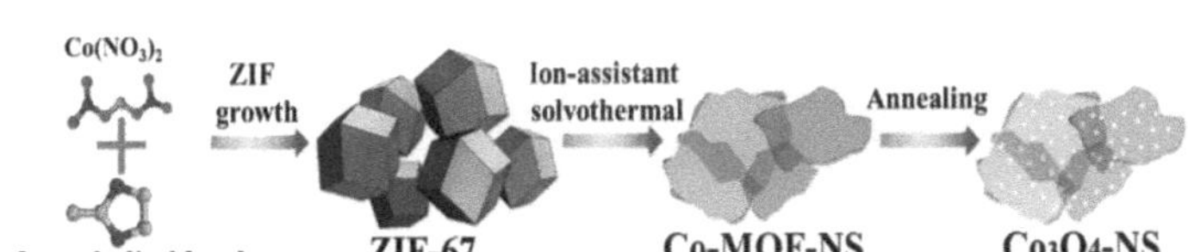

Fig. 11 Schematic illustration of the synthetic process of ultrathin holey Co_3O_4 nanosheets. *(Adapted with the permission from W. Chen, B. Han, C. Tian, X. Liu, S. Liang, H. Deng, Z. Lin, MOFs-derived ultrathin holey Co_3O_4 nanosheets for enhanced visible light CO_2 reduction, Appl. Catal. B Environ. 244 (2019) 996–1003. Copyright 2019, Elsevier.)*

three-component heterojunction $Cu_{2-x}S$ nanotubes coated with a carbon layer and g-C_3N_4 photocatalyst by using HKUST-1 as a precursor [130]. The carbon coat in C-$Cu_{2-x}S$@g-C_3N_4 acted as an electron reservoir, which facilitated electron-hole pair separation, and reached a high reactivity of 1062.6 $\mu mol\,g^{-1}$, which was nearly 7- and 23-fold enhanced compared with bare g-C_3N_4, C-$Cu_{2-x}S$, and was higher than most of the reported Cu_2S or g-C_3N_4 composites under similar conditions.

4.3 Degradation of organic contaminants

MOF-derived porous metal oxides/sulfides, and their porous hybrids, show great potential for the photocatalytic degradation of dyes due to their high tunability, low cost, high stability, excellent light absorption, and strong capability of election-hole separation. Photocatalysts with controllable structure, components, and morphology, such as MOF-derived metal oxides/and their nanohybrids or nanocomposites [131–134], have been shown to be excellent photocatalysts for the degradation of dyes. Zhang and coworkers reported a simple method for the preparation of ZnO-coated three-dimensional graphene composites by utilizing ZIF-8 polyhedron as the precursors through a two-step annealing process [135]. Due to the high surface area and effective interaction between the two composites, the ZnO/3DGN show excellent photocatalytic degradation of methylene blue, exceeding the performance of single 3DGN or ZnO. Lin and coworkers prepared a magnetic cobalt-graphene (MCG) nanocomposite by carbonizing a self-assembly of ZIF-67 and graphene oxide (GO) [136]. The obtained nanocomposite MCG could serve as an effective catalyst to activate peroxymonosulfate (PMS) to decolorize acid yellow dye in the advanced oxidation process, which could be ascribed to the cooperative effect to activate PMS and improve electron transport ability.

Li and coworkers utilized hollow Zn/Co-ZIF matrix consisting of a ZIF-8 shell with some Co-ZIF nanoplates inside the hollow cavity as precursor to fabricate a unique ZnO@C-N-Co core-shell nanocomposite as efficient and recyclable photocatalyst by direct pyrolyzing [131]. The unique ZnO@C-N-Co core-shell nanocomposite showed superior photocatalytic degradation of methyl orange properties, which could be ascribed to the stabilization effect of intermediate ZnO by the core-shell structure, a porous carbon shell that provided high adsorption capability for reactants, as well as the embedded Co NPs that could provide richer electron traps which inhibited the recombination of electrons and holes.

5 MOF-derived porous carbons for electrocatalysis

The consumption of traditional fossil fuels and related environmental problems have made the development of abundant renewable clean energy the topic of extensive research. In recent years, the application of electrocatalysis has been deeply studied, such as oxygen evolution reaction (OER), oxygen reduction reaction (ORR), hydrogen evolution reduction (HER), CO_2 reduction reaction (CO_2RR), and methanol oxidation reaction (MOR). At present, the most widely used catalysts still rely on noble metals such as Ru and Pt. However, the scarcity and stability of precious metals have seriously hampered their large-scale application. Therefore, as the potential alternative catalysts, transition metal-based MOFs-derived porous carbons, metal compounds, and their hybrids have been extensively studied to improve electrochemical performance [54, 137–143]. The advantages of MOF-derived materials as electrocatalysts are summarized as follows: (i) the conductivity of MOFs after carbonization is greatly enhanced relative to the original MOF, and the graphite carbon is significantly beneficial for electron transfer in the electrocatalytic process. (ii) MOF-derived composites inherit the porous structure and large specific surface area, which is more conducive for high-density active sites to access and dispersion. (iii) It is easy to achieve doping of different metals, metal compounds, and heteroatoms, which increases the active sites of the composites. (iv) MOF-derived porous carbon materials have excellent recyclability due to the enhanced stability compared to MOF precursors. From above advantages, MOF-derived porous carbons are promising catalytic materials for electrocatalytic energy-related applications (Table 1).

5.1 Oxygen reduction reaction

The ORR plays a vital role in energy conversion devices (e.g., fuel cells and metal-air batteries). Although Pt and Pt-based catalysts are recognized as excellent catalysts for reducing the energy barriers to accelerate kinetics, their high cost is still not implemented in commercial application. Therefore, as good substitutes, new types of MOFs-derived materials have exploded development in recent years [54, 140, 147, 155, 171–178]. Meanwhile, MOF-derived porous materials exhibit excellent catalytic activity, stability, and methanol durability for ORR [138, 179–183]. In particular, some optimized MOF-derived catalysts achieved consistent or even better performance compared to commercial Pt catalysts. Rong and coworkers reported the N, S, P tri-doped mesoporous carbon (NPSpC) for ORR

Table 1 Summary of MOF-derived porous carbons for electrocatalytic reactions.

Catalyst	MOF precursor	Catalytic reaction	References
Co-N_4	Cobalt imidazolate framework	ORR	[144]
FeIm/ZIF-8	Iron imidazolate framework	ORR	[145]
NPSpC	ZIF-8	ORR	[146]
N-CNTs-650	ZIF-67	ORR	[139]
Z8-Te-1000	ZIF-8	ORR	[147]
H-M_{mix}-N_x-C	ZIF-8	ORR	[148]
Fe SAs-N/C-x	FePc-x@ZIF-8	ORR	[149]
GNPCSs-800	ZIF-8	ORR	[46]
NCNTs-20	Zn-Fe-ZIF	ORR	[150]
Fe-N-C	Fe/Phen/ZIF-8	ORR	[151]
Co/NBCNTs	Zn/CoZIF @ g-C_3N_4	ORR	[152]
CNPs	MIL-88B-NH_2	ORR	[153]
$FeCo_2$-NPC-900	PCN-224-FeCo	ORR	[140]
CoPS-N/C	ZIF-67	HER	[154]
CoP@BCN	Co-MOF	HER	[155]
NiS	Ni-Co PBA	HER	[156]
Co-Fe-P	MIL-88B MOFs	HER	[157]
Co_3O_4/Co-Fe	ZIF-67	OER	[158]
$CoFe_2O_4$/C NRAs	MOF-74-Co/Fe	OER	[143]
NiCoP/C	ZIF-67@LDH	OER	[159]
FeNiP/C	BMM-10	OER	[160]
NGO/Ni_7S_6	Ni-MOF	OER/HER	[161]
Co_3S_4@MoS_2	ZIF-67	OER/HER	[162]
Co_3S_4/EC-MOF	Co-based MOF	OER/HER	[163]
Ni@CoO@CoNC	Ni@CoO@ZIF-67	OER/HER	[164]
Co_3O_4-C	Co-MOF	ORR/OER	[165]
NC@Co-NGC	ZIF-8@ZIF-67	ORR/OER	[166]
NC-Co SA	Co-MOF	ORR/OER	[167]
Pd/ZC-1000	ZIF-8	MOR	[168]
C-AFC@ZIF-8	AFC@ZIF-8	CO_2RR	[169]
Ni/Fe-N-C	ZIF-8	CO_2RR	[170]

via pyrolysis of ZIF-8 treated with dodecyl mercaptan and sodium phytate solution, shown in Fig. 12 [146]. With the large specific surface area and active sites exposed by heteroatoms, the NPSpC possessed the excellent ORR performance with onset potential of 0.923 V and half-wave potential of 0.821 V in 0.1 M KOH. Meanwhile, the stability and methanol tolerance of NPSpC showed the great potential.

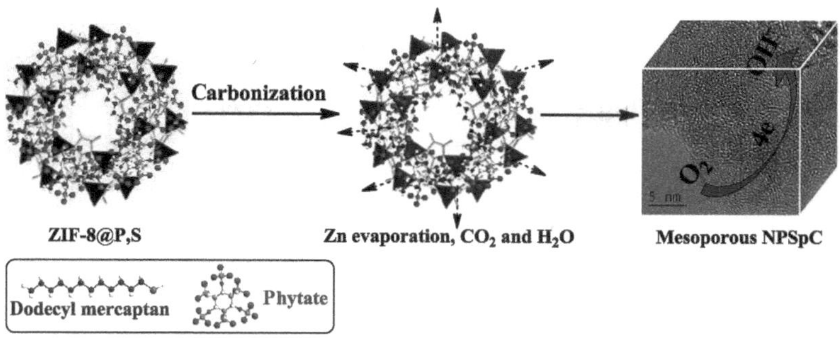

Fig. 12 Schematic illustration for the synthesis of mesoporous NPSpC electrocatalyst. *(Adapted with the permission from H. Rong, T. Zhan, Y. Sun, Y. Wen, X. Liu, H. Teng, ZIF-8 derived nitrogen, phosphorus and sulfur tri-doped mesoporous carbon for boosting electrocatalysis to oxygen reduction in universal pH range, Electrochim. Acta 318 (2019) 783–793. Copyright 2019, Elsevier.)*

As a well-known Co-based MOF that shares the same topology with ZIF-8, ZIF-67 is a potential catalytic material for the preparation of active sites containing M-N. Zhou and coworkers first reported the small diameter nitrogen-doped carbon nanotubes with Co nanoparticles (Co/N-CNTs), which were derived from mesoporous silica ($mSiO_2$)-coated ZIF-67 (Fig. 13A) [184]. The coated $mSiO_2$ effectively prevents the aggregation of Co nanoparticles and plays an important role in inducing CNT growth during annealing. The optimized Co/N-CNTs exhibited higher ORR activity ($E_{1/2}=-0.154\,V$, $j=5.82\,mA\,cm^{-2}$) and stability compared to commercial Pt/C. Meanwhile, with the unique structure and special properties, CNT have attracted the great attention. Recently, Wang et al. demonstrated a simple synthetic strategy for adjusting the structure and size of nitrogen-doped carbon nanomaterials [152]. The fabrication is shown in Fig. 13B, in which nanomaterials with different dimensions are obtained by annealing the composites (different Co/Zn ratios of ZIF arrays and g-C_3N_4). Among them, the obtained Co/NBCNTs from Zn/CoZIF @ g-C_3N_4 showed better ORR activities ($E_{1/2}=0.83\,V$) than commercial Pt/C in alkaline solution. Moreover, the long-term stability and methanol tolerance of Co/NBCNTs were better than commercial Pt catalysts.

The introduction of catalytically active elements (such as Fe) can also improve the ORR activity of the catalyst. The carbon nanocomposites embedded with the combined iron binary alloy via optimized design of bimetallic organic framework (BMOF) were reported by Xiong and coworkers [185]. The electrocatalyst exhibited excellent catalytic activity

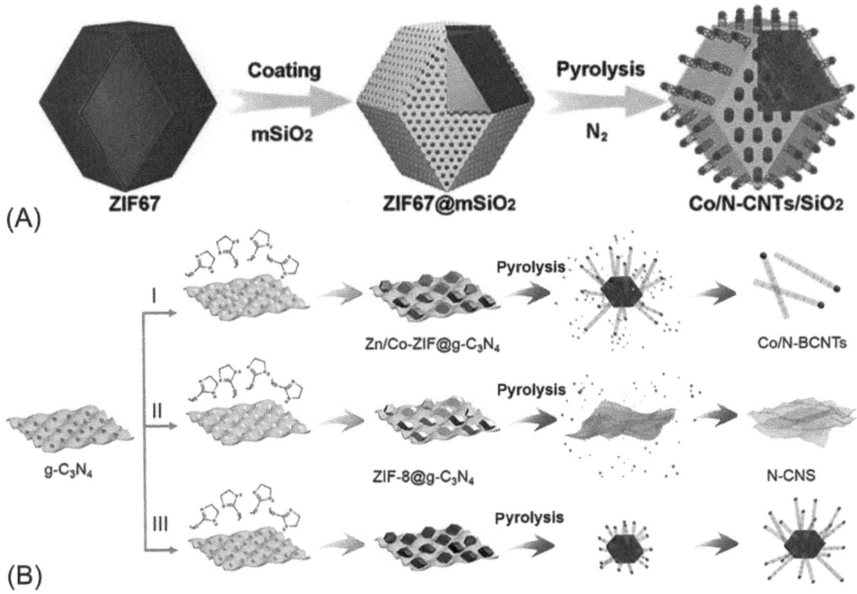

Fig. 13 (A) Synthetic procedure of the Co/N-CNTs by the $mSiO_2$-coated calcination strategy. (B) Schematic illustration of the preparation of nitrogen-doped carbon nanomaterials. *((A) Adapted with the permission from H. Zhou, D. He, A.I. Saana, J. Yang, Z. Wang, J. Zhang, Q. Liang, S. Yuan, J. Zhu, S. Mu, Mesoporous-silica induced doped carbon nanotube growth from metal–organic frameworks, Nanoscale 10 (2018) 6147–6154. Copyright 2018, Royal Society of Chemistry. (B) Adapted with the permission from R. Wang, T. Yan, L. Han, G. Chen, H. Li, J. Zhang, L. Shi, D. Zhang, Tuning the dimensions and structures of nitrogen-doped carbon nanomaterials derived from sacrificial g-C_3N_4/metal–organic frameworks for enhanced electrocatalytic oxygen reduction, J. Mater. Chem. A 6 (2018) 5752–5761. Copyright 2018, Royal Society of Chemistry.)*

for ORR with a high half-wave potential (0.89 V) in 0.1 M NaOH, comparable to commercial Pt/C electrocatalysts. Furthermore, it exhibited robust durability after 30,000 potential cycles.

Single-atom catalysts (SACs) have become a hot research topic due to their special characteristics [186–188]. However, there are still many unsolved challenges in the rational preparation and characterization of SACs [183].

MOFs materials, as the most promising precursors/templates, have been widely used to produce porous and SACs catalysts. MOFs not only provide anchoring sites, but also offer extra coordination sites, which are beneficial to connect single atoms. Recently, Jiao and coworkers reported a series of

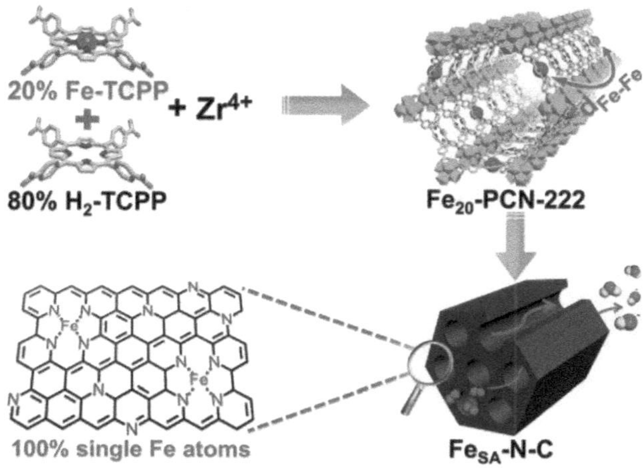

Fig. 14 Illustration of the rational fabrication of single Fe atoms-involved Fe_{SA}-N-C catalyst via a mixed-ligand strategy. *(Adapted with the permission from L. Jiao, G. Wan, R. Zhang, H. Zhou, S.H. Yu, H.L. Jiang, From metal–organic frameworks to single-atom Fe implanted N-doped porous carbons: efficient oxygen reduction in both alkaline and acidic media, Angew. Chem. Int. Ed. 57 (2018) 8525–8529. Copyright 2018, John Wiley and Sons.)*

isomorphic MOFs, prepared by adjusting the ratio between Fe-TCPP and H_2-TCPP. Nitrogen-doped porous carbon anchored with single iron atom (FeSA-N-C) was also obtained after optimized pyrolysis process (Fig. 14) [183]. Combined with highly active single Fe atoms and graded pores that promote mass transfer, optimized FeSA-N-C exhibited excellent activity and stability for ORR, the $E_{1/2}$ of FeSA-N-C is 0.891 V and 0.776 V in 0.1 M $HClO_4$ and KOH, respectively, even higher than that of Pt/C. In order to avoid uncontrolled morphology and additional compounds caused by general synthetic routes, Xiao et al. presented a one-pot synthesis of uniformly dispersed single Fe atom catalysts (Fe-N-C) for ORR by sacrificing ZIF-8 precursors [189]. Monodisperse Fe atoms were uniformly distributed at the edge of porous carbon, and each iron atom was coordinated with four N atoms to form activity sites. The Fe-N-C catalysts exhibited superb ORR performance with the $E_{1/2}$ of 0.9 and 0.81 in alkaline and acidic solutions, respectively.

Apart from the excellent activity of Fe-N_x SACs, Luo and coworkers first reported that pyrolysis could furnish Cr/N/C SACs as the ORR catalysts using MOF as the anchoring matrix [190]. It was confirmed that the monodisperse Cr atom formed a coordination structure with the N atom and the Cr/N/C SACs exhibited excellent activity for ORR, and the half-wave potential was 0.773 V versus RHE. Moreover, the catalysts showed the superb stability due to significantly lower Fenton reaction.

5.2 Hydrogen evolution reaction

Hydrogen is considered to be one of the most efficient and green energy sources owning to its high energy density and low carbon. Currently, the most effective method is to electrolyze the renewable water to produce hydrogen and then store or release energy by oxidizing (such as combustion or fuel cell). Meanwhile, the electrochemical methods are also considered as the most suitable pathways for large-scale production. However, the two half-reactions (HER and OER) in the process of water dissociation, with large reaction energy barriers and slow kinetics, highlight the need for low-cost, highly active electrocatalysts to increase the efficiency of water splitting. At present, a large number of MOF-derived materials have been developed (including metals, metal compounds, and single-atom compounds) and widely recognized as the promising alternative to noble metal catalysts (Pt and Pt-based catalysts) for HER.

Transition metal compounds (e.g., carbides and phosphides) with rational morphologies have proven to be suitable candidates for excellent HER electrocatalysts. Lu and coworkers presented porous nitrogen-doped carbon dodecahedrons (PNCDs) that confine ultrafine carbide nanocrystals via rational annealing of ZIF-8 with molybdate or tungstate, shown in Fig. 15 [191]. The optimized MC-M_2C/PNCD (M=Mo, W) exhibited excellent

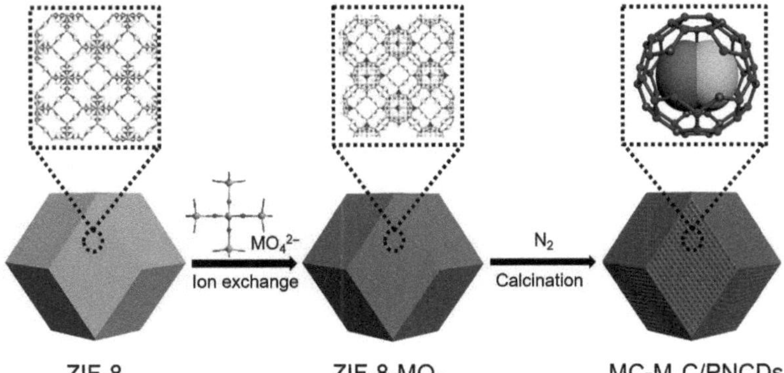

Fig. 15 Schematic illustration of the fabrication of MC-M_2C/PNCDs. *(Adapted with the permission from X.F. Lu, L. Yu, J. Zhang, X.W. Lou, Ultrafine dual-phased carbide nanocrystals confined in porous nitrogen-doped carbon dodecahedrons for efficient hydrogen evolution reaction, Adv. Mater. 31 (2019) 1900699. Copyright 2019, John Wiley and Sons.)*

HER activity due to the large number of active sites exposed by uniformly distributed ultrafine nanocrystals. In addition, the strong coupling between MC and M_2C favors hydrolysis and hydrogen desorption, while PNCD promotes charge transfer. The dual-phased MoC-Mo_2C/PNCDs presented lower overpotential (121 mV@ 10 mA cm^{-2}) and Tafel slope (60 mV dec^{-1}) compared to single-phased MoC/PNCDs or Mo_2C/PNCDs.

Luo's group reported a novel one-dimensional hybrid catalyst fabricated by simple flow and two-step annealing methods, which consisted of CoP and Mo_2C nanoparticles encapsulated in N-doped graphitic carbon (CoP/Mo_2C-NC) [192]. As the electron cloud transfers from the Co-P-Mo bond in CoP/Mo_2C-NC, it led to the high-valent Co and the low-valent Mo, which exhibited abundant HER activity, thereby promoting HER performance over the entire pH range. The overpotential of CoP/Mo_2C-NC was only 55.7 mV @ 10 mA cm^{-2} with favorable reaction kinetics (Tafel slope is 49 mV dec^{-1}).

Apart from the direct modification of MOFs, compositing MOF derivatives with excellent conductivity carriers (such as rGO, carbon cloth, and paper) has also been studied to develop HER catalysts. Recently, Chen et al. reported the grafting of nitrogen-doped carbon nanotubes coated with ultrafine Co nanoparticles onto both sides of rGO (Co@N-CNTs@rGO) [193]. The obtained Co@N-CNTs@rGO exhibited outstanding HER performance, the overpotential requested only 108 and 87 mV@10 mA cm^{-2} under 1 M KOH and 0.5 M H_2SO_4, respectively. High degree of graphitization, uniform distribution of Co nanoparticles, and abundant porosity of Co@N-CNTs@rGO provided a favorable platform for improving HER activity.

Wang and coworkers reported the Co-N-doped carbon nanoflake arrays-encapsulated $CoSe_2$ nanoparticles, which were grown on the carbon cloth via two-step pyrolysis (CS@CNC NAs/CC) [194]. Fig. 16 is schematic illustration of CS@CNC NAs/CC fabrication. The overpotential of CS@CNC NAs/CC requested only 84 mV@10 mA cm^{-2} and the Tafel slope is as small as 38 mV dec^{-1}. The remarkable HER performance is caused by the synergy between $CoSe_2$ nanoparticles and nanoflake arrays, and the unique structure facilitates the diffusion of electrolytes and bubbles. In addition, the carbon cloth provides stable structural support and high electrical conductivity.

Single-atom catalysts have been extensively investigated to improve HER performance owing to their high atomic utilization and unique active sites. Fan' s group presented a Ni-C-based catalyst via the carbonization of MOFs [195]. The obtained catalyst anchored single Ni species,

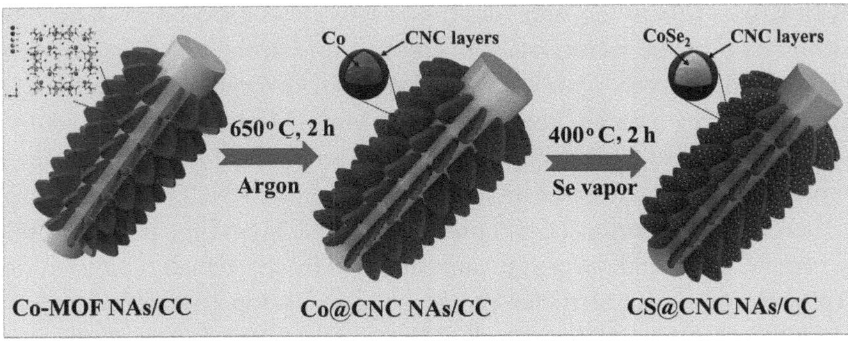

Fig. 16 Schematic illustration of the fabrication of the CS@CNC NAs/CC. *(Adapted with the permission from X. Wang, J. He, B. Yu, B. Sun, D. Yang, X. Zhang, Q. Zhang, W. Zhang, L. Gu, Y. Chen, $CoSe_2$ nanoparticles embedded MOF-derived Co-N-C nanoflake arrays as efficient and stable electrocatalyst for hydrogen evolution reaction, Appl. Catal. B Environ. (2019) 117996. Copyright 2019, Elsevier.)*

which was tuned via electrochemical methods. High activity and durability of activated-Ni-carbon (A-Ni-C) were achieved for HER, which needed a low overpotential of $-34\,mV@10\,mA\,cm^{-2}$ and Tafel slope of $41\,mV\,dec^{-1}$. Moreover, the A-Ni-C showed large exchange current density of $1.2\,mA\,cm^{-2}$ and high stability (>25 h) in acidic conditions.

A MOF-derived nitrogen-doped carbon-anchored single-W-atom catalyst (W-SAC) has been introduced by Chen and coworkers [55]. The single-W-atom contributed greatly to HER activity; consequently, a low overpotential ($85\,mV@10\,mA\,cm^{-2}$) and a small Tafel slope ($53\,mV\,dec^{-1}$) were displayed in 0.1 M KOH.

5.3 Oxygen evolution reaction

OER is a crucial half-reaction involved in metal-air batteries, fuel batteries, and overall water splitting. Due to the four-electron transfer path, OER requires highly efficient electrocatalysts to overcome the sluggish kinetics. Precious metal oxides like IrO_2 and RuO_2 are redeemed as state-of-the-art OER electrocatalysts nowadays. However, the low reserve, high cost, and poor stability obstruct the commercial application of precious metal-based catalysts. Therefore, finding promising alternatives is an urgent issue. Transition metals and their compounds are expected to replace precious metal oxides. MOFs and their derivatives are considered to have preeminent prospect for the preparation of highly active OER catalysts [13, 196–198].

Transition metal oxides (TMOs) have been widely studied for OER in recent years. Wang and coworkers developed an origin MOF (ZIF-67)-hybrid-assisted tactic to fabricate Co_3O_4/Co-Fe oxide with a double-shelled nanobox morphology [158]. The intricate structure and bimetallic composition together contribute to OER in alkaline environment. The obtained nanoboxes exhibited an overpotential of 297 mV to reach $10\,mA\,cm^{-2}$. It is widely known that transition metal oxides may suffer from low conductivity. Zhou's group developed a novel strategy to overcome the difficulty [199]. In fact, 2D MOFs were acted as templates to compose metal oxides/carbon nanosheet arrays as shown in Fig. 17. Benefitting from the large number of active sites exposed by 2D MOFs and porous carbon and the improved conductivity, the optimized Co_3O_4/C_{BDC} displayed high OER activities and excellent durability, outperforming all the reported transition metal oxides so far.

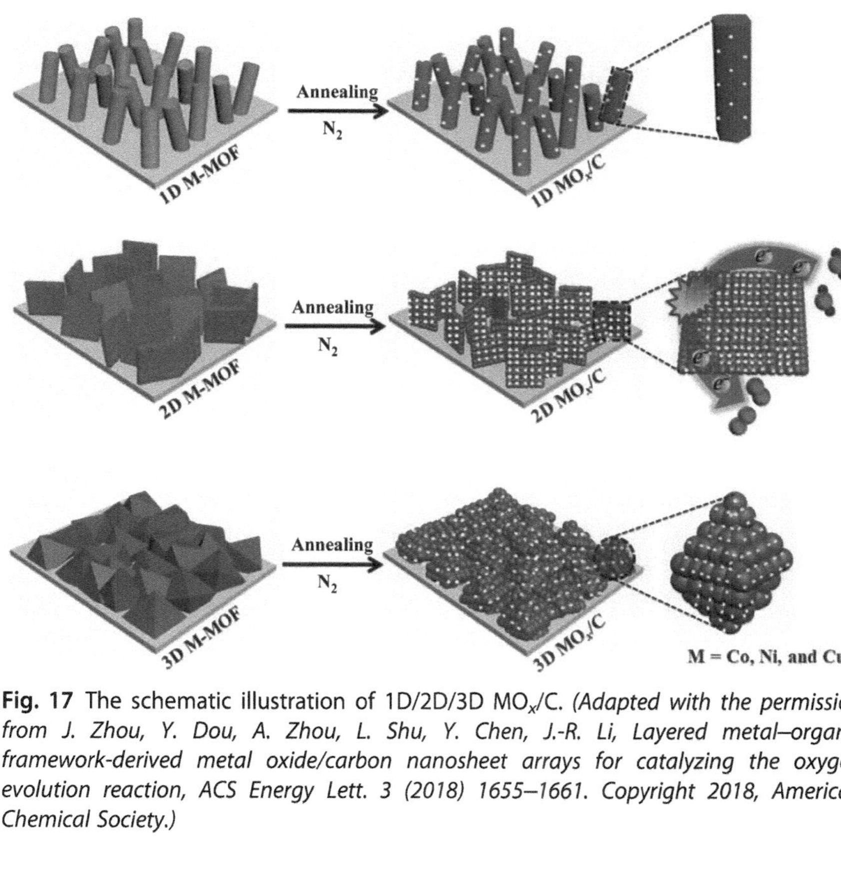

Fig. 17 The schematic illustration of 1D/2D/3D MO_x/C. *(Adapted with the permission from J. Zhou, Y. Dou, A. Zhou, L. Shu, Y. Chen, J.-R. Li, Layered metal–organic framework-derived metal oxide/carbon nanosheet arrays for catalyzing the oxygen evolution reaction, ACS Energy Lett. 3 (2018) 1655–1661. Copyright 2018, American Chemical Society.)*

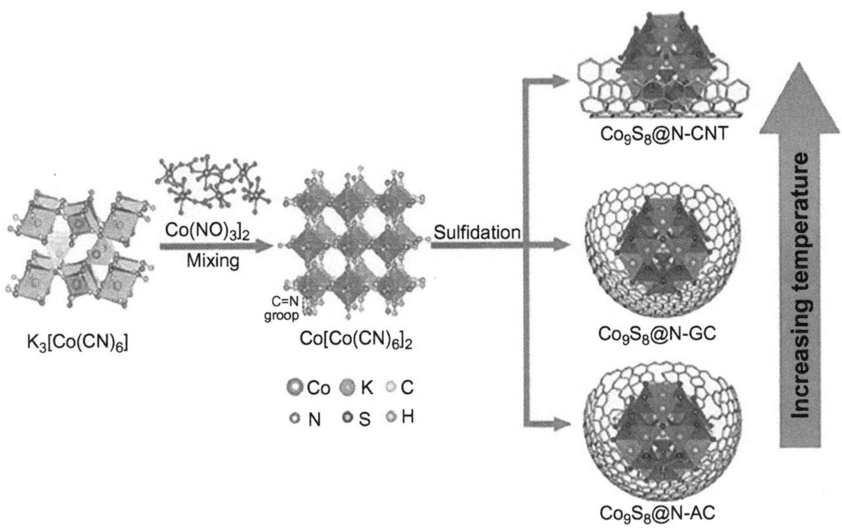

Fig. 18 The schematic illustration of the synthesis of Co_9S_8@N-AC/GC/CNT. *(Adapted with the permission from Z. Chen, R. Wu, M. Liu, Y. Liu, S. Xu, Y. Ha, Y. Guo, X. Yu, D. Sun, F. Fang, Tunable electronic coupling of cobalt sulfide/carbon composites for optimizing oxygen evolution reaction activity, J. Mater. Chem. A 6 (2018) 10304–10312. Copyright 2018, Royal Society of Chemistry.)*

TMSs, especially cobalt sulfides, have been considered as eminent candidates for overall water splitting. Recently, Chen and coworkers developed a Prussian blue analogue-derived Co_9S_8@N-AC/GC/CNT by controlling the annealing temperature as shown in Fig. 18 [200]. The tunable degree of carbonization can generate adjustable electronic coupling between the TMSs and N-doped carbon matrix. The resultant Co_9S_8@NC-800 demonstrated a smaller overpotential (302 mV@10 mA cm^{-2} than Co_9S_8-800) and proved that the coupling effects between N-doped C matrix and Co_9S_8 could greatly enhance the OER catalytic properties. Yang's group recently successfully built MOF-derived CoS_x@MoS_2 microcubes through a novel solvothermal and hydrothermal method and the MoS_2 nanosheet evenly grew on the hollow Co_9S_8 [201]. This heterostructures displayed a small overpotential of 347 mV@ 10 mA cm^{-2} and the excellent OER performance could be ascribed to the synergistic effects between CoS_x and MoS_2.

Transition metal phosphates (TMPs) are a series of superior catalysts in alkaline solution for OER. Wang's group developed a new type of CoP nanoparticle/C nanotubes (denoted as CNT-NC-CoP) through an

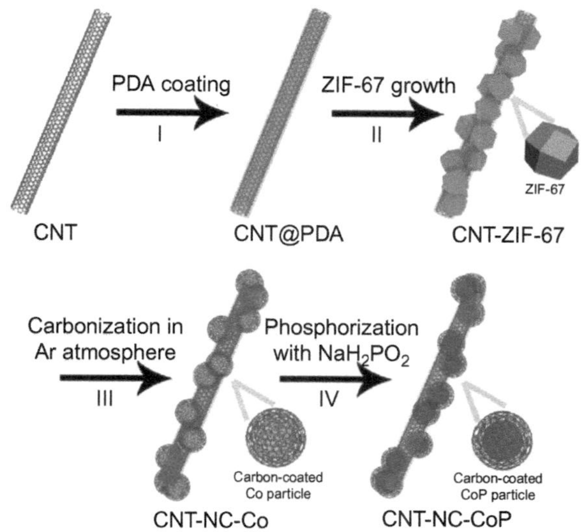

Fig. 19 The schematic illustration of the synthesis of CNT-NC-CoP. *(Adapted with the permission from X. Wang, Z. Ma, L. Chai, L. Xu, Z. Zhu, Y. Hu, J. Qian, S. Huang, MOF derived N-doped carbon coated CoP particle/carbon nanotube composite for efficient oxygen evolution reaction, Carbon 141 (2019) 643–651. Copyright 2019, Elsevier.)*

in situ growth of MOFs onto C nanotubes followed by pyrolysis and phosphorization as shown in Fig. 19 [202]. The CNT-NC-CoP with hierarchical structure performed outstanding OER activities owing to the quick electron diffusion path and high surface area, realizing an ultra-small overpotential of 251 mV@ 10 mA cm^{-2}.

Wang and coworkers imposed another P-containing MOF, BMM-10, integrated with iron salt and ethanol, to provide a precursor to synthesis bimetallic FeNiP/C in different carbonization temperature [160]. Taking advantage of 3D hollow barrel morphology and high specific surface area, the optimized FeNiP/C-900 displayed remarkable OER properties in 1 M solution, with a low overpotential of 229 mV at 10 mA cm^{-2} and 334 mV at 100 mA cm^{-2}.

5.4 Bifunctional electrocatalysis

The dual-function catalysts with HER and OER performance not only save cost, but also achieve overall water splitting to improve efficiency, which are very desirable in recent years. As novel functional catalysts, MOF-derived porous composites (such as transition metal sulfide, phosphide, and selenide) have been widely studied for water splitting. Bu and coworkers developed an adjustable heterostructure of FeNiP/graphene composites via microwave-assisted thermal conversion method [203]. The strategy not only contributes to the doping of phosphorus, but also helps to control the formation of graphitic carbon on the surface of FeNiP compared to the traditional programmed heat treatment. Among them, P doping significantly improved OER and HER activity, while the encapsulated graphite carbon layer effectively prevented electrolyte from eroding the active part. Consequently, (FeNiP)/P-doped graphene delivered an overpotential about 229 and 173 mV@10 mA cm^{-2} for OER and HER in alkaline media, respectively.

Zhang et al. reported a macroporous CoFeP TPAs/Ni, derived from CoFe bi-MOF triangular plate arrays that effectively promote water splitting (Fig. 20) [204]. The large interconnect porous structure of CoFeP TPAs/Ni was attributed to selective etching of the organic ligand in the precursors, which played an important role in both OER and HER for improving performance of overall water splitting. The low overpotentials of CoFeP TPAs/Ni delivered 198 and 43 mV@10 mA cm^{-2} for OER and HER, respectively, and only 1.47 V cell voltage was achieved for overall water splitting.

Liu and coworkers reported an echinops-like Co-based MOF (EC-MOF) well-aligned grown on carbon cloth, and the Co_3S_4/EC-MOF hybrid coating with active Co_3S_4 species obtained after controlled vulcanization method [163]. The obtained Co_3S_4/EC-MOF showed excellent performance in OER and HER, with low overpotentials of 226 and

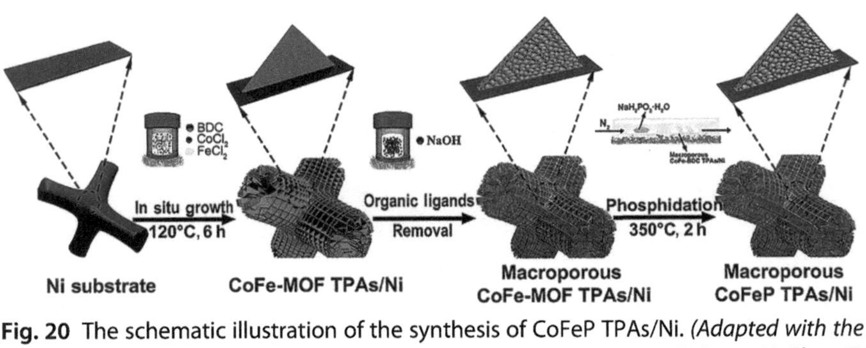

Fig. 20 The schematic illustration of the synthesis of CoFeP TPAs/Ni. *(Adapted with the permission from L. Zhang, X. Wang, A. Li, X. Zheng, L. Peng, J. Huang, Z. Deng, H. Chen, Z. Wei, Rational construction of macroporous CoFeP triangular plate arrays from bimetal–organic frameworks as high-performance overall water-splitting catalysts, J. Mater. Chem. A 7 (2019) 17529–17535. Copyright 2019, Royal Society of Chemistry.)*

84 mV@10 mA cm^{-2}, respectively. In addition, the Co_3S_4/EC-MOF exhibited an excellent cell voltage of 1.55 V@10 mA cm^{-2} and long-term stability.

At present, researche on energy devices, especially metal-air batteries, has become a hot topic. Therefore, the development of electrocatalysts with dual functions of OER and ORR is of great appeal. Shinde and coworkers reported a novel quintet-shelled hollow sphere MOF bifunctional catalyst with a graded porous structure [205]. The catalyst was prepared based on a three-dimensional double-linked hexaiminobenzene MOF and inherited a large surface area (2298 $m^2 g^{-1}$). Thus, it performed exceptionally well in both OER and ORR (the overpotential of 280 mV@10 mA cm^{-2} and half-wave potential 0.883 V). Ma et al. proposed a synthetic strategy for anchoring Fe-Nx on a 2D nitrogen-doped highly graphitic porous carbon layer (Fig. 21) [206]. The high activity of the single-point Fe-Nx species is coupled with the acceleration of charge transfer from the 2D PNC network so that the catalysts exhibited superior performance toward both ORR (half-wave potential is 0.86 V) and OER (390 mV@10 mA cm^{-2}).

In addition, cobalt-based single-atom catalysts have been extensively researched with low cost and high catalytic activity. Zhang and coworkers presented porous N-doped carbon nanosheet arrays anchored with cobalt single atoms derived from Co-MOF precursors [167]. As shown in Fig. 22, removal of the Co metal cluster results in extra pores and active surface area and exposes activities of monodisperse Co atoms that are attached to the carbon network via N—Co bonds. Moreover, single Co atoms catalysts possessed lower overpotential for OER (360 mV@10 mA cm^{-2}) and higher half-wave potential for ORR (0.87 V) compared to Co metal cluster catalysts.

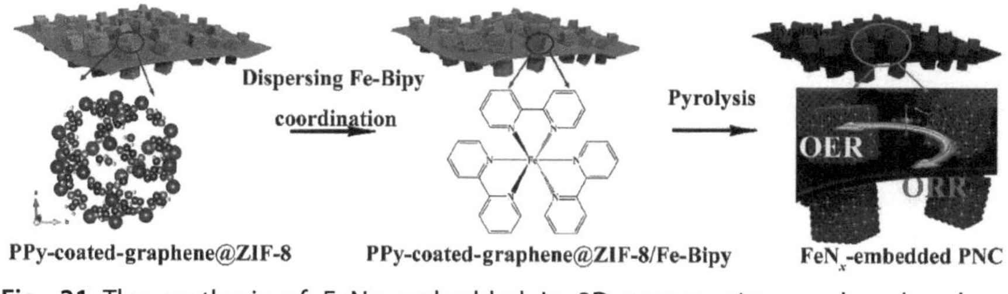

Fig. 21 The synthesis of FeNx embedded in 2D porous nitrogen-doped carbon. *(Adapted with the permission from L. Ma, S. Chen, Z. Pei, Y. Huang, G. Liang, F. Mo, Q. Yang, J. Su, Y. Gao, J.A. Zapien, C. Zhi, Single-site active iron-based bifunctional oxygen catalyst for a compressible and rechargeable zinc–air battery, ACS Nano 12 (2018) 1949–1958. Copyright 2018, American Chemical Society.)*

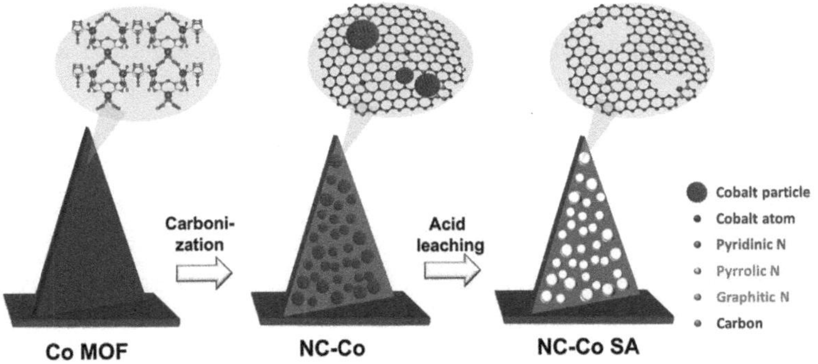

Fig. 22 Schematic illustration of fabrication process for NC-Co SAC nanoflake arrays on carbon cloth. *(Adapted with the permission from W. Zang, A. Sumboja, Y. Ma, H. Zhang, Y. Wu, S. Wu, H. Wu, Z. Liu, C. Guan, J. Wang, S.J. Pennycook, Single Co atoms anchored in porous N-doped carbon for efficient zinc–air battery cathodes,* ACS Catal. *8 (2018) 8961–8969. Copyright 2018, American Chemical Society.)*

5.5 Alcohol oxidation and CO_2 reduction

Apart from the above generally electrocatalytic reactions (ORR, OER, and HER), some other studies of electrocatalytic reactions (e.g., MOR and CO_2RR) based on MOFs-derived porous materials have been carried on. Direct methanol fuel cell has attracted great attention owning to facile preparation method, low cost, and high energy conversion efficiency. This fuel cell is expected to be a promising energy conversion device in the future energy development. Rezaee and coworkers reported a petal-like ultrathin nanoporous carbon rational embedded with NiCo/NiO-CoO/NPCC based on bimetallic MOF, which directly carbonized under nitrogen atmosphere [207]. The unique petal-like porous carbon exhibited excellent MOR performance not only because of superior electrical conductivity, but also due to higher reactivity, lower overpotential, and faster kinetics caused by its free pores structure and large specific surface area. In addition, the synergistic interaction between mixed metals and metal oxides provided favorable activity for MOR.

Electrochemical CO_2RR is also considered to be one of the most promising methods to produce value-added carbon products for reducing greenhouse effect and energy storage [208–212]. The high overpotential in the reaction, the poor selectivity of the catalysts, and the competition of HER seriously hindered the development of CO_2RR [213–216]. Recently, Geng et al. [217] reported Co single-atom catalysts supported on N-doped porous carbon (four-coordinated N and four-coordinated N/C) derived from MOF precursors, through different pyrolysis temperatures. As expected, the single Co atom with four-coordinated N (Co1-N4) showed better CO_2RR performance compared to pure Co atom catalyst, which exhibited a Faradaic efficiency to CO (82%) and a current density of $-15.8\,mA\,cm^{-2}$. Meanwhile, Co1-N4 showed a remarkable stability (10-h constant potential test on CO_2RR). Polynary single-atom structures can expose more atomic interfaces and synergistic effect to promote CO_2RR process. Ren and coworkers fabricated a diatomic metal-nitrogen sites catalyst on N-doped carbon (Ni/Fe-N-C) derived from bimetallic ZIF-8 [170]. The function in synergy of isolated diatomic Ni—Fe sites effectively decreased the reaction barriers to enhanced catalytic efficiency; consequently, the Ni/Fe-N-C exhibited an outstanding Faradaic efficiency, which was up to 98% at $-0.7\,V$. Besides, the excellent switching frequency and robust electrode stability demonstrate the great potential of the diatomic metal-nitrogen sites catalysts in the field of CO_2RR.

6 Conclusions and prospects

To date, as ideal precursors/templates, MOFs have been widely used to investigate and fabricate various carbon, metal-based and metal compound-based porous composites. Despite the early stages, various strategies (e.g., pyrolysis, pyrolysis after solution infiltration) have been developed to fabricate MOF-derived porous carbons. A considerable number of MOF-derived porous carbons have been reported, such as pure carbons, carbon support single-atom catalysts, metal/metal composites/carbons (oxides, nitrides, selenides, sulfides, phosphides, and so on). When MOF-derived porous carbons are used as catalysts, they exhibit remarkable performance in heterogeneous organic catalysis, photocatalysis, and electrocatalysis.

MOF-derived porous carbons possess remarkable features such as: (i) simple fabrication via direct pyrolysis of MOFs precursors or MOFs compounds; (ii) inheriting the morphology and pore structure of the MOFs precursors; (iii) porous structure and large surface areas that ensure the diffusion and dispersion of high-density active sites, efficient mass transfer, and adsorption of reactant molecules during catalytic processes; (iv) the susceptibility of MOFs precursors to modification or doping heteroatoms, providing the necessary catalytically stimulating composition.

Furthermore, some unique advantages are possessed by designing special MOF-derived composites, for example, introducing the substrates such as

GO and NF to improve the conductivity of MOFs-derived materials. The unique structure not only helps mass transfer and improves the efficiency of the active sites, but also facilitates the stripping of bubbles during the catalytic process. In summary, the superior catalytic performance based on efficient mass transfer, effective active sites, and excellent stability can be achieved in the rational design of MOF-derived porous materials.

Since the discovery of MOFs in the 1990s, they have been widely investigated in the fields of sensing, catalysis, and energy storage due to their unique structure with high surface area and high porosity. However, they are limited by their inherent poor conductivity and instability during the catalytic process. Although the development of various composite materials derived from MOFs has made tremendous progress, it is still in the initial stage and some immature problems also need to be solved: (i) more low-cost and large-scale fabrication of MOFs precursors need to be developed to promote basic research in practical industrial production; (ii) more modified and structurally stable MOF precursors should be explored as suitable catalysts; (iii) pore structure and active species of MOF-derived composites need to be precisely designed and controlled, which can facilitate better control of the desired pore characteristics and better understanding of the reaction mechanism; (iv) lack of the techniques to explore the MOF evolution mechanisms during pyrolysis process, which is essential to understand the transformation mechanism of MOFs-derived compounds need to be resolved.

In addition, the research on MOF-derived composites is not limited to the functional design of MOFs precursors and requires an in-depth understanding of the relationship between structure and properties. At the same time, the application of MOF-derived composites for energy conversion/storage should also be concerned with. In view of their unparalleled properties, including flexible tunability of the structure, rich active species, abundant pores, high specific surface area, and so on, MOF-derived porous carbons will open up new horizons in the field of catalysis.

References

[1] S. Chu, A. Majumdar, Opportunities and challenges for a sustainable energy future, Nature 488 (2012) 294.
[2] P.C. Stern, B.K. Sovacool, T. Dietz, Towards a science of climate and energy choices, Nat. Clim. Change 6 (2016) 547.
[3] A. Indra, P.W. Menezes, M. Driess, Uncovering structure–activity relationships in manganese-oxide-based heterogeneous catalysts for efficient water oxidation, ChemSusChem 8 (2015) 776–785.

[4] S.Y. Tee, K.Y. Win, W.S. Teo, L.-D. Koh, S. Liu, C.P. Teng, M.-Y. Han, Recent progress in energy-driven water splitting, Adv. Sci. 4 (2017) 1600337.
[5] A. Indra, A. Acharjya, P.W. Menezes, C. Merschjann, D. Hollmann, M. Schwarze, M. Aktas, A. Friedrich, S. Lochbrunner, A. Thomas, M. Driess, Boosting visible-light-driven photocatalytic hydrogen evolution with an integrated nickel phosphide–carbon nitride system, Angew. Chem. Int. Ed. 56 (2017) 1653–1657.
[6] M.S. Dresselhaus, I.L. Thomas, Alternative energy technologies, Nature 414 (2001) 332–337.
[7] S. Chu, Y. Cui, N. Liu, The path towards sustainable energy, Nat. Mater. 16 (2016) 16.
[8] T. Sadhasivam, K. Dhanabalan, S.-H. Roh, T.-H. Kim, K.-W. Park, S. Jung, M. D. Kurkuri, H.-Y. Jung, A comprehensive review on unitized regenerative fuel cells: crucial challenges and developments, Int. J. Hydrogen Energy 42 (2017) 4415–4433.
[9] D. Aurbach, B.D. McCloskey, L.F. Nazar, P.G. Bruce, Advances in understanding mechanisms underpinning lithium–air batteries, Nat. Energy 1 (2016) 16128.
[10] I. Roger, M.A. Shipman, M.D. Symes, Earth-abundant catalysts for electrochemical and photoelectrochemical water splitting, Nat. Rev. Chem. 1 (2017) 0003.
[11] D.D. Zhu, J.L. Liu, S.Z. Qiao, Recent advances in inorganic heterogeneous electrocatalysts for reduction of carbon dioxide, Adv. Mater. 28 (2016) 3423–3452.
[12] W. Xia, C. Qu, Z. Liang, B. Zhao, S. Dai, B. Qiu, Y. Jiao, Q. Zhang, X. Huang, W. Guo, High-performance energy storage and conversion materials derived from a single metal–organic framework/graphene aerogel composite, Nano Lett. 17 (2017) 2788–2795.
[13] J. Liu, D. Zhu, C. Guo, A. Vasileff, S.Z. Qiao, Design strategies toward advanced MOF-derived electrocatalysts for energy-conversion reactions, Adv. Energy Mater. 7 (2017) 1700518.
[14] Y. Zheng, Y. Jiao, M. Jaroniec, S.Z. Qiao, Advancing the electrochemistry of the hydrogen-evolution reaction through combining experiment and theory, Angew. Chem. Int. Ed. 54 (2015) 52–65.
[15] J. Duan, S. Chen, M. Jaroniec, S.Z. Qiao, Heteroatom-doped graphene-based materials for energy-relevant electrocatalytic processes, ACS Catal. 5 (2015) 5207–5234.
[16] J. Zhang, K. Sasaki, E. Sutter, R. Adzic, Stabilization of platinum oxygen-reduction electrocatalysts using gold clusters, Science 315 (2007) 220–222.
[17] L.C. Seitz, C.F. Dickens, K. Nishio, Y. Hikita, J. Montoya, A. Doyle, C. Kirk, A. Vojvodic, H.Y. Hwang, J.K. Norskov, A highly active and stable Irox/Sriro3 catalyst for the oxygen evolution reaction, Science 353 (2016) 1011–1014.
[18] D. Kim, J. Resasco, Y. Yu, A.M. Asiri, P. Yang, Synergistic geometric and electronic effects for electrochemical reduction of carbon dioxide using gold–copper bimetallic nanoparticles, Nat. Commun. 5 (2014) 4948.
[19] M.S. Faber, S. Jin, Earth-abundant inorganic electrocatalysts and their nanostructures for energy conversion applications, Energy Environ. Sci. 7 (2014) 3519–3542.
[20] M. Shao, Q. Chang, J.-P. Dodelet, R. Chenitz, Recent advances in electrocatalysts for oxygen reduction reaction, Chem. Rev. 116 (2016) 3594–3657.
[21] J. Liu, L. Chen, H. Cui, J. Zhang, L. Zhang, C.-Y. Su, Applications of metal–organic frameworks in heterogeneous supramolecular catalysis, Chem. Soc. Rev. 43 (2014) 6011–6061.
[22] Z. Liang, C. Qu, W. Guo, R. Zou, Q. Xu, Pristine metal–organic frameworks and their composites for energy storage and conversion, Adv. Mater. 30 (2018) 1702891.
[23] L. Zhu, X.-Q. Liu, H.-L. Jiang, L.-B. Sun, Metal–organic frameworks for heterogeneous basic catalysis, Chem. Rev. 117 (2017) 8129–8176.
[24] L. Jiao, Y. Wang, H.-L. Jiang, Q. Xu, Metal–organic frameworks as platforms for catalytic applications, Adv. Mater. 30 (2018) 1703663.

[25] C.-D. Wu, M. Zhao, Incorporation of molecular catalysts in metal–organic frameworks for highly efficient heterogeneous catalysis, Adv. Mater. 29 (2017) 1605446.
[26] S.-N. Zhao, X.-Z. Song, S.-Y. Song, H.-j. Zhang, Highly efficient heterogeneous catalytic materials derived from metal-organic framework supports/precursors, Coord. Chem. Rev. 337 (2017) 80–96.
[27] F.-Y. Yi, D. Chen, M.-K. Wu, L. Han, H.-L. Jiang, Chemical sensors based on metal–organic frameworks, ChemPlusChem 81 (2016) 675–690.
[28] Q. Yang, Q. Xu, H.-L. Jiang, Metal–organic frameworks meet metal nanoparticles: synergistic effect for enhanced catalysis, Chem. Soc. Rev. 46 (2017) 4774–4808.
[29] W. Zhang, Y. Hu, J. Ge, H.-L. Jiang, S.-H. Yu, A facile and general coating approach to moisture/water-resistant metal–organic frameworks with intact porosity, J. Am. Chem. Soc. 136 (2014) 16978–16981.
[30] J. Tang, R.R. Salunkhe, J. Liu, N.L. Torad, M. Imura, S. Furukawa, Y. Yamauchi, Thermal conversion of core–shell metal–organic frameworks: a new method for selectively functionalized nanoporous hybrid carbon, J. Am. Chem. Soc. 137 (2015) 1572–1580.
[31] S. Dang, Q.-L. Zhu, Q. Xu, Nanomaterials derived from metal–organic frameworks, Nat. Rev. Mater. 3 (2017) 17075.
[32] L. Yang, X. Zeng, W. Wang, D. Cao, Recent progress in MOF-derived, heteroatom-doped porous carbons as highly efficient electrocatalysts for oxygen reduction reaction in fuel cells, Adv. Funct. Mater. 28 (2018) 1704537.
[33] Y.-Z. Chen, R. Zhang, L. Jiao, H.-L. Jiang, Metal–organic framework-derived porous materials for catalysis, Coord. Chem. Rev. 362 (2018) 1–23.
[34] K. Shen, X. Chen, J. Chen, Y. Li, Development of Mof-derived carbon-based nanomaterials for efficient catalysis, ACS Catal. 6 (2016) 5887–5903.
[35] Y.V. Kaneti, J. Tang, R.R. Salunkhe, X. Jiang, A. Yu, K.C.W. Wu, Y. Yamauchi, Nanoarchitectured design of porous materials and nanocomposites from metal-organic frameworks, Adv. Mater. 29 (2017) 1604898.
[36] M.H. Yap, K.L. Fow, G.Z. Chen, Synthesis and applications of MOF-derived porous nanostructures, Green Energy Environ. 2 (2017) 218–245.
[37] Y. Qian, I.A. Khan, D. Zhao, Electrocatalysts derived from metal–organic frameworks for oxygen reduction and evolution reactions in aqueous media, Small 13 (2017) 1701143.
[38] S. Fu, C. Zhu, J. Song, D. Du, Y. Lin, Metal-organic framework-derived non-precious metal nanocatalysts for oxygen reduction reaction, Adv. Energy Mater. 7 (2017) 1700363.
[39] L. Liu, Y.-P. Zhu, M. Su, Z.-Y. Yuan, Metal-free carbonaceous materials as promising heterogeneous catalysts, ChemCatChem 7 (2015) 2765–2787.
[40] S. Chen, J. Bi, Y. Zhao, L. Yang, C. Zhang, Y. Ma, Q. Wu, X. Wang, Z. Hu, Nitrogen-doped carbon nanocages as efficient metal-free electrocatalysts for oxygen reduction reaction, Adv. Mater. 24 (2012) 5593–5597.
[41] A.H. Lu, F. Schüth, Nanocasting: a versatile strategy for creating nanostructured porous materials, Adv. Mater. 18 (2006) 1793–1805.
[42] Q.-L. Zhu, Q. Xu, Metal–organic framework composites, Chem. Soc. Rev. 43 (2014) 5468–5512.
[43] Y. Zhao, Z. Song, X. Li, Q. Sun, N. Cheng, S. Lawes, X. Sun, Metal organic frameworks for energy storage and conversion, Energy Storage Mater. 2 (2016) 35–62.
[44] H.-L. Jiang, B. Liu, Y.-Q. Lan, K. Kuratani, T. Akita, H. Shioyama, F. Zong, Q. Xu, From metal–organic framework to nanoporous carbon: toward a very high surface area and hydrogen uptake, J. Am. Chem. Soc. 133 (2011) 11854–11857.
[45] J. Zhang, J. Fang, J. Han, T. Yan, L. Shi, D. Zhang, N, P, S co-doped hollow carbon polyhedra derived from MOF-based core–shell nanocomposites for capacitive deionization, J. Mater. Chem. A 6 (2018) 15245–15252.

[46] H.-x. Zhong, J. Wang, Y.-w. Zhang, W.-l. Xu, W. Xing, D. Xu, Y.-f. Zhang, X.-b. Zhang, Zif-8 derived graphene-based nitrogen-doped porous carbon sheets as highly efficient and durable oxygen reduction electrocatalysts, Angew. Chem. Int. Ed. 53 (2014) 14235–14239.
[47] L. Liu, A. Corma, Metal catalysts for heterogeneous catalysis: from single atoms to nanoclusters and nanoparticles, Chem. Rev. 118 (2018) 4981–5079.
[48] A. Wang, J. Li, T. Zhang, Heterogeneous single-atom catalysis, Nat. Rev. Chem. 2 (2018) 65–81.
[49] M.J. Hülsey, J. Zhang, N. Yan, Harnessing the wisdom in colloidal chemistry to make stable single-atom catalysts, Adv. Mater. 30 (2018) 1802304.
[50] J. Zhang, X. Wu, W.-C. Cheong, W. Chen, R. Lin, J. Li, L. Zheng, W. Yan, L. Gu, C. Chen, Q. Peng, D. Wang, Y. Li, Cation vacancy stabilization of single-atomic-site Pt1/Ni(Oh)X catalyst for diboration of alkynes and alkenes, Nat. Commun. 9 (2018) 1002.
[51] A. Han, W. Chen, S. Zhang, M. Zhang, Y. Han, J. Zhang, S. Ji, L. Zheng, Y. Wang, L. Gu, C. Chen, Q. Peng, D. Wang, Y. Li, A polymer encapsulation strategy to synthesize porous nitrogen-doped carbon-nanosphere-supported metal isolated-single-atomic-site catalysts, Adv. Mater. 30 (2018) 1706508.
[52] M. Piernavieja-Hermida, Z. Lu, A. White, K.-B. Low, T. Wu, J.W. Elam, Z. Wu, Y. Lei, Towards Ald thin film stabilized single-atom Pd1 catalysts, Nanoscale 8 (2016) 15348–15356.
[53] A.J. Howarth, A.W. Peters, N.A. Vermeulen, T.C. Wang, J.T. Hupp, O.K. Farha, Best practices for the synthesis, activation, and characterization of metal–organic frameworks, Chem. Mater. 29 (2017) 26–39.
[54] P. Yin, T. Yao, Y. Wu, L. Zheng, Y. Lin, W. Liu, H. Ju, J. Zhu, X. Hong, Z. Deng, G. Zhou, S. Wei, Y. Li, Single cobalt atoms with precise N-coordination as superior oxygen reduction reaction catalysts, Angew. Chem. Int. Ed. 55 (2016) 10800–10805.
[55] W. Chen, J. Pei, C.-T. He, J. Wan, H. Ren, Y. Wang, J. Dong, K. Wu, W.-C. Cheong, J. Mao, X. Zheng, W. Yan, Z. Zhuang, C. Chen, Q. Peng, D. Wang, Y. Li, Single tungsten atoms supported on MOF-derived N-doped carbon for robust electrochemical hydrogen evolution, Adv. Mater. 30 (2018) 1800396.
[56] Y. Chen, S. Ji, Y. Wang, J. Dong, W. Chen, Z. Li, R. Shen, L. Zheng, Z. Zhuang, D. Wang, Y. Li, Isolated single iron atoms anchored on N-doped porous carbon as an efficient electrocatalyst for the oxygen reduction reaction, Angew. Chem. Int. Ed. 56 (2017) 6937–6941.
[57] A. Kong, C. Mao, Q. Lin, X. Wei, X. Bu, P. Feng, From cage-in-cage Mof to N-doped and co-nanoparticle-embedded carbon for oxygen reduction reaction, Dalton Trans. 44 (2015) 6748–6754.
[58] S.L. Zhang, B.Y. Guan, X.W. Lou, Co–Fe alloy/N-doped carbon hollow spheres derived from dual metal–organic frameworks for enhanced electrocatalytic oxygen reduction, Small 15 (2019) 1805324.
[59] X. Li, D.D. Ma, C. Cao, R. Zou, Q. Xu, X.T. Wu, Q.L. Zhu, Inlaying ultrathin bimetallic Mof nanosheets into 3d ordered macroporous hydroxide for superior electrocatalytic oxygen evolution, Small 15 (2019) 1902218.
[60] W.-F. Chen, J.T. Muckerman, E. Fujita, Recent developments in transition metal carbides and nitrides as hydrogen evolution electrocatalysts, Chem. Commun. 49 (2013) 8896–8909.
[61] X. Zou, Y. Zhang, Noble metal-free hydrogen evolution catalysts for water splitting, Chem. Soc. Rev. 44 (2015) 5148–5180.
[62] Y. Yan, B. Xia, Z. Xu, X. Wang, Recent development of molybdenum sulfides as advanced electrocatalysts for hydrogen evolution reaction, ACS Catal. 4 (2014) 1693–1705.
[63] B.Y. Guan, L. Yu, X.W. Lou, A dual-metal–organic-framework derived electrocatalyst for oxygen reduction, Energy Environ. Sci. 9 (2016) 3092–3096.

[64] H.B. Wu, B.Y. Xia, L. Yu, X.-Y. Yu, X.W. Lou, Porous molybdenum carbide nano-octahedrons synthesized via confined carburization in metal-organic frameworks for efficient hydrogen production, Nat. Commun. 6 (2015) 6512.
[65] W. Gu, L. Hu, J. Li, E. Wang, Hybrid of g-C_3N_4 assisted metal–organic frameworks and their derived high-efficiency oxygen reduction electrocatalyst in the whole Ph range, ACS Appl. Mater. Interfaces 8 (2016) 35281–35288.
[66] J.-S. Li, S.-L. Li, Y.-J. Tang, M. Han, Z.-H. Dai, J.-C. Bao, Y.-Q. Lan, Nitrogen-doped Fe/Fe_3C@graphitic layer/carbon nanotube hybrids derived from Mofs: efficient bifunctional electrocatalysts for Orr and Oer, Chem. Commun. 51 (2015) 2710–2713.
[67] Z.-F. Huang, J. Song, K. Li, M. Tahir, Y.-T. Wang, L. Pan, L. Wang, X. Zhang, J.-J. Zou, Hollow cobalt-based bimetallic sulfide polyhedra for efficient All-Ph-value electrochemical and photocatalytic hydrogen evolution, J. Am. Chem. Soc. 138 (2016) 1359–1365.
[68] Z. Shi, Y. Wang, H. Lin, H. Zhang, M. Shen, S. Xie, Y. Zhang, Q. Gao, Y. Tang, Porous nanomoc@graphite shell derived from a Mofs-directed strategy: an efficient electrocatalyst for the hydrogen evolution reaction, J. Mater. Chem. A 4 (2016) 6006–6013.
[69] Q. Dong, Q. Wang, Z. Dai, H. Qiu, X. Dong, Mof-derived Zn-doped $CoSe_2$ as an efficient and stable free-standing catalyst for oxygen evolution reaction, ACS Appl. Mater. Interfaces 8 (2016) 26902–26907.
[70] B. You, N. Jiang, M. Sheng, S. Gul, J. Yano, Y. Sun, High-performance overall water splitting electrocatalysts derived from cobalt-based metal–organic frameworks, Chem. Mater. 27 (2015) 7636–7642.
[71] Y. Pan, K. Sun, S. Liu, X. Cao, K. Wu, W.-C. Cheong, Z. Chen, Y. Wang, Y. Li, Y. Liu, D. Wang, Q. Peng, C. Chen, Y. Li, Core–shell ZIF-8@ZIF-67-derived cop nanoparticle-embedded N-doped carbon nanotube hollow polyhedron for efficient overall water splitting, J. Am. Chem. Soc. 140 (2018) 2610–2618.
[72] X. Xu, J. Liu, J. Liu, L. Ouyang, R. Hu, H. Wang, L. Yang, M. Zhu, A general metal-organic framework (MOF)-derived selenidation strategy for in situ carbon-encapsulated metal selenides as high-rate anodes for Na-Ion batteries, Adv. Funct. Mater. 28 (2018) 1707573.
[73] C. Xu, R. Fang, R. Luque, L. Chen, Y. Li, Functional metal–organic frameworks for catalytic applications, Coord. Chem. Rev. 388 (2019) 268–292.
[74] Q. Wang, D. Astruc, State of the art and prospects in metal–organic framework (Mof)-based and MOF-derived nanocatalysis, Chem. Rev. 120 (2) (2020) 1438–1511.
[75] C. Wang, B. An, W. Lin, Metal–organic frameworks in solid–gas phase catalysis, ACS Catal. 9 (2018) 130–146.
[76] D. Li, H.-Q. Xu, L. Jiao, H.-L. Jiang, Metal-organic frameworks for catalysis: state-of-the-art, challenges, and opportunities, EnergyChem 1 (2019) 100005.
[77] L. Jiao, H.-L. Jiang, Metal-organic-framework-based single-atom catalysts for energy applications, Chem 5 (2019) 786–804.
[78] J.-K. Sun, Q. Xu, Functional materials derived from open framework templates/precursors: synthesis and applications, Energy Environ. Sci. 7 (2014) 2071–2100.
[79] B.N. Bhadra, S.H. Jhung, Oxidative desulfurization and denitrogenation of fuels using metal-organic framework-based/-derived catalysts, Appl. Catal. B Environ. 259 (2019) 118021.
[80] X.-L. Zhang, D.-X. Zhang, G.-G. Chang, X.-C. Ma, J. Wu, Y. Wang, H.-Z. Yu, G. Tian, J. Chen, X.-Y. Yang, Bimetallic (Zn/Co) Mofs-derived highly dispersed metallic Co/Hpc for completely hydrolytic dehydrogenation of ammonia–borane, Ind. Eng. Chem. Res. 58 (2019) 7209–7216.

[81] T.V. Tran, D.T.C. Nguyen, H.T. Le, L.G. Bach, D.-V.N. Vo, S.S. Hong, T.-Q. T. Phan, T.D. Nguyen, Tunable synthesis of mesoporous carbons from Fe_3O (Bdc)3 for chloramphenicol antibiotic remediation, Nanomaterials 9 (2019) 237.
[82] W. Zhong, H. Liu, C. Bai, S. Liao, Y. Li, Base-free oxidation of alcohols to esters at room temperature and atmospheric conditions using nanoscale co-based catalysts, ACS Catal. 5 (2015) 1850–1856.
[83] Y.-X. Zhou, Y.-Z. Chen, L. Cao, J. Lu, H.-L. Jiang, Conversion of a metal–organic framework to N-doped porous carbon incorporating Co and CoO nanoparticles: direct oxidation of alcohols to esters, Chem. Commun. 51 (2015) 8292–8295.
[84] B.R. Kim, J.S. Oh, J. Kim, C.Y. Lee, Robust aerobic alcohol oxidation catalyst derived from metal–organic frameworks, Catal. Lett. 146 (2016) 734–743.
[85] C. Bai, X. Yao, Y. Li, Easy access to amides through aldehydic C–H bond functionalization catalyzed by heterogeneous co-based catalysts, ACS Catal. 5 (2015) 884–891.
[86] X. Yao, C. Bai, J. Chen, Y. Li, Efficient and selective green oxidation of alcohols by Mof-derived magnetic nanoparticles as a recoverable catalyst, RSC Adv. 6 (2016) 26921–26928.
[87] R. Fang, R. Luque, Y. Li, Selective aerobic oxidation of biomass-derived Hmf to 2, 5-diformylfuran using a Mof-derived magnetic hollow Fe–Co nanocatalyst, Green Chem. 18 (2016) 3152–3157.
[88] N.D. McNamara, J. Kim, J.C. Hicks, Controlling the pyrolysis conditions of microporous/mesoporous Mil-125 to synthesize porous, carbon-supported Ti catalysts with targeted Ti phases for the oxidation of dibenzothiophene, Energy Fuel 30 (2015) 594–602.
[89] H. Tan, C. Ma, L. Gao, Q. Li, Y. Song, F. Xu, T. Wang, L. Wang, Metal–organic framework-derived copper nanoparticle@carbon nanocomposites as peroxidase mimics for colorimetric sensing of ascorbic acid, Chem. Eur. J. 20 (2014) 16377–16383.
[90] S. Li, N. Wang, Y. Yue, G. Wang, Z. Zu, Y. Zhang, Copper doped ceria porous nanostructures towards a highly efficient bifunctional catalyst for carbon monoxide and nitric oxide elimination, Chem. Sci. 6 (2015) 2495–2500.
[91] H. Liu, S. Zhang, Y. Liu, Z. Yang, X. Feng, X. Lu, F. Huo, Well-dispersed and size-controlled supported metal oxide nanoparticles derived from MOF composites and further application in catalysis, Small 11 (2015) 3130–3134.
[92] S. Zhang, H. Liu, C. Sun, P. Liu, L. Li, Z. Yang, X. Feng, F. Huo, X. Lu, Cuo/Cu_2O porous composites: shape and composition controllable fabrication inherited from metal organic frameworks and further application in Co oxidation, J. Mater. Chem. A 3 (2015) 5294–5298.
[93] R. Zhang, L. Hu, S. Bao, R. Li, L. Gao, R. Li, Q. Chen, Surface polarization enhancement: high catalytic performance of Cu/CuO X/C nanocomposites derived from Cu-Btc for Co oxidation, J. Mater. Chem. A 4 (2016) 8412–8420.
[94] F. Zhang, C. Chen, W.-m. Xiao, L. Xu, N. Zhang, CuO/CeO_2 catalysts with well-dispersed active sites prepared from Cu3(BTC)2 metal–organic framework precursor for preferential Co oxidation, Catal. Commun. 26 (2012) 25–29.
[95] C. Zhu, T. Ding, W. Gao, K. Ma, Y. Tian, X. Li, CuO/CeO_2 catalysts synthesized from Ce-Uio-66 metal-organic framework for preferential Co oxidation, Int. J. Hydrogen Energy 42 (2017) 17457–17465.
[96] M. Yuan, H. Zhang, C. Yang, F. Wang, Z. Dong, Co-MOF-derived hierarchical mesoporous yolk-shell-structured nanoreactor for the catalytic reduction of nitroarenes with hydrazine hydrate, ChemCatChem 11 (2019) 3327–3338.
[97] S. Li, J. Pan, X. Wu, Y. Fu, Q. Xiao, F. Zhang, W. Zhu, Metal–organic framework derived Pd/ZrO_2@CN as a stable catalyst for the catalytic hydrogenation of 2, 3, 5-trimethylbenzoquinone, Appl. Organomet. Chem. 33 (2019) e5233.

[98] Y. Jiang, G. Liu, S. Wu, X. Zhang, X. Dai, Q. Sheng, Y. Wang, H. Wang, Enhanced performance of well-dispersed Co species incorporated on porous carbon derived from metal-organic frameworks in 1, 3-butadiene hydrogenation, Microporous Mesoporous Mater. 288 (2019) 109557.
[99] Y. Bai, J. Dong, Y. Hou, Y. Guo, Y. Liu, Y. Li, X. Han, Z. Huang, Co_3O_4@ Pc derived from Zif-67 as an efficient catalyst for the selective catalytic reduction of Nox with Nh3 at low temperature, Chem. Eng. J. 361 (2019) 703–712.
[100] Q. Yang, C.C. Yang, C.H. Lin, H.L. Jiang, Metal–organic-framework-derived hollow N-doped porous carbon with ultrahigh concentrations of single Zn atoms for efficient carbon dioxide conversion, Angew. Chem. Int. Ed. 58 (2019) 3511–3515.
[101] Y.-Z. Chen, G. Cai, Y. Wang, Q. Xu, S.-H. Yu, H.-L. Jiang, Palladium nanoparticles stabilized with N-doped porous carbons derived from metal–organic frameworks for selective catalysis in biofuel upgrade: the role of catalyst wettability, Green Chem. 18 (2016) 1212–1217.
[102] X. Wang, W. Chen, L. Zhang, T. Yao, W. Liu, Y. Lin, H. Ju, J. Dong, L. Zheng, W. Yan, Uncoordinated amine groups of metal–organic frameworks to anchor single Ru sites as chemoselective catalysts toward the hydrogenation of quinoline, J. Am. Chem. Soc. 139 (2017) 9419–9422.
[103] J. Long, Y. Zhou, Y. Li, Transfer hydrogenation of unsaturated bonds in the absence of base additives catalyzed by a cobalt-based heterogeneous catalyst, Chem. Commun. 51 (2015) 2331–2334.
[104] X. Liu, S. Cheng, J. Long, W. Zhang, X. Liu, D. Wei, Mofs-derived Co@ CN bi-functional catalysts for selective transfer hydrogenation of A, B-unsaturated aldehydes without use of base additives, Mater. Chem. Front. 1 (2017) 2005–2012.
[105] R.V. Jagadeesh, K. Murugesan, A.S. Alshammari, H. Neumann, M.-M. Pohl, J. Radnik, M. Beller, Mof-derived cobalt nanoparticles catalyze a general synthesis of amines, Science 358 (2017) 326–332.
[106] X. Li, C. Zeng, J. Jiang, L. Ai, Magnetic cobalt nanoparticles embedded in hierarchically porous nitrogen-doped carbon frameworks for highly efficient and well-recyclable catalysis, J. Mater. Chem. A 4 (2016) 7476–7482.
[107] J. Long, K. Shen, Y. Li, Bifunctional N-doped Co@C catalysts for base-free transfer hydrogenations of nitriles: controllable selectivity to primary amines vs imines, ACS Catal. 7 (2016) 275–284.
[108] X. Ma, Y.-X. Zhou, H. Liu, Y. Li, H.-L. Jiang, A MOF-derived Co–CoO@ N-doped porous carbon for efficient tandem catalysis: dehydrogenation of ammonia borane and hydrogenation of nitro compounds, Chem. Commun. 52 (2016) 7719–7722.
[109] B. Tang, W.-C. Song, E.-C. Yang, X.-J. Zhao, Mof-derived Ni-based nanocomposites as robust catalysts for chemoselective hydrogenation of functionalized nitro compounds, RSC Adv. 7 (2017) 1531–1539.
[110] X. Kang, H. Liu, M. Hou, X. Sun, H. Han, T. Jiang, Z. Zhang, B. Han, Synthesis of supported ultrafine non-noble subnanometer-scale metal particles derived from metal–organic frameworks as highly efficient heterogeneous catalysts, Angew. Chem. Int. Ed. 55 (2016) 1080–1084.
[111] Y. Li, Y.-X. Zhou, X. Ma, H.-L. Jiang, A metal–organic framework-templated synthesis of γ-Fe_2O_3 nanoparticles encapsulated in porous carbon for efficient and chemoselective hydrogenation of nitro compounds, Chem. Commun. 52 (2016) 4199–4202.
[112] J. Long, K. Shen, L. Chen, Y. Li, Multimetal-Mof-derived transition metal alloy Nps embedded in an N-doped carbon matrix: highly active catalysts for hydrogenation reactions, J. Mater. Chem. A 4 (2016) 10254–10262.
[113] K. Shen, L. Chen, J. Long, W. Zhong, Y. Li, MOFs-templated Co@ Pd core–shell NPs embedded in N-doped carbon matrix with superior hydrogenation activities, ACS Catal. 5 (2015) 5264–5271.

[114] H. Niu, S. Liu, Y. Cai, F. Wu, X. Zhao, MOF derived porous carbon supported Cu/Cu_2O composite as high performance non-noble catalyst, Microporous Mesoporous Mater. 219 (2016) 48–53.
[115] X. Li, B. Zhang, Y. Fang, W. Sun, Z. Qi, Y. Pei, S. Qi, P. Yuan, X. Luan, T.W. Goh, Metal–organic-framework-derived carbons: applications as solid-base catalyst and support for Pd nanoparticles in tandem catalysis, Chem. Eur. J. 23 (2017) 4266–4270.
[116] G. Huang, L. Yang, X. Ma, J. Jiang, S.H. Yu, H.L. Jiang, Metal–organic framework-templated porous carbon for highly efficient catalysis: the critical role of pyrrolic nitrogen species, Chem. Eur. J. 22 (2016) 3470–3477.
[117] V.P. Santos, T.A. Wezendonk, J.J.D. Jaén, A.I. Dugulan, M.A. Nasalevich, H.-U. Islam, A. Chojecki, S. Sartipi, X. Sun, A.A. Hakeem, Metal organic framework-mediated synthesis of highly active and stable Fischer-Tropsch catalysts, Nat. Commun. 6 (2015) 6451.
[118] T.A. Wezendonk, V.P. Santos, M.A. Nasalevich, Q.S. Warringa, A.I. Dugulan, A. Chojecki, A.C. Koeken, M. Ruitenbeek, G. Meima, H.-U. Islam, Elucidating the nature of Fe species during pyrolysis of the Fe-Btc Mof into highly active and stable Fischer–Tropsch catalysts, ACS Catal. 6 (2016) 3236–3247.
[119] B. An, K. Cheng, C. Wang, Y. Wang, W. Lin, Pyrolysis of metal–organic frameworks to $Fe_3O_4@Fe_5C_2$ core–shell nanoparticles for Fischer–Tropsch synthesis, ACS Catal. 6 (2016) 3610–3618.
[120] B. Qiu, C. Yang, W. Guo, Y. Xu, Z. Liang, D. Ma, R. Zou, Highly dispersed Co-based Fischer–Tropsch synthesis catalysts from metal–organic frameworks, J. Mater. Chem. A 5 (2017) 8081–8086.
[121] A. Fujishima, K. Honda, Electrochemical photolysis of water at a semiconductor electrode, Nature 238 (1972) 37.
[122] K.E. Dekrafft, C. Wang, W. Lin, Metal-organic framework templated synthesis of Fe_2O_3/TiO_2 nanocomposite for hydrogen production, Adv. Mater. 24 (2012) 2014–2018.
[123] M. Lan, R.-M. Guo, Y. Dou, J. Zhou, A. Zhou, J.-R. Li, Fabrication of porous Pt-doping heterojunctions by using bimetallic MOF template for photocatalytic hydrogen generation, Nano Energy 33 (2017) 238–246.
[124] Y. Qi, J. Xu, Y. Fu, C. Wang, L. Wang, Metal-organic framework templated synthesis Ofg-$C_3N_4/Fe_2O_3@FeP$ composites for enhanced hydrogen production, ChemCatChem 11 (2019) 3465–3473.
[125] D.P. Kumar, H. Park, E.H. Kim, S. Hong, M. Gopannagari, D.A. Reddy, T.K. Kim, Noble metal-free metal-organic framework-derived onion slice-type hollow cobalt sulfide nanostructures: enhanced activity of Cds for improving photocatalytic hydrogen production, Appl. Catal. B Environ. 224 (2018) 230–238.
[126] J.D. Xiao, H.L. Jiang, Thermally stable metal-organic framework-templated synthesis of hierarchically porous metal sulfides: enhanced photocatalytic hydrogen production, Small 13 (2017) 1700632.
[127] H. Zhang, T. Wang, J. Wang, H. Liu, T.D. Dao, M. Li, G. Liu, X. Meng, K. Chang, L. Shi, Surface-plasmon-enhanced photodriven CO_2 reduction catalyzed by metal–organic-framework-derived iron nanoparticles encapsulated by ultrathin carbon layers, Adv. Mater. 28 (2016) 3703–3710.
[128] S. Wang, B.Y. Guan, Y. Lu, X.W.D. Lou, Formation of hierarchical In_2S_3–$CdIn_2S_4$ heterostructured nanotubes for efficient and stable visible light CO_2 reduction, J. Am. Chem. Soc. 139 (2017) 17305–17308.
[129] W. Chen, B. Han, C. Tian, X. Liu, S. Liang, H. Deng, Z. Lin, MOFs-derived ultrathin holey Co_3O_4 nanosheets for enhanced visible light CO_2 reduction, Appl. Catal. B Environ. 244 (2019) 996–1003.
[130] C.Y. Hu, J. Zhou, C.Y. Sun, M.m. Chen, X.L. Wang, Z.M. Su, Hkust-1 derived hollow C-Cu_2–xS nanotube/g-C_3N_4 composites for visible-light CO_2 photoreduction with H_2o vapor, Chem. Eur. J. 25 (2019) 379–385.

[131] H. Chen, K. Shen, J. Chen, X. Chen, Y. Li, Hollow-ZIF-templated formation of a ZnO@ C–N–Co core–shell nanostructure for highly efficient pollutant photodegradation, J. Mater. Chem. A 5 (2017) 9937–9945.
[132] J. Li, X. Xu, X. Liu, W. Qin, M. Wang, L. Pan, Metal-organic frameworks derived cake-like anatase/rutile mixed phase TiO_2 for highly efficient photocatalysis, J. Alloys Compd. 690 (2017) 640–646.
[133] M.Z. Hussain, A. Schneemann, R.A. Fischer, Y. Zhu, Y. Xia, MOF derived porous ZnO/C nanocomposites for efficient dye photodegradation, ACS Appl. Energy Mater. 1 (2018) 4695–4707.
[134] Z. Wu, X. Yuan, J. Zhang, H. Wang, L. Jiang, G. Zeng, Photocatalytic decontamination of wastewater containing organic dyes by metal–organic frameworks and their derivatives, ChemCatChem 9 (2017) 41–64.
[135] X. Cao, B. Zheng, X. Rui, W. Shi, Q. Yan, H. Zhang, Metal oxide-coated three-dimensional graphene prepared by the use of metal–organic frameworks as precursors, Angew. Chem. Int. Ed. 53 (2014) 1404–1409.
[136] K.-Y.A. Lin, F.-K. Hsu, W.-D. Lee, Magnetic cobalt–graphene nanocomposite derived from self-assembly of MOFs with graphene oxide as an activator for peroxymonosulfate, J. Mater. Chem. A 3 (2015) 9480–9490.
[137] E. Proietti, F. Jaouen, M. Lefèvre, N. Larouche, J. Tian, J. Herranz, J.-P. Dodelet, Iron-based cathode catalyst with enhanced power density in polymer electrolyte membrane fuel cells, Nat. Commun. 2 (2011) 416.
[138] B.Y. Xia, Y. Yan, N. Li, H.B. Wu, X.W. Lou, X. Wang, A metal–organic framework-derived bifunctional oxygen electrocatalyst, Nat. Energy 1 (2016) 15006.
[139] J. Meng, C. Niu, L. Xu, J. Li, X. Liu, X. Wang, Y. Wu, X. Xu, W. Chen, Q. Li, Z. Zhu, D. Zhao, L. Mai, General oriented formation of carbon nanotubes from metal–organic frameworks, J. Am. Chem. Soc. 139 (2017) 8212–8221.
[140] X. Fang, L. Jiao, S.-H. Yu, H.-L. Jiang, Metal–organic framework-derived feco-N-doped hollow porous carbon nanocubes for electrocatalysis in acidic and alkaline media, ChemSusChem 10 (2017) 3019–3024.
[141] Y. Xu, W. Tu, B. Zhang, S. Yin, Y. Huang, M. Kraft, R. Xu, Nickel nanoparticles encapsulated in few-layer nitrogen-doped graphene derived from metal–organic frameworks as efficient bifunctional electrocatalysts for overall water splitting, Adv. Mater. 29 (2017) 1605957.
[142] H. Zhang, S. Hwang, M. Wang, Z. Feng, S. Karakalos, L. Luo, Z. Qiao, X. Xie, C. Wang, D. Su, Y. Shao, G. Wu, Single atomic iron catalysts for oxygen reduction in acidic media: particle size control and thermal activation, J. Am. Chem. Soc. 139 (2017) 14143–14149.
[143] X.-F. Lu, L.-F. Gu, J.-W. Wang, J.-X. Wu, P.-Q. Liao, G.-R. Li, Bimetal-organic framework derived $CoFe_2O_4$/C porous hybrid nanorod arrays as high-performance electrocatalysts for oxygen evolution reaction, Adv. Mater. 29 (2017) 1604437.
[144] S. Ma, G.A. Goenaga, A.V. Call, D.-J. Liu, Cobalt imidazolate framework as precursor for oxygen reduction reaction electrocatalysts, Chem. Eur. J. 17 (2011) 2063–2067.
[145] D. Zhao, J.-L. Shui, C. Chen, X. Chen, B.M. Reprogle, D. Wang, D.-J. Liu, Iron imidazolate framework as precursor for electrocatalysts in polymer electrolyte membrane fuel cells, Chem. Sci. 3 (2012) 3200–3205.
[146] H. Rong, T. Zhan, Y. Sun, Y. Wen, X. Liu, H. Teng, ZIF-8 derived nitrogen, phosphorus and sulfur tri-doped mesoporous carbon for boosting electrocatalysis to oxygen reduction in universal Ph range, Electrochim. Acta 318 (2019) 783–793.
[147] W. Zhang, Z.-Y. Wu, H.-L. Jiang, S.-H. Yu, Nanowire-directed templating synthesis of metal–organic framework nanofibers and their derived porous doped carbon nanofibers for enhanced electrocatalysis, J. Am. Chem. Soc. 136 (2014) 14385–14388.

[148] H. Yang, X. Chen, W.-T. Chen, Q. Wang, N. Cantillo Cuello, A. Nafady, A.M. Al-Enizi, G.I.N. Waterhouse, G.A. Goenaga, T.A. Zawodzinski, Tunable synthesis of hollow metal-nitrogen-carbon capsules for efficient oxygen reduction catalysis in proton exchange membrane fuel cells, ACS Nano 13 (2019) 8087–8098.
[149] R. Jiang, L. Li, T. Sheng, G. Hu, Y. Chen, L. Wang, Edge-site engineering of atomically dispersed Fe–N_4 by selective C–N bond cleavage for enhanced oxygen reduction reaction activities, J. Am. Chem. Soc. 140 (2018) 11594–11598.
[150] P. Su, H. Xiao, J. Zhao, Y. Yao, Z. Shao, C. Li, Q. Yang, Nitrogen-doped carbon nanotubes derived from Zn–Fe-ZIF nanospheres and their application as efficient oxygen reduction electrocatalysts with in situ generated iron species, Chem. Sci. 4 (2013) 2941–2946.
[151] A. Zitolo, V. Goellner, V. Armel, M.-T. Sougrati, T. Mineva, L. Stievano, E. Fonda, F. Jaouen, Identification of catalytic sites for oxygen reduction in iron- and nitrogen-doped graphene materials, Nat. Mater. 14 (2015) 937.
[152] R. Wang, T. Yan, L. Han, G. Chen, H. Li, J. Zhang, L. Shi, D. Zhang, Tuning the dimensions and structures of nitrogen-doped carbon nanomaterials derived from sacrificial g-C_3N_4/metal–organic frameworks for enhanced electrocatalytic oxygen reduction, J. Mater. Chem. A 6 (2018) 5752–5761.
[153] S. Zhao, H. Yin, L. Du, L. He, K. Zhao, L. Chang, G. Yin, H. Zhao, S. Liu, Z. Tang, Carbonized nanoscale metal–organic frameworks as high performance electrocatalyst for oxygen reduction reaction, ACS Nano 8 (2014) 12660–12668.
[154] Y. Li, S. Niu, D. Rakov, Y. Wang, M. Cabán-Acevedo, S. Zheng, B. Song, P. Xu, Metal organic framework-derived CoPS/N-doped carbon for efficient electrocatalytic hydrogen evolution, Nanoscale 10 (2018) 7291–7297.
[155] H. Tabassum, W. Guo, W. Meng, A. Mahmood, R. Zhao, Q. Wang, R. Zou, Metal–organic frameworks derived cobalt phosphide architecture encapsulated into B/N Co-doped graphene nanotubes for all Ph value electrochemical hydrogen evolution, Adv. Energy Mater. 7 (2017) 1601671.
[156] X.-Y. Yu, L. Yu, H.B. Wu, X.W. Lou, Formation of nickel sulfide nanoframes from metal–organic frameworks with enhanced pseudocapacitive and electrocatalytic properties, Angew. Chem. Int. Ed. 54 (2015) 5331–5335.
[157] J. Chen, J. Liu, J.-Q. Xie, H. Ye, X.-Z. Fu, R. Sun, C.-P. Wong, Co-Fe-P nanotubes electrocatalysts derived from metal-organic frameworks for efficient hydrogen evolution reaction under wide Ph range, Nano Energy 56 (2019) 225–233.
[158] X. Wang, L. Yu, B.Y. Guan, S. Song, X.W. Lou, Metal–organic framework hybrid-assisted formation of Co_3O_4/Co-Fe oxide double-shelled nanoboxes for enhanced oxygen evolution, Adv. Mater. 30 (2018) 1801211.
[159] P. He, X.-Y. Yu, X.W. Lou, Carbon-incorporated nickel–cobalt mixed metal phosphide nanoboxes with enhanced electrocatalytic activity for oxygen evolution, Angew. Chem. Int. Ed. 56 (2017) 3897–3900.
[160] X. Wang, L. Chai, J. Ding, L. Zhong, Y. Du, T.-T. Li, Y. Hu, J. Qian, S. Huang, Chemical and morphological transformation of Mof-derived bimetallic phosphide for efficient oxygen evolution, Nano Energy 62 (2019) 745–753.
[161] K. Jayaramulu, J. Masa, O. Tomanec, D. Peeters, V. Ranc, A. Schneemann, R. Zboril, W. Schuhmann, R.A. Fischer, Nanoporous nitrogen-doped graphene oxide/nickel sulfide composite sheets derived from a metal-organic framework as an efficient electrocatalyst for hydrogen and oxygen evolution, Adv. Funct. Mater. 27 (2017) 1700451.
[162] Y. Guo, J. Tang, H. Qian, Z. Wang, Y. Yamauchi, One-pot synthesis of zeolitic imidazolate framework 67-derived hollow Co_3S_4@MoS_2 heterostructures as efficient bifunctional catalysts, Chem. Mater. 29 (2017) 5566–5573.

[163] T. Liu, P. Li, N. Yao, T. Kong, G. Cheng, S. Chen, W. Luo, Self-sacrificial template-directed vapor-phase growth of Mof assemblies and surface vulcanization for efficient water splitting, Adv. Mater. 31 (2019) 1806672.
[164] G. Cai, W. Zhang, L. Jiao, S.-H. Yu, H.-L. Jiang, Template-directed growth of well-aligned Mof arrays and derived self-supporting electrodes for water splitting, Chem 2 (2017) 791–802.
[165] T.Y. Ma, S. Dai, M. Jaroniec, S.Z. Qiao, Metal–organic framework derived hybrid Co_3O_4-carbon porous nanowire arrays as reversible oxygen evolution electrodes, J. Am. Chem. Soc. 136 (2014) 13925–13931.
[166] S. Liu, Z. Wang, S. Zhou, F. Yu, M. Yu, C.-Y. Chiang, W. Zhou, J. Zhao, J. Qiu, Metal–organic-framework-derived hybrid carbon nanocages as a bifunctional electrocatalyst for oxygen reduction and evolution, Adv. Mater. 29 (2017) 1700874.
[167] W. Zang, A. Sumboja, Y. Ma, H. Zhang, Y. Wu, S. Wu, H. Wu, Z. Liu, C. Guan, J. Wang, S.J. Pennycook, Single Co atoms anchored in porous N-doped carbon for efficient zinc–air battery cathodes, ACS Catal. 8 (2018) 8961–8969.
[168] J. Li, Q.-L. Zhu, Q. Xu, Pd nanoparticles supported on hierarchically porous carbons derived from assembled nanoparticles of a zeolitic imidazolate framework (ZIF-8) for methanol electrooxidation, Chem. Commun. 51 (2015) 10827–10830.
[169] Y. Ye, F. Cai, H. Li, H. Wu, G. Wang, Y. Li, S. Miao, S. Xie, R. Si, J. Wang, X. Bao, Surface functionalization of ZIF-8 with ammonium ferric citrate toward high exposure of Fe-N active sites for efficient oxygen and carbon dioxide electroreduction, Nano Energy 38 (2017) 281–289.
[170] W. Ren, X. Tan, W. Yang, C. Jia, S. Xu, K. Wang, S.C. Smith, C. Zhao, Isolated diatomic Ni-Fe metal–nitrogen sites for synergistic electroreduction of CO_2, Angew. Chem. Int. Ed. 58 (2019) 6972–6976.
[171] P. Zhang, F. Sun, Z. Xiang, Z. Shen, J. Yun, D. Cao, ZIF-derived in situ nitrogen-doped porous carbons as efficient metal-free electrocatalysts for oxygen reduction reaction, Energy Environ. Sci. 7 (2014) 442–450.
[172] Q. Lai, Y. Zhao, Y. Liang, J. He, J. Chen, In situ confinement pyrolysis transformation of ZIF-8 to nitrogen-enriched meso-microporous carbon frameworks for oxygen reduction, Adv. Funct. Mater. 26 (2016) 8334–8344.
[173] L. Shang, H. Yu, X. Huang, T. Bian, R. Shi, Y. Zhao, G.I.N. Waterhouse, L.-Z. Wu, C.-H. Tung, T. Zhang, Well-dispersed Zif-derived Co,N-Co-doped carbon nanoframes through mesoporous-silica-protected calcination as efficient oxygen reduction electrocatalysts, Adv. Mater. 28 (2016) 1668–1674.
[174] I.S. Amiinu, X. Liu, Z. Pu, W. Li, Q. Li, J. Zhang, H. Tang, H. Zhang, S. Mu, From 3d ZIF nanocrystals to Co–Nx/C nanorod array electrocatalysts for Orr, Oer, and Zn–air batteries, Adv. Funct. Mater. 28 (2018) 1704638.
[175] Q. Ren, H. Wang, X.-F. Lu, Y.-X. Tong, G.-R. Li, Recent progress on MOF-derived heteroatom-doped carbon-based electrocatalysts for oxygen reduction reaction, Adv. Sci. 5 (2018) 1700515.
[176] F.-L. Meng, K.-H. Liu, Y. Zhang, M.-M. Shi, X.-B. Zhang, J.-M. Yan, Q. Jiang, Recent advances toward the rational design of efficient bifunctional air electrodes for rechargeable Zn–air batteries, Small 14 (2018) 1703843.
[177] M. Zhang, Q. Dai, H. Zheng, M. Chen, L. Dai, Novel MOF-derived Co@N-C bifunctional catalysts for highly efficient Zn–air batteries and water splitting, Adv. Mater. 30 (2018) 1705431.
[178] S. Zhao, D.-W. Wang, R. Amal, L. Dai, Carbon-based metal-free catalysts for key reactions involved in energy conversion and storage, Adv. Mater. 31 (2019) 1801526.
[179] Q.-L. Zhu, W. Xia, T. Akita, R. Zou, Q. Xu, Metal-organic framework-derived honeycomb-like open porous nanostructures as precious-metal-free catalysts for highly efficient oxygen electroreduction, Adv. Mater. 28 (2016) 6391–6398.

[180] Z. Li, M. Shao, L. Zhou, R. Zhang, C. Zhang, M. Wei, D.G. Evans, X. Duan, Directed growth of metal-organic frameworks and their derived carbon-based network for efficient electrocatalytic oxygen reduction, Adv. Mater. 28 (2016) 2337–2344.

[181] Y. Wang, X. Chen, Q. Lin, A. Kong, Q.-G. Zhai, S. Xie, P. Feng, Nanoporous carbon derived from a functionalized metal–organic framework as a highly efficient oxygen reduction electrocatalyst, Nanoscale 9 (2017) 862–868.

[182] C.-L. Zhang, B.-R. Lu, F.-H. Cao, Z.-Y. Wu, W. Zhang, H.-P. Cong, S.-H. Yu, Electrospun metal-organic framework nanoparticle fibers and their derived electrocatalysts for oxygen reduction reaction, Nano Energy 55 (2019) 226–233.

[183] L. Jiao, G. Wan, R. Zhang, H. Zhou, S.-H. Yu, H.-L. Jiang, From metal–organic frameworks to single-atom Fe implanted N-doped porous carbons: efficient oxygen reduction in both alkaline and acidic media, Angew. Chem. Int. Ed. 57 (2018) 8525–8529.

[184] H. Zhou, D. He, A.I. Saana, J. Yang, Z. Wang, J. Zhang, Q. Liang, S. Yuan, J. Zhu, S. Mu, Mesoporous-silica induced doped carbon nanotube growth from metal–organic frameworks, Nanoscale 10 (2018) 6147–6154.

[185] Y. Xiong, Y. Yang, F.J. DiSalvo, H.D. Abruña, Metal–organic-framework-derived Co–Fe bimetallic oxygen reduction electrocatalysts for alkaline fuel cells, J. Am. Chem. Soc. 141 (2019) 10744–10750.

[186] H. Zhang, G. Liu, L. Shi, J. Ye, Single-atom catalysts: emerging multifunctional materials in heterogeneous catalysis, Adv. Energy Mater. 8 (2018) 1701343.

[187] B. Qiao, A. Wang, X. Yang, L.F. Allard, Z. Jiang, Y. Cui, J. Liu, J. Li, T. Zhang, Single-atom catalysis of Co oxidation using Pt1/Feox, Nat. Chem. 3 (2011) 634.

[188] P. Liu, Y. Zhao, R. Qin, S. Mo, G. Chen, L. Gu, D.M. Chevrier, P. Zhang, Q. Guo, D. Zang, B. Wu, G. Fu, N. Zheng, Photochemical route for synthesizing atomically dispersed palladium catalysts, Science 352 (2016) 797–800.

[189] F. Xiao, G.-L. Xu, C.-J. Sun, M. Xu, W. Wen, Q. Wang, M. Gu, S. Zhu, Y. Li, Z. Wei, Nitrogen-coordinated single iron atom catalysts derived from metal organic frameworks for oxygen reduction reaction, Nano Energy 61 (2019) 60–68.

[190] E. Luo, H. Zhang, X. Wang, L. Gao, L. Gong, T. Zhao, Z. Jin, J. Ge, Z. Jiang, C. Liu, W. Xing, Single-atom Cr–N4 sites designed for durable oxygen reduction catalysis in acid media. Angew. Chem. Int. Ed. (2019), https://doi.org/10.1002/anie.201906289.

[191] X.F. Lu, L. Yu, J. Zhang, X.W. Lou, Ultrafine dual-phased carbide nanocrystals confined in porous nitrogen-doped carbon dodecahedrons for efficient hydrogen evolution reaction, Adv. Mater. 31 (2019) 1900699.

[192] X. Luo, Q. Zhou, S. Du, J. Li, L. Zhang, K. Lin, H. Li, B. Chen, T. Wu, D. Chen, M. Chang, Y. Liu, One-dimensional porous hybrid structure of Mo_2C-CoP encapsulated in N-doped carbon derived from Mof: an efficient electrocatalyst for hydrogen evolution reaction over the entire Ph range, ACS Appl. Mater. Interfaces 10 (2018) 42335–42347.

[193] Z. Chen, R. Wu, Y. Liu, Y. Ha, Y. Guo, D. Sun, M. Liu, F. Fang, Ultrafine Co nanoparticles encapsulated in carbon-nanotubes-grafted graphene sheets as advanced electrocatalysts for the hydrogen evolution reaction, Adv. Mater. 30 (2018) 1802011.

[194] X. Wang, J. He, B. Yu, B. Sun, D. Yang, X. Zhang, Q. Zhang, W. Zhang, L. Gu, Y. Chen, $CoSe_2$ nanoparticles embedded MOF-derived Co-N-C nanoflake arrays as efficient and stable electrocatalyst for hydrogen evolution reaction, Appl. Catal. B Environ. 258 (2019) 117996.

[195] L. Fan, P.F. Liu, X. Yan, L. Gu, Z.Z. Yang, H.G. Yang, S. Qiu, X. Yao, Atomically isolated nickel species anchored on graphitized carbon for efficient hydrogen evolution electrocatalysis, Nat. Commun. 7 (2016) 10667.

[196] M. Tahir, L. Pan, F. Idrees, X. Zhang, L. Wang, J.-J. Zou, Z.L. Wang, Electrocatalytic oxygen evolution reaction for energy conversion and storage: a comprehensive review, Nano Energy 37 (2017) 136–157.

[197] M. Gong, H. Dai, A mini review of nife-based materials as highly active oxygen evolution reaction electrocatalysts, Nano Res. 8 (2015) 23–39.
[198] N.-T. Suen, S.-F. Hung, Q. Quan, N. Zhang, Y.-J. Xu, H.M. Chen, Electrocatalysis for the oxygen evolution reaction: recent development and future perspectives, Chem. Soc. Rev. 46 (2017) 337–365.
[199] J. Zhou, Y. Dou, A. Zhou, L. Shu, Y. Chen, J.-R. Li, Layered metal–organic framework-derived metal oxide/carbon nanosheet arrays for catalyzing the oxygen evolution reaction, ACS Energy Lett. 3 (2018) 1655–1661.
[200] Z. Chen, R. Wu, M. Liu, Y. Liu, S. Xu, Y. Ha, Y. Guo, X. Yu, D. Sun, F. Fang, Tunable electronic coupling of cobalt sulfide/carbon composites for optimizing oxygen evolution reaction activity, J. Mater. Chem. A 6 (2018) 10304–10312.
[201] L. Yang, L. Zhang, G. Xu, X. Ma, W. Wang, H. Song, D. Jia, Metal–organic-framework-derived hollow CoSx@ MoS_2 microcubes as superior bifunctional electrocatalysts for hydrogen evolution and oxygen evolution reactions, ACS Sustain. Chem. Eng. 6 (2018) 12961–12968.
[202] X. Wang, Z. Ma, L. Chai, L. Xu, Z. Zhu, Y. Hu, J. Qian, S. Huang, MOF derived N-doped carbon coated CoP particle/carbon nanotube composite for efficient oxygen evolution reaction, Carbon 141 (2019) 643–651.
[203] F. Bu, W. Chen, M.F. Aly Aboud, I. Shakir, J. Gu, Y. Xu, Microwave-assisted ultrafast synthesis of adjustable bimetal phosphide/graphene heterostructures from Mofs for efficient electrochemical water splitting, J. Mater. Chem. A 7 (2019) 14526–14535.
[204] L. Zhang, X. Wang, A. Li, X. Zheng, L. Peng, J. Huang, Z. Deng, H. Chen, Z. Wei, Rational construction of macroporous CoFeP triangular plate arrays from bimetal–organic frameworks as high-performance overall water-splitting catalysts, J. Mater. Chem. A 7 (2019) 17529–17535.
[205] S.S. Shinde, C.H. Lee, J.-Y. Jung, N.K. Wagh, S.-H. Kim, D.-H. Kim, C. Lin, S. U. Lee, J.-H. Lee, Unveiling dual-linkage 3d hexaiminobenzene metal–organic frameworks towards long-lasting advanced reversible Zn–air batteries, Energy Environ. Sci. 12 (2019) 727–738.
[206] L. Ma, S. Chen, Z. Pei, Y. Huang, G. Liang, F. Mo, Q. Yang, J. Su, Y. Gao, J. A. Zapien, C. Zhi, Single-site active iron-based bifunctional oxygen catalyst for a compressible and rechargeable zinc–air battery, ACS Nano 12 (2018) 1949–1958.
[207] S. Rezaee, S. Shahrokhian, Facile synthesis of petal-like NiCo/NiO-CoO/nanoporous carbon composite based on mixed-metallic MOFs and their application for electrocatalytic oxidation of methanol, Appl. Catal. B Environ. 244 (2019) 802–813.
[208] S. Lin, C.S. Diercks, Y.-B. Zhang, N. Kornienko, E.M. Nichols, Y. Zhao, A.R. Paris, D. Kim, P. Yang, O.M. Yaghi, Covalent organic frameworks comprising cobalt porphyrins for catalytic CO_2 reduction in water, Science 349 (2015) 1208–1213.
[209] T.-T. Zhuang, Z.-Q. Liang, A. Seifitokaldani, Y. Li, P. De Luna, T. Burdyny, F. Che, F. Meng, Y. Min, R. Quintero-Bermudez, C.T. Dinh, Y. Pang, M. Zhong, B. Zhang, J. Li, P.-N. Chen, X.-L. Zheng, H. Liang, W.-N. Ge, B.-J. Ye, D. Sinton, S.-H. Yu, E.H. Sargent, Steering post-C–C coupling selectivity enables high efficiency electroreduction of carbon dioxide to multi-carbon alcohols, Nat. Catal. 1 (2018) 421–428.
[210] W. Ju, A. Bagger, G.-P. Hao, A.S. Varela, I. Sinev, V. Bon, B. Roldan Cuenya, S. Kaskel, J. Rossmeisl, P. Strasser, Understanding activity and selectivity of metal-nitrogen-doped carbon catalysts for electrochemical reduction of CO_2, Nat. Commun. 8 (2017) 944.
[211] Y. Mun, K. Kim, S. Kim, S. Lee, S. Lee, S. Kim, W. Choi, S.-k. Kim, J.W. Han, J. Lee, A novel strategy to develop non-noble metal catalyst for CO_2 electroreduction: hybridization of metal-organic polymer, Appl. Catal. B Environ. 236 (2018) 154–161.

[212] H. Song, M. Im, J.T. Song, J.-A. Lim, B.-S. Kim, Y. Kwon, S. Ryu, J. Oh, Effect of mass transfer and kinetics in ordered Cu-mesostructures for electrochemical CO_2 reduction, Appl. Catal. B Environ. 232 (2018) 391–396.

[213] X. Li, W. Bi, M. Chen, Y. Sun, H. Ju, W. Yan, J. Zhu, X. Wu, W. Chu, C. Wu, Y. Xie, Exclusive Ni–N_4 sites realize near-unity CO selectivity for electrochemical CO_2 reduction, J. Am. Chem. Soc. 139 (2017) 14889–14892.

[214] H.B. Yang, S.-F. Hung, S. Liu, K. Yuan, S. Miao, L. Zhang, X. Huang, H.-Y. Wang, W. Cai, R. Chen, J. Gao, X. Yang, W. Chen, Y. Huang, H.M. Chen, C.M. Li, T. Zhang, B. Liu, Atomically dispersed Ni(I) as the active site for electrochemical CO_2 reduction, Nat. Energy 3 (2018) 140–147.

[215] C.S. Diercks, Y. Liu, K.E. Cordova, O.M. Yaghi, The role of reticular chemistry in the design of CO_2 reduction catalysts, Nat. Mater. 17 (2018) 301–307.

[216] J. Wu, T. Sharifi, Y. Gao, T. Zhang, P.M. Ajayan, Emerging carbon-based heterogeneous catalysts for electrochemical reduction of carbon dioxide into value-added chemicals, Adv. Mater. 31 (2019) 1804257.

[217] Z. Geng, Y. Cao, W. Chen, X. Kong, Y. Liu, T. Yao, Y. Lin, Regulating the Coordination environment of Co single atoms for achieving efficient electrocatalytic activity in CO_2 reduction, Appl. Catal. B Environ. 240 (2019) 234–240.

CHAPTER 4

The utility of carbon dots for photocatalysis

Samahe Sadjadi
Gas Conversion Department, Faculty of Petrochemicals, Iran Polymer and Petrochemical Institute, Tehran, Iran

1 Introduction

1.1 Definition

Carbon dots (CDs) are an emerging class of carbon nanomaterials [1]. These fluorescent compounds were first discovered in 2004 during the purification of single walled carbon nanotubes (CNTs) [2]. Then, as these carbon materials exhibited spectral features similar to their silicon counterparts, they are called quantum-sized carbon dots [3]. Nowadays, CDs are referred as surface-passivated fine carbon particles that are brightly fluorescent.

CDs are composed of carbon cores that can be amorphous to nanocrystalline and shells with oxygenic functionalities on their surface that can enhance the solubility of CDs in aqueous media [1, 4, 5]. Based on the nature of their cores, CDs can be classified into three types, carbon nanodots (CNDs), graphene quantum dots (GQDs), and polymer dots (PDs) (Fig. 1) [3]. As illustrated in Fig. 1, GQDs can be subdivided into two types, s-GQDs and m-GQDs. In the former, the carbon core is composed of one layer of graphene debris, while the latter features a graphitic nanocrystal structure. PDs are aggregated or cross-linked polymer nanoparticles. CNDs are quasi-spherical carbon nanoparticles without a distinct crystal lattice, which is predominantly composed of sp^2/sp^3 carbons or amorphous carbon together with embedded sp^2-hybridized nanocrystalline regions [3].

Unique properties of CDs such as high surface area, large diameter, and good surface grafting using conjugated π-π network, good biocompatibility, highly tunable photoluminescence (PL), unique spin property, chemical inertness, remarkable UV-blocking property, and high photo stability make them promising candidates for broad range of applications such as catalysis, sensors, biomedicines, biological labeling, drug delivery, photovoltaic devices, and optronics [6].

Emerging Carbon Materials for Catalysis
https://doi.org/10.1016/B978-0-12-817561-3.00004-4

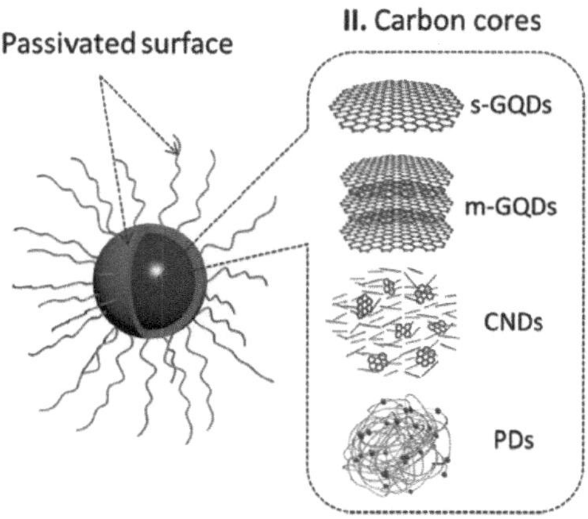

Fig. 1 Schematic diagram of the structure of CDs. *(Reprinted from C. Hu, M. Li, J. Qiu, Y.-P. Sun, Design and fabrication of carbon dots for energy conversion and storage, Chem. Soc. Rev. 48(8) (2019) 2315–2337, with permission from Royal Society of Chemistry.)*

1.2 Synthesis of CDs

Since their discovery, CDs' field of study has witnessed impressive advances [7, 8] and many synthetic methods have been developed for their synthesis. Generally, CDs preparation methods can be categorized into two main approaches, top-down and bottom-up. In the "top-down" approach, the strategy is reducing the size of macroscopic carbon materials such as activated carbon, graphite, CNT, etc. through treatments such as oxygen plasma treatment, electrochemical oxidation, or chemical oxidative cutting/ablation [8] to afford carbon nanomaterials that exhibit the characteristics of fluorescent nanoparticles. It is worth noting that most of the top-down methods are not suitable for the large-scale production of CD because of their low quantum yield (QY) values, special and expensive complicated equipment requirements, and uncontrollable and toxic process conditions [9].

In "bottom-up" approach, however, CDs are prepared by using molecular precursors such as citric acid, urea, glycerol, cysteamine, polyamines, and acrylamide and biological precursors such as carbohydrates, amino acids, etc. through hydrothermal treatment, microwave synthesis, thermal decomposition, and plasma treatment (Fig. 2) [1]. Notably, the type and content of carbon as well as the functionalities of the precursor can influence the optical and/or fluorescence properties of CDs. In recent years, bottom-up approaches have received increasing attention due to their advantages such as precise control of precursor molecules, ease of techniques, low cost, and

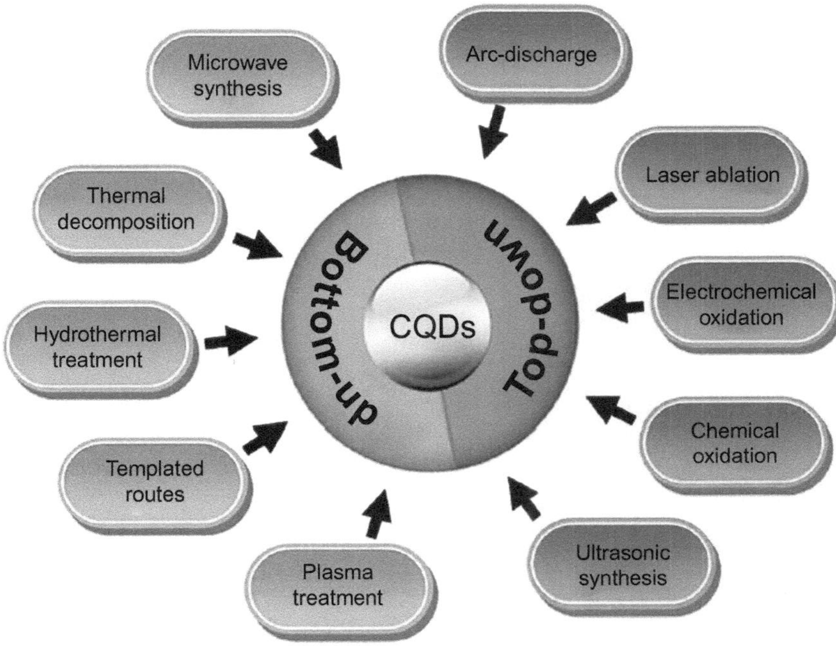

Fig. 2 Schematic illustration of CDs preparation via "top-down" and "bottom-up" approaches. *(Reprinted from R. Wang, K.-Q. Lu, Z.-R. Tang, Y.-J. Xu, Recent progress in carbon quantum dots: synthesis, properties and applications in photocatalysis, J. Mater. Chem. A 5(8) (2017) 3717–3734, with permission from Royal Society of Chemistry.)*

practicality and convenience of the procedure with generally nontoxic precursors [9]. Moreover, in the bottom-up approach, use of passivation agents such as polyethylene glycol, polyethylenimine, for affording fluorescent materials is not required [8]. In Table 1, the advantages and disadvantages of some bottom-up and top-down CD preparation methods are summarized.

1.2.1 Green synthesis of CDs

The green approaches for the synthesis of CDs mostly focused on the use of green starting materials such as biomass [10]. To date, various green raw materials such as chitosan, grass and plant leaves, fruit juices, watermelon or pomelo peels, etc. have been applied as green carbon sources [2]. These precursors not only are readily available and cheap, but also are safer compared to conventional hydrocarbon compounds [10]. In Table 2, some of the reported CDs prepared from bio-based precursors are tabulated. As shown in Table 2, these CDs mostly exhibited relatively high QY. This can be

Table 1 Comparison of the merits and demerits of a representative CD preparation.

	Methods	Merits	Demerits
Top-down	Laser ablation	Controllable morphology and size	Complicated operation, high cost
	Electrochemical oxidation	High purity, high yield, controllable size, good reproducibility	Complicated operation
	Chemical oxidation	Easy operation, large-scale production, no elaborate equipment	Nonuniform size distribution
	Ultrasonic treatment	Easy operation	Instrumental wastage, high energy cost
Bottom-up	Microwave synthesis	Short reaction time, uniform size distribution, easy size control	High energy cost
	Thermal decomposition	Easy operation, solvent-free, low cost, large-scale production	Nonuniform size distribution
	Hydrothermal treatment	High quantum efficiency, low cost, nontoxicity	Low yield

due to the presence of heteroatoms, especially nitrogen, in the bio-based precursors (vide infra). As summarized in Table 2, the fluorescence emission of green CDs is usually in the short wavelength region (blue and green) [10]. CDs from natural resources can be obtained through various methods, such as microwave synthesis, hydrothermal method, solvothermal method, laser ablation technique, and electrochemical method. This topic has been fully reviewed by Das et al. [2] and the readers interested in this topic are referred to this review article.

Apart from bioresources as carbon precursors, developing sustainable synthesis methods such as base catalysis, self-exothermic synthesis, and reduction methods contributes to the green synthesis of CDs. These methods can be achieved at room temperature without any external energy supply and any special equipment [10].

Table 2 The summarized properties and applications of CDs prepared by using varied biomass materials.

Biomass	QY/%	Fluorescence
Sweet pepper	19.3	Blue
Saffron	23.6	Green-blue
Rose-heart radish	13.6	Blue
Lignin	21	Blue
Grass	–	Blue
Chitosan	13	Blue
Peach gum polysaccharide	28.46	Blue
Coriander leaves	6.48	Green
Lotus root	19.0	Blue
Eichhornia crassipes	–	Blue
Wheat straw	9.2	Blue
Silkworm chrysalis	46	Blue
Pumpkin	9.42	Yellow
Potato	6.14	Blue
Lentils	10	–
Fingernail	42.8	Blue
Starch	–	Blue
Garlic	13	Blue
Prawn shells	9	Blue
Orange juice	26	Blue
Ginger juice	13.5	Blue
Grape juice	13.5	Blue
Banana juice	8.95	Green

Adapted from M.L. Liu, B.B. Chen, C.M. Li, C.Z. Huang, Carbon dots: synthesis, formation mechanism, fluorescence origin and sensing applications, Green Chem. 21(3) (2019) 449–471, with permission from Royal Society of Chemistry.

1.2.2 Large-scale synthesis of carbon dots

The large-scale synthesis of CDs is essential for expanding their applications. As mentioned in the previous section, use of molecular precursors such as sucrose as well as both liquid and solid biomass such as orange juice and chicken egg is very appealing for large-scale CD preparation due to their availability and diversity [11]. It is worth noting that reproducibility of synthetic methods for large-scale CD production in terms of particle size, QY, etc. is of great importance. Until now, the large-scale synthesis of CDs by those reported methods only led to yields in gram scale; therefore, it is imperative and challenging to further enlarge the synthesis [11].

1.3 Optical properties of CD

1.3.1 Light absorption

CDs are capable of light absorption in UV region with a tail extending to the visible range [7]. The maximum absorption peak at ~230 nm may correspond to a π-π* transition involving aromatic sp^2 carbons, and the tail starting from ~300 nm can be assigned to an n-π* transition involving functional groups with electron lone pairs [12]. Notably, the absorption properties of CD are influenced by the preparation method. Moreover, the content of oxygen in CD can affect the absorption peak position. Another parameter that can affect the absorbance profile is the presence of a dopant. As an example, the absorbance of N-doped CDs can be enhanced in the near-UV region [13].

1.3.2 Photoluminescence

PL is one of the most important properties of CDs which has been applied for the development of photocatalysts. There are many reports in the literature on the CDs with different PL colors, ranging from the visible into the near-infrared region [14]. One of the influencing factors on the PL is the used precursors and preparation method (Table 3).

Table 3 The optical properties of CDs prepared from different precursors.

Precursor		PL image	PL lifetime (ns)	Quantum yield (%)
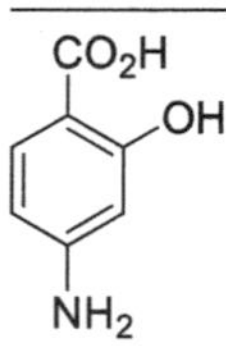		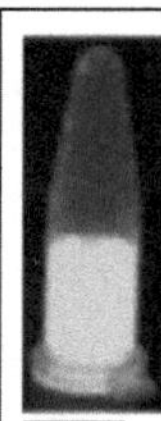	6.7	16.4
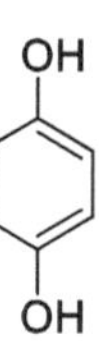	BBr_3		–	14.8
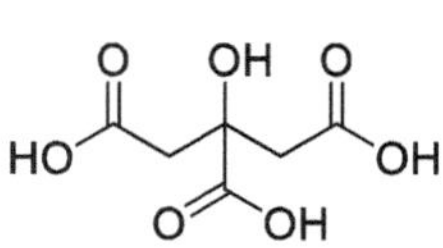	$Tb(NO_3)_{3.5}H_2O$		3.98	–

Table 3 The optical properties of CDs prepared from different precursors—cont'd

Precursor	PL image	PL lifetime (ns)	Quantum yield (%)
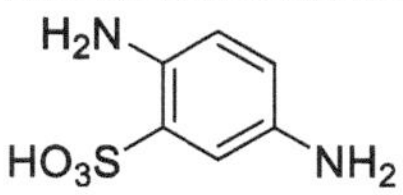 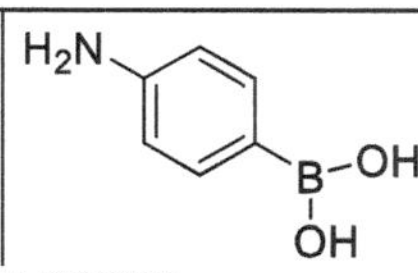		2.3	5.44
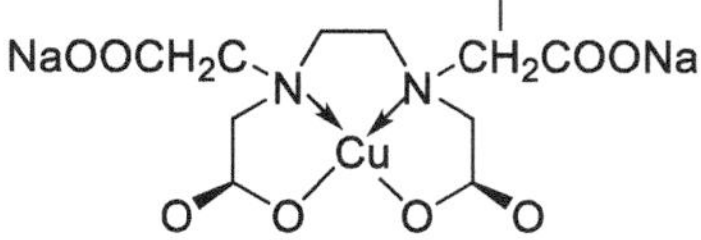		–	–

Adapted from C. Yi, Y. Pan, Y. Fang, Chapter 5—Surface engineering of carbon nanodots (C-dots) for biomedical applications, in: X. Wang, X. Chen (Eds.), Novel Nanomaterials for Biomedical, Environmental and Energy Applications, Elsevier, 2019, pp. 137–188, with permission from Elsevier.

Despite many researches in this field, the mechanism of PL for CDs is not well-understood and need more investigation. One proposed theory regarding the PL of CDs is the quantum confinement effect [12]. The quantum confinement effect occurs when the size of a material is smaller than the Bohr exciton radius and is within the same magnitude as the de Broglie wavelength of the electron, in which there is a discrete size-dependent energy bandgap between valence and conduction bands. Therefore, upon absorption of a photon resulting in an electron from HOMO of the valence band being promoted to LUMO of the conduction band, there is emission of the photon, fluorescence, which blue-shifts with decreasing particle size [15]. In this regard, Chi et al. demonstrated that the fluorescence of oxidized CDs was size-dependent and both excitation and emission spectra had shown a red shift when the molecular weight of the CDs increased [13, 16]. In another research, Choi et al. precisely studied the visible PL of GQDs with different shapes and sizes (5–35 nm) (Fig. 3) and confirmed the size and shape dependency of PL [17]. Upon increasing the size of GQDs, PL emission red shifted. Moreover, within the same size range, the maximum absorption occurred at longer wavelengths and the intensity decreased with increasing GQD size. The shape of the GQDs also dictated their edge states, showing increased armchair edges when the size increased above 17 nm, during which the overall shape was changed from circular to polygonal. As a result, the emission intensity increased with size above 17 nm [15].

Another plausible theory attributes PL to the presence of surface energy traps on surface defects resulting from the passivation step [13]. It was found

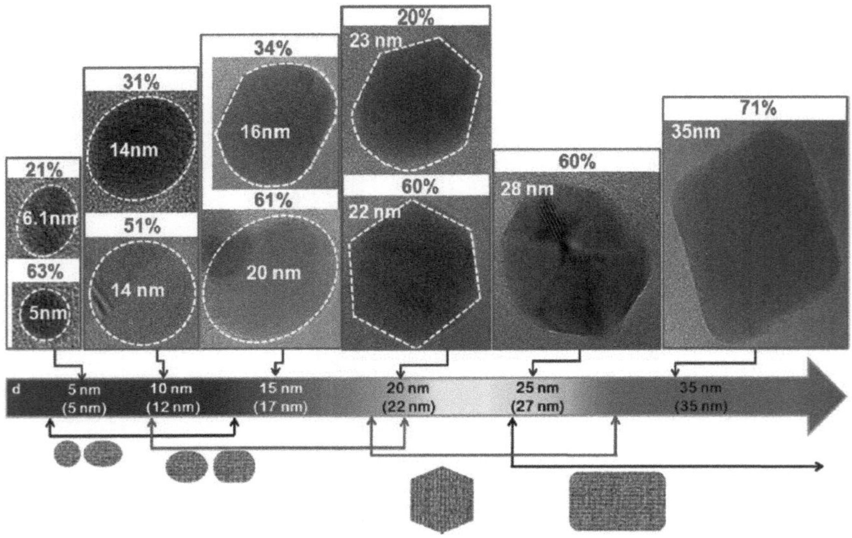

Fig. 3 HRTEM images of GQDs for their major shapes and corresponding populations (p) with increasing average size of GQDs. Here, the *dotted line* indicates the region of a GQD and p is defined as the ratio of number of GQDs with a major shape at each average size. Average sizes (d_a) of GQDs estimated from the HRTEM images at each d are indicated in the parentheses at the bottom of Fig. 2. The connected *arrows* indicate the range of the average size in which GQDs with particular major shapes are found. *(Reprinted from S. Kim, S.W. Hwang, M.-K. Kim, D.Y. Shin, D.H. Shin, C.O. Kim, S.B. Yang, J.H. Park, E. Hwang, S.-H. Choi, G. Ko, S. Sim, C. Sone, H.J. Choi, S. Bae, B.H. Hong, Anomalous behaviors of visible luminescence from graphene quantum dots: interplay between size and shape, ACS Nano 6(9) (2012) 8203–8208, with permission from American Chemical Society.)*

that surface passivation can improve the PL properties of CDs. Studying the PL of raw carbon nanoparticles showed that they exhibit the PL after acid treatment. This observation was attributed to the formation of functional groups such as hydroxyl and carboxylic groups on the surface of CD upon oxidation process. These surface functionalities serve as emission sites for the fluorescence to happen [18]. In an interesting study, Xiong et al. prepared several CDs with similar particle size distribution and graphite structure in their carbon core, but different degree of surface oxidation [19]. The authors attributed the observed red shift in CDs' emission peaks (from 440 to 625 nm) (Fig. 4) to the gradual reduction in their band gaps with the increasing incorporation of oxygen species into their surface structures [19]. In another study, Qiu et al. fabricated CDs with similar size and structure, but different levels of surface oxidation/defect by using similar starting

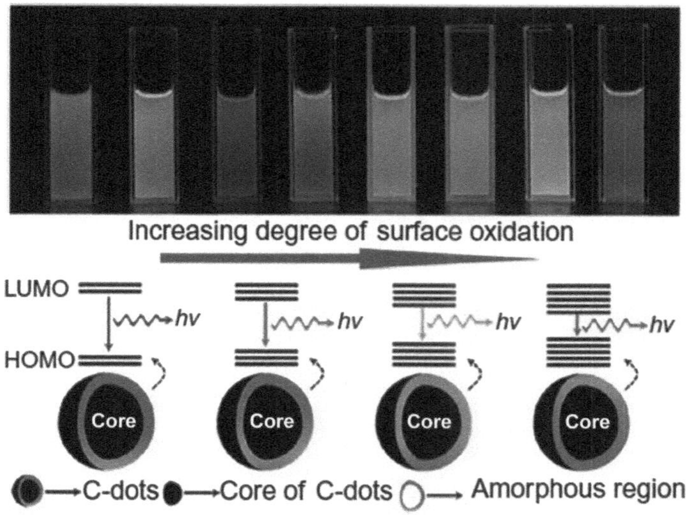

Fig. 4 Model for the tunable PL of CDs with different degrees of oxidation. *(Reprinted from H. Ding, S.-B. Yu, J.-S. Wei, H.-M. Xiong, Full-color light-emitting carbon dots with a surface-state-controlled luminescence mechanism, ACS Nano 10(1) (2016) 484–491, with permission from American Chemical Society.)*

material. The authors confirmed the different fluorescent colors of the prepared samples, indicating role of surface oxidation in the PL [20]. Considering this fact, surface modification was suggested as a tool to control the PL of CDs.

The third theory in this regard is the combination of the two so-called theories, i.e., the possibilities of the quantum confinement effect and the surface states [18]. In this context, Li, Huang et al. prepared a series of full color emitting CDs (with absolute QY > 70%) through one-pot hydrothermal method followed by separation by silica gel column chromatography (Fig. 5). Time-correlated single photon counting measurements suggest that the electron transition takes in effect in the PL progress of the crystalline core-shell structured CDs, and the PL properties could be coarsely adjusted by tuning the size of the crystalline carbon core owing to quantum confinement effects and finely adjusted by changing the surface functional groups consisting shell owing to surface trap states [21].

Recently, it was found that the formation of fluorescent impurities in the course of the bottom-up chemical synthesis contributes to the majority of emission from CDs samples [10]. In this context, Baker et al. prepared a series of CDs through hydrothermal and microwave routes by using citric

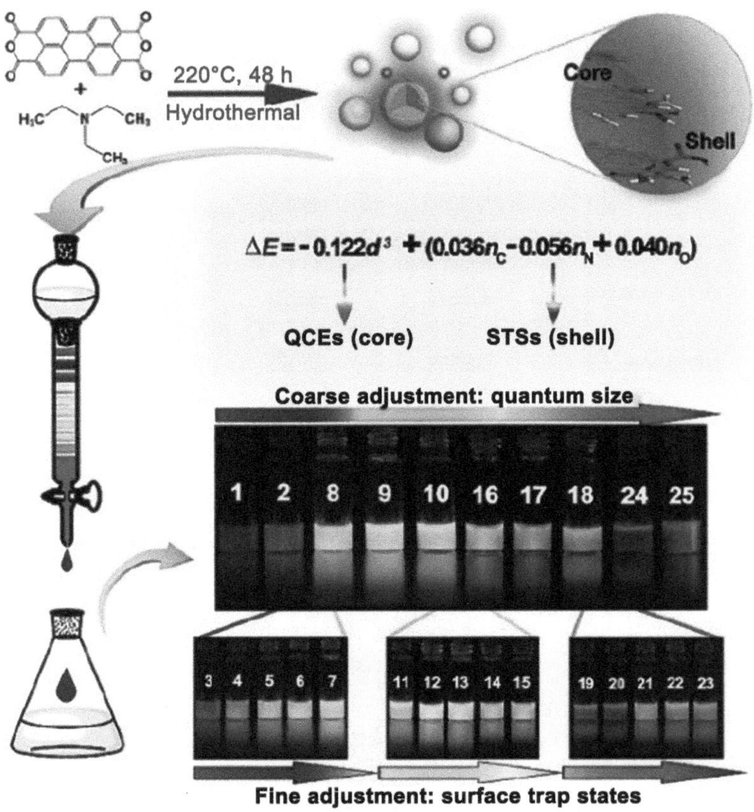

Fig. 5 One-pot synthesis and purification of highly PL CDs. The inserted equation and the photos indicate that the size of the crystalline core in the core-shell nanostructured PL CDs play the leading role owing to the quantum confinement effects, while the functional groups of the shell play a secondary role in the adjustment of PL wavelength owing to the surface trap states. *(Reprinted from Z. Liu, H. Zou, N. Wang, T. Yang, Z. Peng, J. Wang, N. Li, C. Huang, Photoluminescence of carbon quantum dots: coarsely adjusted by quantum confinement effects and finely by surface trap states, Sci. China Chem. 61(4) (2018) 490–496, with permission from Springer Nature.)*

acid (paired with urea or ethylenediamine as a nitrogen source), followed by dialysis or ultrafiltration purification steps [22]. Precise comparison and analysis of the optical properties of the resulting purification products (i.e., dialysate/filtrate vs retentate fractions) confirmed the formation of molecular fluorophores during the bottom-up chemical synthesis which contributes to a majority of the emission from CDs. The authors concluded that the fluorescent impurities produced as by-products of CD synthesis must be rigorously removed to obtain reliable results [22]. Rogach et al. also affirmed this issue by fabrication of three nitrogen-doped CDs using citric acid as

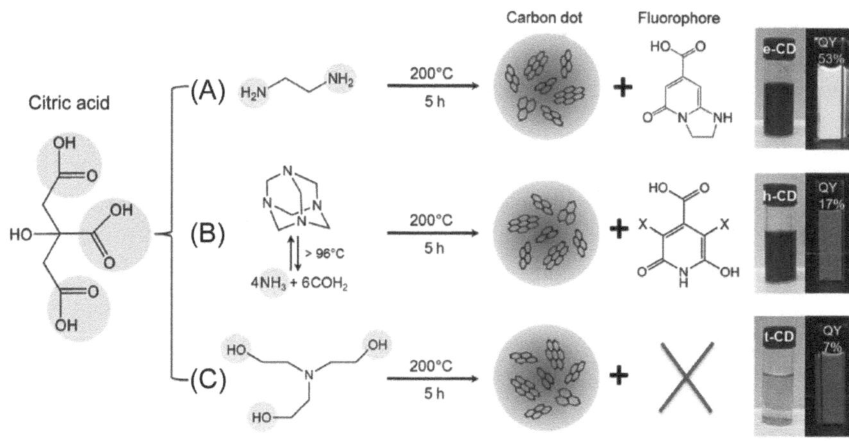

Fig. 6 Synthesis conditions of citric acid-based CDs using three different nitrogen-containing precursors. (A) Reaction of citric acid and ethylenediamine, resulting in e-CDs and the fluorophore 5-oxo-1,2,3,5-tetrahydroimidazo[1,2-α]pyridine-7-carboxylic acid. (B) Reaction of citric acid with hexamethylenetetramine, producing h-CDs and citrazinic acid and/or 3,5 derivatives (marked by −X), due to the decomposition of hexamethylenetetramine to ammonia and formaldehyde at temperatures exceeding 96°C. (C) Reaction of citric acid and triethanolamine, resulting in t-CDs and no derivatives of citrazinic acid since the tertiary amine prohibits their formation. (A–C) Images of the purified reaction products under ambient light and corresponding diluted solutions under UV light excitation, which reveal blue emission with PL QYs as labeled on the graph. *(Reprinted from J. Schneider, C.J. Reckmeier, Y. Xiong, M. von Seckendorff, A.S. Susha, P. Kasák, A.L. Rogach, Molecular fluorescence in citric acid-based carbon dots, J. Phys. Chem. C 121(3) (2017) 2014–2022, with permission from American Chemical Society.)*

carbon precursor and ethylenediamine, hexamethylenetetramine, and triethanolamine as three different nitrogen sources (Fig. 6). The authors fully characterized the CDs and compared their absorption, steady state emission, and PL decays [23]. The results showed that the fluorophores attached to the CDs, instead of just being free in the solution, significantly affected the optical properties of CDs.

1.3.3 Upconverted PL

The upconverted PL (UCPL) of CQDs can be attributed to the multiphoton activation process. In more details, the simultaneous absorption of multiple photons results in the emission of light at a shorter wavelength than the excitation wavelength [7]. This feature of CDs is very interesting for in vivo bioimaging applications since bioimaging at longer wavelengths is usually preferred owing to the improved photon tissue penetration and reduced background autofluorescence [4]. UCPL of CDs has been reported by several research groups. As an example, Lee, Shao et al. reported CD, fabricated through ultrasonic-assisted procedure, with excitation-independent down conversion and upconversion PL properties [24]. In another study, Wang et al. reported preparation of CDs through one-pot low-temperature aqueous heating of mixture of ascorbic acid aqueous and $Cu(Ac)_2{\cdot}H_2O$ solution. The resulting CDs possess excellent aqueous dispersibility, favorable biocompatibility, and unique PL properties such as pH- and polarity-dependent luminescence and upconversion fluorescence [25].

It is worth noting that some scientists believe that some of the apparent UCPL originate from the normal fluorescence excited by the leaking component from the second diffraction in the monochromator of the fluorescence spectrophotometer [26]. It was stated that, by simply inserting a long pass filter into the excitation pathway, the wrong excitation can be removed. Hence, UCPL should be identified by measuring the excitation intensity dependence of the fluorescence.

1.3.4 Quantum yield

QY referred to the efficiency of converting absorbed light into emitted light, which can be in the form of fluorescence. Fluorophores with high QY often emit strong fluorescence, even at low concentration. As high QY decreases the required content of fluorophores, it is desired from economic point of view [13]. The mostly reported QY for bare CDs is below 10%. However, doping of CDs and their surface functionalization could increase their QYs. Another approach for enhancing the QY is doping of as-prepared and passivated CDs with inorganic dopants [4]. As an example, Sun et al. reported photoluminescent CDs doped with ZnO or ZnS to afford C_{ZnO}-Dots or C_{ZnS}-Dots, which were competitive to the commercially available CdSe/ZnS QDs in luminescence brightness [27]. C_{ZnO}-Dots in aqueous solution were slightly less luminescent than C_{ZnS}-Dots (QY higher than 50%), with observed QY about 45%.

1.4 Surface modification

Bare CDs possess many defect sites, such as dangling bonds, nonradiative states, radicals, etc. [12] which render CDs susceptible to the external contaminants that can deteriorate the properties of CDs and resulted in poor photostability and low QY. To circumvent this issue, surface modification is suggested. Surface modification is a potent tool for tuning the properties of CDs for particular applications. Generally, surface modification of CDs can be achieved through three methods: (1) oxidation; (2) functionalization; and (3) passivation [28].

CDs prepared by bottom-up approaches are rich in functional groups such as carbonyl, carboxyl, and hydroxyl groups. These oxygen-containing functional groups can act as surface emissive traps resulting in high PL efficiency. Hence, fluorescent properties of CDs can be adjusted by altering the degree of oxidation. For example, reduction of the oxygenic functionalities can improve the QY, while the oxidation of CDs redshifts the emission [28]. These oxygenic moieties, on the other hand, can be used for further functionalization or even bio-conjugation to tune the solubility and biocompatibility of CDs. Functionalization can proceed through coordination, covalent bonding, π-π interactions, etc. Covalent bonds can be formed via electrophilic addition or substitution mechanism, while physical attraction can occur via hydrogen bonding, physical van der Waals interaction, or electrostatic forces [13].

1.5 Surface passivation

Surface passivation approach is based on introduction of a passivating layer, mostly through attachment of polymeric materials on the acid-treated surface of CDs [4]. Poly(ethylene glycol) (PEG) is the most commonly used polymer for the passivation of the surface of CDs. However, other polymers such as hyperbranched polyethylenimine (PEI) and poly(propionyl ethyleneimine-*co*-ethyleneimine) (PPEI-EI) have also been utilized for this purpose. Moreover, the utility of ionic liquids (ILs) for passivation of CDs is reported. The passivated CDs with organic moieties are strongly photoluminescent both in the solution-like suspension and in the solid state, and the emissions cover the visible wavelength range and extend into the near-infrared (Fig. 7) [4, 29]. As shown in Fig. 8, the PL spectra of the CDs are generally broad and dependent on excitation wavelengths.

The tunable emissions of the passivated CDs can be attributed to the varied fluorescence characteristics of particles of different sizes of the CDs and the distribution of different emissive sites on the surface of the CDs [4].

As surface passivation layer significantly affects the CDs' bright fluorescence emissions, the stability in the fluorescence performance of CDs depends on the structural stability of the surface passivation layer. To achieve desirable fluorescence performance, wise choice of passivation agents and the functionalization chemistry is imperative. Moreover, the cross-linking of the surface passivation layer after the synthesis of CDs could be employed

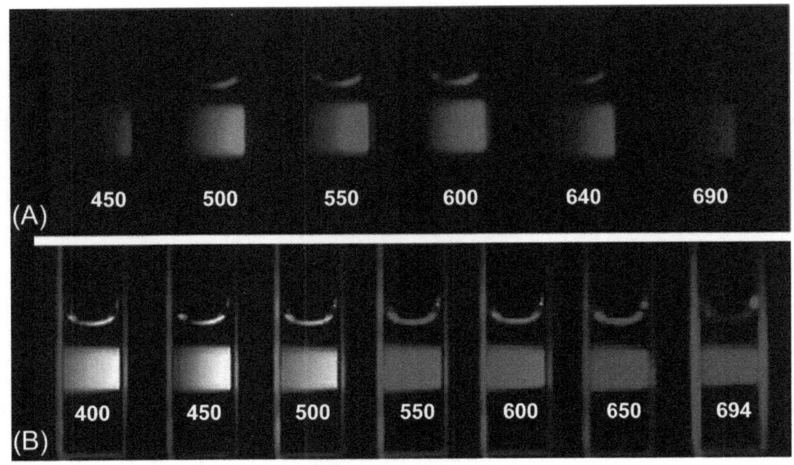

Fig. 7 Aqueous solution of the PEG1500N-attached CDs (A) excited at 400 nm and photographed through band-pass filters of different wavelengths as indicated, and (B) excited at the indicated wavelengths and photographed directly. *(Reprinted from Y.-P. Sun, B. Zhou, Y. Lin, W. Wang, K.A.S. Fernando, P. Pathak, M.J. Meziani, B.A. Harruff, X. Wang, H. Wang, P.G. Luo, H. Yang, M.E. Kose, B. Chen, L.M. Veca, S.-Y. Xie, Quantum-sized carbon dots for bright and colorful photoluminescence, J. Am. Chem. Soc. 128(24) (2006) 7756–7757, with permission from American Chemical Society.)*

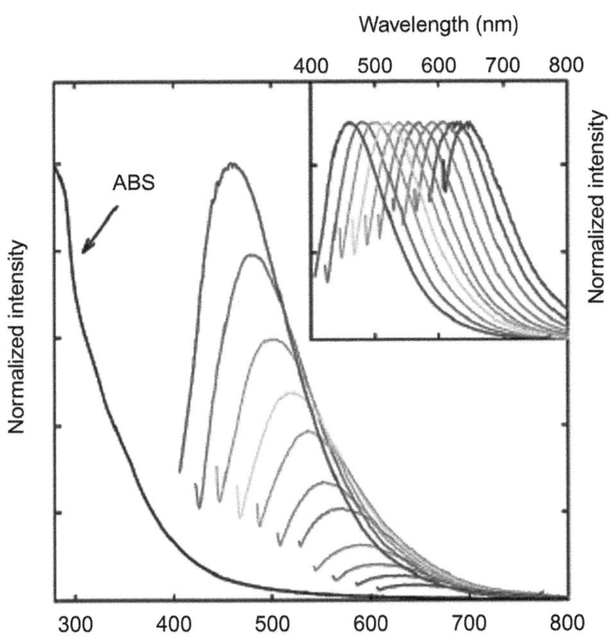

Fig. 8 The absorption and luminescence emission spectra (with progressively longer excitation wavelengths from 400 nm on the left in 20 nm increment) of PPEI-EI CDs in an aqueous solution. *(Reprinted from Y.-P. Sun, B. Zhou, Y. Lin, W. Wang, K.A.S. Fernando, P. Pathak, M.J. Meziani, B.A. Harruff, X. Wang, H. Wang, P.G. Luo, H. Yang, M.E. Kose, B. Chen, L.M. Veca, S.-Y. Xie, Quantum-sized carbon dots for bright and colorful photoluminescence, J. Am. Chem. Soc. 128(24) (2006) 7756–7757, with permission from American Chemical Society.)*

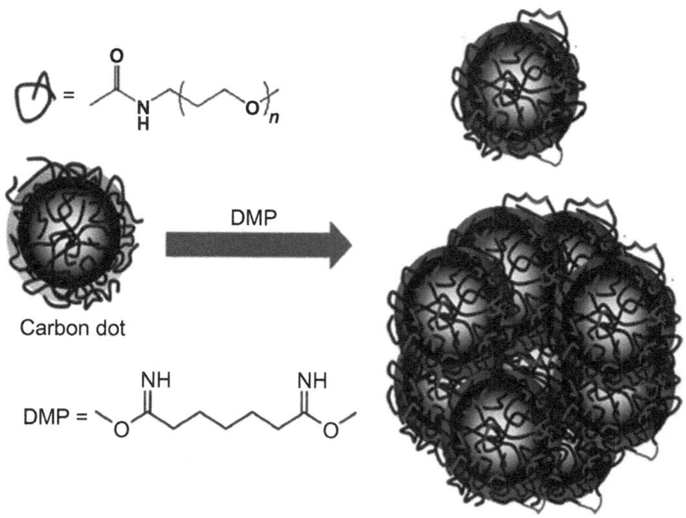

Fig. 9 PEG_{1500N}-CD and cross-linking. *(Reprinted from P. Anilkumar, L. Cao, J.-J. Yu, K.N. Tackett II, P. Wang, M.J. Meziani, Y.-P. Sun, Crosslinked carbon dots as ultra-bright fluorescence probes, Small 9(4) (2013) 545–551, with permission from John Wiley and Sons.)*

to ensure the structural stability [30]. This was reported by Sun et al., who cross-linked PEG-functionalized CD (PEG_{1500N}-CD) (Fig. 9) [30]. As shown in Fig. 9, cross-linking led to the formation of fluorescent particles composed of multiple CDs in covalently bound clusters. The authors confirmed that, for the multiple dots, the fluorescence properties were additive and up to seven CDs exist in a single particle for maximum brightness when one particle is subjected to fluorescence imaging [4].

1.6 Doping

1.6.1 Nitrogen doping

Doping is another approach to tailor the physicochemical and PL properties of CDs and obtain high photostability. Dopants can be classified into two main groups, nonmetallic (heteroatoms such as N, S, P) and metal atoms [31]. Among various nonmetallic dopants, N that has a similar atomic size to carbon atoms is the most common dopant and has been applied for enhancing the emission of CDs by inducing an upward shift in the Fermi level and electrons in the conduction band [28]. The studies confirmed that only the nitrogen bonding to the carbon can really enhance the PL emission of CDs, and the N-doped CDs show nitrogen content-dependent PL intensities with multicolor and two-photon upconversion properties. N-doped

CDs are appealing carbon materials for uses such as sensing, photocatalysis, bioimaging, biochemical bioprobes, and optoelectronic devices. They can be prepared through hydrothermal treatment of a carbon source such as citric acid along with a nitrogen source such as ethanol amine, diethylene amine, ethylene diamine, or urea [32]. The content of N as well as luminescence yield are dependent on the choice of precursors. As an example of N-doped CDs, the synthesis of water-soluble N-doped CD, NCD, through ultrasonic-assisted reaction between glucose as carbon precursor and ammonium hydroxide can be mentioned [33]. The prepared NCD showed stable and strong visible emission and excellent upconversion PL. Moreover, they were highly aqueous dispersible with no need to surface modifications. It is worth mentioning that the study of the photocatalytic activity of the NCD for dye degradation confirmed its superior activity compared to CD. In another research, Feng, Li et al. studied the effect of nitrogen on the PL properties of CD [34]. The authors prepared the NCD (mean particle size of 2.64 nm) via one-step hydrothermal treatment using biomass tar as the carbon precursor and ethylenediamine as N source. The as-prepared NCD was highly stable and water-soluble and emitted bright blue PL under 365 nm ultraviolet light (QY = 26.1%). Very recently, Zhang et al. reported preparation of highly fluorescent and super biocompatible N-doped CD (size of 4–10 nm) using aminated alkali lignin as precursor and disclosed its utility for cellular imaging [35].

1.6.2 Sulfur doping

Stronger electron donors than nitrogen, such as sulfur, have also been doped into CDs to modify their optical properties. Especially, the sulfur atom in CDs plays a role as catalyst for oxidation reduction reaction, which introduces more passivated surface defects to C-dots and further enhances their PL [31]. Similar to N-doped CDs, S-doped CDs can be prepared by hydrothermal treatment of carbon source and sulfur source such as dodecanethiol, sodium hydrosulfide, ethane-sulfonic acid, sodium thiosulfate, waste frying oil, and 2,2′-(ethylenedithio)diacetic acid and sulfuric acid [12]. As an example of S-doped CDs, Dong, Liu et al. reported S-doped GQD through hydrothermal synthesis using 1,3,6-trinitropyrene and Na_2S as precursors [36]. The prepared S-GQD had a broad adsorption in visible region and high separation ratio of photo-generated charges and exhibited high photocatalytic activity for degradation of basic fuchsin under visible light irradiation. Goswami, Pramanik et al. reported simple and scalable synthesis of S-doped CDs by pyrolysis of thiomalic acid as precursor in the presence of sulfuric acid, followed by isolation and polymerization with excess of NH_4OH. The prepared S-CDs showed wide-band gap (4.43 eV) and high QY (11.8%) without any surface passivation [37]. In another research, Kim et al. developed a versatile platform for highly efficient, colorimetric multifunctional sensors that detected different types of information simultaneously, including temperature, pH, and various kinds of metal ions by using blue emitting, thermally responsive block copolymer integrated with green-emitting GQD [38]. The schematic procedure for the synthesis of the block copolymer-integrated GQD (bcp-GQDs) is depicted in Fig. 10. The authors investigated the mechanism of the response of the S-doped CDs via time-resolved fluorescence and dynamic light scattering.

1.6.3 Phosphorus doping

Phosphorus that is larger than C atom is an n-type donor to form substitutional defects in diamond sp^3 thin films, so doping P atoms into CDs can modulate their optical and electronic properties [31]. Various precursors such as monosodium phosphate, phosphorous tribromide, phosphoric acid, and triphenylphosphine can be used for the fabrication of P-doped CDs [12]. As an example, Atabaev et al. reported hydrothermal synthesis of P-doped CD (size distribution of 2.8 ± 0.94 nm) using dextrose and disodium hydrogen phosphate as precursors. The fabricated P-CDs exhibited strong excitation-dependent fluorescence with a maximum emission at 452 nm and applied for low-concentration detection of ferric ions in water [39]. In another attempt, Das et al. prepared biocompatible amino acid-functionalized and P-doped CDs using citric acid, Na-salt of amino acids, and NaH_2PO_4. Notably, P-free CDs were blue emitting, while P-doped ones were green emitting. It was also found that P-doping enhanced the QY and fluorescence intensity [40].

1.6.4 Boron doping

Boron is also capable of serving as an active site for charge transfer. Incorporation of B can lead to the p-type carriers inside CDs, and consequently, alter their electronic structures and optical properties. B-doped CDs can be prepared through various methods such as microwave heating solvothermal method by using B sources such as boric acid and BBr_3. As an example, Feng et al. fabricated B-doped CD through one-pot solvothermal method and using BBr_3 and hydroquinone as the B and C precursors, respectively. The prepared B-CDs with no need to further functionalization were applied as fluorescence sensing system for hydrogen peroxide and glucose detection.

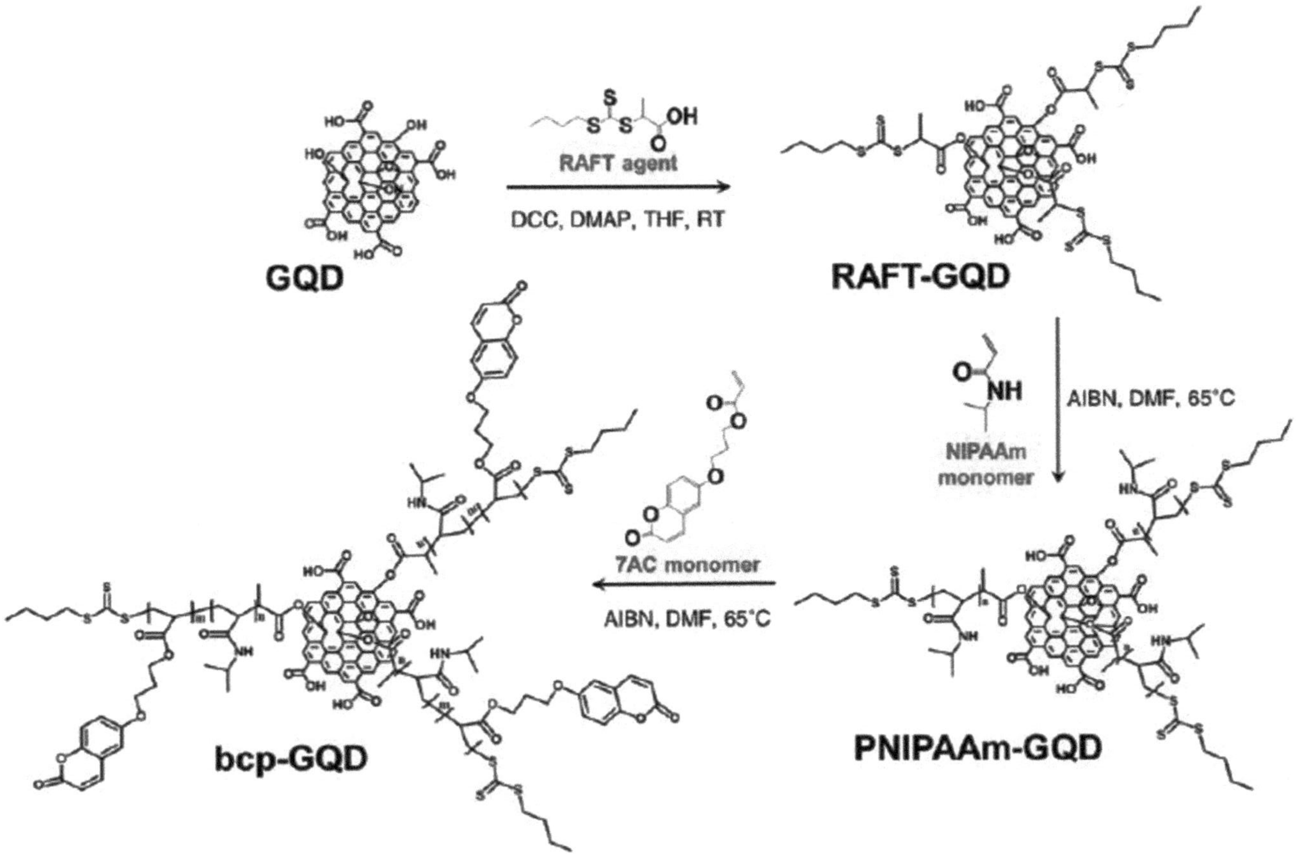

Fig. 10 Synthesis of bcp-GQDs. *(Reprinted from C.H. Park, H. Yang, J. Lee, H.-H. Cho, D. Kim, D.C. Lee, B.J. Kim, Multicolor emitting block copolymer-integrated graphene quantum dots for colorimetric, simultaneous sensing of temperature, pH, and metal ions, Chem. Mater. 27(15) (2015) 5288–5294, with permission from American Chemical Society.)*

Notably, the maximum fluorescence QY of B-CQ-D was 14.8% which was significantly higher compared to bare CD (3.4%). The authors believed that the boron enhancement of fluorescence in B-CDs was relative to CDs because boron could act as an engine for charge transfer, particularly in the excited state [41]. In another approach, biocompatible B-doped GQDs with QY of 21.1% were fabricated via one-pot microwave method by using borax as B precursor [42]. Use of microwave led to shorter reaction time and obviated use of harsh oxidizing acids. B incorporation was achieved through attacking the defects in the graphene structure and resulted in an atomic percentage of 1.44% in the resulting B-GQDs. Hydrothermal method was also applied for the preparation of B-GQDs [43]. As an example, Qui et al. first prepared B-doped graphene by using graphene sheets and boron oxide and then used it as precursor for the preparation of B-GQDs through hydrothermal treatment.

1.6.5 Halogen doping

Considering the fact that halogens were applied to tune the optical properties of graphene and CNTs, some research groups attempted to develop halogen-doped CDs [44]. As an example, Qu et al. fabricated fluorinated GQDs by combining a microwave-assisted technique with hydrothermal treatment and applying fluorinated graphene oxide as starting material [45]. In another attempt, Feng, Qian et al. reported preparation of halogenated CDs with halogen content of 5–46 wt% via solvent-thermal reaction, using carbon tetrachloride and quinol as precursors [46].

1.6.6 Co-doping

Notably, to further tune the CDs' electronic properties, doping of more than one heteroatom (co-doping) can be carried out. In this case, it is expected that synergism between the dopants can affect the electronic structure and properties of CDs and introduces more active cites. To date, various co-doped CDs such as N/S, B/N, P/N, and N/S/P co-doped CDs have been developed [47, 48]. As an instance, Yang et al. reported one-step microwave-assisted synthesis of N- and P-doped CD (QY = 44%) with the utility for dopamine sensing using citric acid, ethylenediamine, and urea phosphate as precursors [49]. The prepared CDs exhibited tremendous optical performance, notable fluorescence, excellent water solubility, and thermosensitivity. N,P-CDs were also prepared through hydrothermal treatment of diethylenetriaminepenta (methylenephosphonic acid) and *m*-phenylenediamine [50]. The authors combined the N,P-CDs with Au nanoparticles as a fluorescent sensing platform for effective determination of carbendazim.

In another attempt, N, S co-doped CD probe was fabricated by using citric acid and L-cysteine as C, N, and S sources through a simple hydrothermal route to detect methotrexate on the basis of the inner filter effect of fluorescence [51]. In this line, N- and S-doped GQD was also reported [52]. S,N-GQD that exhibited high photostability and excitation-dependent emission fluorescence was simply fabricated via hydrothermal method using citric acid and thiourea as the C, N, and S sources, respectively, and successfully applied as a PL probe for detection of ascorbic acid. Using glutathione as precursor, Li et al. prepared N,S-co-doped CDs with good water solubility, QY (17.5%), and optical properties via one-pot hydrothermal approach. The CDs showed good temperature-dependent fluorescence with a sensational linear response from 10°C to 70°C, which indicated the great potential of the N, S co-doped CDs for temperature sensing [48].

As mentioned, another class of dopants is metallic ones. Various metal atoms, including Mg, Zn, Cu, Gd, Si, Se, etc., have been applied for this purpose. These materials can be prepared through various methods such as hydrothermal/solvothermal, chemical reduction, and physical mixing methods. Metal doped shows outstanding photo-induced electron/charge transfer properties as well as UCPL. Therefore, they can be considered as promising candidate for various photochemical and electrochemical applications. This topic is more discussed in the following.

2 Utility of CDs for photocatalysis

As mentioned earlier, CDs are composed of sp^2/sp^3-hybridized carbons and exhibit excellent electron transfer/reservoir properties. In addition, due to the UCPL, CDs can transfer two or more low energy photons to a higher energy photon by sequential absorbing longer-wavelength multiphoton, and hence, the light is emitted at a wavelength shorter than the excitation wavelength. This property expands the light utilization range of CDs, improves photo-induced electron transfer, and provides the possibility to design efficient visible-light-responsive photocatalysts. Furthermore, the light harvesting ability of CDs in the nanocomposites facilitates photo-excited electron transfer phenomena across their interfaces, allowing charge stabilization separation and formation of long-lived holes on the surface. This is attributed mainly to the unique photo-induced electron transfer

(PET) characteristic of CDs [53]. Hence, CDs can play important roles as electron mediator, spectral converter, photosensitizer, and photocatalysts. In the following, these issues are discussed in more details.

2.1 Electron mediator

One of the most effective factors on the photocatalytic performance is separation of photo-generated electrons and holes in the photocatalyst. However, these charge carries can be scattered or trapped by various types of random defects and led to the increase of the chance of the electron-hole recombination [1]. Hence, to improve the photocatalytic performance, prevention from recombination of electron-hole is imperative. Considering the fact that carbon nanomaterials have significant capability of electron-storage, combination of carbons such as CDs with semiconductor photocatalysts can result in the displacement of the photo-excited electrons from semiconductor in the CDs and consequently retard recombination of electron-hole pairs (vide infra) [54].

2.2 Photosensitizer

Taking the capability of CDs for absorption of UV and visible light into account, they can also act as photosensitizers for decorating conventional wide-band gap semiconductors such as TiO_2 to render them visible light photoactive. Compared to the typical photosensitizers such as organic dyes, noble metals, CdS, and CdSe quantum dots that are either toxic or costly, CDs possess low toxicity and can be prepared in relatively large scale economically. Moreover, use of water-soluble CDs can allow photocatalytic processes to take place under environmentally benign conditions (vide infra) [1].

2.3 Spectral converter

As mentioned, one of the challenging issues in the case of conventional semiconductors is that they are photoactive under UV light. To allow these semiconductors use the full spectrum of sunlight, they were coupled with narrow-band gap semiconductors such as CdS (vide infra) [1, 54]. On the contrary of the conventional narrow-band gap semiconductors that are mostly instable and/or toxic, CDs can be exploited to design hybrid photocatalysts with capability of absorbing visible light and consequently improved the photocatalytic performance.

2.4 Reducing agent for metal salt

A common strategy to improve the photocatalytic activity is utilizing metal nanoparticles as cocatalysts. Metals such as Ag and Au that benefit from good chemical, thermal, electronic, and optical properties are used to suppress the recombination rate of electron-hole pairs. This can be attributed to the low-lying Fermi level of metals. The studies revealed that CDs can serve as environmentally benign reducing agents for reduction of metal salts to metallic nanoparticles. As an example, Zheng et al. prepared reduced CDs (r-CD) by using $NaBH_4$ and applied these as both reducing agent and stabilizer for the preparation of Au nanoparticles [55]. CDs prepared from citric acid were also reported to be utilized as reducing agent for the formation of mono and bimetallic Au, Ag, and AuAg nanoparticles [56]. The reduction of metal salts by CDs is not only eco-friendly, but also can ensure the intimate interfacial contact between metal and CDs [54]. This feature is used for the development of metal-CDs nanocomposites with photocatalytic activity [57]. As an instance, CDs were first functionalized by oligomeric poly(ethylene glycol) diamine via conventional amidation reaction and then coated with Au or Pt by simple solution-phase photolysis [58]. In the case of Au-CDs, CD can act both as reducing agent and photosensitizer. The Au-functionalized CDs were applied for the reduction of carbon dioxide in the aqueous media. Upon visible light irradiation, photo-excited CDs generated charge carriers. Au nanoparticles, on the other hand, served as electron sink to concentrate the photo-generated electrons.

2.5 Enhancing adsorption capacity

CDs can increase the adsorption capacity toward reaction substrates. This can be achieved through chemical, physical, or electrostatic interactions. The surface of CD that is rich in functional groups allows it to interact with substrates. On the other hand, the aromatic rings in the CDs provide π-π stacking interactions with proper substrates, and as a result, increase their adsorption. This is beneficial for the enrichment of substrates from the bulk solution onto the surface of photocatalyst. As a result, the adsorbed reaction substrates can effectively react with the photo-generated active species on the surface of composites and result in improvement of the photocatalytic performance [54].

2.6 Sole photocatalyst

In the above-mentioned discussion, it was pointed out that CDs can be used to improve the performance and properties of semiconductors to render them more efficient photocatalysts. However, CDs in their free form can also act as photocatalysts [59].

Despite the advantages of pure CDs such as cost-effectiveness, low-toxicity etc., they suffer from some shortcomings and there are not many reports on the sole CD with satisfying photocatalytic performance, and mostly, doped CDs are applied for this purpose. As an example of sole CDs, Ojha et al. prepared mono-dispersed GQD with well-controlled morphology via in situ hydrothermal procedure, in which graphene oxide (GO) was used as precursor and N_2H_4 as reducing agent. The authors studied the photocatalytic activity of the resulting GQD for the photodegradation of an aqueous solution of RhG dye under sunlight irradiation. The results confirmed decomposition of ~80% of the aforementioned dye after 80 min. The photocatalytic activity of GQD was assigned to the synergistic effect between GQDs and the RhG dye. In other word, the adsorbed dye on GQD surface allowed formation of charge carriers in the GQDs which was followed by charge separation in it. High surface to volume ratio of the spherical GQDs and the presence of localized defect levels could also affect the photocatalytic activity [60]. Regarding the doped CDs, N-doped CDs (size of 10 nm) with high aqueous dispersibility were fabricated through one-pot ultrasonic approach by using glucose and ammonium hydroxide as precursors [33]. Study of the photocatalytic activity of the prepared N-doped CDs confirmed their high activity for photodegradation of methyl orange (MO) that was superior to that of CDs. As an example of co-doped CDs, Li et al. prepared multifunctional N/S-co-doped CDs (3.3 ± 0.5 nm) via one-step hydrothermal treatment of bloom-forming green alga *Dunaliella salina* without using any toxic chemicals [61]. The prepared CDs with negatively charged and functionalized surface showed high photostability, low toxicity, high water solubility, excitation-dependent emission, and enough fluorescent activity. The authors disclosed the photocatalytic activity of the CDs for visible light-induced photodegradation of methylene blue (MB) and methyl violet. It was believed that the carbonyl functionalities on the surface of CDs facilitated the charge separation and transfer in N/S-CDs. Therefore, N/S-CDs generated strongly oxidizing holes on the reaction surface. In another instance, N,S-co-doped GDs with uniform size and strong PL were prepared by hydrothermal treatment of citric acid, thiourea, and urea as precursors [62].

2.7 CDs/semiconductor composite photocatalysts

Considering the above-mentioned discussion, CDs-based photocatalysts can be classified as (1) sole CDs and (2) CDs-based nanocomposites. The latter group can be subdivided into metal-CDs nanocomposites and semiconductor-CDs nanocomposites, as discussed above. Most of the reports on the CDs-based photocatalysts are focused on the nanocomposites, in which CDs are used for enhancement of the photocatalytic performance of semiconductors. The mechanism of excitation and photoactivity of semiconductor-CDs nanocomposites are dependent on band gap, valence band, and conduction band. The main cases can be categorized as: (1) CDs—wide-band gap semiconductor photocatalysts ($Eg \geq 3\,eV$); (2) CDs—narrow-band gap semiconductor photocatalysts ($Eg \leq 3\,eV$); and (3) CDs—*Z*-scheme semiconductor photocatalysts [63]. Considering the importance of this class of photocatalysts, their types as well as their applications for several photocatalytic reactions are discussed in more details in the following.

2.7.1 CDs—Wide-band gap semiconductor photocatalysts

Among this class of semiconductors, TiO_2 and ZnO are the most known and applied ones. Most of the wide-band gap semiconductors possess high photosensitivity and photochemical stability. Moreover, their nontoxicity and low costs expand their applications in various fields such as cosmetics, functional fibers, plastics, coatings, and paints. However, this class of photocatalysts faces some shortcomings that limit their widespread applications. As an example, TiO_2 can just use UV irradiation that is small fraction (~4%) of total solar spectrum. ZnO, on the other hand, has poor quantum efficiency. Although coupling with photosensitizers such as noble metals and quantum dots can furnish a solution to these problems, their toxic nature and high expense encourage more research for disclosing more effective alternatives. In CDs—wide-band gap semiconductors photocatalysts, the improved charge separation reduces charge recombination, leading to the higher photocatalytic activity. There are many reports in the literature on the synthesis and photocatalytic activity of this kind of photocatalysts, especially TiO_2-based ones. Liu et al. reported facile preparation of CDs/TiO_2 nanosheets with highly exposed (001) facets with high stability and superior photocatalytic activity for H_2 evolution compared to CDs/P25. It was believed that CDs loading on the surface of TiO_2-001 acted as electron reservoir to trap photo-generated electrons, which not only facilitated the efficient separation of charge carrier pairs, but also suppressed the recombination of photo-generated charge carriers. Furthermore, CDs acted as photosensitizer and increased visible-light absorption [64].

To prepare CQDs/TiO_2, tedious reactions and posttreatment steps are required. Typically, the as-prepared GDs are combined with semiconductors, where an additional hydrothermal treatment step is typically needed to achieve strong surface integrity between the two components, which cannot be accessible in one-step synthesis process [65]. In recent years, however, facile procedures for the preparation of these nanocomposites have been developed [66]. As an example, Min et al. reported one-pot hydrothermal method for the fabrication of TiO_2-GQD photocatalyst [66]. In the course of hydrothermal treatment, 1,3,6-trinitropyrene that was applied as precursor tolerated an intramolecular fusion to form GQDs, which was simultaneously decorated on the surface of TiO_2 nanoparticles, leading to a strong surface interaction between the two components. This effective coupling between two components effectively extended the light absorption of the TiO_2 to visible region and enhanced the charge separation efficiency of the composites as a result of GQD that acted as a photosensitizer and an excellent electron acceptor [66]. In another attempt, Yang, Jin et al. fabricated CDs-decorated nitrogen (N) and Ti^{3+}/O_v co-doped TiO_2 nanocomposite (CDs-N-TiO_{2-x}) through facile two-step hydrothermal and calcination method and demonstrated its high photocatalytic activity for reduction of Cr(VI) [67]. The CDs' roles were enhancement of light adsorption and charge separation.

In another study, sol-gel and ultrasonic-hydrothermal were employed for the fabrication of CDs-embedded mesoporous TiO_2 composites [68]. It was found that the photocatalytic activity of the composite for photodegradation of MB was superior to that of the commercial sample, Degussa P25, as well as CQDs/P25, CQDs/meso-Ti-450 composites, and pristine mesoporous TiO_2. This composite was also efficient for the degradation of *N*-benzylideneaniline. High photocatalytic activity of the prepared nanocatalyst was attributed to the synergistic effects of the optical properties of CDs and mesoporosity of TiO_2. In more detail, the mesoporous structure of TiO_2 provided more active sites for efficient adsorption of target organic molecules. Meanwhile, the upconversion property and electron acceptor property of CDs facilitated the usage of visible light and hindered the charge carrier recombination.

Apart from TiO_2, other wide-band gap semiconductors such as ZnO were modified by CDs. As an example, ZnO/CDs' heterostructure with enhanced photocatalytic activity for degradation of Rhodamine B (RhB) compared to ZnO was prepared via a sol-gel approach combined with a spin-coating process [69]. It was believed that the interaction between the components could decrease the charge carrier recombination, and consequently, increased the photocatalytic activity. Interestingly, the observed enhancement was dependent on the layers of CDs on ZnO and coating with four layers resulted in the most efficient photocatalyst.

In another study, Krishnan et al. reported synthesis of a heterojunction, ZnO-GQD, active in both UV and visible light region based on decoration of ZnO nanorods with GQDs [70]. The authors confirmed the utility of ZnO-GQD for photodegradation of MB and carbendazim (CZ) fungicide. Noteworthy, the photocatalytic activity of the heterojunction was superior to that of each component. The increase of the activity of ZnO-GQD was not only due to the separation of the charge carrier as well as enhanced light absorption, but also its high specific area (353.447 m^2 g^{-1}) that improved the adsorption of the substrates. The proposed mechanism of photocatalytic degradation of pollutants over ZnO-GQD heterojunctions under sunlight irradiation is illustrated in Fig. 11.

ZnO/CDs were also applied for the degradation of toxic gases. As an example, Kang, Liu et al. employed one-step hydrothermal process for the preparation of ZnO/CDs with utility for the photodegradation of MeOH and benzene at ambient temperature and visible light irradiation [71]. The high photocatalytic activity of the nanocomposite that was

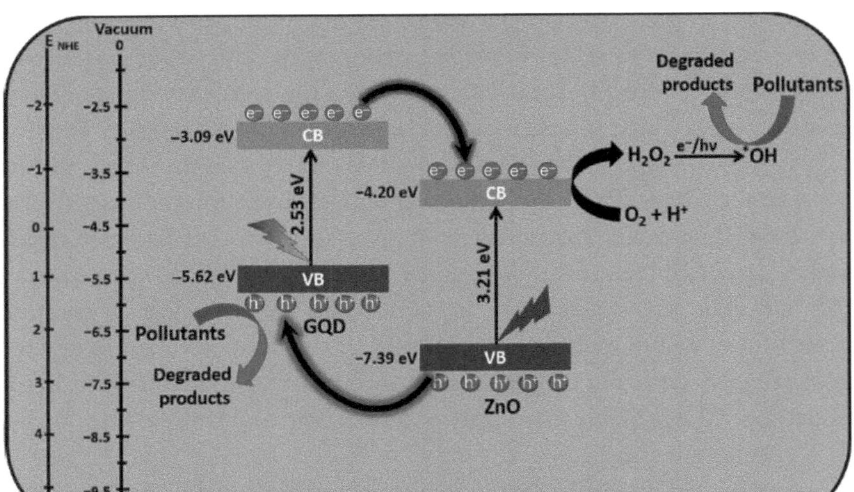

Fig. 11 Schematic illustration of energy band diagram and a proposed mechanism of the charge carrier transitions in ZnO-GQD heterojunction toward photocatalytic pollutant degradation under natural sunlight irradiation. *(Reprinted from S. Kumar, A. Dhiman, P. Sudhagar, V. Krishnan, ZnO-graphene quantum dots heterojunctions for natural sunlight-driven photocatalytic environmental remediation, Appl. Surf. Sci. 447 (2018) 802–815, with permission from Elsevier.)*

superior to that of ZnO and N-doped TiO_2 was assigned to three factors. First, the π-π interaction between aromatic substrate and conjugated structure of CD increased the pollutant adsorption of the photocatalyst surface. Secondly, CDs on the surface of ZnO formed the "dyade" structure providing access to photo-induced charge transfer transitions under visible light irradiation. At the dyade structure, the photo-induced electron was transferred to joint charge transfer states predominately located at the CDs, while the left hole stayed electronically and structurally near ZnO. This process effectively prevented the electron-hole pairs' recombination and guaranteed the high reactivity of photo-generated electron and hole excited by visible light. Thirdly, the CDs with upconversion PL properties could convert long wavelength light to the short wavelength light, which can in turn excite ZnO to form electron-hole pairs [71].

2.7.2 CDs—Narrow-band gap semiconductor photocatalysts

The shortcomings of the narrow-band gap semiconductors such as CdS, Bi_2MoO_6, Bi_2WO_6, Fe_3O_4, and Cu_2O are low light adsorption efficiency, restricted migration, and high recombination of electron and hole pairs [63]. These issues can be circumvented through using CDs [72, 73]. In this section, some advances in this regard are discussed in more detail.

Min et al. reported preparation of CdS/CDs nanocomposite with utility for photocatalytic hydrogen evolution via hydrothermal procedure, in which the crystallization of CdS precursor and coupling of CDs could be accomplished in one-step [74]. This method led to the decoration of CdS with CD with "dot-on-particle" structure. The authors claimed that although CDs did not affect the crystallite structure of CdS significantly, it could improve light adsorption at the wavelength beyond the band edge of CdS. It was found that under the visible light irradiation ($\geq$420 nm), the photocatalytic activity and long-term stability of the hybrid system were superior compared to CdS. Notably, the CD mainly acted as electron acceptor instead of a photosensitizer. The plausible photocatalytic mechanism for enhanced photocatalytic H_2 evolution activity on CdS/GQDs nanohybrids is depicted in Fig. 12.

In another report in this line, Zhang et al. prepared uniform flower-shaped microspheres of CDs-doped CdS, C/CdS, by using glutathione as a template via hydrothermal method [75]. The prepared nanocomposite exhibited high photocatalytic activity for photodegradation of RhB under the solar-simulated light irradiation, superior to that of CdS. The enhanced catalytic activity of the nanocomposite was assigned to the electrons trapped

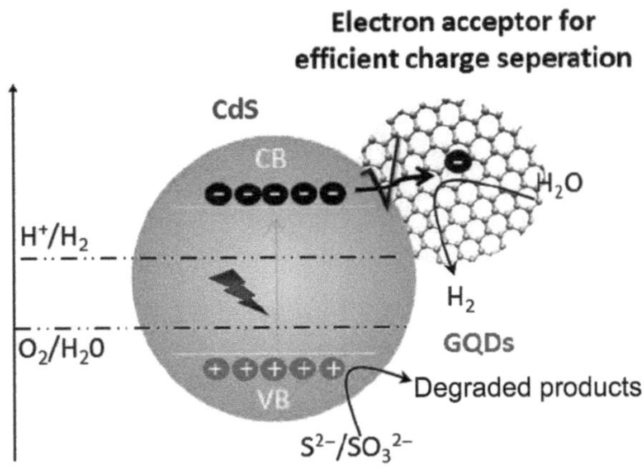

Fig. 12 The plausible photocatalytic mechanism for enhanced photocatalytic H_2 evolution activity on CdS/GQDs nanohybrids. *(Reprinted from Y. Lei, C. Yang, J. Hou, F. Wang, S. Min, X. Ma, Z. Jin, J. Xu, G. Lu, K.-W. Huang, Strongly coupled CdS/graphene quantum dots nanohybrids for highly efficient photocatalytic hydrogen evolution: unraveling the essential roles of graphene quantum dots, Appl. Catal. B Environ. 216 (2017) 59–69, with permission from Elsevier.)*

by the CDs and the hindrance to the recombination of photo-generated electron and hole pairs.

Zhao, Xu, and Xie fabricated (CDs)/Bi_2MoO_6 nanocomposite via one-pot hydrothermal method, in which CD was formed and grafted to Bi_2MoO_6 via hydrothermal treatment of glucose. The study of the photocatalytic activity of the nanocomposite for the degradation of RhB and MB under visible light irradiation confirmed high stability and photocatalytic activity of (CDs)/Bi_2MoO_6. The high photocatalytic activity emerged from formation of efficient junction interface between CDs and Bi_2MoO_6 which prevented the recombination of charge carriers [76].

Photodegradation of pollutants such as MB, RhB, BPA, tetracycline hydrochloride (TC), and CIP under visible light irradiation was also reported by using CD/Bi_2WO_6 nanocomposite [77]. The nanocomposite, in which CD was located on the surface of sphere-like Bi_2WO_6, was prepared through hydrothermal method. The nanocomposite showed higher photocatalytic activity compared to bare Bi_2WO_6. The enhanced activity was attributed to the interfacial transfer of photo-generated electrons from Bi_2WO_6 to CDs, leading to effective charge separation of Bi_2WO_6.

In another study, Zhou et al. reported one-pot solvothermal synthesis of water-dispersible flower-like Fe_3O_4@CDs as stable, efficient, and magnetically separable photocatalysts for MB degradation under visible light irradiation [78]. It is worth noting that the catalytic activity of the magnetic nanocomposite was higher than some previously reported samples such as Fe_3O_4@TiO_2 nanosheets, CNT-confined TiO_2, and P25–grapheme. The observed high catalytic activity was attributed to the strong absorption in the visible light range and upconversion behavior of CDs and production of more charge carriers.

In another research, Kang, Liu et al. reported the utility of Fe_2O_3/CDs nanocomposite for visible-light photodegradation of MeOH and benzene [79]. The authors believed that CD could act as an electron reservoir to trap electrons, thus hindering electron-hole pairs' recombination. Moreover, CDs could absorb longer-wavelength light and then emitted shorter wavelength light as a result of their upconversion PL property, which could in turn excite Fe_2O_3 to form electron-hole pairs. Furthermore, in the case of aromatic substrate, the π-π interactions between the substrate and CD led to the substrate enrichment on the surface of the nanocomposite.

This research group also reported ultrasonic-assisted preparation of CD/Cu_2O photocatalyst with protruding nanostructures on the surface [80]. The photocatalytic activity of the nanocomposite was studied for MB degradation under (N)IR light irradiation. The high photocatalytic activity was justified based on the combined effect of superior light reflecting ability of the Cu_2O protruding nanostructures and the UCPL property of CDs. Noteworthy, the content of CDs played an important role in the observed photocatalytic activity and the optimum amount of CD for achieving the best photocatalytic activity was 7.16 wt%. Interestingly, increase of this amount had a converse effect on the catalytic activity. This observation was due to the fact that surplus CDs could block the electron-hole pairs from reacting with the adsorbed oxidants/reducers to produce active oxygen radicals.

2.7.3 CDs-modified heterojunction photocatalysts

Liu et al. reported fabrication of *Z*-scheme CdS/CDs/BiOCl heterojunction, in which CD was sandwiched between CdS and BiOCl, through region-selective deposition process [81]. The authors studied the photocatalytic activity of the heterojunction for photodegradation of RhB and phenol under visible and UV light, respectively. The results confirmed high photocatalytic activity of CdS/CDs/BiOCl which was superior to that of BiOCl, CdS/BiOCl, and CDs/BiOCl. It was believed that increase of absorption of light as well as migration efficiency of the photo-generated charge carrier and maintaining the high redox ability justify the high photocatalytic activity.

In another research, Li, Zhao, and their coworkers modified BiOCl/BiOBr with CQDs via in situ ionic liquid-induced method [82]. CQDs/BiOCl/BiOBr composite with 5 wt% CD loading was then applied as a photocatalyst for photodegradation of RhB, tetracycline hydrochloride (TC), ciprofloxacin (CIP), and bisphenol A (BPA) under the irradiation of visible light. It was claimed that incorporation of CD could enhance the absorption of light as well as separation of the charge carries. The electron spin resonance and radical quenching experiments showed that holes and O_2^- were the main active species involved in the photocatalytic reactions. The proposed photodegradation mechanism is depicted in Fig. 13.

Li, Cui et al. reported synthesis and photocatalytic activity of Ag/CQDs/$Bi_2O_2CO_3$ for removal of MB and BPA [83]. The nanocomposite could degrade MB and BPA by up to 93.85% and 54.88%, respectively, within 90 min under visible light. Furthermore, it exhibited a photocatalytic activity of 73.75% and 99.25% for MB under UV light and simulated sunlight, respectively. The observed high catalytic activity of the photocatalyst was attributed to the improved spectral absorption range and conversion efficiency as well as rapid charge separation and transfer efficiency.

In another study, Li, Zhang et al. fabricated Bi/BiOCl/TiO_2-CD photocatalyst with lamellar structure via a solvothermal method followed by a hydrothermal process [84] and applied it as an efficient photocatalyst for visible light degradation of methyl orange (MO) and *p*-nitrophenol (PNP). The comparison of the photocatalytic activity of the nanocomposite

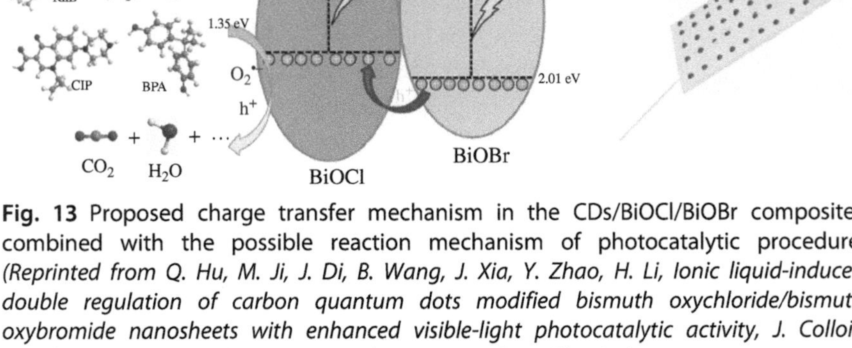

Fig. 13 Proposed charge transfer mechanism in the CDs/BiOCl/BiOBr composites combined with the possible reaction mechanism of photocatalytic procedure. *(Reprinted from Q. Hu, M. Ji, J. Di, B. Wang, J. Xia, Y. Zhao, H. Li, Ionic liquid-induced double regulation of carbon quantum dots modified bismuth oxychloride/bismuth oxybromide nanosheets with enhanced visible-light photocatalytic activity, J. Colloid Interface Sci. 519 (2018) 263–272, with permission from Elsevier.)*

with that of $BiOCl/TiO_2$ confirmed the superior catalytic activity of the former. It was believed that Bi and CDs were served as electron donors, while BiOCl and TiO_2 were acted as electron trappers to extend the lifetime of photo-generated carriers. The enhanced visible light photoactivity was assigned to the strong light absorption and upconversion PL of the CDs, and the surface plasmon resonance effect of the surface Bi also played an important role in this system (Fig. 14).

In another instance, Yan et al. utilized solvothermal-precipitation method for the fabrication of N-doped $CD/Ag_3PO_4/BiVO_4$ *Z*-scheme photocatalyst [85]. The study of the photocatalytic activity of the nanocomposite for the degradation of tetracycline (TC) under visible light irradiation confirmed high photocatalytic activity of $BiVO_4$/N-CQDs/Ag_3PO_4 (88.9% removal efficiency of TC (10 mg/L) in 30 min and 59.8% mineralization in 90 min). The authors attributed the observed catalytic activity to the following parameters: (1) the molecular oxygen activation ability of N-CQDs; (2) enhancement of visible-light absorption by coating $BiVO_4$

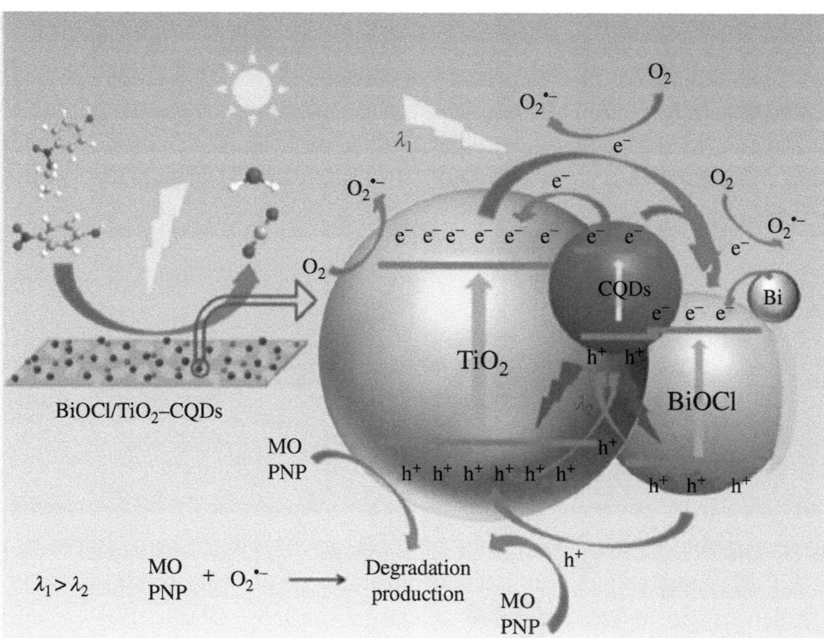

Fig. 14 Schematic illustration of the proposed photocatalytic mechanism of the Bi/$BiOCl/TiO_2$-CD photocatalyst under visible light irradiation. *(Reprinted from W. Li, C. Zhao, Q. Zhang, Synthesis of Bi/BiOCl-TiO_2-CQDs quaternary photocatalyst with enhanced visible-light photoactivity and fast charge migration, Catal. Commun. 107 (2018) 74–77, with permission from Elsevier.)*

with N-CD and Ag_3PO_4; and (3) improved separation efficiency of electron-hole pairs and maintenance of the strong redox ability of these photo-generated charges.

N-doped CD was also applied for the modification of dibismuth tetraoxide microrods to furnish a photocatalyst, m-Bi_2O_4/NCDs, for photodegradation of MO and phenol under visible-light illumination [86]. The photocatalytic activity of the nanocomposite was superior to that of m-Bi_2O_4, confirming the contribution of N-doped CDs to the photocatalysis. In fact, the formation of Z-scheme heterojunctions between m-Bi_2O_4 and NCDs improved the charge carrier separation and enhances the molecular oxygen activation ability. Moreover, enhancement of light harvesting resulted in higher catalytic activity. Notably, m-Bi_2O_4/NCDs preserved its activity upon four photodegradation cycles.

Wan et al. embedded CDs chemically onto the exterior surface of TiO_2 nanotubes (TNT) via modified hydrothermal procedure and used the resulting photocatalyst CD/TNT for the photodegradation of MB under visible light irradiation [87]. The high photocatalytic activity that was superior to bare TNT was attributed to several parameters. First, the π-π bonds between CD and MB increased its absorption of the photocatalyst. Additionally, CDs with excellent upconversion PL properties as cocatalysts converted long wavelength light (>600 nm) absorbed from infrared light into visible light (<600 nm), which activated the TNTs to generated e^-/h^+ pairs. Furthermore, the CDs also could capture the photo-excited electrons and prolong the lifetime of the photo-excited charges.

3 Conclusion

The unique physiochemical and optical features of CDs led to the increasing research on this new class of carbon nanomaterials in diverse fields of studies, such as delivery systems, photovoltaics, biosensors, bioimaging, and (photo) catalysis. To date, many valuable review articles and book chapters have been published, covering general or specific aspects of CDs. Considering the remarkable optical properties of CDs, such as high light harvesting, PL, UCPL, tunable QY, etc., and importance of photocatalysis for photodegradation and environmental remediation, the scope of this chapter is limited to the use of CDs for photocatalysis. Considering the discussed issues, it can be concluded that the properties of CDs can be tuned by varying the degree of oxidation, surface passivation, doping, and using different starting materials and preparation methods. CDs can play important roles as electron mediator, spectral converter, photosensitizer, and photocatalysts. Moreover,

CDs can be used in combination with semiconductors to develop efficient photocatalysts. To date, various CDs—wide-band gap semiconductor photocatalysts, CDs—narrow-band gap semiconductor photocatalysts, and CDs—*Z*-scheme semiconductor photocatalysts have been reported that were successfully applied for photo-induced reactions such as photodegradation, hydrogen evolution, etc. High surface area of CD can improve the adsorption of the substrates. On the other hand, the roles of CD in enhancing light harvesting and preventing the recombination of charge carriers can justify the superior activity of the CD-based photocatalysts.

Considering the promising properties of CDs as well as the possibility of using biomass precursors for their (large-scale) synthesis, it is expected that use of CDs not only in photocatalysis but also in various applications will witness growing advances.

Acknowledgment

The author appreciates the support of Iran Polymer and Petrochemical Institute.

References

[1] R. Wang, K.-Q. Lu, Z.-R. Tang, Y.-J. Xu, Recent progress in carbon quantum dots: synthesis, properties and applications in photocatalysis, J. Mater. Chem. A 5 (8) (2017) 3717–3734.

[2] R. Das, R. Bandyopadhyay, P. Pramanik, Carbon quantum dots from natural resource: a review, Mater. Today Chem. 8 (2018) 96–109.

[3] C. Hu, M. Li, J. Qiu, Y.-P. Sun, Design and fabrication of carbon dots for energy conversion and storage, Chem. Soc. Rev. 48 (8) (2019) 2315–2337.

[4] S.Y. Lim, W. Shen, Z. Gao, Carbon quantum dots and their applications, Chem. Soc. Rev. 44 (1) (2015) 362–381.

[5] A. Sciortino, A. Cannizzo, F. Messina, Carbon nanodots: a review—from the current understanding of the fundamental photophysics to the full control of the optical response, C J. Carbon Res. 4 (2018) 67.

[6] M. Kaur, M. Kaur, V.K. Sharma, Nitrogen-doped graphene and graphene quantum dots: a review onsynthesis and applications in energy, sensors and environment, Adv. Colloid Interf. Sci. 259 (2018) 44–64.

[7] Y. Wang, A. Hu, Carbon quantum dots: synthesis, properties and applications, J. Mater. Chem. C 2 (34) (2014) 6921–6939.

[8] K. Hola, Y. Zhang, Y. Wang, E.P. Giannelis, R. Zboril, A.L. Rogach, Carbon dots—emerging light emitters for bioimaging, cancer therapy and optoelectronics, Nano Today 9 (5) (2014) 590–603.

[9] S. Sagbas, N. Sahiner, 22—Carbon dots: preparation, properties, and application, in: A. - Khan, M. Jawaid, A.M.A. Inamuddin (Eds.), Nanocarbon and Its Composites, Woodhead Publishing, 2019, pp. 651–676.

[10] M.L. Liu, B.B. Chen, C.M. Li, C.Z. Huang, Carbon dots: synthesis, formation mechanism, fluorescence origin and sensing applications, Green Chem. 21 (3) (2019) 449–471.

[11] J. Zhang, S.-H. Yu, Carbon dots: large-scale synthesis, sensing and bioimaging, Mater. Today 19 (7) (2016) 382–393.
[12] Y. Park, J. Yoo, B. Lim, W. Kwon, S.W. Rhee, Improving the functionality of carbon nanodots: doping and surface functionalization, J. Mater. Chem. A 4 (30) (2016) 11582–11603.
[13] J.F.Y. Fong, Y.H. Ng, S.M. Ng, Chapter 7—Carbon dots as a new class of light emitters for biomedical diagnostics and therapeutic applications, in: A.M. Grumezescu (Ed.), Fullerens, Graphenes and Nanotubes, William Andrew Publishing, 2018, pp. 227–295.
[14] Z. Kang, Y. Liu, Catalytic applications of carbon dots, in: N. Yang, X. Jiang, D.-W. Pang (Eds.), Carbon Nanoparticles and Nanostructures, Springer International Publishing, Cham, 2016, pp. 257–298.
[15] J. Hong, M. Kim, C. Cha, 17—Multimodal carbon dots as biosensors, in: W. Cui, X. Zhao (Eds.), Theranostic Bionanomaterials, Elsevier, 2019, pp. 377–400.
[16] Y. Dong, N. Zhou, X. Lin, J. Lin, Y. Chi, G. Chen, Extraction of electrochemiluminescent oxidized carbon quantum dots from activated carbon, Chem. Mater. 22 (21) (2010) 5895–5899.
[17] S. Kim, S.W. Hwang, M.-K. Kim, D.Y. Shin, D.H. Shin, C.O. Kim, S.B. Yang, J. H. Park, E. Hwang, S.-H. Choi, G. Ko, S. Sim, C. Sone, H.J. Choi, S. Bae, B. H. Hong, Anomalous behaviors of visible luminescence from graphene quantum dots: interplay between size and shape, ACS Nano 6 (9) (2012) 8203–8208.
[18] S.M. Ng, 12—Carbon dots as optical nanoprobes for biosensors, in: S.C.B. Gopinath, T. Lakshmipriya (Eds.), Nanobiosensors for Biomolecular Targeting, Elsevier, 2019, pp. 269–300.
[19] H. Ding, S.-B. Yu, J.-S. Wei, H.-M. Xiong, Full-color light-emitting carbon dots with a surface-state-controlled luminescence mechanism, ACS Nano 10 (1) (2016) 484–491.
[20] J. Wei, J. Qiu, Unveil the fluorescence of carbon quantum dots, Adv. Eng. Mater. 17 (2) (2015) 138–142.
[21] Z. Liu, H. Zou, N. Wang, T. Yang, Z. Peng, J. Wang, N. Li, C. Huang, Photoluminescence of carbon quantum dots: coarsely adjusted by quantum confinement effects and finely by surface trap states, Sci. China Chem. 61 (4) (2018) 490–496.
[22] J.B. Essner, J.A. Kist, L. Polo-Parada, G.A. Baker, Artifacts and errors associated with the ubiquitous presence of fluorescent impurities in carbon nanodots, Chem. Mater. 30 (6) (2018) 1878–1887.
[23] J. Schneider, C.J. Reckmeier, Y. Xiong, M. von Seckendorff, A.S. Susha, P. Kasák, A. L. Rogach, Molecular fluorescence in citric acid-based carbon dots, J. Phys. Chem. C 121 (3) (2017) 2014–2022.
[24] S. Zhuo, M. Shao, S.-T. Lee, Upconversion and downconversion fluorescent graphene quantum dots: ultrasonic preparation and photocatalysis, ACS Nano 6 (2) (2012) 1059–1064.
[25] X. Jia, J. Li, E. Wang, One-pot green synthesis of optically pH-sensitive carbon dots with upconversion luminescence, Nanoscale 4 (18) (2012) 5572–5575.
[26] X. Wen, P. Yu, Y.-R. Toh, X. Ma, J. Tang, On the upconversion fluorescence in carbon nanodots and graphene quantum dots, Chem. Commun. 50 (36) (2014) 4703–4706.
[27] Y.-P. Sun, X. Wang, F. Lu, L. Cao, M.J. Meziani, P.G. Luo, L. Gu, L.M. Veca, Doped carbon nanoparticles as a new platform for highly photoluminescent dots, J. Phys. Chem. C 112 (47) (2008) 18295–18298.
[28] B. De, Chapter 7—Carbon dots and their polymeric nanocomposites, in: N. Karak (Ed.), Nanomaterials and Polymer Nanocomposites, Elsevier, 2019, pp. 217–260.
[29] Y.-P. Sun, B. Zhou, Y. Lin, W. Wang, K.A.S. Fernando, P. Pathak, M.J. Meziani, B. A. Harruff, X. Wang, H. Wang, P.G. Luo, H. Yang, M.E. Kose, B. Chen, L.M. Veca, S.-Y. Xie, Quantum-sized carbon dots for bright and colorful photoluminescence, J. Am. Chem. Soc. 128 (24) (2006) 7756–7757.

[30] P. Anilkumar, L. Cao, J.-J. Yu, K.N. Tackett II, P. Wang, M.J. Meziani, Y.-P. Sun, Crosslinked carbon dots as ultra-bright fluorescence probes, Small 9 (4) (2013) 545–551.

[31] C. Yi, Y. Pan, Y. Fang, Chapter 5—Surface engineering of carbon nanodots (C-dots) for biomedical applications, in: X. Wang, X. Chen (Eds.), Novel Nanomaterials for Biomedical, Environmental and Energy Applications, Elsevier, 2019, pp. 137–188.

[32] A. Abbas, L.T. Mariana, A.N. Phan, Biomass-waste derived graphene quantum dots and their applications, Carbon 140 (2018) 77–99.

[33] Z. Ma, H. Ming, H. Huang, Y. Liu, Z. Kang, One-step ultrasonic synthesis of fluorescent N-doped carbon dots from glucose and their visible-light sensitive photocatalytic ability, New J. Chem. 36 (4) (2012) 861–864.

[34] X. Deng, Y. Feng, H. Li, Z. Du, Q. Teng, H. Wang, N-doped carbon quantum dots as fluorescent probes for highly selective and sensitive detection of Fe^{3+} ions, Particuology 41 (2018) 94–100.

[35] Y. Shi, X. Liu, M. Wang, J. Huang, X. Jiang, J. Pang, F. Xu, X. Zhang, Synthesis of N-doped carbon quantum dots from bio-waste lignin for selective irons detection and cellular imaging, Int. J. Biol. Macromol. 128 (2019) 537–545.

[36] B. Huang, J. He, S. Bian, C. Zhou, Z. Li, F. Xi, J. Liu, X. Dong, S-doped graphene quantum dots as nanophotocatalyst for visible light degradation, Chin. Chem. Lett. 29 (11) (2018) 1698–1701.

[37] S. Chandra, P. Patra, S.H. Pathan, S. Roy, S. Mitra, A. Layek, R. Bhar, P. Pramanik, A. Goswami, Luminescent S-doped carbon dots: an emergent architecture for multimodal applications, J. Mater. Chem. B 1 (18) (2013) 2375–2382.

[38] C.H. Park, H. Yang, J. Lee, H.-H. Cho, D. Kim, D.C. Lee, B.J. Kim, Multicolor emitting block copolymer-integrated graphene quantum dots for colorimetric, simultaneous sensing of temperature, pH, and metal ions, Chem. Mater. 27 (15) (2015) 5288–5294.

[39] A. Molkenova, T.S. Atabaev, Phosphorus-doped carbon dots (P-CDs) from dextrose for low-concentration ferric ions sensing in water, Optik 187 (2019) 70–73.

[40] S. Sarkar, K. Das, M. Ghosh, P.K. Das, Amino acid functionalized blue and phosphorous-doped green fluorescent carbon dots as bioimaging probe, RSC Adv. 5 (81) (2015) 65913–65921.

[41] X. Shan, L. Chai, J. Ma, Z. Qian, J. Chen, H. Feng, B-doped carbon quantum dots as a sensitive fluorescence probe for hydrogen peroxide and glucose detection, Analyst 139 (10) (2014) 2322–2325.

[42] X. Hai, Q.-X. Mao, W.-J. Wang, X.-F. Wang, X.-W. Chen, J.-H. Wang, An acid-free microwave approach to prepare highly luminescent boron-doped graphene quantum dots for cell imaging, J. Mater. Chem. B 3 (47) (2015) 9109–9114.

[43] L. Zhang, Z.-Y. Zhang, R.-P. Liang, Y.-H. Li, J.-D. Qiu, Boron-doped graphene quantum dots for selective glucose sensing based on the "abnormal" aggregation-induced photoluminescence enhancement, Anal. Chem. 86 (9) (2014) 4423–4430.

[44] J. Zhou, H. Zhou, J. Tang, S. Deng, F. Yan, W. Li, M. Qu, Carbon dots doped with heteroatoms for fluorescent bioimaging: a review, Microchim. Acta 184 (2) (2017) 343–368.

[45] H. Sun, H. Ji, E. Ju, Y. Guan, J. Ren, X. Qu, Synthesis of fluorinated and nonfluorinated graphene quantum dots through a new top-down strategy for long-time cellular imaging, Chem. Eur. J. 21 (9) (2015) 3791–3797.

[46] J. Zhou, P. Lin, J. Ma, X. Shan, H. Feng, C. Chen, J. Chen, Z. Qian, Facile synthesis of halogenated carbon quantum dots as an important intermediate for surface modification, RSC Adv. 3 (25) (2013) 9625–9628.

[47] J.-m. Wei, B.-t. Liu, X. Zhang, C.-c. Song, One-pot synthesis of N, S co-doped photoluminescent carbon quantum dots for Hg^{2+} ion detection, New Carbon Mater. 33 (4) (2018) 333–340.

[48] C. Zhao, Y. Jiao, Z. Gao, Y. Yang, H. Li, N, S co-doped carbon dots for temperature probe and the detection of tetracycline based on the inner filter effect, J. Photochem. Photobiol. A Chem. 367 (2018) 137–144.
[49] S.K. Tammina, D. Yang, S. Koppala, C. Cheng, Y. Yang, Highly photoluminescent N, P doped carbon quantum dots as a fluorescent sensor for the detection of dopamine and temperature, J. Photochem. Photobiol. B 194 (2019) 61–70.
[50] Y. Yang, D. Huo, H. Wu, X. Wang, J. Yang, M. Bian, Y. Ma, C. Hou, N, P-doped carbon quantum dots as a fluorescent sensing platform for carbendazim detection based on fluorescence resonance energy transfer, Sensors Actuators B Chem. 274 (2018) 296–303.
[51] Y. Zhao, S. Zou, D. Huo, C. Hou, M. Yang, J. Li, M. Bian, Simple and sensitive fluorescence sensor for methotrexate detection based on the inner filter effect of N, S co-doped carbon quantum dots, Anal. Chim. Acta 1047 (2019) 179–187.
[52] H. Safardoust-Hojaghan, O. Amiri, M. Hassanpour, M. Panahi-Kalamuei, H. Moayedi, M. Salavati-Niasari, S,N co-doped graphene quantum dots-induced ascorbic acid fluorescent sensor: design, characterization and performance, Food Chem. 295 (2019) 530–536.
[53] M. Pirsaheb, A. Asadi, M. Sillanpää, N. Farhadian, Application of carbon quantum dots to increase the activity of conventional photocatalysts: a systematic review, J. Mol. Liq. 271 (2018) 857–871.
[54] K.-Q. Lu, Q. Quan, N. Zhang, Y.-J. Xu, Multifarious roles of carbon quantum dots in heterogeneous photocatalysis, J. Energy Chem. 25 (6) (2016) 927–935.
[55] X. Wang, Y. Long, Q. Wang, H. Zhang, X. Huang, R. Zhu, P. Teng, L. Liang, H. Zheng, Reduced state carbon dots as both reductant and stabilizer for the synthesis of gold nanoparticles, Carbon 64 (2013) 499–506.
[56] J.-C. Jin, Z.-Q. Xu, H.-F. Zou, Z.-Q. Zhou, Q.-Q. Yang, B.-B. Wang, F.-L. Jiang, Y. Liu, Carbon dots reduced and stabilized silver nanoclusters: synthesis and formation mechanisms, RSC Adv. 6 (80) (2016) 76989–76995.
[57] S. Sahu, Y. Liu, P. Wang, C.E. Bunker, K.A.S. Fernando, W.K. Lewis, E.A. Guliants, F. Yang, J. Wang, Y.-P. Sun, Visible-light photoconversion of carbon dioxide into organic acids in an aqueous solution of carbon dots, Langmuir 30 (28) (2014) 8631–8636.
[58] L. Cao, S. Sahu, P. Anilkumar, C.E. Bunker, J. Xu, K.A.S. Fernando, P. Wang, E. A. Guliants, K.N. Tackett, Y.-P. Sun, Carbon nanoparticles as visible-light photocatalysts for efficient CO_2 conversion and beyond, J. Am. Chem. Soc. 133 (13) (2011) 4754–4757.
[59] M. Han, S. Zhu, S. Lu, Y. Song, T. Feng, S. Tao, J. Liu, B. Yang, Recent progress on the photocatalysis of carbon dots: classification, mechanism and applications, Nano Today 19 (2018) 201–218.
[60] B. Ahmed, S. Kumar, A.K. Ojha, F. Hirsch, S. Riese, I. Fischer, Facile synthesis and photophysics of graphene quantum dots, J. Photochem. Photobiol. A Chem. 364 (2018) 671–678.
[61] Y. Li, F. Liu, J. Cai, X. Huang, L. Lin, Y. Lin, H. Yang, S. Li, Nitrogen and sulfur co-doped carbon dots synthesis via one step hydrothermal carbonization of green alga and their multifunctional applications, Microchem. J. 147 (2019) 1038–1047.
[62] D. Qu, M. Zheng, P. Du, Y. Zhou, L. Zhang, D. Li, H. Tan, Z. Zhao, Z. Xie, Z. Sun, Highly luminescent S, N co-doped graphene quantum dots with broad visible absorption bands for visible light photocatalysts, Nanoscale 5 (24) (2013) 12272–12277.
[63] S. Sharma, V. Dutta, P. Singh, P. Raizada, A. Rahmani-Sani, A. Hosseini-Bandegharaei, V.K. Thakur, Carbon quantum dot supported semiconductor photocatalysts for efficient degradation of organic pollutants in water: a review, J. Clean. Prod. 228 (2019) 755–769.

[64] Y. Sui, L. Wu, S. Zhong, Q. Liu, Carbon quantum dots/TiO_2 nanosheets with dominant (001) facets for enhanced photocatalytic hydrogen evolution, Appl. Surf. Sci. 480 (2019) 810–816.
[65] D. Pan, C. Xi, Z. Li, L. Wang, Z. Chen, B. Lu, M. Wu, Electrophoretic fabrication of highly robust, efficient, and benign heterojunction photoelectrocatalysts based on graphene-quantum-dot sensitized TiO_2 nanotube arrays, J. Mater. Chem. A 1 (11) (2013) 3551–3555.
[66] S. Min, J. Hou, Y. Lei, X. Ma, G. Lu, Facile one-step hydrothermal synthesis toward strongly coupled TiO_2/graphene quantum dots photocatalysts for efficient hydrogen evolution, Appl. Surf. Sci. 396 (2017) 1375–1382.
[67] L. Xu, X. Bai, L. Guo, S. Yang, P. Jin, L. Yang, Facial fabrication of carbon quantum dots (CDs)-modified N-TiO2-x nanocomposite for the efficient photoreduction of Cr(VI) under visible light, Chem. Eng. J. 357 (2019) 473–486.
[68] R. Miao, Z. Luo, W. Zhong, S.-Y. Chen, T. Jiang, B. Dutta, Y. Nasr, Y. Zhang, S. L. Suib, Mesoporous TiO_2 modified with carbon quantum dots as a high-performance visible light photocatalyst, Appl. Catal. B Environ. 189 (2016) 26–38.
[69] Y. Li, B.-P. Zhang, J.-X. Zhao, Z.-H. Ge, X.-K. Zhao, L. Zou, ZnO/carbon quantum dots heterostructure with enhanced photocatalytic properties, Appl. Surf. Sci. 279 (2013) 367–373.
[70] S. Kumar, A. Dhiman, P. Sudhagar, V. Krishnan, ZnO-graphene quantum dots heterojunctions for natural sunlight-driven photocatalytic environmental remediation, Appl. Surf. Sci. 447 (2018) 802–815.
[71] H. Yu, H. Zhang, H. Huang, Y. Liu, H. Li, H. Ming, Z. Kang, ZnO/carbon quantum dots nanocomposites: one-step fabrication and superior photocatalytic ability for toxic gas degradation under visible light at room temperature, New J. Chem. 36 (4) (2012) 1031–1035.
[72] W. Huang, S. Wang, Q. Zhou, X. Liu, X. Chen, K. Yang, C. Yu, D. Li, Constructing novel ternary composites of carbon quantum dots/Bi_2MoO_6/graphitic nanofibers with tunable band structure and boosted photocatalytic activity, Sep. Purif. Technol. 217 (2019) 195–205.
[73] S. Sajjadi, A. Khataee, R.D.C. Soltani, A. Hasanzadeh, N, S co-doped graphene quantum dot–decorated Fe_3O_4 nanostructures: preparation, characterization and catalytic activity, J. Phys. Chem. Solids 127 (2019) 140–150.
[74] Y. Lei, C. Yang, J. Hou, F. Wang, S. Min, X. Ma, Z. Jin, J. Xu, G. Lu, K.-W. Huang, Strongly coupled CdS/graphene quantum dots nanohybrids for highly efficient photocatalytic hydrogen evolution: unraveling the essential roles of graphene quantum dots, Appl. Catal. B Environ. 216 (2017) 59–69.
[75] Y. Liu, Y.-X. Yu, W.-D. Zhang, Carbon quantum dots-doped CdS microspheres with enhanced photocatalytic performance, J. Alloys Compd. 569 (2013) 102–110.
[76] C. Sun, Q. Xu, Y. Xie, Y. Ling, J. Jiao, H. Zhu, J. Zhao, X. Liu, B. Hu, D. Zhou, High-efficient one-pot synthesis of carbon quantum dots decorating Bi_2MoO_6 nanosheets heterostructure with enhanced visible-light photocatalytic properties, J. Alloys Compd. 723 (2017) 333–344.
[77] J. Di, J. Xia, Y. Ge, H. Li, H. Ji, H. Xu, Q. Zhang, H. Li, M. Li, Novel visible-light-driven CQDs/Bi_2WO_6 hybrid materials with enhanced photocatalytic activity toward organic pollutants degradation and mechanism insight, Appl. Catal. B Environ. 168–169 (2015) 51–61.
[78] H. Wang, Z. Wei, H. Matsui, S. Zhou, Fe_3O_4/carbon quantum dots hybrid nanoflowers for highly active and recyclable visible-light driven photocatalyst, J. Mater. Chem. A 2 (38) (2014) 15740–15745.
[79] H. Zhang, H. Ming, S. Lian, H. Huang, H. Li, L. Zhang, Y. Liu, Z. Kang, S.-T. Lee, Fe_2O_3/carbon quantum dots complex photocatalysts and their enhanced photocatalytic activity under visible light, Dalton Trans. 40 (41) (2011) 10822–10825.

[80] H. Li, R. Liu, Y. Liu, H. Huang, H. Yu, H. Ming, S. Lian, S.-T. Lee, Z. Kang, Carbon quantum dots/Cu_2O composites with protruding nanostructures and their highly efficient (near) infrared photocatalytic behavior, J. Mater. Chem. 22 (34) (2012) 17470–17475.
[81] J. Pan, J. Liu, S. Zuo, U.A. Khan, Y. Yu, B. Li, Structure of Z-scheme CdS/CQDs/BiOCl heterojunction with enhanced photocatalytic activity for environmental pollutant elimination, Appl. Surf. Sci. 444 (2018) 177–186.
[82] Q. Hu, M. Ji, J. Di, B. Wang, J. Xia, Y. Zhao, H. Li, Ionic liquid-induced double regulation of carbon quantum dots modified bismuth oxychloride/bismuth oxybromide nanosheets with enhanced visible-light photocatalytic activity, J. Colloid Interface Sci. 519 (2018) 263–272.
[83] Z. Zhang, S. Lin, W. Cui, X. Li, H. Li, Enhanced photocatalytic activity of Ag/CQDs/$Bi_2O_2CO_3$ composite photocatalyst under full-spectrum light, Mater. Lett. 234 (2019) 264–268.
[84] W. Li, C. Zhao, Q. Zhang, Synthesis of bi/BiOCl-TiO_2-CQDs quaternary photocatalyst with enhanced visible-light photoactivity and fast charge migration, Catal. Commun. 107 (2018) 74–77.
[85] J. Zhang, M. Yan, X. Yuan, M. Si, L. Jiang, Z. Wu, H. Wang, G. Zeng, Nitrogen doped carbon quantum dots mediated silver phosphate/bismuth vanadate Z-scheme photocatalyst for enhanced antibiotic degradation, J. Colloid Interface Sci. 529 (2018) 11–22.
[86] X. Yue, X. Miao, Z. Ji, X. Shen, H. Zhou, L. Kong, G. Zhu, X. Li, S. Ali Shah, Nitrogen-doped carbon dots modified dibismuth tetraoxide microrods: a direct Z-scheme photocatalyst with excellent visible-light photocatalytic performance, J. Colloid Interface Sci. 531 (2018) 473–482.
[87] F. Zhao, Y. Rong, J. Wan, Z. Hu, Z. Peng, B. Wang, High photocatalytic performance of carbon quantum dots/TNTs composites for enhanced photogenerated charges separation under visible light, Catal. Today 315 (2018) 162–170.

CHAPTER 5

Catalytic carbon materials from biomass

Gareth Davies, Ahmed El Sheikh, Catherine Collett, Ibrahim Yakub, and James McGregor
Department of Chemical & Biological Engineering, University of Sheffield, Sheffield, United Kingdom

1 Introduction

Carbonaceous materials have a long history of application in heterogeneous catalysis. Their primary application has been as a support to disperse an active material—normally a metal or metal oxide. However, carbons can also act directly as catalysts, known as "carbocatalysts," in a range of industrial processes [1].

Catalysts made from carbon, either with or without metals present, may provide a more sustainable and economic alternative to costly and scarce platinum group metal (PGM)-based materials. In particular, the production of carbon from waste and fresh biomass sources (Section 2) provides an opportunity to embed catalyst synthesis within the framework of the so-called "circular economy." Carbon has been shown to catalyze a range of hydrocarbon transformation reactions: hydrogenations, dehydrogenations, hydrocarbon transfer, oxidations, and reductions. In electrocatalysis, carbon is also used for electrodes, where it has been found to outperform silver and platinum. A variety of possible active sites, such as oxygen-containing functional groups or delocalized electrons in graphitic defects, make carbonaceous materials versatile catalysts.

Carbon surfaces typically possess functional groups (with O, N, H heteroatoms) and imperfections or defects from an ordered carbon structure. The structure of a number of key functional groups is shown in Fig. 1. These provide anchoring sites for metal (oxide) nanoparticles during catalyst synthesis [2,3]. Additionally, surface functional groups may act to decrease the hydrophobicity of the carbon and increase water uptake [4]. The interaction between the precursor and carbon varies, with strong anchoring sites beneficial for enhancing catalyst dispersion. In the case of direct catalytic applications, surface functionalities (which may also be introduced through

Emerging Carbon Materials for Catalysis
https://doi.org/10.1016/B978-0-12-817561-3.00005-6

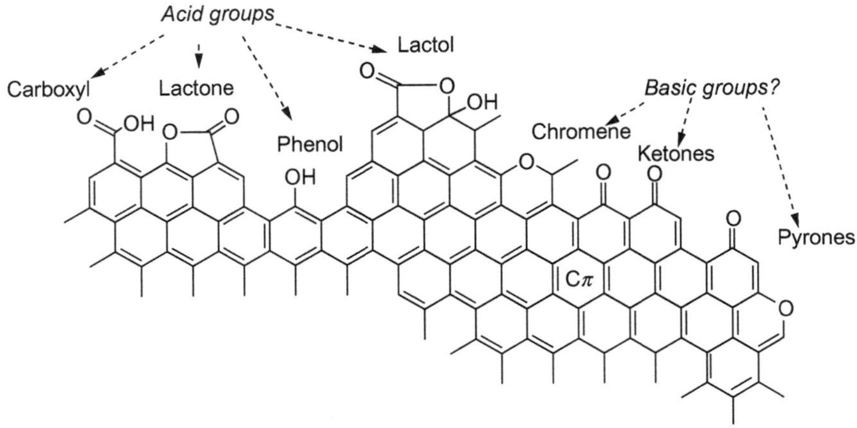

Fig. 1 Surface functional groups commonly found on carbonaceous catalysts and supports. *(Reprinted from M.A. Montes-Morán, D. Suárez, J.A. Menéndez, E. Fuente, On the nature of basicsites on carbon surfaces: an overview, Carbon 42 (7) (2004) 1219–1225, with permission of Elsevier.)*

post-synthesis treatments) can play a critical role in catalytic performance, in particular hydroxyl and carboxyl moieties [5].

2 Carbon materials from biomass

Carbonaceous materials can be categorized as naturally occurring (e.g., charcoals and graphites) and synthetic carbons (such as carbon nanostructures and biochars). These carbons can further be categorized through their degree of defined structure; for example, contrasting highly regular graphites with amorphous carbons [6]. The categorization of different carbons in this way for applications in catalysis is shown in Fig. 2. This degree of structure, or graphiticity, does not, however, have a clearly defined and metricized scale, with different analytical techniques probing different properties (Section 3). The carbonization of biomass precursors produces mixtures of solid carbon, liquid biofuels, gas, and water depending on the temperature and residence time [7]. The different precursors or processing mechanisms influence the structure, and therefore, lead to different carbon products. For example, the gas-phase pyrolysis of a hydrocarbon precursor produces carbon black (or soot), but that of chemical vapor deposition (CVD) on metallic surfaces leads to vapor-grain fibers or nanotubes. Many synthetic carbons are produced on an industrial scale for a range of applications; for example, activated carbons as adsorbents for wastewater treatment [8]. This chapter

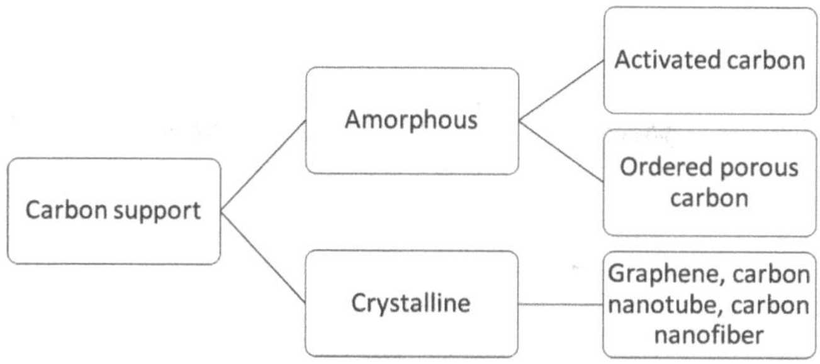

Fig. 2 Applications of different carbon polymorphs sorbents and catalyst supports. *(Adapted from Y. Yang, K. Chiang, N. Burke, Porous carbon-supported catalysts for energy and environmental applications: a short review, Catal. Today 178 (2011) 197–205, with permission of Elsevier.)*

presents the different types of carbon materials with potential catalytic applications that can be derived from biomass sources.

2.1 Activated carbon

Several types of carbon have found applications as catalytic materials. The most commonly encountered is activated carbon, i.e., carbon that has been chemically or thermally activated to increase its surface area or functionality for catalytic applications [9]. Activated carbon is produced by pyrolysis of agricultural wastes such as wood chips, peat, pits, coal, and nutshells. After pyrolysis, an activation process, using CO_2 for example, is pursued to develop the porous structure of the carbon. The synthesis of activated carbons is well-established and a number of reviews on the topic exist [10,11].

Carbon black (CB) is a related material comprising elemental carbon in an almost-spherical particulate graphitic shape with a diameter from 10 nm to 400 nm [12]. These small particles can coalesce in the form of either an aggregate (85–500 nm) or agglomerate (1–100 μm). Industrial CB is produced using the furnace black process in which a highly aromatic fossil oil is sprayed onto a reactor. Meanwhile, the oil droplets meet a high-temperature (>1800°C) gas, which is produced by a secondary feedstock. The oil droplets vaporize and pyrolyze at first. Then, the primary CB nodules are formed, followed by aggregation. Finally, agglomeration of aggregates occurs when there is no more surface growth coagulation [13].

2.2 Biochar

Biochar is a by-product of biomass pyrolysis—heating in the absence of oxygen—and provides a sustainable and carbon-neutral source of

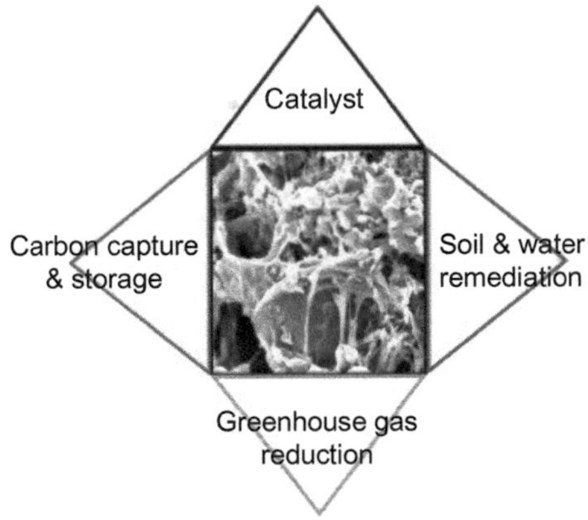

Fig. 3 Current and potential applications of biochar. *(Reprinted from J. Lee, K.-H. Kim, E.E. Kwon, Biochar as a catalyst, Renew. Sust. Energ. Rev. 77 (2017) 70–79, with permission of Elsevier.)*

carbonaceous materials, in contrast to graphite and coal. Biochar is generally applied as a soil additive; however, the study of biochar as a catalytic material is a rapidly growing field, with a number of excellent reviews covering this [14–17]. Applications of biochars include fuel cells, adsorbents, gas storage, and catalysts [18], as shown in Fig. 3 [17]. The synthesis of biochar produces a highly porous, carbon-rich material ideally suited for catalytic applications [19,20]. The properties of biochar can be easily tailored by altering feedstock or pyrolysis conditions, or applying post-synthetic surface treatments, providing excellent versatility.

Biochar has the further advantage that the biomass feedstock can be sourced from a wide range of materials, such as food, animal and municipal waste, as well as plant materials. Additionally, the pyrolysis of biomass also produces non-condensable gases, which can be burnt for fuel, as well as condensable organic liquids [21].

2.3 Hydrochar

Hydrochar is a carbonaceous material with coal-like properties produced from the pressurized low-temperature thermal conversion of a feedstock, e.g., domestic municipal waste, agricultural residues or other biomass sources, or waste plastics [22]. This process is known as hydrothermal carbonization (HTC) and was first studied by Bergius and Specht [23]. For HTC, the feedstock is submerged in water and heated in the range 180–260°C in a closed system at 20–100 bar, i.e., subcritical conditions [24]. The reaction time for

HTC typically ranges from minutes to a few days, although there are significant findings to show that the greatest extent of reaction occurs within the first 30 min [25,26]. During HTC, subcritical water effects change to the physical and chemical structure of the feedstock predominately via hydrolysis reactions; however, aromatization, dehydration, decarboxylation, and recondensation reactions also take place [25,27]. When the severity of the reaction conditions is increased, liquefaction and gasification dominate over carbonization [28]. Fig. 4 shows a phase diagram indicating the P/T range of these processes as compared to HTC.

HTC is performed in water; hence, the feedstock does not require predrying. This, therefore, eliminates the need for energy-intensive and costly pretreatment steps which are required in conventional thermal processes such as pyrolysis [29]. In addition, this presents opportunities to utilize feedstocks with very high moisture contents (>30%), e.g., sewage sludge which is ubiquitous worldwide [30].

Compared to the raw feedstock, the synthesized hydrochars have a higher percentage of carbon [31]. This is because dehydration and decarboxylation reactions remove hydrogen and oxygen from the feedstock as water and carbon dioxide, respectively [25]. Hydrochar also has reduced nitrogen (N) and sulfur (S) content compared to the raw feedstock because these are lost as oxides into the liquid phase during processing [32]. Hydrochar has a lower ash content (incombustible material) compared to other chars (including biochars) because the inorganic compounds which normally form the ash after combustion are extracted into the liquid phase during HTC [33].

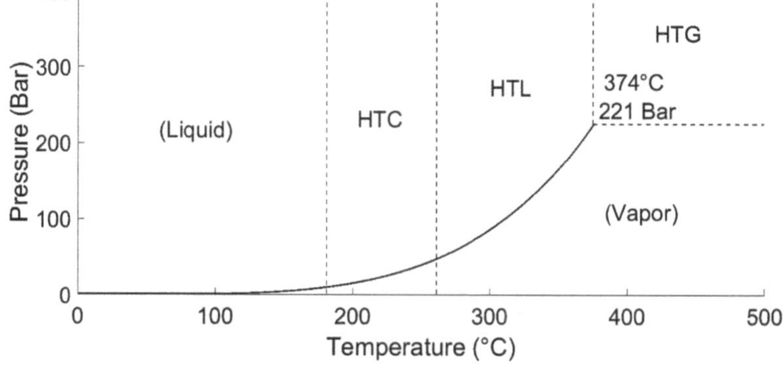

Fig. 4 Water phase diagram showing the operational range of hydrothermal carbonization (HTC), hydrothermal liquefaction (HTL), and hydrothermal gasification (HTG).

2.4 Other carbon materials

Graphitic and nanostructured carbons are also applied in catalysis, although their preparation process is more complex than other carbons, often requiring a metal template and carbon vapor deposition processes [34]. A schematic for the production of carbon nanotubes (CNTs), fullerenes, and carbon blacks from hydrocarbons is shown in Fig. 5. Although not typically derived from biomass, the carbon sources could be sourced in that way. For instance, CNTs may be grown using CO as the carbon source with CO readily available from biomass via gasification [35], while the pyrolysis oils produced from biomass can provide the hydrocarbonaceous materials for carbon black production. Carbon nanofibers are prepared via the catalytic pyrolysis of hydrocarbons over oxide-supported catalysts and range in surface area from 70 to 300 m^2/g and in pore size from 5 to 50 nm. Carbon nanotubes are similar to nanofibers except that the structure of the former is hollow and the latter is rod-like; however, both are famous filamentous carbonaceous materials [36].

3 Characterization of catalytic carbon material from biomass

There are a variety of physicochemical properties which influence the performance of biomass-derived carbons in catalysis. These include, but are not limited to: surface area; pore size and structure; surface functionality; graphiticity and electron conductivity; hydrophobicity/philicity; and elemental

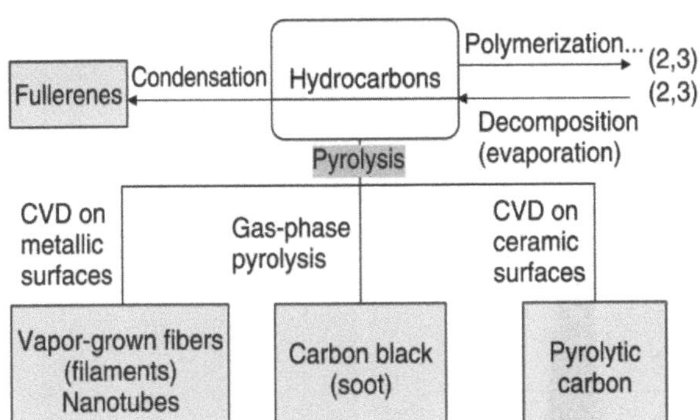

Fig. 5 Processes of carbon in the gas phase and the formation of carbon macrostructures. *(Reprinted from L.R. Radovic, Physicochemical properties of carbon materials: a brief overview, in: Carbon Materials for Catalysis, John Wiley & Sons, Inc., Hoboken, NJ, USA, 2008, pp. 1–44, with permission of John Wiley & Sons.)*

composition. It is therefore necessary to employ several techniques in order to develop structure-performance relationships as no one technique can capture all of these information [37]. Techniques vary in their ability to measure bulk or surface properties, and some may be invasive or destructive. The importance of using a wide range of characterization methods, ideally conducted simultaneously on the same sample, cannot be overstated. This section initially presents some additional considerations when characterizing biomass-derived carbons (Section 3.1), followed by a description of the most relevant characterization techniques.

3.1 Additional considerations for the characterization of biomass-derived carbons

The characterization of biomass-derived carbons such as biochars and hydrochars poses its own challenges. There are many properties of these carbons which affect their suitability for catalytic applications, and there is little consistency in how these properties are quantified. There is also a difference in approach between characterization of biochar for catalysis and characterization for soil remediation. Studies of biochar for catalytic applications will often focus on one or two source materials and aim to demonstrate their suitability for a given application. The aim of characterization is to improve the activity of a specific biochar, for example, to understand the effect of activation or functionalization treatments. Examples of this are the development of a sulfonated acid catalyst from pine wood chips for hemicellulose hydrolysis [38] and the use of karanja seeds pyrolyzed at different temperatures for glycerol esterification [39]. In contrast, it is typical in soil remediation applications to characterize multiple biochars, in order to choose the biochar with the properties best suited to the application. This can be seen in Table 1, where up to 12 biochars are investigated in a single study, in order to select the feedstock with the most suitable properties. This application-centered approach could be valuable in catalysis research, where characterization is performed in order to select the best material for the application, rather than the best application for the material.

The characterization methods themselves vary depending on the application. Some representative examples are given in Table 1. In catalytic applications, the majority of studies will characterize the surface area, porosity, and surface acidity or basicity of a sample; however, characterization of carbon structure is not always performed. Studies often quantify the trace metal content through ICP analysis of elements digested in aqua regia; however, this technique is less effective at removing K and Na from the biochar matrix [53].

Table 1 Characterization techniques used in the literature for biochars for

			Composition				
Source	**Biochars studied**	**Target application**	**Proximate (TGA)**	**Ultimate (CHNOS)**	**Elemental (EDX/XPS/ICP/XRF)**	**Surface area (BET/N2)**	**Porosity (BJH)**
Catalytic applications							
[40]	Commercial biomass char (unspecified), pinewood char, pinewood ash	Tar reduction	X	X		X	X
[41]	Three commercial hardwood chars	Biodiesel production (solid acid catalyst)			X	X	
[42]	Pelletized peanut hulls, pine pellets, pine chip char	Esterification of fatty acids (solid acid catalyst)	X	X	X	X	X
[43]	Pistachio hull biochar	Ozonation of water recalcitrant concentrations	X	X		X	
[38]	Pine chip, wood-based AC	Catalyst (solid acid) for hemicellulose hydrolysis	X	X		X	X
[39]	Karanja seed shells	Esterification of glycerol with acetic acid	X	X		X	
[44]	Rice husk char	Conversion of tar using rice husk char-supported nickel-iron catalysts.	X	X	X	X	
[45]	Shengli brown coal	Pyrolysis and gasification of biomass			X	X	
[46]	Woody biomass	(Solid acid) Transesterification of canola oil	X	X		X	X
[47]	Modified cotton biochar	Low-temperature selective catalytic reduction (SCR) of NO			X	X	
[48]	Calcium oxide-based catalyst from palm kernel shell biochar	Transesterification of sunflower oil with methanol to produce biodiesel	X	X	X	X	
[49]	Fly ash and egg shell-derived solid Catalysts	Solid base catalyst, transesterification of soybean oil to biodiesel				X	X
[50]	Ash from cocoa pod husks (supported and unsupported)	Transesterification of soybean oil to biodiesel. (Supported and unsupported catalysts)			X	X	X
[51]	Palm bunch ash, support for KOH	Biodiesel synthesis (simultaneous ozonolysis and transesterification)			X	X	
[52]	Rice husk biochar	Tar reforming	X	X	X	X	

different applications.

Techniques/properties								
Structure		Surface chemistry			Other			
Crystal structure (XRD)	Imaging (SEM/TEM)	Functional groups (FTIR)	Acidity, basicity (TPD/titration)	Carbon content (^{13}C NMR/Raman/EPR/PAH)	Conductivity (CEC/EC)	Calorimetry	Adsorbance capacity	pH
			X					
		X	X					X
		X	X	X				X
		X	X					
X		X	X	X				
X	X			X				
X		X	X					
X		X	X					
X	X		X					
X	X		X					
			X					
	X							

Continued

Table 1 Characterization techniques used in the literature for biochars for

			Composition				
Source	Biochars studied	Target application	Proximate (TGA)	Ultimate (CHNOS)	Elemental (EDX/XPS/ICP/XRF)	Surface area (BET/N2)	Porosity (BJH)
Soil remediation applications							
[53]	Wood chip, blend of paper sludge and wheat husks, sewage sludge	Soil amendment		X		X	
[54]	Pine wood chips, wheat straw, and wheat straw pellets	Soil enhancement/C sequestration					
[55]	Rice husk, rice straw, apple tree wood chips (*Malus pumila*), oak tree wood chips (*Quercus serrata*)	Carbon sequestration/soil fertility	X	X		X	X
[56]	Animal manure (cow, pig), wood waste (sawdust), crop residue (wheat straw, grass), food waste (peanut shell, shrimp hull), aquatic plants (waterweeds, chlorella), municipal waste (wastewater sludge, waste paper, bone dregs)	Soil: carbon sink, contaminant sorbent, soil nutrient amendment	X		X	X	
[57]	corn stover, switchgrass x 3 (different preparation methods)	Soil remediation/carbon sequestration	X	X		X	
Other applications							
[58]	Primary sewage sludge	Production of hydrochar as fuel source/soil additive	X	X	X		
[59]	Rice husk biochar	Adsorbent for acetanilide herbicide metolachlor		X		X	
[60]	Manchurian walnut	Catalyzing pyrolysis (studying effect of pyrolysis temperature on alkali andalkaline earth metallic species)	X	X	X		
[61]	Rice husks, corn cobs	Adsorbent for phenol (model organic pollutant)	X		X	X	
[62]	Corn straw, poplar leaf	Adsorbent for dodecylbenzene sulfonic acid			X	X	

Representative examples are given for catalytic and soil remediation applications, with other notable different applications—cont'd

Techniques/properties								
Structure		Surface chemistry			Other			
Crystal structure (XRD)	Imaging (SEM/TEM)	Functional groups (FTIR)	Acidity, basicity (TPD/titration)	Carbon content (^{13}C NMR/Raman/EPR/PAH)	Conductivity (CEC/EC)	Calorimetry	Adsorbance capacity	pH
				X	X			X
				X	X			X
		X		X				X
		X		X	X			
		X		X	X			X
						X		
	X	X	X				X	
				X				
	X	X						X
	X	X		X				

examples also listed.

XPS would also allow surface concentrations to be determined, which is more relevant for catalytic applications. While some factors analyzed for soil remediation are less relevant for catalysis, such as the cation exchange capacity, properties such as carbon structure are more often characterized. Carbon structure, and particularly graphitic content, may, however, exert a significant influence over catalytic activity; however, this is rarely considered in the biochar characterization literature.

It is also worth noting that there are no formal standards for the characterization of biochar. A recent "round robin" study found that when three standard biochars were characterized by 22 laboratories in 12 countries, using methods of their choice, reproducibility between laboratories was generally poor, with mean reproducibility standard deviation values over 20% for most of the parameters studied [53]. Similar problems can be expected in the characterization of biochars for catalytic applications. In the future, standardized procedures for preparation and characterization of biochar may be beneficial for comparison of results.

3.2 Surface area and porosity

Various properties impact the role of carbon as a catalyst and support, but the most crucial are surface area, porosity, and surface chemical properties. In the majority of applications, the highest surface area and most well-developed porosity are beneficial for the catalytic reaction. The highest surface area will present the greatest number of reaction sites. The porosity of the catalyst determines the mass transfer of reactants to, and products from, the active centers which may play a key role in determining the observed rate of reaction. Diffusion constraints are likely to be of particular concern in liquid-phase reactions. The measurement of surface area and porosity is most commonly via adsorption of an inert gas, e.g., N_2 [63]. N_2 physisorption, applying The Brunauer-Emmett-Teller (BET) method, is used to calculate the specific surface area. It is based on the physical adsorption of N_2 gas molecules, through van de Waals forces. Scanning electron microscopy (SEM) can additionally be used to analyze the pore size, shape, and overall surface morphology of the carbon [64–66].

As an example, the surface area and pore volume of orange peel waste have been measured by N_2 adsorption and compared with the corresponding hydrochar formed via HTC at 200°C for 10h. The surface area of the orange peel waste was $0.4\,m^2/g$ and that of the hydrochar was $117\ m^2/g$, while the pore volume was $0.0003\,cm^3/g$ and $0.059\,cm^3/g$, respectively

[65]. Elsewhere, in the context of electrocatalysts, a commercial carbon black, Vulcan carbon (XC72), has a pore size of 20 nm, and another, selectivity from Cabot, has a pore size of 12 nm [67]. These materials have applications as supports for electrocatalysts in, e.g., alcohol fuel cells, an application where biomass-derived carbons could play a key role.

3.3 Elemental analysis

Elemental analysis (ultimate analysis) is typically performed on carbon materials rich in hydrogen and heteroatoms, such as biochars and hydrochars. It allows for the determination of oxygen, nitrogen, carbon, hydrogen, and sulfur content [30,68]. These data can then be used to plot a van Krevelen diagram, wherein the H/C atomic ratio is plotted against the O/C atomic ratio (Fig. 6). van Krevelen diagrams are typically used to assess the quality of the chars produced for use as solid fuels where lower ratios are considered favorable because of reduced smoke, water vapor, and energy losses during combustion [71]. During HTC, oxygen and hydrogen are removed in a higher ratio than carbon via decarboxylation and dehydration reactions. Taking the HTC of oak wood as an example, with processing parameters of 250°C and 1 h reaction time, the O/C and H/C ratios for the oak wood hydrochar were 0.16 and 1.10, respectively. This compares to untreated oak wood with O/C and H/C ratios of 0.74 and 1.63. The O/C and H/C ratios

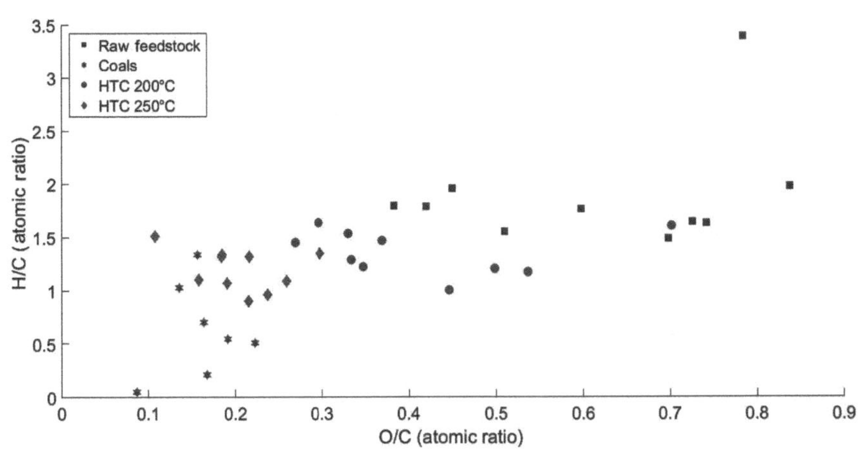

Fig. 6 An example of a van Krevelen diagram. *(Data from A.M. Smith, S. Singh, A.B. Ross, Fate of inorganic material during hydrothermal carbonisation of biomass: influence of feedstock on combustion behaviour of hydrochar, Fuel 169 (2016) 135–145; A. Kozlov, D. Svishchev, I. Donskoy, V. Shamansky, A. Ryzhkov, A technique proximate and ultimate analysis of solid fuels and coal tar, J. Therm. Anal. Calorim. 122 (2015) 1213–1220.)*

of the hydrochar produced from oak wood is very close to some types of brown coal (Haranutsky), which has ratios of 0.16 and 1.34, respectively [69,70]. The same analysis is directly applicable to materials employed in catalytic applications and provides a useful mechanism to compare and contrast different materials.

3.4 Thermal methods

Thermal methods involve heating the sample under interrogation in a controlled atmosphere and either monitoring the species desorbed, for example by mass spectrometry, or monitoring the mass change of the sample [37]. Among the most commonly utilized thermal methods in the analysis of carbon materials are: temperature programmed oxidation (TPO), which can probe resistance to oxidation and reveal information on the degree of structuring or graphiticity of the carbon [72–76]; temperature programmed desorption (TPD) which can yield information on the functional groups present on the surface of the carbonaceous materials, including acid sites [77,78]; and thermal gravimetric analysis (TGA) which can characterize the composition of carbonaceous materials in terms of their fixed carbon content, moisture, ash content, and volatiles—this is often termed "proximate analysis" [22,30,79]. Other variations include temperature programmed hydrogenation (TPH) and reduction (TPR).

Thermogravimetric analysis (TGA) measures weight decrease or increase due to temperature increase in certain atmospheres such as N_2 and O_2. Loss of water or solvent, decarboxylation, pyrolysis, ash, and residues can all be analyzed using this technique. TGA is extensively used as tool to analyze the different constituents of carbon-supported catalysts (moisture, volatiles, carbon, metal, and ash) by increasing the temperature in an oxidative atmosphere [80–82]. Standardized methods for thermogravimetric analysis of carbonaceous materials (coals and cokes) have been developed by the American Society for Testing and Materials, providing the basis of most methods applied to carbocatalysts.

3.5 Vibrational spectroscopy

Vibrational spectroscopy techniques analyze the interactions between photons or particles with a surface and the resulting excitation or de-excitation. This category of techniques includes infrared (IR) spectroscopy, Raman spectroscopy, THz-time domain spectroscopy (THz-TDS), and ultraviolet-visible (UV-vis) spectroscopy, and they have many and varied applications in the study of biomass-derived carbons.

3.5.1 Fourier transform infrared spectroscopy (FTIR)

IR spectroscopy is used to identify the functional groups present in a sample, provided the vibration mode produces a change in the dipole moment of the bond. Functional groups can be identified based on the wavenumber of absorption for specific covalent bonds, e.g., hydroxide groups can be identified by stretching (3500–3300 cm^{-1}) or scissoring (1460–1000 cm^{-1}). The most relevant common functional groups which are readily identified by this method are: O—H, C=O, C—O, N—H, C—H, N—O (Table 2). A primary application in carbonaceous samples is the detection of changes in surface chemistry following surface treatment [83]. This can range from treatments with acids and bases [84–87], to oxidation [88,89], heat treatment [90], microwave treatment [91], and heteroatom doping [92]. Adsorption and reaction studies have been coupled with in situ IR spectroscopy, allowing the species formed and adsorbed during reaction to be studied [93–97]. This provides valuable information on the adsorption species and sites on the surface of the carbonaceous sample. However, the highly absorbing nature

Table 2 Infrared spectroscopy absorptions useful in biomass-derived carbon surface analysis.

Absorbance (cm^{-1})	Strength (s, m, w)	Absorption bond	Functional group	Type of absorbance
3700–3584	m	O—H	Stretching	Alcohol
3550–3200	s	O—H	Stretching	Alcohol
3350–3310	m	N—H	Stretching	Secondary amine
3300–2500	s	O—H	Stretching	Carboxylic acid
3200–2700	w	O—H	Stretching	Alcohol
3333–3267	s	C—H	Stretching	Alkyne
3100–3000	m	C—H	Stretching	Alkene
2830–2695	m	C—H	Stretching	Aldehyde
2000–1650	w	C—H	Bending	Aromatic compound
1760	s	C=O	Stretching	Carboxylic acid
1740–1720	s	C=O	Stretching	Aldehyde
1720–1706	s	C=O	Stretching	Carboxylic acid
1680	s	C=O	Stretching	Secondary amide and tertiary amide
1550–1500	s	N—O	Stretching	Nitro compound

of carbon samples, which are often black in color, can obscure key features of the spectrum.

3.5.2 Raman spectroscopy

Raman spectroscopy is a complementary technique to IR spectroscopy and is widely used for the study of carbonaceous samples. This is because the Raman scattering effect is dependent on the polarisability of the species, and thus can be used to indicate the degree of graphiticity of a carbon network, where high densities of polarisable electrons are located [98,99]. The most commonly studied region for carbonaceous samples is from 800 to 2000 cm^{-1}, where disordered and graphitic carbon bands are observed. Deconvolution of these bands can yield information on the ratio of graphitic to disordered carbon, a key indicator of the extent of graphiticity in carbonaceous and coked catalyst samples. While a number of key Raman bands are known for carbonaceous materials and assigned to physical characteristics of the material [100–103], the heterogeneity of biochar and similar materials can lead to highly complex spectral deconvolution [56,60,104,105].

3.5.3 Other vibrational spectroscopies

The degree of graphiticity in carbonaceous samples can also be examined quantitatively through terahertz-time domain spectroscopy (THz-TDS) [100,106,107]. Compared to IR and Raman spectroscopy, THz-TDS probes a lower energy region of the electromagnetic spectrum and is, hence, ideally suited for characterizing low energy modes in extended graphitic-like networks. THz-TDS can be used to quantify graphiticity in terms of electron density and mobility, parameters which can in some cases be directly correlated with catalytic performance. Fig. 7 shows such a correlation for (non-biomass-derived) carbon nanofibers in the oxidative dehydrogenation of ethylbenzene [108].

Other spectroscopic techniques commonly used in the study of carbons include UV-vis. Typical functionalities which can be identified by UV-vis spectroscopy include conjugated double bonds, aromatics, and unsaturated carbenium cations [109].

3.6 NMR

NMR techniques have many applications in the study of coke deposits on heterogeneous catalysts such as zeolites and supported metals. ^{13}C NMR played a key role in the discovery of the hydrocarbon pool mechanism by which carbon deposits play a catalytic role in the conversion of methanol

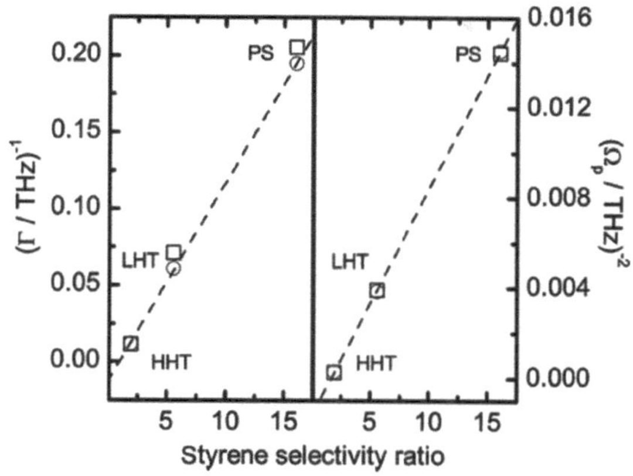

Fig. 7 Correlations of electronic properties (denoted Ω_p and Γ) with the selectivity of styrene to other hydrocarbon products for the three CNF samples: *circles*, Maxwell-Garnett model parameters; *squares*, Bruggeman model parameters. *(Reprinted from E.P.J. Parrott, J. Axel Zeitler, J. McGregor, S.-P. Oei, H.E. Unalan, S.-C.Tan, W.I. Milne, J.-P. Tessonnier, R. Schlogl, L.F. Gladden, Understanding the dielectric properties of heat-treated carbon nanofibers at terahertz frequencies: a new perspective on the catalytic activity of structured carbonaceous materials, J. Phys. Chem. C 113 (2009) 10554–10559, with permission of American Chemical Society.)*

to higher hydrocarbons [110]. In studying carbonaceous materials, however, ^{13}C NMR spectra can be composed of very complicated overlapping signals, e.g., C, CH, CH_2, and CH_3 peaks [111]. NMR has not, therefore, been widely applied in the study of carbons to-date. However, there have been some innovations towards using ^1H and ^{13}C nuclear magnetic resonance to provide semiquantitative data on organics present in the structure of hydrochars and biochars [112]. When using NMR for analysis of hydrochars, 2D $^{13}C—^1H$ heteronuclear single quantum correlation (HSQC) spectra are beneficial to correctly identify the structures present [113].

3.7 XRD

Interference or diffraction methods, such as X-ray diffraction (XRD) or neutron diffraction, can be used to study the geometry and symmetry of a surface [37]. Since XRD relies on the study of diffraction patterns, it is suitable for the study of materials with long-range order which can produce such diffraction patterns, for example carbon fibers [114] and chars and activated carbons with graphitic character [46,89,115]. XRD is conversely unsuitable for materials which are polycrystalline or amorphous. sp^2 graphitic carbons are the most common applied support materials for electrocatalysis, and hence, XRD finds wide application in this area. For instance, semi-amorphous graphitic carbon blacks (such as Vulcan XC72) show broad peak at 25°, and tubular structures such as carbon nanofibers (CNFs) and carbon nanotubes (CNTs) present more crystalline characteristic sharp peak at 25° [116,117].

3.8 Electron spectroscopy

Electron spectroscopy techniques such as Auger electron spectroscopy (AES) and X-ray photoelectron spectroscopy (XPS) are highly surface-sensitive and can provide detailed information on the chemical environment of surface atoms. They are commonly used for the study of carbonaceous materials.

In both techniques, electrons are emitted with kinetic energies related to the atomic or molecular environment of the atom of origin, yielding approximate elemental surface composition data and oxidation states. XPS, for example, can quantify the C/H ratio or degree or aromaticity in a carbon sample [118]. XPS is also a useful tool to study the carbon surface functionalities by measuring the peaks of C1s, N1s, and O1s. After oxidation, the C1s peak of graphitic carbon splits into different peaks such as graphitic carbon (284.6–285.1 eV), carbidic carbon (282.6–288.1 eV), alcohol (286.3–287.0 eV), carboxyl (289.33–290.0 eV), and π-π transitions (291.2–292.1 eV) [119,120]. Additionally, where carbon is employed as the support material, XPS can probe the oxidation state of the active metal and quantify the presence of each oxidation state [121,122]. One limitation of XPS is the calibration of the electron binding energies. The position of the adventitious carbon peak is often used, even for carbonaceous samples [104,123,124]. A detailed study of the C1s peak identified seven components, based on DFT and experimental data [125]. This implies that the position of the C1s peak will vary, depending on the relative contribution of these components. As with Raman curve deconvolution, there is frequently little consistency between researchers on curve deconvolution and calibration methods for C1s spectra.

Electron spectroscopy also forms the basis of a number of imaging techniques, which can be applied for investigating carbonaceous catalysts. Scanning electron microscopy (SEM) and transmission electron microscopy (TEM) use a beam of focused electrons to produce an image of the catalyst surface and have been applied to image biochar and carbocatalyst samples [104,126,127]. These techniques are also particularly valuable in electrocatalysis where the employed carbonaceous materials are usually electronically conductive, and therefore, interact strongly with the electron beam revealing information about surface morphology, particle size, and shape. High-resolution electron microscopy (HRTEM), equipped with phase contrast imaging, can reveal the structure of lamellar carbon blacks, giving a wealth of localized information other techniques cannot provide [13]. Nanoscale details such as fringe length, curvature, lamella stacking, paracrystallinity, and spacing can be easily interpreted using HRTEM.

In combination with energy dispersive X-ray spectroscopy (EDX) and electron energy loss spectroscopy (EELS), TEM can be used to identify the chemical elements present on a catalyst surface, producing a compositional map on top of the microscope image. Atomic force microscopy (AFM) can also be used to obtain high-resolution images with minimal sample preparation, with applications in studies of carbon films [98] and graphene samples [128].

4 Applications of carbonaceous catalysts from biomass

This section provides an introduction to the application of carbonaceous materials in catalysis, before considering specific applications of biochar (Section 4.2) and hydrochar (Section 4.3) in conventional chemical catalysis and finally discussing the specific case of electrocatalysis (Section 4.4).

4.1 Carbon in catalysis

Carbon finds a number of applications in catalysis due to its physical and chemical properties. Carbon structures can be stable at temperatures as high as 1000 K in the absence of oxygen, while the high electron conductivity of graphitic structures can provide advantages [18]. Carbonaceous materials often possess high surface areas and porosities, which can be tailored easily by altering the feedstock or method of preparation [6]. These features lend carbon to application as a support material for active metals or metal oxides. In this case, an active catalyst is synthesized through a variety of methods including impregnation (excess-solution or incipient-wetness) and deposition-precipitation (DP). Biomass-derived carbons present a sustainable and tailorable source of carbonaceous catalysts and supports for a wide range of applications. Among the many applications of carbon supports in catalysis, some of the most investigated are hydrodesulfurization (HDS) and hydrodenitrogenation (HDN) [129–131], hydrogenation (including Fischer-Tropsch synthesis) [132,133], and ammonia synthesis [134,135].

Many of the same features that make carbonaceous materials suitable as a support material are also beneficial for catalysts in both liquid- and gas-phase applications; namely, high surface area, tailorable porosity, corrosion resistance, affordability, and chemical, mechanical, and thermal stability. The ability of carbon to act as a catalyst is dependent on the availability of active sites. Functional groups containing oxygen and nitrogen may be naturally present, or can be introduced through chemical treatments, and can act as acidic or basic active sites. This is a particular advantage of carbon from biomass where such heteroatoms are present due to the composition of the source material. In graphitic materials, delocalized unpaired electrons at the edge of graphene sheets can act as active sites, while the conductivity of graphite can facilitate electron transfer between reagents [136]. The charge distribution of graphite and CNTs can also be modified by doping with heteroatoms, such as nitrogen, sulfur, and boron [137]. The variety of active sites, and the ease of modification, makes carbonaceous materials versatile catalysts. Among the many reactions in which carbocatalysts are active are oxidations, reductions, hydrogenations, dehydrogenations, dehydration, and other bond-forming and bond-cleaving reactions (Fig. 8). Uses in environmental catalysis include the oxidation of SOx and NOx [138–140]. In such applications, the presence of naturally occurring nitrogen heteroatoms in biomass-derived carbons may be beneficial for catalyst activity. Industrial applications of activated carbons include the catalysis of phosgene synthesis [141,142] and flue gas cleaning [143]. Graphite and fullerenes have been shown to catalyze the reduction of substituted nitrobenzenes to aniline [144], while carbon nanotubes catalyze dehydrogenation of *n*-butane to 1-butene [145]. Graphene oxide, in particular, has demonstrated ability to catalyze hydration and oxidation reactions, such as alcohols to ketones and aldehydes, alkenes to diones, and alkynes to hydrates [146].

4.2 Applications of biochar in catalysis

Biochar has found numerous applications in catalytic research, principally as a support and as a functionalized carbon. This is a fast-growing subject of research interest, with a number of reviews available [5,14,15,17]. One of the key challenges in the application of biomass-derived carbon as a catalyst is a lack of reproducibility in experimental studies. This is most likely due to the inherent variation in the biomass feedstock, affecting the properties of the catalyst. A key note in the use of biomass-derived carbons is that they should not, as carbonaceous catalysts often are, be described as "metal-free"

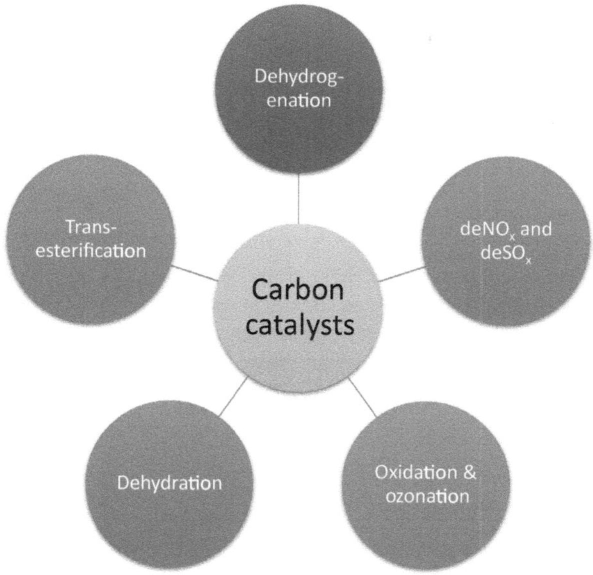

Fig. 8 A selection of the many classes of reaction which can be catalyzed by carbonaceous materials.

due to trace metal content present in a biomass feedstock [6]. This provides advantages and disadvantages insofar as either carbon or the metal content may be the active species, but discerning which can present a significant challenge.

Unlike activated carbons, biochars do not require extensive chemical and thermal treatments. As a support material, biochar has found applications in a wide range of reactions including hydrogenation of phenol to cyclohexanol [147], low-temperature selective catalytic reduction of NO [148], oxidation of glycerol, hydrogenation of levulinic acid [149], and dry reforming using biochar-supported tungsten carbide [17]. Recently, biomass precursors, such as wood and starch, have been applied as sources to produce carbonaceous catalyst supports [150].

Considering the application of biochars directly as catalysts, a key area of research is their use for processes closely related to biochar production. These include the reduction of tar [40,61,148] and the conversion of biomass to chemical products [18]. The activity of biochar in these processes is often attributed to the mineral or ash content, particularly the presence of potassium and alkali/alkali earth metals (AAEM). For example, the CO_2 gasification rate of a spruce wood char was found to increase linearly

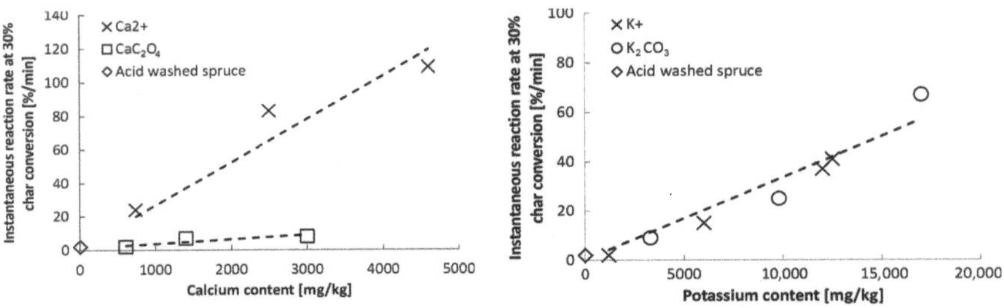

Fig. 9 Instantaneous reaction rate at 30%, char conversion vs. initial Ca and K-content for Ca- and K-loaded spruce samples and acid washed spruce. *(Reprinted from M. Perander, N. DeMartini, A. Brink, J. Kramb, O. Karlström, J. Hemming, A. Moilanen, J. Konttinen, M. Hupa, Catalytic effect of Ca and K on CO_2 gasification of spruce wood char, Fuel 150 (2015) 464–472, with permission of Elsevier.)*

with Ca and K content (Fig. 9) [151]. Surface alkali metals in biochars may also be involved in the catalytic decomposition of tar compounds, such as toluene [52,152]. Biochar may, therefore, provide a more sustainable source of catalysts for these processes, in place of the currently used iron- and nickel-based catalysts.

The activity of ash has also been demonstrated in other cases, such as the formation of glycerol from glycerol carbonate over boiler ash [153], while potassium is a known promoter in conventional heterogeneous catalysts; for example, potassium-promoted iron catalysts are used in dehydrogenation of hydrocarbons [137]. In methane decomposition, ash content was found to catalyze C—C and C—H bond breakage, with carbon acting as both an active site for methane cracking and as a support for inorganic metals such as Fe, Ca, and K; oxygen-containing groups were not found to be involved in the reaction [154]. In biochar samples, the potassium most likely exists as ions, while other alkali earth metals such as calcium and magnesium are more likely bound in organic compounds [60]. The role of ash content should, therefore, not be neglected in studies of the catalytic activity of biochar.

Other reactions employing biochar catalysts include the production of biodiesel via the transesterification of vegetable oils [14,155]. Elsewhere, biochars have been employed in applications such as syngas production [156], biomass hydrolysis [156], hydrotreating [156], and selective catalytic reduction of NO_x [156].

Biochars have been used for the production of both solid acid catalysts [42,46] and solid bases [42,48], and as supports for traditional homogeneous catalysts such as KOH [51]. Biochars with high AAEM contents are particularly suitable for the formation of solid base catalysts, containing active species such as calcium oxide and potassium carbonate. Solid acid catalysts are generally prepared by treating carbonaceous materials with a sulfonating agent, usually fuming or concentrated sulfuric acid.

4.3 Applications of hydrochar in catalysis

While not as widely investigated as biochars, hydrochars have the potential to be used in the same manner and are potentially more tailorable for different applications through variation of the synthesis parameters such as temperature, pressure, solvent, etc. The most heavily research application of hydrochar is as a renewable solid fuel [157]. This is because hydrochar has heavily improved combustion properties compared to the raw biomass such as improved combustion efficiency and reduced emissions of water vapor, smoke, and pollutants during combustion. Another significant area of application for hydrochars is as a low cost adsorbent for aqueous and volatile contaminants [31]. As-synthesized hydrochars often have lower than desired surface area for catalytic applications (20–50 m^2/g), alongside a low pore volume. These characteristics can be altered by changing the process conditions; however, it is often more effective to increase the surface area and other physicochemical characteristics of the hydrochars via post-processing. This is either done by physical activation where the hydrochar is heated at temperatures in excess of 700°C under an inert gas such as CO_2, or alternatively hydrochars can also undergo chemical activation where an activation agent, e.g., KOH, is added to the hydrochar and heated at temperatures between 600°C and 800°C [27]. Falco et al. investigated the potential to tailor the porosity of hydrochars using KOH to increase the surface area and pore volume [158]. Hydrochars were synthesized from glucose, sucrose, and rye straw at 180°C, 240°C, and 280°C. The highest surface area was achieved at 240°C where each hydrochar had a surface area over 2200 m^2/g. Pari et al. produced and tested adsorbents via a similar method using cassava and tapioca flour raw material as the feedstock.

A variety of different methods have been used to produce hydrochar-supported heterogeneous catalysts with examples such as nanoparticles supported on hydrochar synthesized via a one step in situ processes, nanoparticle encapsulation into hydrochars, and acid-treating hydrochar to form a sulfonated catalyst [159–161]. Tavasoli et al. produced a porous hydrochar-supported nickel catalyst from canola stalks through HTC (240°C, 1 h) followed by chemical activation using $ZnCl_2$ and heating under argon at 700°C. Finally, nickel was deposited onto the hydrochar via wet impregnation. This nickel hydrochar catalyst was then used in the hydrothermal gasification of cattle manure and found to improve H_2 yields by a factor of up to 2.27 when using blended catalyst systems as compared to non-catalytic hydrothermal gasification [162].

4.4 Electrocatalytic applications

Carbon materials are crucially important for energy storage devices such as batteries, fuel cells, and supercapacitors. In electrocatalytic systems, carbons find application as electrode materials and supports for the electrocatalytically active metal. This has been reviewed by Liu and Dai [137]. Key processes of interest include the oxygen reduction reaction (ORR), oxygen evolution reaction (OER), and hydrogen evolution reaction (HER). Carbon nanofibers have been reported to demonstrate activity for CO_2 reduction, outperforming silver electrodes, with a negligible overpotential of 0.17 V [163]. This activity was attributed to the higher binding energy of intermediates to the carbon nanofiber surface, rather than to the electronegativity of nitrogen heteroatoms.

The carbon supports required for fuel cell electrocatalysis should have high surface area, reasonable porosity, and high electronic conductivity [116]. Similar but not identical carbon characteristics are required in carbon-based electrodes in supercapacitors and batteries [164]. An emerging application of hydrochars is as sustainable energy storage devices; hydrochars can be used as electric double layer capacitors (EDLC)—a type of supercapacitor [165]. Jain et al. have optimized the stoichiometry and pretreatment of hydrochar and were able to produce mesoporous carbons with surface areas in excess 2400 m^2/g (Fig. 10) [166]. These have been successfully applied as a supercapacitor electrode material. Biochars have also found application in microbial fuel cells, exhibiting higher maximum power densities than Pt/C and graphite electrodes [167,168].

5 Future outlook

One of the key challenges in studying carbon, particularly from biomass, is reproducibility. Commercial activated charcoals and biochars are usually produced for the purpose of soil remediation, and so, reproducibility in catalytic applications is not considered. Properties such as porosity and distribution of elements at the surface of the material can have a great impact on the catalytic activity of a sample. The development of robust structure-performance relationships where the catalytic properties of biomass-derived

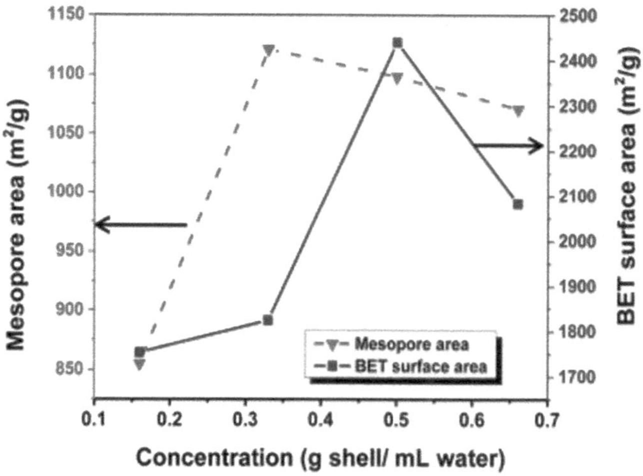

Fig. 10 Porosity of ACs obtained by successive hydrothermal treatment with H_2O_2 and $ZnCl_2$. *(Reprinted from A. Jain, C. Xu, S. Jayaraman, R. Balasubramanian, J.Y. Lee, M.P. Srinivasan, Mesoporo mesoporous activated carbons with enhanced porosity by optimal hydrothermal pre-treatment of biomass for supercapacitor applications, Microporous Mesoporous Mater. 218 (2015) 55–61, with permission of Elsevier.)*

catalysts are correlated with distinct physicochemical properties is, therefore, crucial.

Addressing these challenges may require further development of characterization techniques, which are identified as a limiting factor. A wide range of techniques should be applied, as no one technique can fully characterize a sample. Combined techniques are continually being developed; however, these entail a compromise between the quality of data sets and the benefits of a combined data set. Standardization of characterization techniques, particularly for biochar in catalytic applications, would facilitate comparisons between different studies; round robin studies have shown interlaboratory reproducibility to be poor for standard biochars. In Raman and XPS analysis of carbon samples in particular, there was little similarity in curve deconvolution methods used between studies, making comparisons of results difficult.

Additional factors which are often overlooked in the study of biomass-derived catalysts are practical considerations, such as the long-term durability and mechanical strength of the catalysts. Particle attrition, for example, can be a problem in stirred tank reactors, due to agitation. In addition, chemical activations and surface treatments used in the literature are not always sustainable at larger scales; any sustainability advantages from the improved catalyst may be canceled out by the volume of waste chemicals produced. The goal of tailor-made catalysts for desired applications can only be realized if the resulting catalysts are cost-effective.

However, in spite of the challenges, it is clear that biomass-derived catalysts and catalyst supports can be designed and synthesized such that they have high efficacy in a range of important reactions; in particular, in key "green" processes such as deNO_x and deSO_x processes, biodiesel synthesis, and as electrocatalysts in fuel cells. If the scalable and standardized production of catalytic materials from biomass can be realized, then this will provide a significant advancement in the development of a future sustainable chemical industry.

References

[1] X. Mo, D.E. López, K. Suwannakarn, Y. Liu, E. Lotero, J.G. Goodwin, L. Changqing, Activation and deactivation characteristics of sulfonated carbon catalysts, J. Catal. 254 (2) (2008) 332–338.
[2] M.M. Antxustegi, A.R. Pierna, N. Ruiz, Chemical activation of Vulcan® XC72R to be used as support for NiNbPtRu catalysts in PEM fuel cells, Int. J. Hydrog. Energy 39 (8) (2014) 3978–3983.
[3] E.L. da Silva, M.R.O. Vega, P. dos Santos Correa, A. Cuña, N. Tancredi, C. de Fraga Malfatti, Influence of activated carbon porous texture on catalyst activity for ethanol electro-oxidation, Int. J. Hydrog. Energy 39 (27) (2014) 14760–14767.
[4] J.H. Kim, J.Y. Cheon, T.J. Shin, J.Y. Park, S.H. Joo, Effect of surface oxygen functionalization of carbon support on the activity and durability of Pt/C catalysts for the oxygen reduction reaction, Carbon 101 (2016) 449–457.
[5] K. Qian, A. Kumar, H. Zhang, D. Bellmer, R. Huhnke, Recent advances in utilization of biochar, Renew. Sust. Energ. Rev. 42 (2015) 1055–1064.
[6] D.R. Dreyer, C.W. Bielawski, Carbocatalysis: heterogeneous carbons finding utility in synthetic chemistry, Chem. Sci. 2 (7) (2011) 1233.
[7] A.V. Bridgwater, Review of fast pyrolysis of biomass and product upgrading, Biomass Bioenergy 38 (2012) 68–94.
[8] A. Bhatnagar, W. Hogland, M. Marques, M. Sillanpää, An overview of the modification methods of activated carbon for its water treatment applications, Chem. Eng. J. 219 (2013) 499–511.
[9] E. Auer, A. Freund, J. Pietsch, T. Tacke, Carbons as supports for industrial precious metal catalysts, Appl. Catal. A Gen. 173 (2) (1998) 259–271.
[10] J.J. Freeman, Active Carbon Edited by R. C. Bansal, J.-B. Donnet and F. Stoeckli. Marcel Dekker, New York, 1988, pp. xiv + 482, US$150.00. ISBN 0-8247-7842-1, J. Chem. Technol. Biotechnol. 48 (2) (2007) 240–241.
[11] F. Rodríguez-Reinoso, M. Molina-Sabio, Activated carbons from lignocellulosic materials by chemical and/or physical activation: an overview, Carbon 30 (7) (1992) 1111–1118.
[12] H.P. Boehm, Zur Struktur Der Rußteilchen, Z. Anorg. Allg. Chem. 297 (5–6) (1958) 315–322.
[13] P. Toth, T. Vikström, R. Molinder, H. Wiinikka, Structure of carbon black continuously produced from biomass pyrolysis oil, Green Chem. 20 (17) (2018) 3981–3992.

[14] S.H.Y.S. Abdullah, N.H.M. Hanapi, A. Azid, R. Umar, H. Juahir, H. Khatoon, A. Endut, A review of biomass-derived heterogeneous catalyst for a sustainable biodiesel production, Renew. Sust. Energ. Rev. 70 (2017) 1040–1051.
[15] X. Cao, S. Sun, R. Sun, Application of biochar-based catalysts in biomass upgrading: a review, RSC Adv. 7 (77) (2017) 48793–48805.
[16] J.S. Cha, S.H. Park, S.-C. Jung, C. Ryu, J.-K. Jeon, M.-C. Shin, Y.-K. Park, Production and utilization of biochar: a review, J. Ind. Eng. Chem. 40 (2016) 1–15.
[17] J. Lee, K.-H. Kim, E.E. Kwon, Biochar as a catalyst, Renew. Sust. Energ. Rev. 77 (2017) 70–79.
[18] E. Lam, J.H.T. Luong, Carbon materials as catalyst supports and catalysts in the transformation of biomass to fuels and chemicals, ACS Catal. 4 (10) (2014) 3393–3410.
[19] S.-H. Kong, S.-K. Loh, R.T. Bachmann, S.A. Rahim, J. Salimon, Biochar from oil palm biomass: a review of its potential and challenges, Renew. Sust. Energ. Rev. 39 (2014) 729–739.
[20] A. Sharma, V. Pareek, D. Zhang, Biomass pyrolysis—a review of modelling, process parameters and catalytic studies, Renew. Sust. Energ. Rev. 50 (2015) 1081–1096.
[21] K. Crombie, O. Mašek, S.P. Sohi, P. Brownsort, A. Cross, The effect of pyrolysis conditions on biochar stability as determined by three methods, GCB Bioenergy 5 (2) (2013) 122–131.
[22] J. Cai, B. Li, C. Chen, J. Wang, M. Zhao, K. Zhang, Hydrothermal carbonization of tobacco stalk for fuel application, Bioresour. Technol. 220 (2016) 305–311.
[23] F. Bergius, Production of hydrogen from water and coal from cellulose at high temperatures and pressures, J. Soc. Chem. Ind. 32 (9) (1913) 462–467.
[24] S. Nizamuddin, H.A. Baloch, G.J. Griffin, N.M. Mubarak, A.W. Bhutto, R. Abro, S. A. Mazari, B.S. Ali, An overview of effect of process parameters on hydrothermal carbonization of biomass, Renew. Sust. Energ. Rev. 73 (December 2015) (2017) 1289–1299.
[25] A. Funke, F. Ziegler, Hydrothermal carbonization of biomass: a summary and discussion of chemical mechanisms for process engineering, Biofuels Bioprod. Biorefin. 6 (4) (2010) 160–177.
[26] J.G. Lynam, C.J. Coronella, W. Yan, M.T. Reza, V.R. Vasquez, Acetic acid and lithium chloride effects on hydrothermal carbonization of lignocellulosic biomass, Bioresour. Technol. 102 (10) (2011) 6192–6199.
[27] H.S. Kambo, A. Dutta, A comparative review of biochar and hydrochar in terms of production, physico-chemical properties and applications, Renew. Sust. Energ. Rev. 45 (2015) 359–378.
[28] M. Kumar, A.O. Oyedun, A. Kumar, A review on the current status of various hydrothermal technologies on biomass feedstock, Renew. Sust. Energ. Rev. 81 (May 2017) (2018) 1742–1770.
[29] G.K. Parshetti, S. Chowdhury, R. Balasubramanian, Hydrothermal conversion of urban food waste to chars for removal of textile dyes from contaminated waters, Bioresour. Technol. 161 (2014) 310–319.
[30] P. Zhao, Y. Shen, S. Ge, Z. Chen, K. Yoshikawa, Clean solid biofuel production from high moisture content waste biomass employing hydrothermal treatment, Appl. Energy 131 (2014) 345–367.
[31] J. Fang, Z. Lu, Y. Sik, B. Gao, Journal of industrial and engineering chemistry minireview of potential applications of hydrochar derived from hydrothermal carbonization of biomass, J. Ind. Eng. Chem. 57 (2018) 15–21.
[32] S. Kang, X. Li, J. Fan, J. Chang, Characterization of hydrochars produced by hydrothermal carbonization of lignin, cellulose, d-xylose, and wood meal, Ind. Eng. Chem. Res. 51 (26) (2012) 9023–9031.

[33] V. Benavente, E. Calabuig, A. Fullana, Upgrading of moist agro-industrial wastes by hydrothermal carbonization, J. Anal. Appl. Pyrolysis 113 (2015) 89–98.
[34] J. Phillips, T. Shiina, M. Nemer, K. Lester, Graphitic structures by design, Langmuir 22 (23) (2006) 9694–9703.
[35] A. Thaib, G.A. Martin, P. Pinheiro, M.C. Schouler, P. Gadelle, Formation of carbon nanotubes from the carbon monoxide disproportionation reaction over Co/Al_2O_3 and Co/SiO_2 catalysts, Catal. Lett. 63 (3/4) (1999) 135–141.
[36] P. Serp, M. Corrias, P. Kalck, Carbon nanotubes and nanofibers in catalysis, Appl. Catal. A Gen. 253 (2) (2003) 337–358.
[37] C.H. Collett, J. McGregor, Things go better with coke: The beneficial role of carbonaceous deposits in heterogeneous catalysis, Catal. Sci. Technol. 6 (2) (2016) 363–378.
[38] R. Ormsby, J.R. Kastner, J. Miller, Hemicellulose hydrolysis using solid acid catalysts generated from biochar, Catal. Today 190 (1) (2012) 89–97.
[39] J. Rafi, A. Mahammad, M. Rajashekar, B.V.S. Srinivas, K. Rao, R.B.N. Prasad, N. Lingaiah, Esterification of glycerol over a solid acid biochar catalyst derived from waste biomass, RSC Adv. 5 (55) (2015) 44550–44556.
[40] Z. Abu El-Rub, E.A. Bramer, G. Brem, Experimental comparison of biomass chars with other catalysts for tar reduction, Fuel 87 (10 – 11) (2008) 2243–2252.
[41] A.M. Dehkhoda, A.H. West, N. Ellis, Biochar based solid acid catalyst for biodiesel production, Appl. Catal. A Gen. 382 (2) (2010) 197–204.
[42] J.R. Kastner, J. Miller, D.P. Geller, J. Locklin, L.H. Keith, T. Johnson, Catalytic esterification of fatty acids using solid acid catalysts generated from biochar and activated carbon, Catal. Today 190 (1) (2012) 122–132.
[43] G. Moussavi, R. Khosravi, Preparation and characterization of a biochar from pistachio hull biomass and its catalytic potential for ozonation of water recalcitrant contaminants, Bioresour. Technol. 119 (2012) 66–71.
[44] Y. Shen, P. Zhao, Q. Shao, D. Ma, F. Takahashi, K. Yoshikawa, In-situ catalytic conversion of tar using rice husk char-supported nickel-Iron catalysts for biomass pyrolysis/gasification, Appl. Catal. B Environ. 152–153 (2014) 140–151.
[45] F.-J. Wang, S. Zhang, Z.-D. Chen, C. Liu, Y.-G. Wang, Tar reforming using char as catalyst during pyrolysis and gasification of Shengli Brown coal, J. Anal. Appl. Pyrolysis 105 (2014) 269–275.
[46] J.T. Yu, A.M. Dehkhoda, N. Ellis, Development of biochar-based catalyst for transesterification of canola oil, Energy Fuel 25 (1) (2011) 337–344.
[47] B. Shen, J. Chen, S. Yue, G. Li, A comparative study of modified cotton biochar and activated carbon based catalysts in low temperature SCR, Fuel 156 (2015) 47–53.
[48] A. Bazargan, M.D. Kostić, O.S. Stamenković, V.B. Veljković, G. McKay, A calcium oxide-based catalyst derived from palm kernel shell gasification residues for biodiesel production, Fuel 150 (2015) 519–525.
[49] R. Chakraborty, S. Bepari, A. Banerjee, Transesterification of soybean oil catalyzed by fly ash and egg shell derived solid catalysts, Chem. Eng. J. 165 (3) (2010) 798–805.
[50] C. Ofori-Boateng, K.T. Lee, The potential of using cocoa pod husks as green solid base catalysts for the transesterification of soybean oil into biodiesel: effects of biodiesel on engine performance, Chem. Eng. J. 220 (2013) 395–401.
[51] L. Riadi, E. Purwanto, H. Kurniawan, R. Oktaviana, Effect of bio-based catalyst in biodiesel synthesis, Procedia Chem. 9 (2014) 172–181.
[52] D. Feng, Z. Yu, Y. Zhao, S. Sun, J. Gao, Improvement and maintenance of biochar catalytic activity for in-situ biomass tar reforming during pyrolysis and H_2O/CO_2 gasification, Fuel Process. Technol. 172 (2018) 106–114.
[53] H.J. Bachmann, T.D. Bucheli, A. Dieguez-Alonso, D. Fabbri, H. Knicker, H.-P. Schmidt, A. Ulbricht, R. Becker, A. Buscaroli, D. Buerge, A. Cross, D. Dickinson, A. Enders, V.I. Esteves, M.W.H. Evangelou, G. Fellet, K. Friedrich,

G.G. Guerrero, B. Glaser, U.M. Hanke, K. Hanley, I. Hilber, D. Kalderis, J. Leifeld, O. Masek, J. Mumme, M.P. Carmona, R.C. Pereira, F. Rees, A.G. Rombolà, J.M. de la Rosa, R. Sakrabani, S. Sohi, G. Soja, M. Valagussa, F. Verheijen, F. Zehetner, Toward the standardization of biochar analysis: the COST action TD1107 Interlaboratory comparison, J. Agric. Food Chem. 64 (2) (2016) 513–527.

[54] K. Crombie, O. Mašek, A. Cross, S. Sohi, Biochar – synergies and trade-offs between soil enhancing properties and C sequestration potential, GCB Bioenergy 7 (5) (2015) 1161–1175.

[55] K. Jindo, H. Mizumoto, Y. Sawada, M.A. Sanchez-Monedero, T. Sonoki, Physical and chemical characterization of biochars derived from different agricultural residues, Biogeosciences 11 (23) (2014) 6613–6621.

[56] L. Zhao, X. Cao, O. Mašek, A. Zimmerman, Heterogeneity of biochar properties as a function of feedstock sources and production temperatures, J. Hazard. Mater. 256–257 (2013) 1–9.

[57] C.E. Brewer, R. Unger, K. Schmidt-Rohr, R.C. Brown, Criteria to select biochars for field studies based on biochar chemical properties, BioEnergy Res. 4 (4) (2011) 312–323.

[58] E. Danso-Boateng, G. Shama, A.D. Wheatley, S.J. Martin, R.G. Holdich, Hydrothermal carbonisation of sewage sludge: effect of process conditions on product characteristics and methane production, Bioresour. Technol. 177 (2015) 318–327.

[59] L. Wei, Y. Huang, Y. Li, L. Huang, N.N. Mar, Q. Huang, Z. Liu, Biochar characteristics produced from rice husks and their sorption properties for the acetanilide herbicide metolachlor, Environ. Sci. Pollut. Res. 24 (5) (2017) 4552–4561.

[60] Y. Zhao, D. Feng, Z. Yu, Y. Huang, S. Sun, Effect of pyrolysis temperature on char structure and chemical speciation of alkali and alkaline earth metallic species in biochar, Fuel Process. Technol. 141 (2016) 54–60.

[61] W.-J. Liu, F.-X. Zeng, H. Jiang, X.-S. Zhang, Preparation of high adsorption capacity bio-chars from waste biomass, Bioresour. Technol. 102 (17) (2011) 8247–8252.

[62] N. Zhao, X. Yang, J. Zhang, L. Zhu, Y. Lv, N. Zhao, X. Yang, J. Zhang, L. Zhu, Y. Lv, Adsorption mechanisms of dodecylbenzene sulfonic acid by corn straw and poplar leaf biochars, Materials 10 (10) (2017) 1119.

[63] J. Fang, B. Gao, J. Chen, A.R. Zimmerman, Hydrochars derived from plant biomass under various conditions: characterization and potential applications and impacts, Chem. Eng. J. 267 (2015) 253–259.

[64] A. Dutta, J. Datta, Outstanding catalyst performance of PdAuNi nanoparticles for the anodic reaction in an alkaline direct ethanol (with anion-exchange membrane) fuel cell, J. Phys. Chem. C 116 (49) (2012) 25677–25688.

[65] M.E. Fernandez, B. Ledesma, S. Román, P.R. Bonelli, A.L. Cukierman, Development and characterization of activated hydrochars from Orange peels as potential adsorbents for emerging organic contaminants, Bioresour. Technol. 183 (2015) 221–228.

[66] S. Kabir, A. Serov, K. Artyushkova, P. Atanassov, Design of novel graphene materials as a support for palladium nanoparticles: highly active catalysts towards ethanol Electrooxidation, Electrochim. Acta 203 (2016) 144–153.

[67] A. Elsheikh, V.L. Martins, J. McGregor, Influence of physicochemical characteristics of carbon supports on Pd ethanol oxidation catalysts, Energy Procedia 151 (2018) 79–83.

[68] M. Sevilla, A.B. Fuertes, Chemical and structural properties of carbonaceous products obtained by hydrothermal carbonization of saccharides, Chem. Eur. J. 15 (16) (2009) 4195–4203.

[69] A. Kozlov, D. Svishchev, I. Donskoy, V. Shamansky, A. Ryzhkov, A technique proximate and ultimate analysis of solid fuels and coal tar, J. Therm. Anal. Calorim. 122 (3) (2015) 1213–1220.

[70] A.M. Smith, S. Singh, A.B. Ross, Fate of inorganic material during hydrothermal carbonisation of biomass: influence of feedstock on combustion behaviour of hydrochar, Fuel 169 (2016) 135–145.
[71] C. He, A. Giannis, J.Y. Wang, Conversion of sewage sludge to clean solid fuel using hydrothermal carbonization: hydrochar fuel characteristics and combustion behavior, Appl. Energy 111 (2013) 257–266.
[72] O. Bayraktar, E.L. Kugler, Characterization of coke on equilibrium fluid catalytic cracking catalysts by temperature-programmed oxidation, Appl. Catal. A Gen. 233 (1–2) (2002) 197–213.
[73] K. Chen, Z. Xue, L. He, A. Guo, Z. Wang, A temperature-programmed oxidation method for quantitative characterization of the thermal cokes morphology, Fuel 113 (2013) 274–279.
[74] W. Qi, D. Su, Metal-free carbon catalysts for oxidative dehydrogenation reactions, ACS Catal. 4 (9) (2014) 3212–3218.
[75] C.A. Querini, S.C. Fung, Coke characterization by temperature programmed techniques, Catal. Today 37 (3) (1997) 277–283.
[76] B. Sánchez, M.S. Gross, B.D. Costa, C.A. Querini, Coke analysis by temperature-programmed oxidation: morphology characterization, Appl. Catal. A Gen. 364 (1–2) (2009) 35–41.
[77] S. Haydar, C. Moreno-Castilla, M.A. Ferro-Garcia, F. Carrasco-Marin, J. Rivera-Utrilla, A. Perrard, J.P. Joly, Regularities in the temperature-programmed desorption spectra of CO_2 and CO from activated carbons, Carbon 38 (9) (2000) 1297–1308.
[78] H. Muckenhuber, H. Grothe, The heterogeneous reaction between soot and NO_2 at elevated temperature, Carbon 44 (3) (2006) 546–559.
[79] A. Bagreev, T.J. Bandosz, H2S adsorption/oxidation on unmodified activated carbons: Importance of Prehumidification, Carbon 39 (15) (2001) 2303–2311.
[80] L. Li, M. Chen, G. Huang, N. Yang, L. Zhang, H. Wang, L. Yu, W. Wang, J. Gao, A green method to prepare Pd–Ag nanoparticles supported on reduced graphene oxide and their electrochemical catalysis of methanol and ethanol oxidation, J. Power Sources 263 (2014) 13–21.
[81] S.-S. Li, Y.-Y. Hu, J.-J. Feng, Z.-Y. Lv, J.-R. Chen, A.-J. Wang, Rapid room-temperature synthesis of Pd nanodendrites on reduced graphene oxide for catalytic oxidation of ethylene glycol and glycerol, Int. J. Hydrog. Energy 39 (8) (2014) 3730–3738.
[82] F. Nitze, Synthesis and Characterization of Palladium Based Carbon Nanostructure-Composites and their Clean-Energy Application, Umeå University, Sweden, 2013.
[83] L. Zhang, Q. Wang, B. Wang, G. Yang, L.A. Lucia, J. Chen, Hydrothermal carbonization of corncob residues for hydrochar production, Energy Fuels 29 (2) (2015) 872–876.
[84] R.J.J. Jansen, H. van Bekkum, Amination and ammoxidation of activated carbons, Carbon 32 (8) (1994) 1507–1516.
[85] C. Moreno-Castilla, F. Carrasco-Marín, A. Mueden, The creation of acid carbon surfaces by treatment with (NH4)2S2O8, Carbon 35 (10–11) (1997) 1619–1626.
[86] H. ShamsiJazeyi, T. Kaghazchi, Investigation of nitric acid treatment of activated carbon for enhanced aqueous mercury removal, J. Ind. Eng. Chem. 16 (5) (2010) 852–858.
[87] C. Toles, S. Rimmer, J.C. Hower, Production of activated carbons from a Washington lignite using phosphoric acid activation, Carbon 34 (11) (1996) 1419–1426.
[88] A. Koch, A. Krzton, G. Finqueneisel, O. Heintz, J.-V. Weber, T. Zimny, A study of carbonaceous char oxidation in air by semi-quantitative FTIR spectroscopy, Fuel 77 (6) (1998) 563–569.

[89] B.K. Pradhan, N.K. Sandle, Effect of different oxidizing agent treatments on the surface properties of activated carbons, Carbon 37 (8) (1999) 1323–1332.
[90] S. Shin, J. Jang, S.H. Yoon, I. Mochida, A study on the effect of heat treatment on functional groups of pitch based activated carbon fiber using FTIR, Carbon 35 (12) (1997) 1739–1743.
[91] J.M. Valente Nabais, P.J.M. Carrott, M.M.L. Ribeiro Carrott, J.A. Menéndez, Preparation and modification of activated carbon fibres by microwave heating, Carbon 42 (7) (2004) 1315–1320.
[92] C. Gai, Y. Guo, N. Peng, T. Liu, Z. Liu, N-doped biochar derived from co-hydrothermal carbonization of rice husk and *Chlorella pyrenoidosa* for enhancing copper ion adsorption, RSC Adv. 6 (59) (2016) 53713–53722.
[93] S. Al-Khattaf, C. D'Agostino, M.N. Akhtar, N. Al-Yassir, N.Y. Tan, L.F. Gladden, The effect of coke deposition on the activity and selectivity of the HZSM-5 zeolite during ethylbenzene alkylation reaction in the presence of ethanol, Catal. Sci. Technol. 4 (4) (2014) 1017.
[94] J. Guo, H. Lou, L. Mo, X. Zheng, The reactivity of surface active carbonaceous species with CO_2 and its role on hydrocarbon conversion reactions, J. Mol. Catal. A Chem. 316 (1–2) (2010) 1–7.
[95] E. Raymundo-Piñero, D. Cazorla-Amorós, A. Linares-Solano, Temperature programmed desorption study on the mechanism of SO_2 oxidation by activated carbon and activated carbon fibres, Carbon 39 (2) (2001) 231–242.
[96] J. Zawadzki, M. Wiśniewski, K. Skowrońska, Heterogeneous reactions of NO2 and NO–O2 on the surface of carbons, Carbon 41 (2) (2003) 235–246.
[97] J. Zawadzki, M. Wiśniewski, J. Weber, O. Heintz, B. Azambre, IR study of adsorption and decomposition of Propan-2-Ol on carbon and carbon-supported catalysts, Carbon 39 (2) (2001) 187–192.
[98] P. Ayala, M.E.M. da Costa, R. Prioli, F. Freire, Nano- and micro-scale wear of fluorinated carbon films, Surf. Coat. Technol. 182 (2–3) (2004) 335–341.
[99] F. Haghseresht, G.Q. Lu, A.K. Whittaker, Carbon structure and porosity of carbonaceous adsorbents in relation to their adsorption properties, Carbon 37 (9) (1999) 1491–1497.
[100] G. Sanz, L.M.M. Sara, M.G. James, J. Axel Zeitler, N. Al-Yassir, S. Al-Khattaf, L. F. Gladden, A new perspective on catalytic dehydrogenation of ethylbenzene: the influence of side-reactions on catalytic performance, Catal. Sci. Technol. 5 (7) (2015) 3782–3797.
[101] R.J. Nemanich, S.A. Solin, First- and second-order Raman scattering from finite-size crystals of graphite, Phys. Rev. B 20 (2) (1979) 392–401.
[102] M.A. Pimenta, G. Dresselhaus, M.S. Dresselhaus, L.G. Cançado, A. Jorio, R. Saito, Studying disorder in graphite-based systems by Raman spectroscopy, Phys. Chem. Chem. Phys. 9 (11) (2007) 1276–1290.
[103] J. Schwan, S. Ulrich, V. Batori, H. Ehrhardt, S.R.P. Silva, Raman spectroscopy on amorphous carbon films, J. Appl. Phys. 80 (1) (1998) 440.
[104] C.H. Chia, B. Gong, S.D. Joseph, C.E. Marjo, P. Munroe, A.M. Rich, Imaging of mineral-enriched biochar by FTIR, Raman and SEM–EDX, Vib. Spectrosc. 62 (2012) 248–257.
[105] D.M. Quyn, H. Wu, J.-i. Hayashi, C.-Z. Li, Volatilisation and catalytic effects of alkali and alkaline earth metallic species during the pyrolysis and gasification of Victorian Brown coal. Part IV. Catalytic effects of NaCl and ion-exchangeable Na in coal on char reactivity, Fuel 82 (5) (2003) 587–593.
[106] R. Arrigo, M. Hävecker, S. Wrabetz, R. Blume, M. Lerch, J. McGregor, E.P. J. Parrott, J.A. Zeitler, L.F. Gladden, A. Knop-Gericke, R. Schlögl, D.S. Su, Tuning the acid/base properties of Nanocarbons by functionalization via amination, J. Am. Chem. Soc. 132 (28) (2010) 9616–9630.

[107] J. McGregor, Z. Huang, E.P.J. Parrott, J.A. Zeitler, K.L. Nguyen, J.M. Rawson, A. Carley, T.W. Hansen, J.-P. Tessonnier, D.S. Su, Active coke: carbonaceous materials as catalysts for alkane dehydrogenation, J. Catal. 269 (2) (2010) 329–339.
[108] E.P.J. Parrott, J. Axel Zeitler, J. McGregor, S.-P. Oei, H.E. Unalan, S.-C. Tan, W. I. Milne, J.-P. Tessonnier, R. Schlögl, L.F. Gladden, Understanding the dielectric properties of heat-treated carbon nanofibers at terahertz frequencies: a new perspective on the catalytic activity of structured carbonaceous materials, J. Phys. Chem. C 113 (24) (2009) 10554–10559.
[109] H. Hu, J.H. Xin, H. Hu, X. Wang, Structural and mechanistic understanding of an active and durable graphene carbocatalyst for reduction of 4-nitrophenol at room temperature, Nano Res. 8 (12) (2015) 3992–4006.
[110] W. Wang, Y. Jiang, M. Hunger, Mechanistic investigations of the methanol-to-olefin (MTO) process on acidic zeolite catalysts by in situ solid-state NMR spectroscopy, Catal. Today 113 (1–2) (2006) 102–114.
[111] J.Z. Hu, M.S. Solum, C.M.V. Taylor, R.J. Pugmire, D.M. Grant, Structural'determination in carbonaceous solids using advanced solid state NMR techniques, Energy Fuels 15 (1) (2001) 14–22.
[112] Q. Wu, S. Yu, N. Hao, T.W. Jr, X. Meng, M. Li, P. Yunqiao, S. Liu, A.J. Ragauskas, Characterization of products from hydrothermal carbonization of pine, Bioresour. Technol. 244 (July) (2017) 78–83.
[113] M. Foston, G.A. Nunnery, X. Meng, Q. Sun, F.S. Baker, A. Ragauskas, NMR a critical tool to study the production of carbon fiber from lignin, Carbon 52 (2013) 65–73.
[114] S.-J. Park, M.-K. Seo, Y.-S. Lee, Surface characteristics of fluorine-modified PAN-based carbon fibers, Carbon 41 (4) (2003) 723–730.
[115] Y.-J. Liou, W.-J. Huang, Quantitative analysis of graphene sheet content in wood char powders during catalytic pyrolysis, J. Mater. Sci. Technol. 29 (5) (2013) 406–410.
[116] M.J. Lázaro, L. Calvillo, V. Celorrio, J. Pardo, S. Perathoner, R. Moliner, Study and Application of Vulcan XC-72 in Low Temperature Fuel Cells, 2011, pp. 41–68.
[117] T. Maiyalagan, K. Scott, Performance of carbon nanofiber supported Pd–Ni catalysts for electro-oxidation of ethanol in alkaline medium, J. Power Sources 195 (16) (2010) 5246–5251.
[118] W. Grünert, Auger electron, X ray and UV photoelectron spectroscopies, in: Characterization of Solid Materials and Heterogeneous Catalysts, Wiley-VCH Verlag GmbH & Co. KGaA., Weinheim, Germany, 2012, pp. 537–583.
[119] T.J. Bandosz, Surface chemistry of carbon materials, in: Carbon Materials for Catalysis, John Wiley & Sons, Inc., Hoboken, NJ, USA, 2008, pp. 45–92.
[120] A.B. Fuertes, M. Camps Arbestain, M. Sevilla, J.A. Maciá-Agulló, S. Fiol, R. López, R.J. Smernik, W.P. Aitkenhead, F. Arce, F. Macías, Chemical and structural properties of carbonaceous products obtained by pyrolysis and hydrothermal carbonisation of corn Stover, Soil Res. 48 (7) (2010) 618.
[121] S. Hu, F. Munoz, J. Noborikawa, J. Haan, L. Scudiero, S. Ha, Carbon supported Pd-based bimetallic and trimetallic catalyst for formic acid electrochemical oxidation, Appl. Catal. B Environ. 180 (2016) 758–765.
[122] Y.-H. Qin, Y. Li, R.-L. Lv, T.-L. Wang, W.-G. Wang, C.-W. Wang, Pd-Au/C catalysts with different alloying degrees for ethanol oxidation in alkaline media, Electrochim. Acta 144 (2014) 50–55.
[123] R.J.J. Jansen, H. van Bekkum, XPS of nitrogen-containing functional groups on activated carbon, Carbon 33 (8) (1995) 1021–1027.
[124] Y. Nakayama, F. Soeda, A. Ishitani, XPS study of the carbon fiber matrix interface, Carbon 28 (1) (1990) 21–26.
[125] M. Smith, L. Scudiero, J. Espinal, J.-S. McEwen, M. Garcia-Perez, Improving the deconvolution and interpretation of XPS spectra from chars by ab initio calculations, Carbon 110 (2016) 155–171.

[126] X. Ma, B. Zhou, A. Budai, A. Jeng, X. Hao, D. Wei, Y. Zhang, D. Rasse, Study of biochar properties by scanning electron microscope – Energy dispersive X-Ray spectroscopy (SEM-EDX), Commun. Soil Sci. Plant Anal. 47 (5) (2016) 593–601.
[127] A.O. Neto, M.M. Tusi, N.S. de Oliveira Polanco, S.G. da Silva, M.C. dos Santos, E. V. Spinacé, PdBi/C electrocatalysts for ethanol electro-oxidation in alkaline medium, Int. J. Hydrog. Energy 36 (17) (2011) 10522–10526.
[128] J. Hong, M.K. Park, E.J. Lee, D.E. Lee, D.S. Hwang, S. Ryu, Origin of new broad Raman D and G peaks in annealed graphene, Sci. Rep. 3 (1) (2013) 2700.
[129] G.M.K. Abotsi, A.W. Scaroni, A review of carbon-supported Hydrodesulfurization catalysts, Fuel Process. Technol. 22 (2) (1989) 107–133.
[130] M. Breysse, P. Afanasiev, C. Geantet, M. Vrinat, Overview of support effects in hydrotreating catalysts, Catal. Today 86 (1–4) (2003) 5–16.
[131] K. Sakanishi, T. Nagamatsu, I. Mochida, D.D. Whitehurst, Hydrodesulfurization kinetics and mechanism of 4,6-dimethyldibenzothiophene over NiMo catalyst supported on carbon, J. Mol. Catal. A Chem. 155 (1–2) (2000) 101–109.
[132] M.C. Bahome, L.L. Jewell, K. Padayachy, D. Hildebrandt, D. Glasser, A.K. Datye, N.J. Coville, Fe-Ru small particle bimetallic catalysts supported on carbon nanotubes for use in Fischer–Tröpsch synthesis, Appl. Catal. A Gen. 328 (2) (2007) 243–251.
[133] R.C. Reuel, C.H. Bartholomew, Effects of support and dispersion on the CO hydrogenation activity/selectivity properties of cobalt, J. Catal. 85 (1) (1984) 78–88.
[134] Z. Kowalczyk, S. Jodzis, J. Sentek, Studies on kinetics of ammonia synthesis over ruthenium catalyst supported on active carbon, Appl. Catal. A Gen. 138 (1) (1996) 83–91.
[135] X. Zheng, S. Zhang, J. Xu, K. Wei, Effect of thermal and oxidative treatments of activated carbon on its surface structure and suitability as a support for barium-promoted ruthenium in Ammonia synthesis catalysts, Carbon 40 (14) (2002) 2597–2603.
[136] J.W. Larsen, M. Freund, K.Y. Kim, M. Sidovar, J.L. Stuart, Mechanism of the carbon catalyzed reduction of nitrobenzene by hydrazine, Carbon 38 (5) (2000) 655–661.
[137] X. Liu, L. Dai, Carbon-based metal-free catalysts, Nat. Rev. Mater. 1 (11) (2016) 16064.
[138] A. Lisovskii, R. Semiat, C. Aharoni, Adsorption of sulfur dioxide by active carbon treated by nitric acid: I. Effect of the treatment on adsorption of SO_2 and extractability of the acid formed, Carbon 35 (10–11) (1997) 1639–1643.
[139] I. Martin-Gullon, R. Andrews, M. Jagtoyen, F. Derbyshire, PAN-based activated carbon fiber composites for sulfur dioxide conversion: influence of fiber activation method, Fuel 80 (7) (2001) 969–977.
[140] B. Xia, J. Phillips, C.K. Chen, L.R. Radovic, I.F. Silva, J. Angel Menéndez, Impact of pretreatments on the selectivity of carbon for NOx adsorption/reduction, Energy Fuels 13 (4) (1999) 903–906.
[141] K.L. Dunlap, Phosgene, in: Kirk-Othmer Encyclopedia of Chemical Technology, John Wiley & Sons, Inc., Hoboken, NJ, USA, 2010, pp. 1–13.
[142] W. Schneider, W. Diller, Phosgene, in: Ullmann's Encyclopedia of Industrial Chemistry, Wiley-VCH Verlag GmbH & Co. KGaA, Weinheim, Germany, 2000.
[143] K. Knoblauch, E. Richter, H. Jüntgen, Application of active coke in processes of SO2- and NOx-removal from flue gases, Fuel 60 (9) (1981) 832–838.
[144] B. Li, Z. Xu, A nonmetal catalyst for molecular hydrogen activation with comparable catalytic hydrogenation capability to noble metal catalyst, J. Am. Chem. Soc. 131 (45) (2009) 16380–16382.
[145] J. Zhang, X. Liu, R. Blume, A. Zhang, R. Schlögl, D.S. Su, Surface-modified carbon nanotubes catalyze oxidative dehydrogenation of n-butane, Science (New York, N.Y.) 322 (5898) (2008) 73–77.
[146] D.R. Dreyer, H.-P. Jia, C.W. Bielawski, Graphene oxide: a convenient carbocatalyst for facilitating oxidation and hydration reactions, Angew. Chem. Int. Ed. 49 (38) (2010) 6813–6816.

[147] P. Makowski, R.D. Cakan, M. Antonietti, F. Goettmann, M.-M. Titirici, Selective partial hydrogenation of hydroxy aromatic derivatives with palladium nanoparticles supported on hydrophilic carbon, Chem. Commun. 8 (2008) 999–1001.
[148] Y. Shen, Chars as carbonaceous adsorbents/catalysts for tar elimination during biomass pyrolysis or gasification, Renew. Sust. Energ. Rev. 43 (2015) 281–295.
[149] L. Prati, D. Bergna, A. Villa, P. Spontoni, C.L. Bianchi, T. Hu, H. Romar, U. Lassi, Carbons from second generation biomass as sustainable supports for catalytic systems, Catal. Today 301 (2018) 239–243.
[150] C.M. Long, M.A. Nascarella, P.A. Valberg, Carbon black vs. black carbon and other airborne materials containing elemental carbon: Physical and chemical distinctions, Environ. Pollut. 181 (2013) 271–286.
[151] M. Perander, N. DeMartini, A. Brink, J. Kramb, O. Karlström, J. Hemming, A. Moilanen, J. Konttinen, M. Hupa, Catalytic effect of Ca and K on CO_2 gasification of spruce wood char, Fuel 150 (2015) 464–472.
[152] S. Mani, J.R. Kastner, A. Juneja, Catalytic decomposition of toluene using a biomass derived catalyst, Fuel Process. Technol. 114 (2013) 118–125.
[153] V.P. Indran, A.S.H. Saud, G.P. Maniam, M.M. Yusoff, Y.H. Taufiq-Yap, M.H. A. Rahim, Versatile boiler ash containing potassium silicate for the synthesis of organic carbonates, RSC Adv. 6 (41) (2016) 34877–34884.
[154] N.B. Klinghoffer, M.J. Castaldi, A. Nzihou, Influence of char composition and inorganics on catalytic activity of char from biomass gasification, Fuel 157 (2015) 37–47.
[155] M. Li, Y. Zheng, Y. Chen, X. Zhu, Biodiesel production from waste cooking oil using a heterogeneous catalyst from pyrolyzed rice husk, Bioresour. Technol. 154 (2014) 345–348.
[156] S. Ren, H. Lei, W. Lu, B. Quan, S. Chen, W. Joan, Hydrocarbon and hydrogen-rich syngas production by biomass catalytic pyrolysis and bio-oil upgrading over biochar catalysts, RSC Adv. 4 (21) (2014) 10731–10737.
[157] Z. Liu, Q. Augustine, S. Kent Hoekman, R. Balasubramanian, Production of solid biochar fuel from waste biomass by hydrothermal carbonization, Fuel 103 (2013) 943–949.
[158] C. Falco, J.P. Marco-Lozar, D. Salinas-Torres, E. Morallón, D. Cazorla-Amorós, M. M. Titirici, D. Lozano-Castelló, Tailoring the porosity of chemically activated hydrothermal carbons: influence of the precursor and hydrothermal carbonization temperature, Carbon 62 (2013) 346–355.
[159] Y. Chen, X. Ai, B. Huang, M. Huang, Y. Huang, L. Yi, Consecutive preparation of hydrochar catalyst functionalized in situ with sulfonic groups for efficient cellulose hydrolysis, Cellulose 24 (7) (2017) 2743–2752.
[160] C. Gai, F. Zhang, T. Yang, Z. Liu, W. Jiao, N. Peng, T. Liu, Q. Lang, X. Yu, Hydrochar supported bimetallic Ni-Fe nanocatalysts with tailored composition, size and shape for improved biomass steam reforming performance, Green Chem. 20 (12) (2018) 2788–2800.
[161] Q. Ma, C. Lin, S. Zhou, Y. Li, W. Shi, S. Ai, Iron nanoparticles in situ encapsulated in lignin-derived hydrochar as an effective catalyst for phenol removal, Environ. Sci. Pollut. Res. 25 (21) (2018) 20833–20840.
[162] A. Tavasoli, M. Aslan, M. Salimi, B. Salar, S.M. Pirbazari, H. Hashemi, K. Kohansal, Influence of the blend nickel/porous hydrothermal carbon and cattle manure hydrochar catalyst on the hydrothermal gasification of cattle manure for H2 production, Energy Convers. Manag. 173 (April) (2018) 15–28.
[163] B. Kumar, M. Asadi, D. Pisasale, S. Sinha-Ray, B.A. Rosen, R. Haasch, J. Abiade, A. L. Yarin, A. Salehi-Khojin, Renewable and metal-free carbon nanofibre catalysts for carbon dioxide reduction, Nat. Commun. 4 (1) (2013) 2819.

[164] X. Chen, R. Paul, L. Dai, Carbon-based supercapacitors for efficient energy storage, Natl. Sci. Rev. 4 (3) (2017) 453–489.
[165] M. Härmas, T. Thomberg, T. Romann, A. Jänes, E. Lust, Carbon for energy storage derived from granulated white sugar by hydrothermal carbonization and subsequent zinc chloride activation, J. Electrochem. Soc. 164 (9) (2017) A1866–A1872.
[166] A. Jain, C. Xu, S. Jayaraman, B. Rajasekhar, J.Y. Lee, M.P. Srinivasan, Mesoporous activated carbons with enhanced porosity by optimal hydrothermal pre-treatment of biomass for supercapacitor applications, Microporous Mesoporous Mater. 218 (2015) 55–61.
[167] T.M. Huggins, J.J. Pietron, H. Wang, Z.J. Ren, J.C. Biffinger, Graphitic biochar as a cathode electrocatalyst support for microbial fuel cells, Bioresour. Technol. 195 (2015) 147–153.
[168] Y. Yuan, T. Liu, F. Peng, J. Tang, S. Zhou, Conversion of sewage sludge into high-performance bifunctional electrode materials for microbial energy harvesting, J. Mater. Chem. A 3 (16) (2015) 8475–8482.

Further readings

M.A. Montes-Moran, D. Suarez, J.A. Menendez, E. Fuente, On the nature of basic sites on carbon surfaces: an overview, Carbon 42 (7) (2004) 1219–1225.
L.R. Radovic, Physicochemical properties of carbon materials: a brief overview, Carbon Materials for Catalysis, John Wiley & Sons, Inc., Hoboken, NJ, USA, 2008, pp. 1–44
Y. Yang, K. Chiang, N. Burke, Porous carbon-supported catalysts for energy and environmental applications: a short review, Catal. Today 178 (1) (2011) 197–205.

CHAPTER 6

Electrospun carbon (nano) fibers for catalysis

Samahe Sadjadi[a] and Sodeh Sadjadi[b]
[a]Gas Conversion Department, Faculty of Petrochemicals, Iran Polymer and Petrochemical Institute, Tehran, Iran
[b]Nuclear Science and Technology Research Institute, Tehran, Iran

1 Introduction

Electrospinning is a facile and versatile approach to fabricate one-dimensional nanostructures such as nanofibers from different types of raw materials with controllable diameters (submicron diameters down to nanometer diameters), compositions, and morphologies, patterns, and functionalities [1, 2].

The term "Electrospinning" is derived from "electrostatic spinning." This is a relatively new term [3]. However, the original idea of applying high electric potentials for the production of liquid drops dated back to the last century [4]. The first patent in this regard was filed by John Francis Cooley in 1900 [5]. The required set-up for the fabrication of polymer filaments by an electrostatic force was patented by Formalas [3]. Sir Geoffrey Ingram Taylor developed the theoretical foundation of electrospinning between 1964 and 1969. In the early 1990s, different groups of researchers (most notably, Reneker, who popularized the name electrospinning) demonstrated that many polymers could be used to produce nanofibers [6].

Although several other methods including phase separation, melt or solution blowing, self-assembly, and template synthesis have been developed for nanofibers fabrication [2], nowadays, electrospinning has attracted considerable attention from different scientific sectors, as well as industry. This attraction stems from the low cost, relative easiness, high speed, and more importantly, high versatility of this technique, allowing control over the fibers' diameters, structures and arrangements, and the possibility of a wide selection of raw materials (e.g., polymers, ceramics, and metals) [7]. Therefore, various nanofiber morphologies and arrangements with unique properties, such as large surface-to-volume ratio, high porosity with excellent pore interconnectivity, flexibility in surface functionalities, and superior mechanical performance, can be fabricated by electrospinning technique [7].

Emerging Carbon Materials for Catalysis
https://doi.org/10.1016/B978-0-12-817561-3.00006-8

The specified structural features of nanofibers result in flexibility in their applicability, and the nonwoven fabrics produced by this technique have made their way into several scientific areas, such as tissue engineering, biomedical, environmental protection, composite reinforcements, energy storage, supercapacitors, nanosensors, catalyst supports, and so forth [8].

Considering the importance of electrospinning, many researchers reviewed various aspect of this issue such as its set-up, utility, technical issues, and mechanism [1, 9–12]. This chapter covers a short overview of recent advances of electrospun carbon materials and their applications in catalysis. After a short introduction to the basic principles of electrospinning, the material classes, the influencing factors on the properties of the electrospun nanofibers, and the advanced techniques for tuning the porosity, composition, structure, and surface properties of electrospun nanofibers are discussed. Finally, some examples of the electrospun nanofibers in catalysis are highlighted.

2 Basic set-up and principle for needle-based electrospinning

One of the classifications of electrospinning is based on the nature of the applied polymer. According to this classification, electrospinning can be classified into two classes: solution electrospinning and melt electrospinning [2].

Environmental concerns due to involvement of organic solvents, the need for the additional solvent extraction process, and low productivity are the main drawbacks of the solution electrospinning process. Melt electrospinning approach, on the other hand, is free from the above-mentioned challenges. However, it has not been studied extensively. This may be because of the limiting constraints associated with the process including: the difficulties inherent in finer fiber formation, the complex equipment used, the intrinsic difficulties associated with the polymer, such as high viscosity and low electrical conductivity, and the electrical discharge issues associated with the use of high voltage to polymeric melt [2, 13].

A basic needle electrospinning set-up is depicted in Fig. 1, which comprises a high voltage power supply, a syringe pump, a spinneret (or a needle nozzle), and a grounded collecting plate (usually a metal screen, plate, or rotating mandrel).

Direct current (DC) is mostly employed as the source of power supply. However, use of alternating current (AC) is also feasible [2].

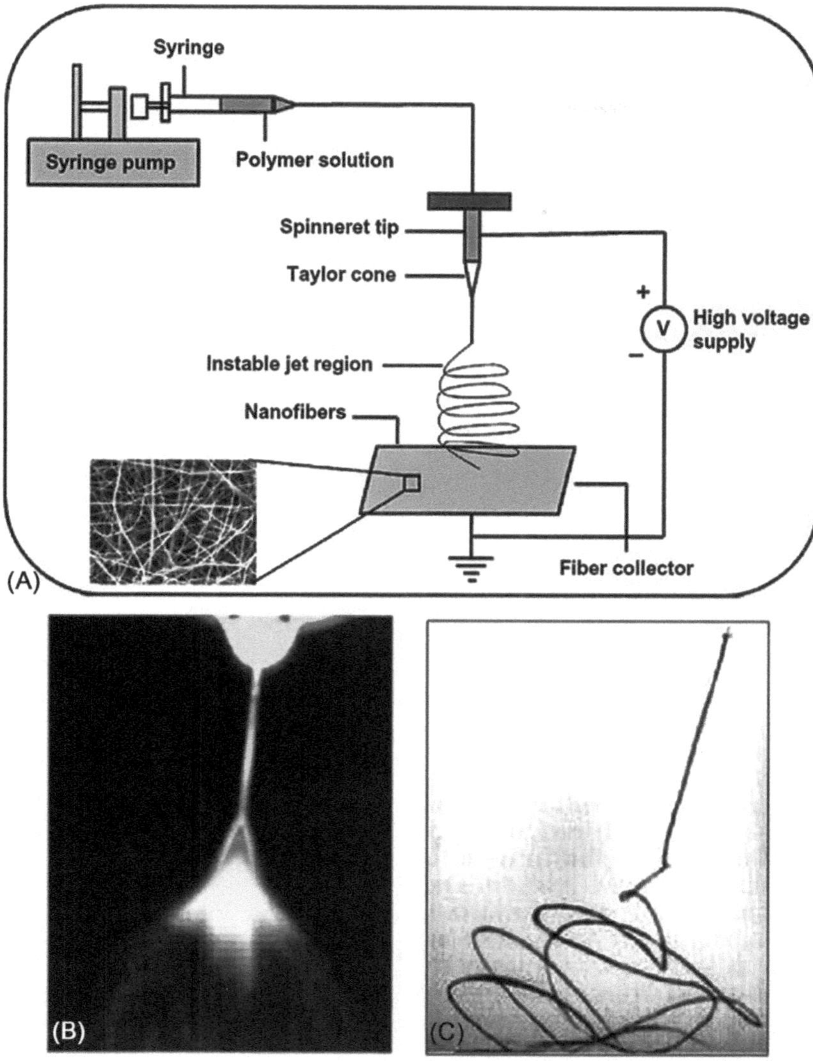

Fig. 1 (A) An illustration of a typical set-up for electrospinning. Photographs of the jet obtained with (B) a digital video camera using the interference color technique, and (C) a high-speed camera at an exposure time of 0.1 ms. *((A) Reproduced from M. Ramalingam, S. Ramakrishna, 1—Introduction to nanofiber composites, in: M. Ramalingam, S. Ramakrishna (Eds.), Nanofiber Composites for Biomedical Applications, Woodhead Publishing, 2017, pp. 3–29, with permission from Elsevier). (B) and (C) Reproduced from J. Xue, et al., Electrospun nanofibers: new concepts, materials, and applications, Acc. Chem. Res. 50(8) (2017) 1976–1987, with permission from American Chemical Society.)*

When a polymer solution with a moderate viscosity value is pumped through a syringe at a constant and controllable rate, a spherical droplet is formed at the end of the needle because of the confinement of surface tension [14]. During electrospinning, charge is induced on the liquid surface of polymer solution by a DC or AC high voltage due to electrical potential difference between the spinneret and grounded collecting plate (fiber collector). The reciprocal repulsion of these charges produces a force that opposes the surface tension, deforming the spherical droplet into a conical shape known as the Taylor cone. When the repulsive electric force overcomes the surface tension force, the charged jets are eventually sprinkled from the tip of the Taylor cone to the collector. At the onset of jetting, the droplet instantly is involved in a process that is referred as the "cone-jet" regime. The combination of the repulsion among surface charges and the effect of the electric field leads to constant decrease in diameter of jet till it bends. Subsequently, the jet starts the "whipping instability" regime, in which it accelerates and fluctuates rapidly in a "whipping" motion. Hence, the jet diameter significantly decreases over time while the solvent evaporates, leaving behind a charged polymer fiber which lays itself randomly on a collector [14–20].

Both negative and positive electrical charges can be applied for electrospinning. However, most of electrospinning experiments have been performed by using a positive potential [21].

Fibers were also deposited on the nonconducting surface by applying AC high voltage electrospinning.

To address prime problem associated with needle electrospinning (i.e., low productivity) and to scale-up nanofiber production, researchers have devoted a great deal of efforts to modify the general electrospinning setup. Currently, different types of electrospinning systems are available for large-scale production of nanofiber such as multijets from a single needle and multijets from multiple needles, and needleless systems as alternative electrospinning technologies. Large-scale production of nanofibers can be fulfilled by using needleless electrospinning approach. This approach benefits from the significant capacity to launch multiple-jets from a charged liquid surface without the influence of capillary effect that is normally associated with needle-like nozzles [2, 7, 22].

3 Theoretical background

Although the process may appear simple, the physics behind it is extremely complicated [5]. Significant efforts have been made for developing a theoretical framework for explaining the physical mechanisms that control the formation of the fibers via electrospinning. The theory has two components: a stability analysis of a cylinder of fluid with a static charge density in an external electric field and a theory for how these properties vary along the jet as it thins away from the needle [23, 24].

Many researches have been devoted to the study of the early stage of electrospun jets, which have emphasized on the importance of viscoelastic rheology [25]. For example, Yarin et al. modeled the jet as a series of charged beads connected by viscoelastic dumbbell elements. Spivak et al. suggested a 1D model for the electrospun jet, with fluid behavior represented by a non-linear power law rheological constitutive equation [26].

By contrast, a few studies have considered the unstable part of the jet, in part because of the difficulties inherent to its 3D and unsteady character [25]. As an example, to explain the phenomena detected in the course of electrospinning, Hohman et al. suggested instability models. Yarin and his coworkers also suggested models related to higher order instabilities [26].

These proposed models have had validation to some extent against experimental data and offer in-depth insights into the physical understanding of many complex phenomena that cannot be fully explained experimentally [25].

4 Controlling parameters on electrospinning

There are several parameters that significantly influence the electrospinning process and the quality of the resulting nanofibers. These parameters included process parameters, solution parameters, and ambient parameters.

Solution parameters include dielectric constant, surface tension, electrical conductivity, solvent (vapor pressure, diffusivity in air, etc.), viscosity of the polymer solution—depending strongly on the polymer (its concentration, molecular weight, and architecture), and additives (surfactants, salts, etc.). Process parameters include strength of the electric field and distance between the spinneret tip and the collector, collector shape, diameter of the needle, and feeding rate.

Ambient parameters include the gas used (such as air, N_2, Ar, vacuum) and its flow rate, humidity, and temperature [15].

Detailed discussions of the correlations between individual processing parameters and morphology of the electrospun nanofibers can be found in many references [2, 5]. However, giving a brief description of the controlling parameters on electrospinning is necessary for the better understanding of how to achieve nanofibers of a controlled structure and desirable functions. It should be noted that the morphology of the electrospun fibers usually differs from the predictions, implying the complex interplay of numerous parameters on electrospun fiber morphology [27].

4.1 Solution parameters

Surface tension and viscosity of a polymer have significant effects on the rheological properties of the polymer solution, which ultimately decide the spinnability of the solution into nanofibers [28]. Several polymer properties such as molecular weight, molecular weight distribution, and architecture of the polymer (linear, branched, etc.) could affect the concentration range that is suitable for electrospinning fibers [29].

The concentration of the polymer solution is of great importance. It needs to be high enough to have an adequate number of polymer entanglements, yet not so high that the viscosity prevents sufficient polymer flow being induced by the pump and sufficient stretching being induced by the electrical field.

The criteria of the solution are as follows: high enough charge density, low enough surface tension, and enough viscosity to prevent the jet from coalescing into droplets before the solvent has evaporated [29].

Fig. 2 shows the effect of polymer concentration on the morphology of polystyrene (PS) fibers electrospun from various concentrations of (a) 5 wt%, (b) 10 wt%, (c) 20 wt%, and (d) 30 wt% [30].

When the polymer concentration is too low, electrospraying occurs due to the effect of the applied voltage and surface tension of the polymeric solution, which causes the entangled polymer chains to break into fragments before reaching the collector and beads formation instead of fibers [19, 28, 31].

Increasing the concentration of the polymeric solution leads to a decrease in the beaded fiber and a mixture of beads and fibers is obtained.

At high concentration of the polymeric solution, the viscoelastic force is dominant and overcomes the surface tension, resulting in nearly bead-free electrospun nanofibers [2].

As the concentration is little higher, higher viscosity results in a larger fiber diameter.

Further increase in the concentration (beyond a critical value at which bead-free uniform nanofibers are formed) hampers the flow of the solution through the needle tip that results in the deformation of fiber structure into spindle-like structure.

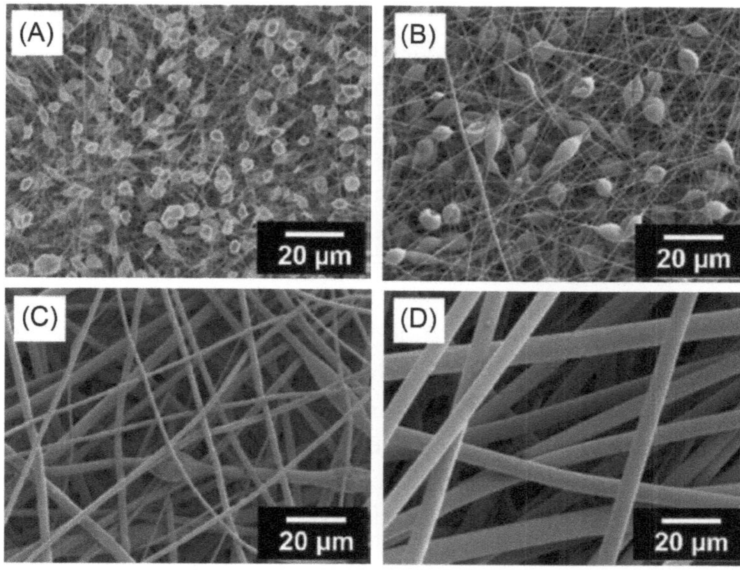

Fig. 2 The effect of polymer concentration on the morphology of polystyrene (PS) fibers electrospun from various concentrations of (A) 5 wt%, (B) 10 wt%, (C) 20 wt%, and (D) 30 wt%. *(Reproduced from B. Ding, J. Yu (Eds.), Electrospun Nanofibers for Energy and Environmental Applications, Springer-Verlag, Berlin, Heidelberg, 2014, with permission of Springer Nature; J. Lin, et al., Direct fabrication of highly nanoporous polystyrene fibers via electrospinning, ACS Appl. Mater. Interfaces 2(2) (2010) 521–528, with permission from American Chemical Society and Springer.)*

Generally, at the constant concentration, high molecular weight polymer solutions provide desired entanglement and viscosity for fiber formation. However, low molecular weight solutions tend to form beads rather than fibers. Increasing the molecular weight above the critical value could lead to the formation of micro-ribbon morphology. It is also important to note that high molecular weights are not considered necessary for electrospinning process if sufficient intermolecular interactions can provide a substitute for the interchain connectivity obtained through chain entanglements [32].

The solvent volatility, solvent composition, solvent dipole moment, and conductivity are other important factors affecting the morphology of the electrospun nanofibers. Generally, volatile solvents are mostly preferred because of easy evaporation of the solvent from the nanofibers during their flight from the spinneret to collector (the residual solvent will cause the formation of beaded nanofibers).

Furthermore, the solvent also plays an important role in the fabrication of highly porous nanofibers. It is well-documented that the porous fibers are favored to form if the solvent applied in the electrospinning process is volatile and sensitive to moisture absorption. One possible account is that the solvent quality can be reduced by the surrounding moisture, which eventually would trigger the liquid-liquid phase separation during electrospinning [33–38].

Solution conductivity not only affects the Taylor cone formation, but also helps in controlling the diameter of the nanofibers [18]. The electrical conductivity of the solution is mainly determined by the polymer type and can be tuned by addition of salts or changing the polarity of a solvent mixture. In general, an increase in solution conductivity gives rise to a smaller fiber diameter with less beads. However, highly conductive solutions are not stable in the presence of strong electric fields, which causes a dramatic bending instability and a wider range of diameter distribution [32].

Surface tension, which has a key role in beads formation on electrospun nanofibers, is a function of solvent compositions. Surface tension and solution viscosity can be adjusted by changing the mass ratio of solvents mixture or adding surfactant [5, 32].

4.2 Process parameters

In the electrospinning process, only after applying a threshold voltage to the solution, deposition of nanofibers on the collector can be observed [5, 32]. The applied voltage affects the diameter and morphology of electrospun fibers. Several researchers reported that increasing the applied voltage could lead to the formation of beads [2, 39–41]. A short overview of the literature concerning the effect of the applied voltages on the diameter of electrospun fibers indicates that different or contradictory results have been published in this regard. Several studies found that higher voltages caused an increase in the electrostatic repulsive force on the charged jet, favoring the narrowing of fiber diameter [2]. However, other authors demonstrated a different trend, a preliminary decrease followed by an increase in fiber diameter [2, 28]. Gu et al. [42] demonstrated that there is not much effect of applied voltages on the diameter of electrospun polyacrylonitrile (PAN) fibers [2].

For obtaining the small diameter fiber, lower flow rate is more recommended as the polymer solution will get enough drying time and stretching force. By increasing the flow rate above the critical value, the pore size and fiber diameter increase and the bead formation or even ribbon-like defects and unspun droplets are highly probable [18].

The type of the collectors has a significant impact on the structure and alignment of nanofiber (see following examples). Nanofibers may be collected on a static collector or on dynamic collectors, for example, rotating drum. In the latter case, fiber alignments and diameter depend on the rotation speed of the drum.

In a study, Kim et al. proved that by altering the types of the collector, the morphology of the nylon 6 nanofibers can be adjusted. Moreover, high degree of morphological change can be achieved by varying some parameters such as the conductivity of the water bath and the temperature. The results showed that the change of the morphology of the fibers stemmed from the change in the physical state [43].

Adomaviciute and Stanys investigated the influence of the type of grounded electrode on electrospun poly(vinyl alcohol) (PVA) nanofibers collected on different support materials. It was found that the type of grounding electrodes had a greater influence on the structure of electrospun mats than the support material. However, the type of grounded electrode and nature of the support material did not have important effects on the diameter of nanofibers [44].

Most recently, several groups demonstrated that electrospun nanofibers could be collected as uniaxially aligned arrays by using specially designed collectors [45]. As an instance, Xia and his coworkers reported preparation of uniaxially aligned nanofibers by using a collector consisting of two pieces of electrically conductive substrates separated by a gap [46]. This research group also investigated the effects of the area and geometric shape of the insulating gap on the deposition of fibers [45]. In another attempt, Zhang et al. showed that electrospun mats with different patterned architectures could be prepared by wise design of the collectors. The authors suggested that the influencing factors for controlling the arrangement of the fibers and the architectures of the electrospun mats were protrusions and the diameter of the electroconductive wires in the collector, the dimensions and spacings of the protrusions as well as the wire spacings in the woven collectors [47, 48]. Schematic illustration of some electrospinning set-ups using different collector designs is depicted in Fig. 3. The central points are to change the electric field distributions and control the jet path in the course of electrospinning [47]. More examples can be found in the literature, illustrating the exploring various collector designs to achieve different nanofiber patterns [47, 49–54].

In addition to the modification of the collector, the spatial orientation of the nanofibers can also be controlled by manipulating externally exerted

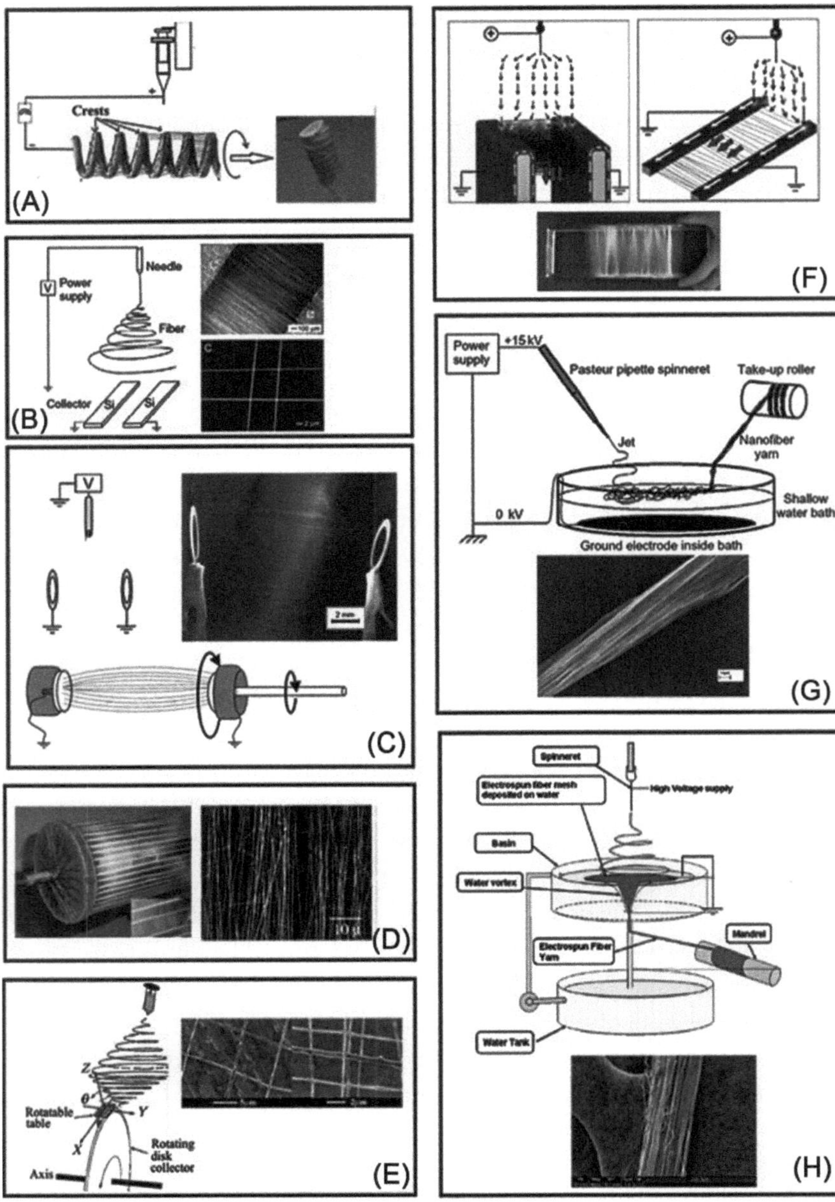

Fig. 3 Schematic illustration of an electrospinning set-up using different collector designs. (A) Helical spring collector. (B) A parallel collector containing two pieces of conductive silicon stripes separated by a gap and the collected aligned nanofibers . (C) The dual collection rings and the collected aligned nanofibers. (D) A rotating drum with parallel wires. (E) Typical SEM image of crossed arrays of nanofibers collected on an aluminum table, which can be rotated about the *z*-axis. (F) The

forces (e.g., magnetic field, electric field, postdrawing, centrifugal force, and gas force, etc.). (See Ref. [2] for more details.)

Finally, optimal distances between collector and spinneret tip give the fibers sufficient time to dry and ensure the nanofibers without beads.

4.3 Ambient parameters

The last factor that may affect the properties of electrospun nanofibers is the ambient parameters such as temperature and humidity. Temperature has an effect on the average diameter and morphology of the nanofibers because it is related to evaporation rate of the solvent, the viscosity of the polymer solution, and rigidity of the polymer chain [31]. For example, Mit-uppatham et al. [55] demonstrated that higher temperature favors the thinner fiber diameter with polyamide-6 fibers.

Amiraliyan et al. investigated the effect of temperature (ranging from 25 to 75°C) on the morphology of electrospun silk fibers. Circular morphology was observed at 25°C, whereas flat fibers were obtained by increasing the temperature. The obtained fibers were within the same range of diameters at different temperatures [5, 56].

Kocbek et al. explored the morphology and size distribution of nanofiber obtained from electrospinning of aqueous single polymer solutions (PVA or

---

automated parallel tracks. (G) Yarn-spinning set-up with water bath grounded collector electrode. (H) A dynamic liquid support system for continuous electrospun yarn fabrication. *((A) Reproduced from F. Hejazi, et al., Novel class of collector in electrospinning device for the fabrication of 3D nanofibrous structure for large defect load-bearing tissue engineering application, J. Biomed. Mater. Res. A 105(5) (2017) 1535–1548, with permission from John Wiley and Sons). (B) Redrawn from D. Li, Y. Wang, Y. Xia, Electrospinning nanofibers as uniaxially aligned arrays and layer-by-layer stacked films, Adv. Mater.* ***16****(4) (2004) 361–366, with permission from Wiley. (C) Reproduced from P.D. Dalton, D. Klee, M. Möller, Electrospinning with dual collection rings, Polymer 46(3) (2005) 611–614, with permission from Elsevier. (D) Reproduced from P. Katta, et al., Continuous electrospinning of aligned polymer nanofibers onto a wire drum collector, Nano Lett. 4(11) (2004) 2215–2218, with permission from American Chemical Society. (E) Reproduced from E. Zussman, A. Theron, A.L. Yarin, Formation of nanofiber crossbars in electrospinning, Appl. Phys. Lett. 82(6) (2003) 973–975, with permission from AIP Publishing. (F) Reproduced from V. Beachley, et al., A novel method to precisely assemble loose nanofiber structures for regenerative medicine applications, Adv. Healthc. Mater. 2(2) (2013) 343–351, with permission from Wiley. (G) Reproduced from E. Smit, U. Bűttner, R.D. Sanderson, Continuous yarns from electrospun fibers, Polymer 46(8) (2005) 2419–2423, with permission from Elsevier. (H) Reproduced from W.-E. Teo, et al., A dynamic liquid support system for continuous electrospun yarn fabrication, Polymer 48(12) (2007) 3400–3405, with permission from Elsevier.)*

poly(ethylene oxide) (PEO)) and blend polymer solutions (PVA/hyaluronic acid (HA) or PEO/chitosan (CS)) as a function of relative humidity (RH). They observed thinner nanofibers with a more heterogeneous size distribution at higher humidity and for the blend; the decrease of the fiber diameter was even more. At very high humidity, however, beaded fibers were formed for individual polymers, while almost no electrospinning was carried out from blends [18, 27]. The thinning of the electrospun nanofibers with increasing RH can be successfully explained by the velocity of solvent evaporation and jet solidification. At higher RH conditions, the velocity of the solvent evaporation decreases and solidification occurs more slowly and the polymer jet is consequently exposed to voltage-induced stretching for longer time, resulting in the formation of thinner fibers. This phenomenon is, however, dependent on the chemical nature of the polymer; hence, for different polymers electrospun nanofibers produced from, variations in fiber diameter are different.

Humidity also has a key role in the fabrication of porous nanofibers when the nanofibers are electrospun from organic solvents (water (RH) acts as a nonsolvent) [18]. In this case, the fiber diameter can be regulated through the control of solvent vapor pressure in the environment [27].

Higher water vapor pressure results in more water molecules between the needle and collector, which decreases the amount of excess charges on the electrospinning jet. As a result, the intensity of the electric field is decreased and the jets would experience a smaller drawdown force and thus undergo reduced elongation. Furthermore, the existence of water vapor that acts as a nonsolvent makes precipitation of the fiber easier before deposition of the collector surface. This hastened precipitation locks in the fiber diameter before the rest of the solvent has a chance to evaporate. Therefore, increasing humidity during spinning generally results in increased diameter of nanofibers [57]. Although the observed trends are the same for many polymers, there are some differences in fiber size distribution and surface morphology [57].

Because the diameter of nanofibers correlates with their rigidity, the mechanical properties of nanofibers can be tuned by precise regulation of the relative humidity [27].

The effects of RH are also strongly coupled to other parameters. For example, Putti et al. reported effects of humidity and temperature on the porous structure of polycaprolactone fibers, as depicted in Fig. 4 [31, 58].

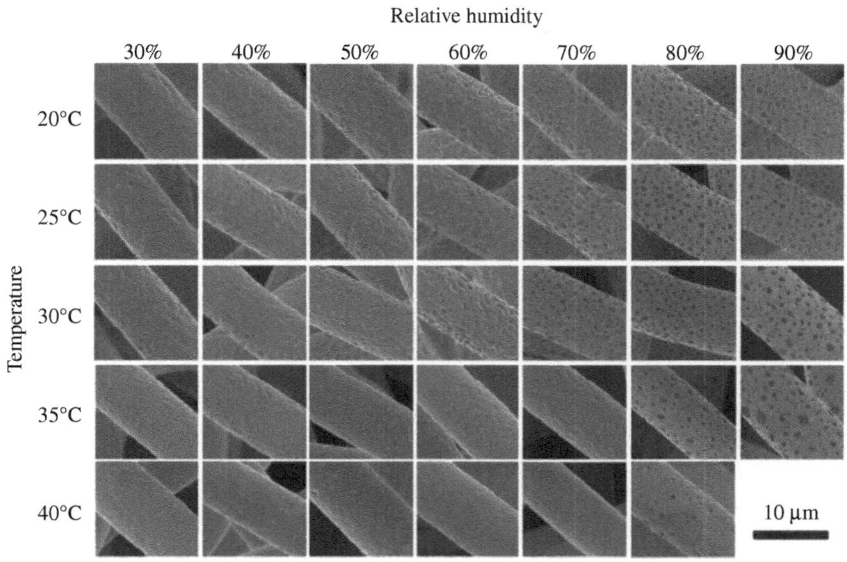

Fig. 4 SEM images of polycaprolactone fibers spun from a 15 wt% solution at different temperatures and relative humidity. *(Reproduced from M. Putti, et al., Electrospinning poly (ε-caprolactone) under controlled environmental conditions: influence on fiber morphology and orientation, Polymer 63 (2015) 189–195, with permission from Elsevier.)*

5 Materials classes

5.1 Polymeric fibers (synthetic and natural polymers)

Electrospinning technique can be used for synthetic and natural polymers. The most commonly used natural polymers for electrospun nanofibers are polysaccharides and proteins such as alginate, cellulose, chitosan, dextran, chitin, hyaluronic acid, zein, silk fibroin, fibrinogen, gelatin, elastin, wheat gluten, soybean proteins, collagen, casein, and fibrinogen [2, 5]. Nanofibers have also been produced from DNA and lignin [2].

Electrospun nanofibers derived from synthetic polymers have been extensively investigated in the last decade. As this class of polymers can be tuned to furnish desired properties, they are more appealing compared to natural polymers.

The more commonly used synthetic polymers with the corresponding used solvents that are able to yield beadless nanofibers are listed in Table 1 [5].

Table 1 Different synthetic polymers used in electrospinning [3, 5].

Polymer	Solvent	Polymer	Solvent
Polyurethanes (PU)	Dimethyl formamide/*N,N*-dimethylacetamide	Polyethylene terephtalate (PET)	Dichlormethane/trifluoracetic acid
Polybenzimidazole (PBI)	*N,N*-Dimethylacetamide	Polyvinylchloride (PVC)	Tetrahydrofuran/dimethylformamide
Polyvinil alcohol	Water	Polyamide (PA)	*N,N*-Dimethylacetamide
Nylon-6	Formic acid	Polyethylene oxide	Water
Polyacrylonitrile	Dimethyl formamide	Poly vinyl phenol	Tetrahydrofuran/ethanol
Polysulfone (PSF)	*N,N*-Dimethylacetamide/acetone	Polystyrene	Tetrahydrofuran/dimethylformamide/methylethylketone
Polyimide (PI)	Dimethyl formamide	Polypyrrole (PPy)	Water
Polylactic acid	Dichlormethane	Polycarbonate (PC)	Dimethyl formamide/tetrahydrofuran
Polyvinyl butyral (PVB)	Water/ethanol	Polymethyl methacrylate (PMMA)	Dimethyl formamide/dichlormethane

Adapted from Z.-M. Huang, et al., A review on polymer nanofibers by electrospinning and their applications in nanocomposites, Compos. Sci. Technol. 63 (15) (2003) 2223–2253, with permission of Elsevier.

In order to obtain the best combination of properties according to the intended application for the nanofibers, electrospinning of a mixture of synthetic and natural polymers is commonly used. These polyblend nanofibers retain the mechanical strength of their synthetic polymer component as well as the biological functionality of the natural polymer component [2].

5.2 Carbon (nano)fibers

Carbon (nano)fibers, CNFs, one example of one-dimensional carbon materials, possess high mechanical strengths and moduli, excellent electrical and thermal conductivities, and strong corrosion resistance [59]. These properties combined with high strength-to-weight ratio and superior stiffness rendered CNFs promising candidates for the development of high performance composite structures in sensors, lithium-ion batteries, supercapacitors, high temperature catalysis, tissue engineering, filters, heat management materials in aircraft, nanoscale electronics and photonics, and other industries [60–62].

CNFs are fabricated either by melt spinning from various organic precursors (e.g., polyacrylonitrile, or alternatively pitch) or by chemical vapor deposition (CVD). Spinning method can only be used for the production of microscale CNFs, while CVD method can be applied for the fabrication of CNFs with diameters from several microns down to less than 100 nm. However, the CVD method involves a complicated chemical and physical process, and thus, the associated cost is inevitably high; additionally, the CVD method is not suitable for producing long fibers and nonwoven fabrics [59, 60, 63, 64]. Electrospinning technique provides an efficient and inexpensive approach to fabricate carbon nanofibers in micro-, even nanoscale, in different forms such as nonwoven mats, yarns, etc. [15, 59, 60].

The preparation of electrospun CNFs involves electrospinning of its polymer precursor and conversion of the polymer fibers to CNFs through thermal treatments, including stabilization and carbonization processes [59, 65]. The initial fiber formation step has a remarkable effect on the structure and features of the final CNFs. Thus, the judicious selection of the precursors combined with the optimized spinning processes (spatially solution properties) could be one of the ways to tailor the performance of the resultant fibers [66, 67].

Treatment at elevated temperatures can improve electrical characteristics and induce some structural changes due to the structural development from disordered to graphitic carbon (three-dimensionally ordered structure). However, graphitizability of a carbon is dramatically dependent on the used precursor [15].

The polymer fibers, which have been converted to CNFs, are rather limited, such as PAN, PI, PVA, poly(vinyliden fluoride) (PVDF), and pitch [15].

In the case of some of the precursors, such as PAN and pitches, to preserve fibrous morphology, stabilization is necessary prior to carbonization. The essential reaction of stabilization is oxidation [15]. Generally, the diameter of the electrospun polymer fibers decreases during thermal treatments.

Multifunctional composite nanofibers can be provided using different nanofillers added into the electrospinning precursor solution [68]. Carbon nanotubes are promising nanofillers to increase electrical conductivity of electrospun carbon nanofibers and reinforce their mechanical properties [67, 69, 70]. Embedding of platinum to carbon nanofiber webs was also performed in relation to fuel cell applications [71]. To improve the performance of electrochemical capacitors and lithium-ion rechargeable, various nanoparticles were included in CNFs via electrospinning [61, 72–78].

In order to attain electrospun CNFs with larger surface areas, many studies applying electrospinning are aiming to produce porous carbon fibers.

Previous methods of producing porous fibers heavily relied on the varying polymer architecture and processing parameters. An approach to modify and tune the content of micropores is activation. Typically, this can be achieved by using either steam, KOH, or H_3PO_4 on electrospun nanofibers after carbonization [15, 79].

In addition, the use of different additives into polymer solution was reported for tuning pore structure without activation process [80–82]. For example, reduction of the oxidative stabilization time, the improved carbon yield, and no necessity to use an activation process are the advantages provided by the use of zinc chloride in PAN/DMF electrospinning solutions because of its catalytic function [80].

Alternatively, electrospinning of two immiscible polymer solutions can be performed. This furnishes nanofibers containing two separate phases. Subsequent thermal treatment gives pores after selective pyrolyze of one of the components. For instance, Kim et al. reported the fabrication of porous, fibrous carbon materials with many hollow pores developed along the fiber by electrospinning of polymethyl methacrylate (PMMA) emulsions in a continuous phase of PAN and DMF and removal of the elongated rod-like PMMA phase within a single nanofiber through thermal treatment.

As shown in Fig. 5, blend ratio of the constituent polymers strongly influences the number, diameter, and length of the hollow cores in a single CNFs. The higher the fraction of PAN, the smaller the diameter of the

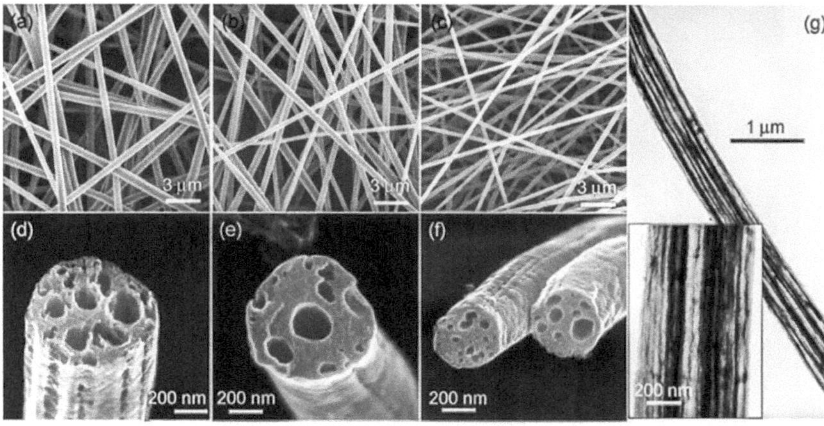

Fig. 5 (A–C) Macromorphology of nanofibers containing two polymer phases. PAN:PMMA = (A) 5:5, (B) 7:3, and (C) 9:1. (D–F) Cross-sectional FESEM images of thermally treated nanofibers at 1000°C. PAN:PMMA = (D) 5:5, (E) 7:3, and (F) 9:1. (G) TEM images (D), showing linearly developed hollow cores along the fiber length. The inset is a magnified TEM image. *(Reprinted from C. Kim, et al., Synthesis and characterization of porous carbon nanofibers with hollow cores through the thermal treatment of electrospun copolymeric nanofiber webs, Small 3(1) (2007) 91–95, with permission from John Wiley and Sons.)*

nanofibers; this is due to the high spinnability of PAN. During thermal treatment, the PAN and PMMA polymers show different thermal behavior: PAN easily transforms into residual carbon, while PMMA decomposes without carbon residue, resulting hollow cores within a carbon fiber. It is, therefore, expected that the number of hollow cores increases with an increase in PMMA concentration [83].

Another strategy to synthesize porous CNFs with high surface areas is use of a porogen such as nanoparticles. For example, carbonization of electrospun PAN/SiO_2 composite nanofiber, followed by etching of SiO_2 nanoparticles, led to the formation of porous CNFs [84].

More complex architectures such as hollow nanofibers can be fabricated by *co*-electrospinning of two polymer solutions and subsequent thermal treatment [85]. As heat is applied, volatile constituents such as decomposed polymer and residual solvent diffuse to the outer surface of the shell, where they evaporate. As gas transport through the shell layer is blocked due to either relatively high shell thickness or low permeability, the internal pressure in the core increases, resulting in internal circumferential stresses in the shell. As can be seen in Fig. 6, this type of pressure build-up causes different shell-fracture patterns [85].

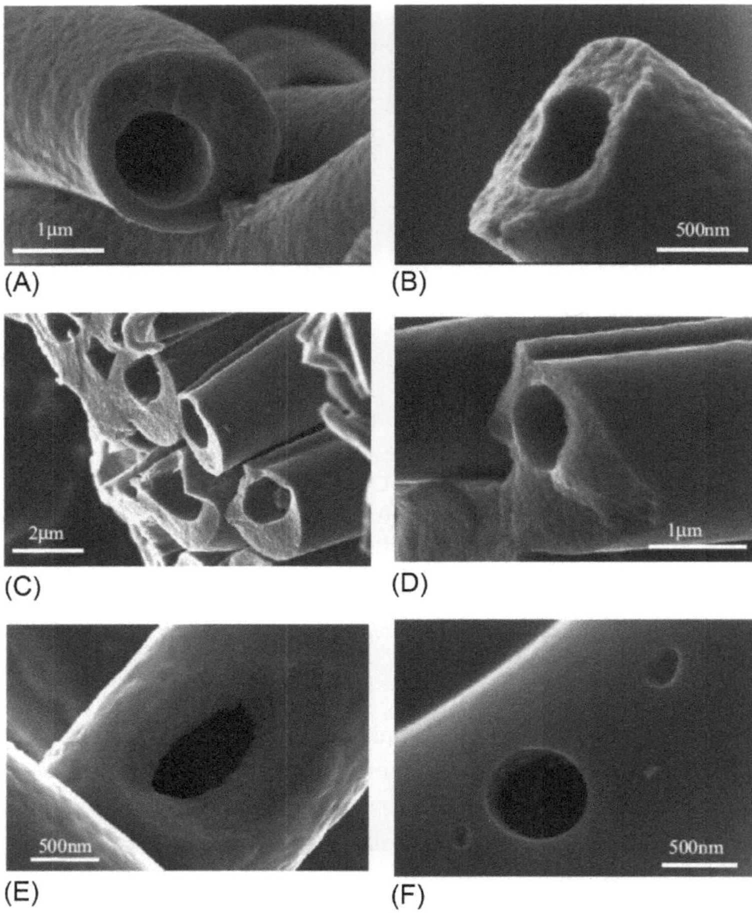

Fig. 6 (A and B) SEM images of fractured surfaces of carbonized hollow nanofibers, (C) aligned array of carbonized hollow nanofibers, (D) carbonized nanotube with an arch on the surface, (E) carbonized nanotube with a brittle fracture, and (F) as-spun fiber with almost perfect holes. *(Reproduced from E. Zussman, et al., Electrospun polyaniline/poly(methyl methacrylate)-derived turbostratic carbon micro-/nanotubes, Adv. Mater. 18(3) (2006) 348–353, with permission from John Wiley and Sons.)*

6 Application of electrospun carbon nanofibers for catalysis

Electrospun carbon nanofibers have been extensively used for catalysis. Mostly, CNFs have been applied as electrocatalysts. This issue has been addressed in many articles, reviews, and book chapters [86]. In this section, some examples of the utility of CNFs for other types of chemical transformations are addressed.

6.1 Photocatalysis

Taking advantage of electrospinning technique and hydrothermal treatment, Chen, Shao et al. fabricated ZnO-CNFs' heteroarchitectures through initial formation of CNFs via electrospinning of PAN and carbonization followed by hydrothermal treatment of as-prepared CNFs with zinc acetate. Loading of ZnO was tunable by controlling the mass ratio of CNFs and zinc precursor in the course of hydrothermal treatment. ZnO-CNFs' heteroarchitecture exhibited high photocatalytic activity for degradation of rhodamine B [87]. This was attributed to the improved separation of photo-induced charge carriers. Another merit of the prepared heteroarchitecture was its facile recovery.

Shao, Chen et al. reported fabrication of one-dimensional In_2O_3 nanocubes/CNFs through electrospinning and solvothermal approach [88]. The authors studied the photocatalytic activity of the resulting heterostructure for rhodamine B degradation under visible high irradiation and compared it with that of bare In_2O_3. The results confirmed superior activity of In_2O_3 nanocubes/CNFs. The increase of the catalytic activity of In_2O_3 nanocubes/CNFs was assigned to the better separation of photo-induced electrons and holes due to the formation of heterostructures. Notably, the catalyst could be easily recycled with no loss of the catalytic activity.

Combining electrospinning and solvothermal methods, this research group also developed Bi_2MoO_6/CNF hierarchical heterostructures and studied their photocatalytic activity for degradation of rhodamine B under visible light irradiation [89]. Similarly, the comparison of the catalytic activity with that of bare Bi_2MoO_6 confirmed higher activity of the Bi_2MoO_6/CNF hierarchical heterostructures. It was believed that improved separation of photogenerated electrons and holes and absorption led to the enhanced activity of Bi_2MoO_6/CNF. Notably, the morphology of the secondary Bi_2MoO_6 could be tuned by altering the reaction variables such as temperature and concentration of raw materials and solvent. Furthermore, Bi_2MoO_6/CNF was recyclable with preserving its activity.

TiO_2-loaded graphene/carbon composite nanofibers (CCNFs) were prepared by electrospinning of graphene-DMF-PAN mixture, followed by treating with titanium *n*-butoxide and carbonization [90]. Investigation of the photocatalytic activity of the prepared composite for degradation of methylene blue under visible light irradiation confirmed its high photocatalytic activity and recyclability. It was believed that graphene could serve as an electron acceptor and a photosensitizer that retarded recombination

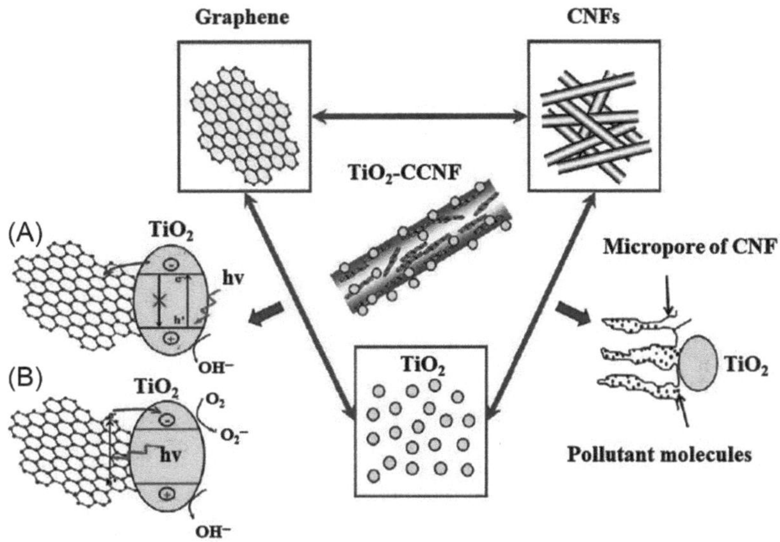

Fig. 7 In the TiO_2-CCNF materials with a mixture of anatase and rutile structures, the graphene may act as an electron acceptor (a) for rutile and as a photosensitizer (b) for anatase which prevents the recombination of electron-hole pairs. *(Reprinted from C.H. Kim, B.-H. Kim, K.S. Yang, TiO_2 nanoparticles loaded on graphene/carbon composite nanofibers by electrospinning for increased photocatalysis, Carbon 50(7) (2012) 2472–2481, with permission of Elsevier.)*

rate of charge carriers and consequently accelerated the photodegradation rate. High specific surface area of CNFs, on the other hand, improved the activity of TiO_2 by enhancing the physical adsorption of the substrate (Fig. 7).

Electrospun CdS-TiO_2-doped carbon nanofibers were also prepared and applied for the photocatalytic purposes. To prepare the photocatalyst, a mixture of titanium tetraisopropoxide, cadmium acetate dehydrate, poly (vinyl pyrrolidone), and a few amount of ammoniumsulfide was used as spinning dope. After electrospinning, the resulting nanofibers were calcined under inert atmosphere. Electrospun CdS-TiO_2-doped carbon nanofibers that benefited from high surface area, favorable electrons-transfer, and adsorption properties as well as synergism between CdS and TiO_2 exhibited high catalytic activity for visible light hydrolytic dehydrogenation of ammonia borane [91].

Chen et al. reported visible light-induced photocatalytic effect of electrospun carbon nanofiber/tin(IV) sulfide (CNF@SnS_2) core/sheath fibers for treating chromium(VI)-containing wastewater. CNFs were synthesized via electrospinning of PAN precursor and subsequent carbonization. SnS_2 nanoparticles were then uniformly anchored on the CNFs by chemical bath deposition, which yielded hierarchical nanostructures. A schematic of the mechanism for the photocatalytic reduction of Cr(VI) is shown in Fig. 8. The conducting CNF cores could quickly transport photoelectrons produced by the SnS_2 sheath, under visible light irradiation, enhancing the photocatalytic efficiency of CNF@SnS_2 fibers compared with the individual components, CNFs and SnS_2. In addition, CNFs prevented the SnS_2 nanoparticles from aggregating during the photocatalytic degradation cycles and yielded good cycling stability [92].

Li et al. loaded Au and TiO_2 nanoparticles simultaneously on the CNFs to prepare Au-TiO_2/CNFs composite [93]. In more detail, using PAN, nanofibers were first prepared through electrospinning and then calcined to afford CNFs. Then, using one-step solvothermal synthesis, Au and TiO_2 were loaded on CNFs. The authors studied the photocatalytic activity

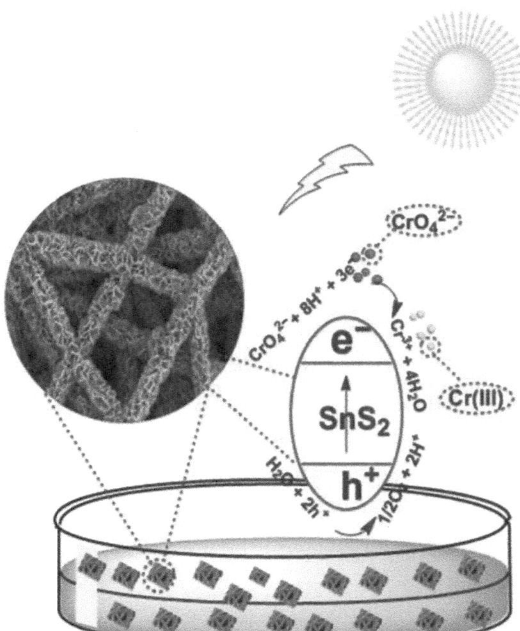

Fig. 8 Schematic illustration of the reaction mechanism for Cr(VI) removal. *(Reproduced from Y. Zhong, et al., Flexible electrospun carbon nanofiber/tin(IV) sulfide core/sheath membranes for photocatalytically treating chromium(VI)-containing wastewater, ACS Appl. Mater. Interfaces 8(42) (2016) 28671–28677, with permission from American Chemical Society.)*

of the catalyst for photodegradation of methyl orange, methylene blue, and acid red 18 under UV light irradiation. Noteworthy, the activity of Au-TiO_2/CNFs was higher than that of TiO_2/CNFs. This was assigned to the presence of Au nanoparticles. Furthermore, the catalyst could be easily separated and recycled.

Recently, Park, Kim et al. decorated PAN-derived carbon nanofibers with Ag-ZnO nanoparticles via hydrothermal approach and used them as photocatalysts for adsorption and degradation of methylene blue under visible and UV irradiation [94]. The catalyst exhibited high photocatalytic activity, superior to that of pristine ZnO. The higher photocatalytic activity of Ag-ZnO/CNF composite was attributed to the reduced recombination rate of the photo-induced electrons and holes, enhanced electron transport, mobility of charge carriers, and adsorption capacity of the carbon fibers. Furthermore, the adsorption property of carbon fibers also accelerates the decolorization process. Notably, it was found that incorporation of Ag nanoparticles enhanced the visible light absorption ability of ZnO.

In another attempt, Mu et al. combined electrospinning technique and solvothermal method to couple the two-dimensional (2D) rectangular iron phthalocyanine nanosheets with one-dimensional (1D) electrospun CNFs to afford a photocatalyst, TNFePc/CNFs, with utility for visible light photodegradation of rhodamine B and methyl orange [95]. It is worth mentioning that the activity of TNFePc/CNFs was superior to that of bare TNFePc. In fact, the higher activity of the catalyst emerged from the improved charge separation in the composite. Moreover, TNFePc/CNFs could be easily recycled by sedimentation due to the large length to diameter ratio of the one-dimensional nanofibers morphology without affecting the catalyst activity.

In an example, Shao, Li et al. [96] reported synthesis and photocatalytic application of molybdenum diselenide nanosheet/carbon nanofiber ($MoSe_2$/CNF) heterojunctions, prepared via electrospinning and solvothermal approaches for photodegradation of rhodamine B under visible and infrared irradiation. Noteworthy, $MoSe_2$ particles were well-dispersed on carbon nanofibers and their loading could be easily controlled by adjusting the synthetic parameters. It was shown that $MoSe_2$/CNF exhibited high catalytic activity superior to bare $MoSe_2$. It was believed that the formation of heterojunction could improve separation of photogenerated electrons and holes. On the other hand, $MoSe_2$/CNF that possessed ultralong one-dimensional and self-supporting structure could be easily separated via sedimentation with no negative effect on the photocatalytic activity.

Lee et al. fabricated CNFs via electrospinning of PAN followed by carbonization and then used the resulting one-dimensional carbon nanofibers for the preparation of graphitic carbon nitride (GCN) [97]. In more detail, to a solution of urea and thiourea certain amount of CNF powder was added, and after solvent evaporation and sintering, GCN-coated CNFs were furnished. Photoelectrochemical results confirmed a superior photocurrent of 3 μA for the hybrid compared with that of 1 μA for the pristine GCN. The high photocurrent for the hybrid structures indicated the formation of heterojunctions that resulted from a lower recombination rate of charge carriers. Furthermore, the sample prepared by using 0.075 g of urea and 0.075 g of thiourea exhibited the highest performance of hydrogen generation with its numerical value of 437 $\mu mol\,g^{-1}$. This higher hydrogen production could be explained again with successful formation of heterojunctions between GCN and CNFs.

In another example, Dai, Chen et al. developed a core-shell photocatalyst for photocatalytic reduction of CO_2 through fabrication of carbon nanofibers core via electrospinning and in situ growth of TiO_2 shell [98]. The catalytic activity of the catalyst was 2.3-fold superior compared to bare TiO_2. The authors believed that the improved catalytic activity emerged from higher specific surface area of the core-shell composite. CNFs that possessed good electrical conductivity could improve separation of electrons and holes. As shown in Fig. 9, the photogenerated electrons could transfer to CNFs, which remarkably retarded the charges recombination. The remaining electrons would react with the adsorbed CO_2 and water molecules to furnish CH_4. Furthermore, the adsorption

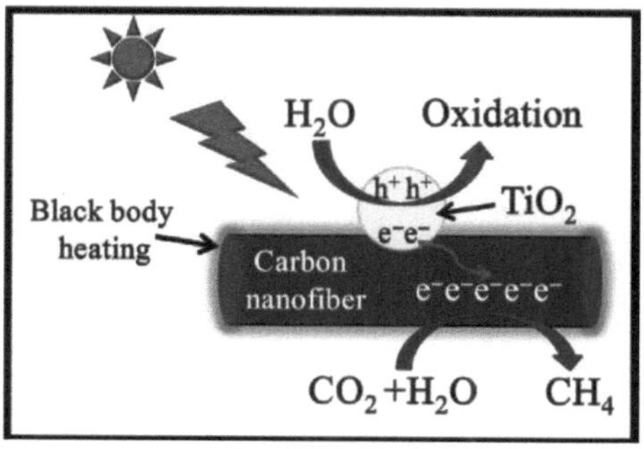

Fig. 9 A schematic illustration of photocatalytic mechanism over the carbon nanofiber@TiO_2 system. *(Reprinted from J. Zhang, et al., 1D carbon nanofibers@TiO_2 core-shell nanocomposites with enhanced photocatalytic activity toward CO_2 reduction, J. Alloys Compd. 746 (2018) 168–176, with permission from Elsevier.)*

of water molecules would consume the photogenerated holes to produce some oxidation products. On the other hand, the presence of black CNFs in the structure of the catalyst improved light absorption. As the light energy could be converted into heat energy, it accelerated the diffusion of reactants and products.

Combining electrospinning and carbonization, Zheng et al. developed TiO_2/SiO_2/carbon electrospun nanofiber mat (TSC NFM) with flexibility and porous hierarchy [99]. The composite consists of carbon with interconnected meso- and macropores, amorphous SiO_2, and anatase nanocrystals that were uniformly distributed. TSC NFM was successfully applied for degradation of rhodamine B and 4-nitrophenol. Interestingly, the catalyst was recyclable and its activity was superior to that of TiO_2/SiO_2 nanofiber mat. The enhanced activity of TSC NFM was ascribed to the improved adsorption of substrates and enhanced interfacial charge separation between TiO_2 nanoparticles and carbon.

Recently, Du, Ruan, Zhao et al. fabricated electrospun-reduced graphene oxide/TiO_2/poly(acrylonitrile-*co*-maleic acid) composite nanofibers (Espun RGO/TiO_2/PANCMA NFs) and applied it for adsorption and photocatalytic degradation of malachite green and leucomalachite green from aqueous solution under UV irradiation [100]. Notably, RGO/TiO_2/PANCMA NFs exhibited superior adsorption and degradation efficiency compared to TiO_2/PANCMA NFs and GO/TiO_2/PANCMA NFs. The enhanced performance was attributed to the improvement of charge migration and carrier separation in photocatalytic degradation process. Moreover, the catalyst was recyclable for long-term adsorption and irradiation.

6.2 Coupling reaction

Hou et al. developed an efficient catalyst for Sonogashira coupling reaction by palladating carbon nanofibers. More precisely, electrospun polyacrylonitrile/$Pd(OAc)_2$ composite nanofibers were prepared and then carbonized in furnace by the following steps: first, annealing at 210°C in air was performed for oxidation of PAN; secondly, annealing was pursed at 400°C in H_2 and Ar mixture for reduction of Pd^{2+}; and finally, heating at 550°C was carried out in Ar for the formation of metallic nanoparticles with diameter range from 4 to 40 nm on CNFs. The catalyst not only exhibited high catalytic activity, but also showed high recyclability with low Pd leaching in liquid phase [101].

Using a three-step procedure, including electrospinning of $PdCl_2$/PAN/DMF, gas-phase hydrogenation reaction, and carbonization, Bai et al. reported preparation of Pd/CNFs [102]. It was shown that Pd nanoparticles were highly dispersed and placed both on the exterior surface and inner space of carbon matrix. Notably, the carbonization temperature significantly affected the size and distribution of Pd nanoparticles. The authors examined the catalytic activity of the as-prepared catalyst for Heck coupling reaction and confirmed high catalytic activity and recyclability of the catalyst. The study of the effect of carbonization temperature on the catalytic activity of the catalyst confirmed that the best results were achieved by carbonization at 600°C. In fact, after five reaction runs, no Pd aggregation was observed and the catalyst preserved its initial morphology. Moreover, the Pd leaching was insignificant and the catalyst almost preserved its catalytic activity.

Song et al. also applied $CaCo_3$ nanoparticles for fabrication of hierarchical macro/mesopores inside carbon nanofibers [103]. To prepare the porous CNFs, PAN and $CaCo_3$ nanoparticles were electrospun. Subsequently, carbonization and acid leaching led to the hierarchical macro/mesopores CNFs. $CaCo_3$ was served as a dual purpose template. In the course of carbonization, decomposition of $CaCO_3$ produced CO_2 gas which formed mesopores when being discharged out of the nanofibers, and then the acid washing removed the as-formed CaO nanoparticles to produce macropores (Fig. 10). The as-prepared porous CNF was then applied as a support for the immobilization of Pd nanoparticles to furnish a catalyst for Suzuki coupling reaction. The catalyst showed high catalytic activity and recyclability. It was believed that the macropores in the CNFs framework could facilitate the mass transport and the attached mesopores provided high surface area for the contact of the reagents and Pd nanoparticles. On the other hand, small size of Pd nanoparticles and the improved anchoring of Pd nanoparticles by the residual nitrogen in CNFs could justify the high catalytic activity of the catalyst.

In another example, Li et al. reported the utility of CNFs-supported copper nanoparticles for Ullmann-type coupling reactions in the absence of additional ligand [104]. To fabricate the catalyst, the authors prepared β-cyclodextrins/poly (acrylonitrile) composite fibers via electrospinning and then loaded copper nanoparticles on the surface of the as-prepared fibers. Subsequently, the composite was thermal-treated to form CuNPs/CNFs. Investigation of the catalytic activity of the catalyst confirmed high catalytic activity and selectivity of the catalyst. The broad substrate scope of

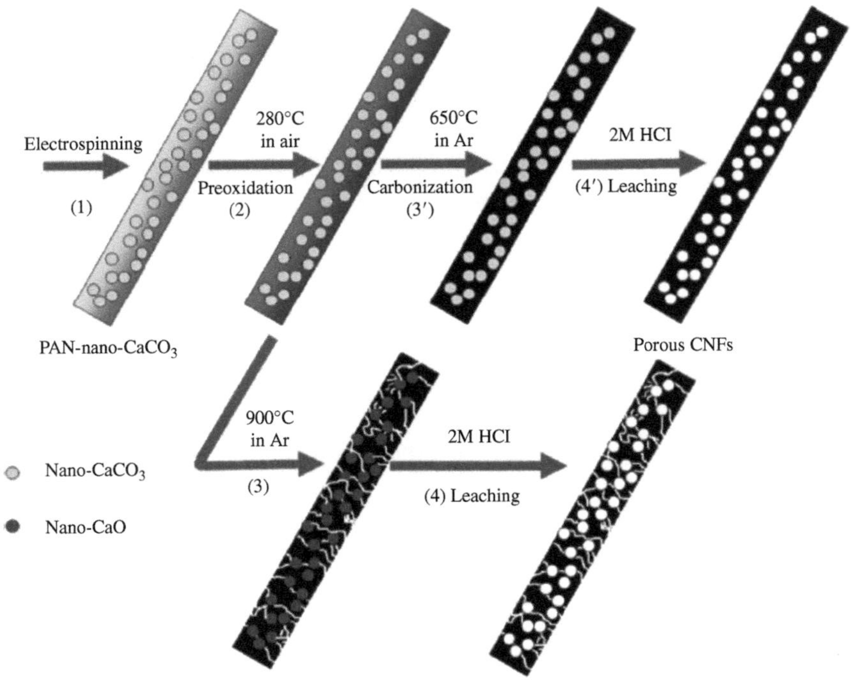

Fig. 10 Schematic illustration of the fabrication process of the hierarchical macro/mesoporous carbon nanofibers. *(Reprinted from H. Liu, et al., Fabrication of macroporous/mesoporous carbon nanofiber using $CaCO_3$ nanoparticles as dual purpose template and its application as catalyst support, J. Phys. Chem. C 117(41) (2013) 21426–21432, with permission from American Chemical Society.)*

the protocol and high recyclability of the catalyst up to five reaction runs were other merits of this catalyst.

This research group also fabricated several Pd nanoparticles/highly porous carbon nanofibers via initial preparation of $PdCl_2$/PAN/PS nanofibers with different molar ratio of PAN: PS, followed by reduction and carbonization [105]. In the course of heat treatment, PS that is a thermoplastic polymer acted as a pore-forming polymer, while PAN converted to carbon matrix. This procedure led to the formation of a porous structure with high specific surface area (250.9 m^2 g^{-1}) with a majority of meso- or nanopores/nanochannels in the range of several to tens of nanometer. Characterization of the catalyst confirmed that well-dispersed Pd nanoparticles were anchored both on the exterior surface and inner space of CNFs. It was shown that the prepared catalyst with PAN:PS molar ratio of 40 could efficiently promote Heck coupling reaction and led to high conversion and yields of the desired products.

In another example, Bai et al. supported Pd nanoparticles on electrospun carbon fibers and used them as catalysts for Suzuki and Heck reactions [106]. The procedure for the preparation of the catalyst included electrospinning $PdCl_2$/PAN/DMF solution followed by reduction with hydrazine hydrate and carbonization. Using this procedure, Pd nanoparticles were uniformly supported on CNFs. The authors confirmed high catalytic activity and recyclability of the catalyst up to five reaction runs.

Bai et al. also reported the utility of Pd/CNFs for Suzuki coupling reaction [107]. In the synthetic procedure of the catalyst, $PdCl_2$/PAN nanofibers were first obtained via electrospinning technology and then treated with glucose. Finally, carbonization under inert atmosphere furnished Pd/CNFs, in which Pd nanoparticles were uniformly dispersed inner and outer of CNFs with no aggregation. The study of the catalytic activity of Pd/CNFs confirmed high catalytic activity and recyclability of the catalyst (up to 10 reaction runs).

In a related work, electrospinning was exploited for the preparation of $CuCl_2$/poly (acrylonitrile) nanofibers [108]. The resulting fibers were then carbonized to furnish zero valance copper on carbon nanofibers, Cu/CNFs. Cu/CNFs exhibited high catalytic activity for Ullmann coupling of C—O and C—N reactions. Moreover, Cu/CNFs were recyclable and could be recovered and reused for five reaction runs with insignificant loss of the catalytic activity.

6.3 Reduction

As an example of the utility of carbon nanofibers for development of catalysts for reduction reactions, Shao et al. reported synthesis of carbon nanofibers/silver nanoparticles composite through fabrication of nanofibers via electrospinning and thermal treatment, treatment with HNO_3 followed by hydrothermal growth of silver nanoparticles. It was found that treatment with acid could provide large number of active sites that favored homogeneous growth of silver nanoparticles. Moreover, the silver loading could be adjusted by altering the concentration of the reactants in the course of hydrothermal treatment. Characterization of the as-prepared composite confirmed that silver nanoparticles have been well-distributed on carbon nanofibers with no aggregation [109]. The catalyst could successfully promote reduction of 4-nitrophenol in the presence of $NaBH_4$. High activity of the catalyst was attributed to the synergistic effect on delivery of electrons between carbon nanofibers and silver nanoparticles as well as high surface area of silver nanoparticles.

Li et al. reported synthesis of three-dimentional free-standing network composed of cross-linked carbon@Au core-shell nanofibers and investigated its catalytic activity for reduction of 4-nitrophenol [110]. To prepare the catalyst, PAN-derived CNFs were fabricated via electrospinning technology and carbonization. Subsequently, as-prepared CNFs were treated with acidic solution of $SnCl_2$ to furnish activated CNFs that were further reacted with $HAuCl_4$. Au nanoparticles, obtained through in situ reduction, uniformly formed a layer around CNFs with thickness of 5 nm. The synergism between Au and CNFs improved the catalytic activity of CNFs@Au network. Moreover, the free-standing 3D nanofibrous cross-linked network structure improved separation and reuse of the catalyst.

Kang, Chen et al. reported hallow structured Ag/CNFs with high specific surface area via co-electrospinning and in situ reduction [111]. The authors studied the catalytic activity of the prepared Ag/CNFs for the reduction of methylene blue in the presence of $NaBH_4$. High specific surface area of the hallow nanofibers assured increase of content of active sites. On the other hand, the waste of Ag nanoparticles was prevented by hallow structure of the catalyst. These features led to high catalytic activity of hallow structured Ag/CNFs for decolorization of methylene blue.

In another instance, electrospinning and carbonization methods were exploited for the fabrication of electrospun Ag/g-C_3N_4-loaded composite carbon nanofibers. In more detail, PAN and polyvinyl pyrrolidone were dissolved and mixed and then $AgNO_3$ as silver precursor and melamine were introduced to prepare the spinning dope [112]. After electrospinning, the nanofibers were preoxidized and then carbonized. Examining the catalytic activity of the resulting catalyst for reduction of 4-nitrophenol and benzylamine confirmed high catalytic activity of the catalyst. The observed catalytic activity was attributed to the high adsorption capability of carbon nanofibers as well as synergism between silver nanoparticles and g-C_3N_4. Facile recycling and high stability and need for low dosage of the membrane-like composite were other advantages of the developed catalyst.

Recently, electrospinning method has been utilized for the synthesis of N-doped carbon-supported FeO_x with excellent catalytic activity for hydrogenation of nitroarenes in the presence of hydrazine hydrate [113]. As shown in Fig. 11, the synthetic procedure for the fabrication of the catalyst (FeO_x@CN-hpes) included formation of Fe@HMTA metal organic framework (MOF), electrospinning of Fe@HMTA, and polymethylmethacrylate and PAN mixture in DMF, followed by pyrolysis. It was found that the pyrolysis temperature could affect the catalytic activity

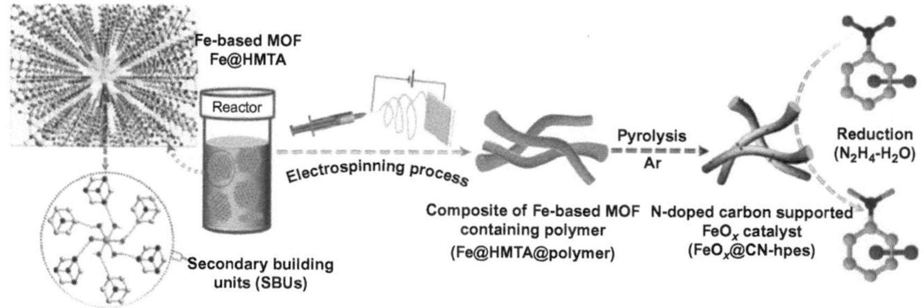

Fig. 11 Schematic illustration of preparing FeO_x-based (FeO_x@CN-hpes) catalyst for hydrogenation of nitroarenes to functional arylamines. *(Reprinted from X. Li, et al., Enhanced catalytic performance of nitrogen-doped carbon supported FeOx-based catalyst derived from electrospun nanofiber crosslinked N, Fe-containing MOFs for efficient hydrogenation of nitroarenes, Mol. Catal. 477 (2019) 110544, with permission from Elsevier.)*

and the highest activity was observed for the sample that was fabricated at 400°C. It is worth mentioning that apart from high catalytic activity, the catalyst exhibited high recyclability without any loss in both activity and selectivity. On the other hand, the catalytic activity of the catalyst was superior compared to control catalysts prepared by using $Fe(NO_3)_3 \cdot 6H_2O$, instead of the Fe@HMTA MOFs and sample prepared without using the electrospun technique.

In another example, Li, Liu et al. fabricated hierarchical porous carbon nanofibers-encapsulated Au nanoparticles, PCNFs-Au [114], and used it for the reduction of 4-nitrophenol in the presence of $NaBH_4$. To prepare the catalyst, MOF (ZIF-8-Au) was embedded into PAN via electrospinning and then the resulting composite was carbonized. The reduction reaction proceeded via pseudo-first-order kinetic with a rate constant $k \sim 0.10455\,s^{-1}$. Noteworthy, the catalyst could be recycled for five reaction runs with insignificant loss of the catalytic activity. Moreover, the characterization of the recycled catalyst confirmed that the recycling did not induce aggregation of Au nanoparticles. This observation was attributed to the high surface area of PCNFs that provided large amounts of active sites for the load of Au nanoparticles.

6.4 Oxidation reaction

Jiang et al. reported synthesis of electrospun CeO_2/Ag@carbon nanofiber through electrospinning and carbonization by using PAN, polyvinylpyrrolidone, $Ce(NO_3)_3{\cdot}6H_2O$, and $AgNO_3$ as precursors [86]. The electrospun nanofiber hybrid exhibited high catalytic performance (high conversion and selectivity) for selective oxidation of *n*-butyl and benzyl alcohol to the corresponding aldehydes. Notably, CeO_2/Ag@carbon nanofiber could be easily separated and recycled.

In another attempt, Lee et al. supported Pt catalyst on PAN-derived electrospun carbon fibers with diameters of 39, 158, and 309 nm [87]. The resulting compound was applied as a heterogeneous catalyst for glucose oxidation. The authors proved that, among three catalysts, the one that was prepared by using the fibers with higher curvature and smaller diameter (Pt/CF_{39nm}) had higher number of Pt atoms on its surface and led to the highest catalytic activity. It was also shown that the prepared Pt/CF could be used as a nonenzymatic glucose sensor with significant tolerance to foreign substrates. Among three systems, Pt/CF_{39nm} showed a higher sensitivity ($2.03\,\mu A mM^{-1}\,cm^{-2}$), detection limits (33 μM), and linear range (0.3–17 mM).

In another example, Bai et al. electrospun $AgNO_3$/PAN/DMF solutions to obtain $AgNO_3$/PAN nanofibers [115]. The resulting fibers were then treated with H_2 gas to reduce silver salt to silver nanoparticles. Finally, the carbonization of Ag/NFs led to the formation of Ag/CNFs. The characterization of Ag/CNFs showed that Ag nanoparticles with sizes in the range of 10–30 nm were uniformly dispersed. The study of the catalytic activity of Ag/CNFs for styrene epoxidation also confirmed high catalytic activity of Ag/CNFs. High activity of the catalyst was attributed to high surface areas of Ag nanoparticles and synergistic effect on delivery of electrons between Ag nanoparticles and CNFs.

6.5 Hydrogen production

Electrospun carbon nanofibers containing Co-TiC nanoparticles-like superficial protrusions were also prepared from electrospinning of solution of cobalt acetate tetrahydrate, titanium (IV) isopropoxide, and polyvinylpyrrolidone, followed by carbonization. The resulting catalyst was then applied for promoting hydrolysis of an aqueous ammonia borane solution [116]. It was found that the activity of the catalyst (turn over frequency of $32.18\ mol_{H2}\ min^{-1}\ (mol_{metal}^{-1})$) was superior to that of Co NPs, Co/CNFs, and CoTiC/C NPs. Moreover, it was shown that the catalyst was recyclable and the rate of ammonia borane hydrolysis depended on the catalyst concentration and temperature. The low activation energy of the catalyst was attributed to the synergistic effects of the Co and TiC-like superficial protrusions.

Hydrolysis of aqueous ammonia borane was also catalyzed by electrospun $CoCr_7C_3$-supported C nanofibers [117]. The catalyst was simply fabricated by carbonization of nanofibers prepared from electrospinning of cobalt acetate tetrahydrate, chromium (II) acetate dimermonohydrate, and polyvinylpyrrolidone. The comparison of the activity of the nanocomposite with that of cobalt/carbon nanofibers (Co/CNFs) and cobalt nanoparticles confirmed superior activity of the catalyst (turn over frequency and activation energy of $25.78\,mol_{H2}\,min^{-1}(mol_{metal}^{-1})$ and $\sim 24.2\,kJ\,mol^{-1}$). The study of the recyclability of the catalyst also proved high recyclability of the catalyst up to six reaction runs. It was also shown that the reaction temperature and catalyst concentration could affect the rate of ammonia borane hydrolysis.

Brooks, Nafady, Yousef et al. reported an efficient catalyst, NiS@CNFs, for hydrogen production from hydrolytic dehydrogenation of sodium borohydride [118]. The catalyst that possessed high specific surface area ($650.92\,m^2\,g^{-1}$) was prepared via electrospinning of a mixture of PAN, nickel acetate, and ammonium sulfide, followed by carbonization (Fig. 12). It is worth mentioning that NiS@CNFs exhibited high

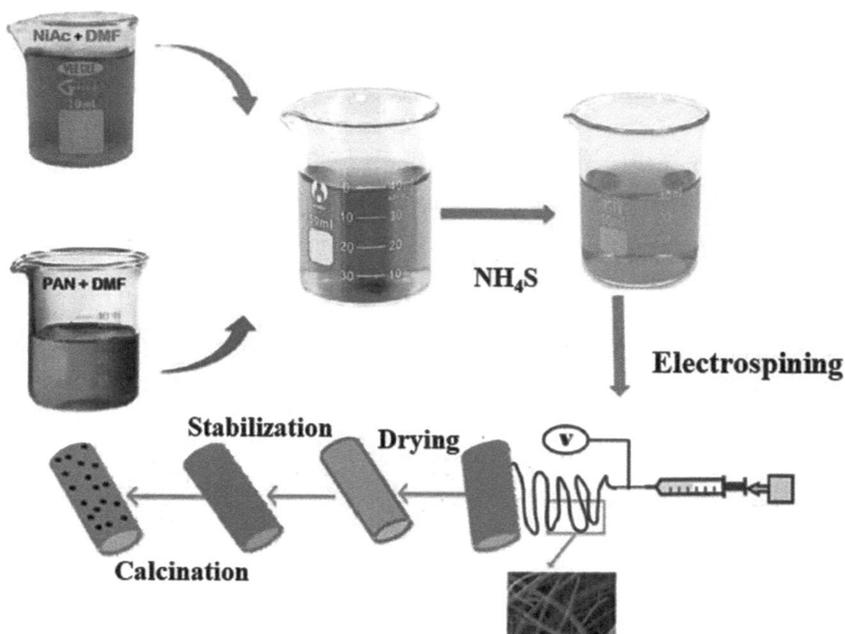

Fig. 12 Fabrication processes of NiS@CNFs. *(Reprinted from A.M. Al-Enizi, et al., Electrospun carbon nanofiber-encapsulated NiS nanoparticles as an efficient catalyst for hydrogen production from hydrolysis of sodium borohydride, Int. J. Hydrogen Energy 44(39) (2019) 21716–21725, with permission from Elsevier.)*

recyclability and catalytic activity ($E_a = 25.11576\,kJ\,mol^{-1}$, $\Delta S = 0.02374\,kJ\,mol^{-1}$, $\Delta H = 22.56004\,kJ\,mol^{-1}$) that was superior to that of Ni@CNFs. The reaction followed half-order kinetics with respect to the catalyst concentration.

Maafa, Yousef et al. reported fabrication of electrospun bimetallic NiCr nanoparticles@CNFs that served as a catalyst for hydrolytic dehydrogenation of ammonia borane [119]. To prepare the catalyst, a solution of chromium acetate dimer, nickel acetate tetrahydrate, and polyvinyl alcohol was electrospun and the resulting nanofibers were then sintered under inert atmosphere. The catalyst with 15 wt% Cr exhibited high catalytic activity (activation energy, ΔS, and ΔH of 37.6, 0.094, and 35.03 $kJ\,mol^{-1}$), higher than that of Cr-free Ni-CNFs. Moreover, it was recyclable for six runs with slight decrease of the catalytic activity.

7 Conclusion

Recently, breakthroughs in electrospun nanofibers, particularly in the electrospinning techniques and the possibility of preparing nanofibers of diverse morphologies and architectures, have opened up great opportunities for the specific design and fabrication of different nanofibers. Electrospun nanofibers can be transformed to carbon nanofibers via carbonization process. Carbon nanofibers can be decorated with catalytic species such as nanoparticles, alloys, and bimetallic nanoparticles either by electrospinning of a dope containing the catalyst precursor or by posttreatment. Carbon nanofiber-based catalysts can be applied for catalyzing various photo/electro/chemical transformations. From some discussed examples in this chapter, it can be concluded that use of carbon nanofibers as catalyst support can lead to the catalyst with improved catalytic activity and recyclability. The enhanced catalytic activity can be attributed to the chemical and physical properties of carbon nanofibers such as specific surface area as well as the possible synergism between carbon nanofiber and the immobilized catalytic species. Furthermore, the N-doped carbon nanofibers can potentially improve anchoring of the catalytic species and assure their high dispersion. In the case of photocatalysis, separation of photo-induced charge carriers can also be observed. Moreover, the porosity of the carbon nanofibers can be controlled through adjusting the spinning precursors. This feature paves the way for design and synthesis of the catalysts with improved mass transfer. Considering the advantages of carbon nanofibers for the catalysis, it can be expected that this issue will witness new advances in the future. In this regard, use of new spinning dopes for the development of doped carbon nanofiners with high specific surface area, combining electrospinning technology with other approaches for the preparation of nanocomposites, immobilization of novel catalytic species on carbon nanofibers, etc. can be forecasted.

It is hoped that the results presented in this chapter will encourage and inspire researchers to develop novel catalysts based on electrospun carbon nanofibers.

References

[1] Y. Dai, et al., Ceramic nanofibers fabricated by electrospinning and their applications in catalysis, environmental science, and energy technology, Polym. Adv. Technol. 22 (3) (2011) 326–338.

[2] M. Afshari (Ed.), Electrospun Nanofibers, In: Woodhead Publishing Series in Textiles, Woodhead Publishing, 2017.

[3] Z.-M. Huang, et al., A review on polymer nanofibers by electrospinning and their applications in nanocomposites, Compos. Sci. Technol. 63 (15) (2003) 2223–2253.

[4] L.A. Bosworth, S. Downes (Eds.), Electrospinning for Tissue Regeneration, Woodhead Publishing, 2011.

[5] B. Ding, J. Yu (Eds.), Electrospun Nanofibers for Energy and Environmental Applications, Springer-Verlag, Berlin, 2014.

[6] N. Tucker, et al., The history of the science and technology of electrospinning from 1600 to 1995, J. Eng. Fiber Fabr. **7** (2_suppl) (2012) 63–72.

[7] H. Niu, T. Lin, Fiber generators in needleless electrospinning, J. Nanomater. 2012 (2012) 13.

[8] S. Homaeigohar, et al., The electrospun ceramic hollow nanofibers, Nanomedicine 7 (11) (2017) 383.

[9] K. Yoon, B.S. Hsiao, B. Chu, Functional nanofibers for environmental applications, J. Mater. Chem. 18 (44) (2008) 5326–5334.

[10] C. Feng, K.C. Khulbe, T. Matsuura, Recent progress in the preparation, characterization, and applications of nanofibers and nanofiber membranes via electrospinning/interfacial polymerization, J. Appl. Polym. Sci. 115 (2) (2010) 756–776.

[11] X. Lu, C. Wang, Y. Wei, One-dimensional composite nanomaterials: synthesis by electrospinning and their applications, Small 5 (21) (2009) 2349–2370.

[12] S. Tan, X. Huang, B. Wu, Some fascinating phenomena in electrospinning processes and applications of electrospun nanofibers, Polym. Int. 56 (11) (2007) 1330–1339.

[13] R. Nayak, et al., Recent advances in nanofibre fabrication techniques, Text. Res. J. 82 (2) (2012) 129–147.

[14] H. Ara Begum, M.K. Rahman Khan, Study on the various types of needle based and needleless electrospinning system for nanofiber production, Int. J. Text. Sci. 6 (2017) 110–117.

[15] M. Inagaki, Y. Yang, F. Kang, Carbon nanofibers prepared via electrospinning, Adv. Mater. 24 (19) (2012) 2547–2566.

[16] J. Doshi, D.H. Reneker, Electrospinning process and applications of electrospun fibers, J. Electrost. 35 (2) (1995) 151–160.

[17] H.S. Yoo, T.G. Kim, T.G. Park, Surface-functionalized electrospun nanofibers for tissue engineering and drug delivery, Adv. Drug Deliv. Rev. 61 (12) (2009) 1033–1042.

[18] A. Haider, S. Haider, I.-K. Kang, A comprehensive review summarizing the effect of electrospinning parameters and potential applications of nanofibers in biomedical and biotechnology, Arab. J. Chem. 11 (8) (2018) 1165–1188.
[19] S.S. Ray, et al., A comprehensive review: electrospinning technique for fabrication and surface modification of membranes for water treatment application, RSC Adv. 6 (88) (2016) 85495–85514.
[20] J. Xue, et al., Electrospun nanofibers: new concepts, materials, and applications, Acc. Chem. Res. 50 (8) (2017) 1976–1987.
[21] H. Wu, et al., Electrospinning of ceramic nanofibers: fabrication, assembly and applications, J. Adv. Ceram. 1 (1) (2012) 2–23.
[22] I. Bhattacharyya, et al., Free surface electrospinning of aqueous polymer solutions from a wire electrode, Chem. Eng. J. 289 (2016) 203–211.
[23] C.J. Angammana, S.H. Jayaram, A theoretical understanding of the physical mechanisms of electrospinning, in: Proc. ESA Annual Meeting on Electrostatics, 2011.
[24] M.M. Hohman, et al., Electrospinning and electrically forced jets. I. Stability theory, Phys. Fluids 13 (8) (2001) 2201–2220.
[25] T.A. Kowalewski, S. Barral, T. Kowalczyk, Modeling Electrospinning of Nanofibers, Springer Netherlands, Dordrecht, 2009.
[26] R. Sahay, V. Thavasi, S. Ramakrishna, Design modifications in electrospinning setup for advanced applications, J. Nanomater. 2011 (2011) 17.
[27] J. Pelipenko, et al., The impact of relative humidity during electrospinning on the morphology and mechanical properties of nanofibers, Int. J. Pharm. 456 (1) (2013) 125–134.
[28] A.D. Sekar, M. Manickam, Current trends of electrospun nanofibers in water and wastewater treatment, in: X.-T. Bui et al., (Ed.), Water and Wastewater Treatment Technologies, Springer Singapore, Singapore, 2019, pp. 469–485.
[29] J. Huang, T. You, Electrospun nanofibers: from rational design, fabrication to electrochemical sensing applications, in: Advances in Nanofibers, InTech, 2013, pp. 35–83.
[30] J. Lin, et al., Direct fabrication of highly nanoporous polystyrene fibers via electrospinning, ACS Appl. Mater. Interfaces 2 (2) (2010) 521–528.
[31] P.S. Kumar, S. Jayaraman, G. Singh, Polymer and composite nanofiber: electrospinning parameters and rheology properties, in: S. Thomas, R. Muller, J. Abraham (Eds.), Rheology and Processing of Polymer Nanocomposites, Wiley, 2016.
[32] N. Bhardwaj, S.C. Kundu, Electrospinning: a fascinating fiber fabrication technique, Biotechnol. Adv. 28 (3) (2010) 325–347.
[33] E.S. Medeiros, et al., Effect of relative humidity on the morphology of electrospun polymer fibers, Can. J. Chem. 86 (6) (2008) 590–599.
[34] Z. Qi, et al., Highly porous fibers prepared by electrospinning a ternary system of nonsolvent/solvent/poly(l-lactic acid), Mater. Lett. 63 (3) (2009) 415–418.
[35] P. Dayal, et al., Experimental and theoretical investigations of porous structure formation in electrospun fibers, Macromolecules 40 (21) (2007) 7689–7694.
[36] C.L. Casper, et al., Controlling surface morphology of electrospun polystyrene fibers: effect of humidity and molecular weight in the electrospinning process, Macromolecules 37 (2) (2004) 573–578.
[37] S. Megelski, et al., Micro- and nanostructured surface morphology on electrospun polymer fibers, Macromolecules 35 (22) (2002) 8456–8466.
[38] C.-L. Pai, M.C. Boyce, G.C. Rutledge, Morphology of porous and wrinkled fibers of polystyrene electrospun from dimethylformamide, Macromolecules 42 (6) (2009) 2102–2114.
[39] C.J. Buchko, et al., Processing and microstructural characterization of porous biocompatible protein polymer thin films, Polymer 40 (26) (1999) 7397–7407.

[40] M.M. Demir, et al., Electrospinning of polyurethane fibers, Polymer 43 (11) (2002) 3303–3309.
[41] J.M. Deitzel, et al., The effect of processing variables on the morphology of electrospun nanofibers and textiles, Polymer 42 (1) (2001) 261–272.
[42] S.Y. Gu, J. Ren, G.J. Vancso, Process optimization and empirical modeling for electrospun polyacrylonitrile (PAN) nanofiber precursor of carbon nanofibers, Eur. Polym. J. 41 (11) (2005) 2559–2568.
[43] B.K.C. Remant, et al., Effect of collector temperature and conductivity on the chain conformation of nylon 6 electrospun nanofibers, Fiber. Polym. 8 (2) (2007) 186–191.
[44] E. Adomavičiūtė, S. Stanys, Formation of electrospun PVA mats on different types of support materials using various kinds of grounded electrodes, Fibres Text. East. Eur. 19 (2011) 34–40.
[45] D. Li, et al., Collecting electrospun nanofibers with patterned electrodes, Nano Lett. 5 (5) (2005) 913–916.
[46] D. Li, Y. Wang, Y. Xia, Electrospinning of polymeric and ceramic nanofibers as uniaxially aligned arrays, Nano Lett. 3 (8) (2003) 1167–1171.
[47] X.-Y. Ye, X.-J. Huang, Z.-K. Xu, Nanofibrous mats with bird's nest patterns by electrospinning, Chin. J. Polym. Sci. 30 (1) (2012) 130–137.
[48] D. Zhang, J. Chang, Patterning of electrospun fibers using electroconductive templates, Adv. Mater. 19 (21) (2007) 3664–3667.
[49] F. Hejazi, et al., Novel class of collector in electrospinning device for the fabrication of 3D nanofibrous structure for large defect load-bearing tissue engineering application, J. Biomed. Mater. Res. A 105 (5) (2017) 1535–1548.
[50] D. Li, Y. Wang, Y. Xia, Electrospinning nanofibers as uniaxially aligned arrays and layer-by-layer stacked films, Adv. Mater. 16 (4) (2004) 361–366.
[51] P.D. Dalton, D. Klee, M. Möller, Electrospinning with dual collection rings, Polymer 46 (3) (2005) 611–614.
[52] P. Katta, et al., Continuous electrospinning of aligned polymer nanofibers onto a wire drum collector, Nano Lett. 4 (11) (2004) 2215–2218.
[53] E. Zussman, A. Theron, A.L. Yarin, Formation of nanofiber crossbars in electrospinning, Appl. Phys. Lett. 82 (6) (2003) 973–975.
[54] V. Beachley, et al., A novel method to precisely assemble loose nanofiber structures for regenerative medicine applications, Adv. Healthc. Mater. 2 (2) (2013) 343–351.
[55] C. Mit-uppatham, M. Nithitanakul, P. Supaphol, Ultrafine electrospun polyamide-6 fibers: effect of solution conditions on morphology and average fiber diameter, Macromol. Chem. Phys. 205 (17) (2004) 2327–2338.
[56] N. Amiraliyan, M. Nouri, M.H. Kish, Effects of some electrospinning parameters on morphology of natural silk-based nanofibers, J. Appl. Polym. Sci. 113 (1) (2009) 226–234.
[57] L. Huang, et al., Controlling electrospun nanofiber morphology and mechanical properties using humidity, J. Polym. Sci. B Polym. Phys. 49 (24) (2011) 1734–1744.
[58] M. Putti, et al., Electrospinning poly(ε-caprolactone) under controlled environmental conditions: influence on fiber morphology and orientation, Polymer 63 (2015) 189–195.
[59] Z. Zhou, et al., Development of carbon nanofibers from aligned electrospun polyacrylonitrile nanofiber bundles and characterization of their microstructural, electrical, and mechanical properties, Polymer 50 (13) (2009) 2999–3006.
[60] Y. Wang, J.J. Santiago-Avilés, Large negative magnetoresistance and two-dimensional weak localization in carbon nanofiber fabricated using electrospinning, J. Appl. Phys. 94 (3) (2003) 1721–1727.

[61] L. Wang, et al., Electrospinning synthesis of C/Fe_3O_4 composite nanofibers and their application for high performance lithium-ion batteries, J. Power Sources 183 (2) (2008) 717–723.
[62] J. Liu, Z. Yue, H. Fong, Continuous nanoscale carbon fibers with superior mechanical strength, Small 5 (5) (2009) 536–542.
[63] E. Zussman, et al., Mechanical and structural characterization of electrospun PAN-derived carbon nanofibers, Carbon 43 (10) (2005) 2175–2185.
[64] K. Suzuki, et al., Preparation of carbon fiber fabrics from phenolic resin by electrospray deposition, Polym. J. 39 (2007) 1128.
[65] C. Kim, K.S. Yang, Electrochemical properties of carbon nanofiber web as an electrode for supercapacitor prepared by electrospinning, Appl. Phys. Lett. 83 (6) (2003) 1216–1218.
[66] C. Kim, et al., Fabrications and structural characterization of ultra-fine carbon fibres by electrospinning of polymer blends, Solid State Commun. 142 (1) (2007) 20–23.
[67] S. Prilutsky, E. Zussman, Y. Cohen, The effect of embedded carbon nanotubes on the morphological evolution during the carbonization of poly(acrylonitrile) nanofibers, Nanotechnology 19 (16) (2008) 165603.
[68] M. Bayat, H. Yang, F. Ko, Electromagnetic properties of electrospun Fe_3O_4/carbon composite nanofibers, Polymer 52 (7) (2011) 1645–1653.
[69] H. Hou, et al., Electrospun polyacrylonitrile nanofibers containing a high concentration of well-aligned multiwall carbon nanotubes, Chem. Mater. 17 (5) (2005) 967–973.
[70] F. Ko, et al., Electrospinning of continuous carbon nanotube-filled nanofiber yarns, Adv. Mater. 15 (14) (2003) 1161–1165.
[71] R. Ruiz-Rosas, et al., The production of submicron diameter carbon fibers by the electrospinning of lignin, Carbon 48 (3) (2010) 696–705.
[72] L. Wang, et al., Electrospun carbon–cobalt composite nanofiber as an anode material for lithium ion batteries, Scr. Mater. 58 (5) (2008) 405–408.
[73] L. Ji, A.J. Medford, X. Zhang, Porous carbon nanofibers loaded with manganese oxide particles: formation mechanism and electrochemical performance as energy-storage materials, J. Mater. Chem. 19 (31) (2009) 5593–5601.
[74] L. Ji, X. Zhang, Fabrication of porous carbon/Si composite nanofibers as high-capacity battery electrodes, Electrochem. Commun. 11 (6) (2009) 1146–1149.
[75] Y. Yu, et al., Encapsulation of Sn@carbon nanoparticles in bamboo-like hollow carbon nanofibers as an anode material in lithium-based batteries, Angew. Chem. Int. Ed. 48 (35) (2009) 6485–6489.
[76] L. Zou, et al., A film of porous carbon nanofibers that contain Sn/SnOx nanoparticles in the pores and its electrochemical performance as an anode material for lithium ion batteries, Carbon 49 (1) (2011) 89–95.
[77] H.S. Choi, et al., Effects of surrounding confinements of Si nanoparticles on Si-based anode performance for lithium ion batteries, Electrochim. Acta 56 (2) (2010) 790–796.
[78] O. Toprakci, et al., Fabrication and electrochemical characteristics of electrospun $LiFePO_4$/carbon composite fibers for lithium-ion batteries, J. Power Sources 196 (18) (2011) 7692–7699.
[79] J.S. Im, et al., Improved capacitance characteristics of electrospun ACFs by pore size control and vanadium catalyst, J. Colloid Interface Sci. 327 (1) (2008) 115–119.
[80] C. Kim, et al., Self-sustained thin webs consisting of porous carbon nanofibers for supercapacitors via the electrospinning of polyacrylonitrile solutions containing zinc chloride, Adv. Mater. 19 (17) (2007) 2341–2346.
[81] J. Li, et al., Nickel/carbon nanofibers composite electrodes as supercapacitors prepared by electrospinning, J. Alloys Compd. 478 (1) (2009) 371–374.

[82] S.K. Nataraj, et al., Effect of added nickel nitrate on the physical, thermal and morphological characteristics of polyacrylonitrile-based carbon nanofibers, Mater. Sci. Eng. B 162 (2) (2009) 75–81.
[83] C. Kim, et al., Synthesis and characterization of porous carbon nanofibers with hollow cores through the thermal treatment of electrospun copolymeric nanofiber webs, Small 3 (1) (2007) 91–95.
[84] L. Ji, et al., Porous carbon nanofibers from electrospun polyacrylonitrile/SiO_2 composites as an energy storage material, Carbon 47 (14) (2009) 3346–3354.
[85] E. Zussman, et al., Electrospun polyaniline/poly(methyl methacrylate)-derived turbostratic carbon micro-/nanotubes, Adv. Mater. 18 (3) (2006) 348–353.
[86] I.I. Ponomarev, et al., Preparation and thermal treatment influence on Pt-decorated electrospun carbon nanofiber electrocatalysts, RSC Adv. 9 (47) (2019) 27406–27418.
[87] J. Mu, et al., High photocatalytic activity of ZnO – carbon nanofiber heteroarchitectures, ACS Appl. Mater. Interfaces 3 (2) (2011) 590–596.
[88] J. Mu, et al., In_2O_3 nanocubes/carbon nanofibers heterostructures with high visible light photocatalytic activity, J. Mater. Chem. 22 (5) (2012) 1786–1793.
[89] M. Zhang, et al., Hierarchical heterostructures of Bi_2MoO_6 on carbon nanofibers: controllable solvothermal fabrication and enhanced visible photocatalytic properties, J. Mater. Chem. 22 (2) (2012) 577–584.
[90] C.H. Kim, B.-H. Kim, K.S. Yang, TiO_2 nanoparticles loaded on graphene/carbon composite nanofibers by electrospinning for increased photocatalysis, Carbon 50 (7) (2012) 2472–2481.
[91] B. Pant, et al., Electrospun CdS–TiO_2 doped carbon nanofibers for visible-light-induced photocatalytic hydrolysis of ammonia borane, Cat. Com. 50 (2014) 63–68.
[92] Y. Zhong, et al., Flexible electrospun carbon nanofiber/tin(IV) sulfide core/sheath membranes for photocatalytically treating chromium(VI)-containing wastewater, ACS Appl. Mater. Interfaces 8 (42) (2016) 28671–28677.
[93] Y. Gu, et al., One-step solvothermal synthesis of Au-TiO_2 loaded electrospun carbon fibers to enhance photocatalytic activity, Vacuum 130 (2016) 1–6.
[94] B. Pant, et al., Ag-ZnO photocatalyst anchored on carbon nanofibers: synthesis, characterization, and photocatalytic activities, Synth. Met. 220 (2016) 533–537.
[95] Z. Guo, et al., Constructing iron phthalocyanine nanosheets/electrospun carbon nanofibers heterostructures with enhanced photocatalytic activity under visible light irradiation, J. Alloys Compd. **690** (2017) 160–168.
[96] S. Yang, et al., Molybdenum diselenide nanosheet/carbon nanofiber heterojunctions: controllable fabrication and enhanced photocatalytic properties with a broad-spectrum response from visible to infrared light, J. Colloid Interface Sci. 518 (2018) 1–10.
[97] J. Jang, et al., Electrospun one-dimensional graphitic carbon nitride-coated carbon hybrid nanofibers (GCN/CNFs) for photoelectrochemical applications, Curr. Appl. Phys. 18 (9) (2018) 1006–1012.
[98] J. Zhang, et al., 1D carbon nanofibers@TiO_2 core-shell nanocomposites with enhanced photocatalytic activity toward CO_2 reduction, J. Alloys Compd. 746 (2018) 168–176.
[99] X.-Q. Wu, et al., Flexible and porous TiO_2/SiO_2/carbon composite electrospun nanofiber mat with enhanced interfacial charge separation for photocatalytic degradation of organic pollutants in water, J. Colloid Interface Sci. 553 (2019) 156–166.
[100] F. Du, et al., Electrospun reduced graphene oxide/TiO_2/poly(acrylonitrile-co-maleic acid) composite nanofibers for efficient adsorption and photocatalytic removal of malachite green and leucomalachite green, Chemosphere 239 (2020) 124764.
[101] L. Chen, et al., Novel Pd-carrying composite carbon nanofibers based on polyacrylonitrile as a catalyst for Sonogashira coupling reaction, Cat. Com. 9 (13) (2008) 2221–2225.

[102] L. Guo, et al., Fabrication of palladium nanoparticles-loaded carbon nanofibers catalyst for the Heck reaction, New J. Chem. 37 (12) (2013) 4037–4044.
[103] H. Liu, et al., Fabrication of macroporous/mesoporous carbon nanofiber using $CaCO_3$ nanoparticles as dual purpose template and its application as catalyst support, J. Phys. Chem. C 117 (41) (2013) 21426–21432.
[104] H. Li, et al., Carbon nanofiber supported copper nanoparticles catalyzed Ullmann-type coupling reactions under ligand-free conditions, RSC Adv. 4 (89) (2014) 48362–48367.
[105] L. Guo, et al., Fabricating series of controllable-porosity carbon nanofibers-based palladium nanoparticles catalyst with enhanced performances and reusability, J. Mol. Catal. Chem. A 400 (2015) 95–103.
[106] S.-J. Guo, et al., The controllable preparation of electrospun carbon fibers supported Pd nanoparticles catalyst and its application in Suzuki and Heck reactions, Chin. Chem. Lett. 27 (3) (2016) 459–463.
[107] D. Yu, et al., Assembling formation of highly dispersed Pd nanoparticles supported 1D carbon fiber electrospun with excellent catalytic active and recyclable performance for Suzuki reaction, Appl. Surf. Sci. 399 (2017) 185–191.
[108] H. Mu, et al., In situ synthesis of Cu/CNFs composite catalyst by electrospun nanofibers wrapped copper chloride and applied for Ullmann coupling reaction, J. Mol. Struct. 1165 (2018) 90–100.
[109] P. Zhang, et al., In situ assembly of well-dispersed Ag nanoparticles (AgNPs) on electrospun carbon nanofibers (CNFs) for catalytic reduction of 4-nitrophenol, Nanoscale 3 (8) (2011) 3357–3363.
[110] P. Zhang, et al., An electron-rich free-standing carbon@Au core–shell nanofiber network as a highly active and recyclable catalyst for the reduction of 4-nitrophenol, Phys. Chem. Chem. Phys. 15 (25) (2013) 10453–10458.
[111] F. Li, et al., Preparation and catalytic behavior of hollow Ag/carbon nanofibers, Cat. Com. 69 (2015) 150–153.
[112] B. Yu, et al., Preparation of electrospun Ag/g-C_3N_4 loaded composite carbon nanofibers for catalytic applications, Mater. Res. Express 4 (2017) 015603.
[113] X. Li, et al., Enhanced catalytic performance of nitrogen-doped carbon supported FeOx-based catalyst derived from electrospun nanofiber crosslinked N, Fe-containing MOFs for efficient hydrogenation of nitroarenes, Mol. Catal. 477 (2019) 110544.
[114] J. Wang, et al., Electrospun metal-organic frameworks with polyacrylonitrile as precursors to hierarchical porous carbon and composite nanofibers for adsorption and catalysis, Chemosphere 239 (2020) 124833.
[115] H. Liu, et al., The preparation of silver nanoparticles/carbon nanofibers as catalyst in the styrene epoxidation, Colloids Surf. A Physicochem. Eng. Asp. 448 (2014) 154–159.
[116] A.M. Al-Enizi, et al., Electrospun carbon nanofibers containing Co-TiC nanoparticles-like superficial protrusions as a catalyst for H_2 gas production from ammonia borane complex, Ceram. Int. 43 (17) (2017) 15735–15742.
[117] A. Yousef, et al., Electrospun $CoCr_7C_3$-supported C nanofibers: effective, durable, and chemically stable catalyst for H_2 gas generation from ammonia borane, Mol. Catal. 434 (2017) 32–38.
[118] A.M. Al-Enizi, et al., Electrospun carbon nanofiber-encapsulated NiS nanoparticles as an efficient catalyst for hydrogen production from hydrolysis of sodium borohydride, Int. J. Hydrogen Energy 44 (39) (2019) 21716–21725.
[119] R.M. Brooks, et al., Electrospun bimetallic NiCr nanoparticles@carbon nanofibers as an efficient catalyst for hydrogen generation from ammonia borane, Nanomedicine 9 (2019) 1082.

CHAPTER 7

Pristine, transition metal and heteroatom-doped carbon aerogels for catalytic and electrocatalytic applications

Naveen Chandrasekaran and K.S. Adarsh
Electroplating Metal Finishing and Technology Division, CSIR-Central Electrochemical Research Institute, Karaikudi, Tamil Nadu, India

1 Introduction

Open-cell, three-dimensional assemblies of organic or inorganic nanoparticles with ultra low-density and high surface area are termed as aerogels [1–5]. Aerogels are obtained by supercritical carbon dioxide drying of wet gels, which preserves the tenuous network from collapse and maintains the attractive physical properties such as high surface area and porosity. The wet gels dried under ambient temperature and pressure are termed as xerogels. Organic aerogels are a class of materials most appropriate for their conversion to electrically conducting three-dimensional porous carbon interconnects [6–15]. Hitherto, various organic aerogels such as resorcinol-formaldehyde, melamine-formaldehyde, polyurethane, polyimide, polyamide polybenzoxazine, and polyisocyanurates have been employed to synthesize nanoporous carbon networks [16–25]. Additionally, graphene oxide (GO) has been utilized to prepare three-dimensional graphene networks using cross-linkers or by π-π stacking [26–30]. Graphene oxide is obtained by oxidation of graphite, resulting in oxygen functionalities (hydroxyl, epoxide, carbonyl, and carboxylic acid) on the surface of the carbon. These surface functionalities can be used to react with various cross-linkers (such as isocyanates) to form covalent three-dimensional GO networks. In another approach, GO is either hydrothermally reacted or freeze-dried to obtain physically interlinked three-dimensional networks through π-π interaction. To tap the outstanding physicochemical properties of rGO (reduced graphene oxide) and aerogels, rigorous synthetic approaches have been formulated to synthesize porous three-dimensional

Emerging Carbon Materials for Catalysis
https://doi.org/10.1016/B978-0-12-817561-3.00007-X

assemblies of rGO and aerogels. For example, the mesoporous carbon, derived from resorcinol-formaldehyde networks, has been used as interconnects to synthesize graphene aerogels. In another attempt, ascorbic acid was used as reducing agent to form porous networks of rGO. However, very few reports are available on GO-induced gelation of polymeric or cellulose networks. For example, Zhang et al. reported graphene oxide-induced gelation of cellulose networks with exceptional mechanical properties. Drawing from our experience, GO can also be used to induce the gelation of resorcinol-formaldehyde [31] and poly (urethane-amide) networks [32]. In the former, the acidity offered by deprotonation of oxygenated species in acetonitrile played the key role in the formation of resorcinol-formaldehyde network; in the latter, the hydroxyl and carboxylic acid groups present in GO reacted with aromatic triisocyanates to form poly (urethane-amide) network. These covalent or physically linked GO networks are further subjected to thermal treatment under inert conditions to form electrically conducting rGO/graphene networks.

Catalysis and electrocatalysis remained the core for sustainable and green energy conversion devices and other related applications. Platinum group metals (PGM) and their oxides are commonly used electrocatalysts for common energy conversion reactions. Nevertheless, the paucity and high cost have limited their industrial applications. The pristine carbon aerogels can be prepared as monoliths and thin films and can be employed for multifarious technological applications such as catalysts supports, adsorption, insulation, and energy storage and capacitive deionization reactions.

Different synthetic protocols have been reported for the synthesis of transition metal or alloy aerogels by various groups. For example, Leventis et al. reported the synthesis of porous transition metal aerogels and rare earth carbides by carbothermal reduction [33–36]. Here, metal ions can be utilized as acid catalysts for gelation of R-F networks and ring opening of epoxides followed by condensation to form metal oxide (M-O-M) networks. The surface hydroxyl functionalities of the monolithic M-O-M networks can be conformably coated with polymeric derivatives of dopamine (pDA) or urethane (PU). The metal ion-doped R-F and M-O-M coated with pDA and PU can be further pyrolyzed under inert atmosphere to yield nanoporous metal/alloy/oxide-decorated carbon by carbothermal reduction. Tappan et al. [37] and Echymuller group prepared metal/alloy aerogels without using carbothermal reduction [38]. The metal/alloy porous networks combine the unique properties of the individual metal and aerogels. Due to the limited mass transfer through the nanochannels, efficient catalysis or electrocatalysis can be hampered through manipulation of metal foams. Careful design and engineering with balanced active centers and micro-*meso* and macroporous carbon networks may lead to efficient carbon-based catalysts for energy storage and conversion reactions.

It is, thus, imperative that: (i) the metal aerogels/monoliths are reinforced on an electrically conducting matrix; and (ii) the chemical interconnects should act both as a template and a carbon source. For example, high porosity offers improved mass transfer, high specific surface area, and more active sites, and the metallic or the carbonaceous backbone provides superior electrical conductivity, necessary for the effective electrocatalytic activity. In situ or posttreatment doping of carbon nanostructures with heteroatoms (sulfur, nitrogen, phosphorous) facilitates catalytic reactions through electronic modulation of their adjacent carbon atoms. The heteroatom-doped nanoporous carbon and transition metal/oxide or alloy-doped carbon aerogels are most sought-after alternates for precious metal catalysts for energy conversion reactions such as direct conversion of carbon dioxide to fuels, I^{3-} to I^- reduction in dye-sensitized solar cells, direct four electron ($4e^-$) oxygen reduction, two-electron ($2e^-$) transfer ORR to produce H_2O_2, nitrogen reduction, and oxygen and hydrogen evolution reactions [39–45]. For example, certain mono- or dual heteroatom-doped carbon materials were found to catalyze the anodic OER and cathodic ORR and HER for rechargeable air batteries and water splitting. By introducing one or multiple heteroatoms into the carbon matrix, variety of active sites for catalyzing different chemical reactions can be achieved. It is well-known that physical mixing of metal and carbon allotropes has been a common practice for fabrication of electrode materials for electrocatalytic reactions which may lead to lack of stability during the electrocatalytic process. Due to the nonavailability of metal in the heteroatom-doped carbon matrix and since the transition metal ions catalyse the reaction of the polymers and take integral part in the polymer formation, the metal/alloy formed during the carbothermal reduction remains intact into the porous carbon matrix. Hence, the heteroatom-doped and transition metal/oxide-decorated nanoporous carbons may offer good stability and do not suffer from dissolution of metal during the electrocatalytic process.

In this chapter, we summarize the topical advances, challenges, and opportunities in developing pristine, metal, and heteroatom-doped carbon aerogels for the catalytic and electrocatalytic applications.

2 Types of nanostructured carbons

Carbon allotropes are classified according to the arrangement of carbon atoms and are classified into amorphous carbon, graphite, and diamond. C_{60}, Carbon nanotubes (CNTs), and graphene led to advanced technologies based on carbon nanomaterials. Carbon aerogels are nanoporous carbon materials obtained from carbonization of R-F, M-F (Melamine-Formamide), and other organic aerogels. Carbon aerogels have networks of interconnected nanosize primary particles which are connected to form secondary particles. The nanoporous structures of the carbon aerogels are made up of intra-connected micropores (<2 nm) and interconnected meso- (2–50 nm) and macropores (>50 nm). The concentration of micropores and mesopores can be independently controlled in the carbon aerogels. The nature of the carbon material and concentration of pores and other physicochemical properties can be controlled by varying the nature of precursors, the drying methods, and the conditions during the carbonization process. Recently, carbon aerogels have been classified as nanostructured carbon due to the ability to control the structure and texture at the nanoscale.

3 Preparation of carbon aerogels

Carbon aerogels can be obtained by carbonization of organic aerogels. Organic aerogels are obtained by polymerization of certain organic monomers such as resorcinol, melamine, formaldehyde and isocyanates, etc. R-F and M-F aerogels were first reported by R.W. Pekala and these classes of material were extensively studied for almost two decades. R-F gels can be prepared by either acid or base catalysis by dissolving the reactants in water or organic solvents such as acetonitrile and *N,N′* dimethyl formamide. Typically, R-F gels are prepared within a week using Na_2CO_3 as the catalyst. Leventis et al. synthesized R-F gels using acetonitrile as the solvent and HCl as the catalyst and the gelation time was reduced to 2 h at room temperature and 10 min at 80°C [14]. The resultant gel was found to be akin in the physicochemical properties attained by the classic base-catalyzed R-F gels. By changing the concentrations of monomers and catalysts, the microstructure of the gels can be tuned. For example, small particles with well-connected networks were formed when the R/C (Reactant/Catalyst) ratio was low, whereas high R/C ratios led to colloidal gels with spherical particles with narrow necks. Unlike in basic conditions, where the mechanism of formation of the RF network is initiated by the activation of the aromatic ring leading to an increase in its electron density, the initiation of polymerization under acidic conditions occurs by protonation of formaldehyde preceded by hydroxymethyl formation in the aromatic ring. This step is followed by condensation of hydroxymethyl groups leading to the formation of CH_2 bridges between aromatic groups, resulting in a three-dimensional network of resorcinol-formaldehyde. The red color of the R-F gels can be due to the formation of *o*-quinonemethide during either acid- or base-catalyzed mechanism. Recently, our research group disclosed the polymerization of R-F using graphene oxide (GO). Here, the surface hydroxyl and oxygen functionalities of the GO play a crucial role in the increasing of the acidity of the acetonitrile solution containing R and F, leading to network formation similar to acid catalysis. In this approach, a catalytic amount of GO (0.2–0.8 wt%) is added to the resorcinol-formaldehyde sol where H^+ ions obtained from deprotonation of GO functionalities act as the acid catalyst for the gelation of the resorcinol-formaldehyde network, which takes only a few minutes for gelation at room temperature.

Scheme 1 illustrates the synthesis of heteroatom- and metal-doped carbon aerogels from pyrolysis of polymeric aerogels. The main stages of preparing aerogels include mixing of the reactants to form a sol, gelation at either ambient or elevated temperatures, and solvent exchange to remove the unreacted precursors subsequently followed by supercritical drying. Supercritical drying is employed to preserve the tenuous porous network

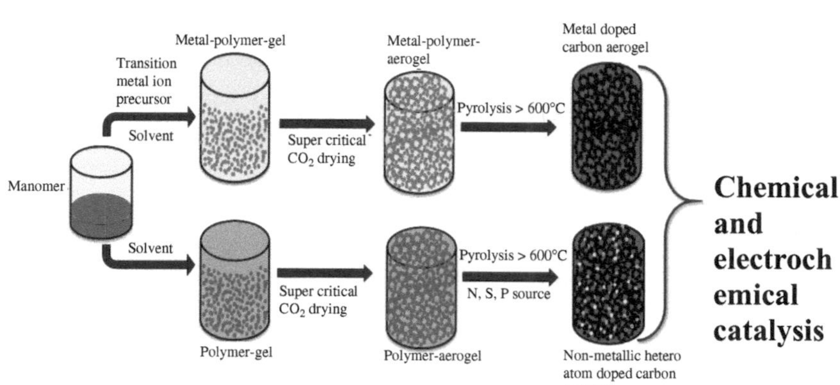

Scheme 1 Schematic representation of synthesis of heteroatom and metal-doped carbon aerogel from polymeric aerogels.

of the gels by subsiding the vapor-liquid interface. The wet gels dried under ambient temperature and pressure can be termed as xerogels. Low boiling point/surface tension solvents such as pentane and hexane were also employed to remove the solvent to preserve the volume of the gel. When the solvent is removed by freeze-drying, the gels are termed as cryogels. Supercritical or freeze-drying techniques can be used to preserve the volume of the gel. Once the three-dimensional networks are preserved, the organic aerogels are transformed to carbon aerogels by carbonizing at elevated temperatures (>600°C) under inert conditions. During the carbonization process, the organic gel loses the oxygen functionalities leaving high pure porous carbon structure.

Thermal treatment of the gels leads to shrinkage of the gels due to the loss of oxygen functionalities and mesopores tend to form at these temperatures. Once the carbonization temperature is increased, gas evolutions take place and lead to the formation of micropores and consequently increase in surface area. At higher temperatures, the gel undergoes maximum shrinkage with complete transformation of the organic moieties to carbon with very low surface oxygen functionalities. However, the porous structure of the carbon aerogels can be controlled by carbonization temperatures. Leventis et al. reported a method by which R-F gels can be conformably coated with polyurethane by reacting with the surface hydroxyl groups present in the R-F gels with aliphatic isocyanates [15]. The processed gels can be converted to electrically conducting macroporous carbon aerogels. The cross-linker polyurethane was found to be responsible for the conversion of mesopores to macropores by melting, followed by exerting surface tension forces on the RF framework leading to partial structural collapse. The bulk resistivity of these macroporous networks was found to lesser by 7 manifolds relative to its native counterparts. The surface area and pore volume of the carbon aerogels can also be increased by activating with steam and CO_2 at temperatures between 800°C and 900°C, making them potential candidates for many adsorption-related applications. Other than R-F gels, organic aerogels such as polyurea, polyurethane, polyamide, polyimide, polybenzoxazine, polyacrylonitrile, and polysaccharides have been pursued for transformation to carbon aerogels. Recently, various biomass-derived porous organic materials have been used as precursors for carbon aerogels [46–49]. When carbonized, the polymeric gels prepared with aliphatic carbon chain monomer lead to char, whereas the carbon obtained from polymers with aromatic chains was found to contain more sp^2 hybridized carbon with enhanced electrical conductivity.

4 Heteroatom-doped carbon aerogels for electrocatalytic applications

The plethora of free flowing electrons in carbon networks are too inert to be applied as electrodes in electrocatalysis. Carbon allotropes doped with heteroatom (nitrogen, sulfur, boron, and phosphorous) are most sought-after replacements for precious metal catalysts for energy conversion (hydrogen evolution, oxygen evolution, and reduction) and storage reactions. The enhanced electroactivity of these materials on par with benchmark (precious metal/metal oxide) catalysts can be reasoned to the activation of π electrons of carbon by the lone pair electrons of nitrogen, enabling the more positively charged carbon to effectively reduce oxygen molecules. Nitrogen-doped carbon catalysts have been found to act as a metal-free ORR electrocatalyst, but the type of nitrogen doping such as pyridinic-N, pyrrolic-N, and graphitic-N responsible for the electroactivity is quiet unclear. Nitrogen-doped carbon showed enhanced catalytic activity, extended stability, and good methanol resistance than platinum in alkaline solutions. Various studies have reported that nitrogen, phosphorous, and co-doping have improved the electrocatalytic activity towards ORR and are comparable with the commercial Pt/C catalyst, with the electron transfer per oxygen molecule estimated close to 4. The N-, P-doped porous carbon catalytic materials were also employed as air electrode in Zn-air batteries, providing an open circuit potential of 1.48 V and a specific capacity of 735 $mAhgZn^{-1}$. Apart from the doping, the hierarchal porous structure of carbon frameworks was related with ORR activity. N-doped carbons with the pore size distribution in the mesoporous regime show enhanced catalytic activity, facilitating the mass transport. But only the presence of mesopores limits the increase in active site density, and hence, the design of N-doped carbon with balanced micropores and mesopores is highly desirable for improving the active site density and mass transport towards oxygen reduction reactions. The heteroatom can be introduced into the carbon networks by either an in situ or ex-situ process. In a typical ex-situ process, the heteroatom precursor, likely NH_3 for incorporation of nitrogen or elemental sulfur/H_2S in the case of sulfur, will be employed during the pyrolysis of the organic networks. On the other hand, the sulfur or nitrogen containing organic molecules will be reacted with the carbon yielding monomer, forming porous networks which can be transformed to heteroatom-doped carbon during pyrolysis. Jasiniki et al. first demonstrated the synthesis of nitrogen-doped carbons from Co-pthalocyanines for ORR [50].

Yang et al. developed an approach for the synthesis of sulfur-doped graphene for ORR by pyrolyzing graphene oxide and benzyl disulfide as carbon and sulfur precursor. S-doped carbon aerogels derived upon pyrolysis of RF networks in the presence of H_2S displayed $4e^-$ transfer with a more positive onset potential and excellent methanol tolerance for ORR [51]. The electrocatalytic activity of the heteroatom-doped carbon allotropes can be enhanced by incorporating two or more heteroatom into the carbon backbone. For example, the S- and N-doped carbon showed remarkably better ORR activity than its mono-doped and pristine counterparts. Antonieetti et al. reported the synthesis of S- and N-doped carbon aerogels by pyrolyzing a mixture of glucose (as carbon source), S-(2-thienyl)-L-cysteine (TC) and 2-thienyl carboxaldehyde (TCA) (sulfur precursors), and ovalbumin (nitrogen source). The resultant N-, S-doped carbon networks revealed 4e- transfer for ORR [52]. The mechanism behind enhanced electrocatalytic activity of N-, S-doped carbon networks is still unclear and yet to be investigated.

We have also reported a simple and efficient method for the synthesis of simultaneous N- and S-doped carbon aerogels by forming a three-dimensional network of polyisocyanurate aerogels (PIRs) and pyrolyzing PIR in the presence of elemental sulfur yield N- and S-doped carbon monolithic structures [53]. The presence of nitrogen in the isocyanurate linkages acts as the source for N-doping into the carbon. The sulfur content can be changed by modulating the amount of elemental sulfur during thermal treatment. This approach eradicates the use of numerous precursors, meticulous methods, and harsh chemicals as reported in the previous methods. The electrocatalytic performance of the above-mentioned materials was examined towards ORR in alkaline medium. The analyzed physicochemical properties of these materials displayed that the ORR activity were extensively reliant on the doping concentration of elemental sulfur. The relatively more electrocatalytic activity of the optimized concentration of elemental sulfur into the carbon matrix than their native counterparts can be attributed to the introduction of micro- and mesopores into the existing macropores. It enhances the external surface area facilitating considerably more catalytic sites and enhanced inward and outward movement of ions in the electrolyte, presence of more —C—S—C— active species (oxygen adsorption sites), and monolayer formation on the micropores of carbon providing improved electronic conduction. N/P co-doping in carbon frameworks was also explored largely; Li et al. synthesized N, P co-doped carbon nanosheets from graphene oxide, polyaninile, and phytic acid by pyrolysis [54]. The material showed a low overpotential of 340 mV at 10 mA cm^{-1}, which is lesser than N and P individually doped carbon, indicating the synergistic advantage of multi-heteroatom doping. Combined doping metal and nonmetal on carbon has also proven excellent catalytic activity. Among this class of materials, metal- and nitrogen-doped materials showed significant OER activity. Zhao et al. reported the Fe—N—C and Co—N—C catalyst systems for OER with an overpotentials of ~360 mV and 380 mv at the current density 10 mA cm^{-1} [55]. The role of metal ions in these systems is not exactly known, as the onset potentials are similar to nonmetal-doped systems. Single atom-doped carbon catalytic systems are also known for OER catalysis. Chen et al. studied the single Fe-atom-doped N/S-doped carbon system and evaluated its OER activity with an onset potential of ~370 mV [56]. Nonmetal elements such as N, P, B, S also have been used to improve the activity of carbon towards HER. Among this, the N-doped systems were extensively researched; Huang et al. reported the N-doped graphene carbon with an overpotential of 239 mV at 10 mA cm^{-2} [57], and in acidic electrolyte, slightly higher overpotential of 290 mV was displayed by S-doped system [58]. Shervedani et al. studied the sulfur-doped graphene as a catalyst support and the influences of carbon black and ruthenium nanoparticles on the hydrogen evolution reaction performance [58]. Sathe et al. reported B-doping into carbon using borane tetrahydrofuran and the electrocatalytic analysis showed an overpotential of 430 mV at 10 mA cm^{-2}, which is slightly less than its parent graphene [59]. These results proved that B doping is not as effective as nitrogen and sulfur towards HER.

5 Metal-doped carbon aerogels

Introducing metal species in the carbon matrix leads to the change in the structure, physicochemical properties, and the catalytic activity of the carbon aerogels [60]. Four main strategies have been employed to introduce metal ion species into the carbon aerogels matrix: (1) dissolving the metal precursors in the organic sol before gelation; (2) Use of an organic monomer (resorcinol) derivative containing ion exchange moiety, which can be ion-exchanged with other metal ions; (3) Physical, chemical, or electrodeposition of the metal or its derivatives on the polymer or carbon aerogel; (4) Dipping the organic gels in the metal ion solution. Different metal containing carbon aerogels have been prepared for the catalysis and energy storage applications. Earlier attempts of incorporating metal species into the carbon matrix were focused only on RF gels [61,62]. Cerium and Zirconia were doped in the carbon aerogels by dissolving its respective nitrate salts in the RF sol [63]. Introduction of these salts resulted in the increase in the initial pH. These aerogels were pyrolyzed under Argon at 1000°C to yield Ce- and Zr-doped carbon aerogels [64,65]. The carbon aerogels with Zr was found to have relatively more micropores compared to the Ce-doped carbon aerogels [65]. The nature of the dopant was found to influence the physicochemical properties of the porous networks. In another approach, Pt-, Pd-, and Ag-doped carbon aerogels were obtained using $[Pt(NH_3)_4]Cl_2$, $PdCl_2$, or $AgOOC—CH_3$ as the catalysts for polymerizing the RF gels. Here, the precious metal salts were expected to act as an acid catalyst in driving the polymerization of RF networks. The RF gels containing the precious metal ions were pyrolyzed at higher temperatures (1050°C) and activated by steam at 900°C. The Pt containing carbon aerogels were found to display large meso- and macropore volumes, which can be attributed to the high degree of activation offered by Pt (gasification catalyst) during the activation step. Mayer et al. synthesized Pt/carbon aerogels by pyrolyzing the Pt black-RF aerogels [66]. In this case, the aqueous dispersion of Pt black was introduced to the initial RF sol. In another approach, the Pt-doped carbon aerogels were obtained by impregnation of the carbon aerogel with hexachloroplatinic acid. The composite was pyrolyzed at higher temperatures in inert Ar atmosphere to yield Pt/C aerogels. Supercritical deposition method was employed by Erkey et al. to synthesize Pt-doped carbon aerogels. In this approach, monolithic organic and carbon aerogels were impregnated with dimethyl (1,5-cyclooctadiene) platinum(II) dissolved in supercritical CO_2 [67]. Dresslhaus et al. synthesized Cu-doped carbon aerogels by pyrolyzing the Cu metal ion exchanged 2,4-dihydroxybenzoic acid formaldehyde aerogels [68]. Miller and Dunn reported ruthenium containing carbon aerogels by sublimation method [69]. The higher amount of metal loading decreased the BET surface area. It was found that even at high metal loading and repeated thermal treatment at 300°C, the Ru metal particles remained homogeneous with no aggregation. It led to a conclusion that the Ru metal particles are anchored at the surface of the carbon impeding the growth or agglomeration of the particles, establishing a good contact between the substrate and the particles. Moreno-Catilla et al. studied the possibilities of graphitization of carbon aerogels at relatively lower temperatures using transition metal (Cr, Fe, Co, Ni, W) ions [70,71]. These aerogels were synthesized by mixing metal acetate salts in the initial RF sol. The resulting carbon nanostructures revealed that the carbon atoms grew into a highly ordered crystalline solid by the catalytic activity of metal nanoparticles. Leventis et al. reported synthesis of interpenetrating networks of RF-Metal oxide (n-RF-MOx) gels. Upon pyrolysis at inert conditions, the gels were transformed to metal or metal carbide/carbon aerogels additionally; the polyurethane was conformably coated on the RF-MOx networks (X-RF-MOx) [33,34]. Thermal treatment of these gels resulted in macroporous metal or metal carbide/carbon aerogels. The polyurethane-coated gels were found to be thermodynamically efficient to yield the metal nanoparticles at lower temperatures. Pure metallic iron aerogels were also synthesized by pyrolyzing the interpenetrating networks of polybenzoxaine and iron oxide. Here, the formation of polybenzoxazine was catalyzed by $[Fe(H_2O)_6]^{3+}$, a bronsted acid catalyst [23,35].

We reported the synthesis of Ni, Fe alloy carbon aerogels by conformably coating polydopamine and polyurea on the NiFeOx particles; here, the NiFeOx particles were formed by ring opening of epoxides by using respective metal salts followed by condensation to form NiFeOx networks [72]. The NiFeOx networks were soaked in the polydopamine (pDA) solution. Subsequently, the surface hydroxyl groups were reacted with the aliphatic isocyanates to yield polyurethane (PU)-coated NiFeOx@pDA@pU. These gels were pyrolyzed to yield Ni, Fe alloy-decorated macroporous carbon networks. In another approach, GO was incorporated into the metal oxide sol and allowed to form gel at room temperature. Pyrolysis of these gels resulted in Ni, Fe alloy-reduced graphene oxide (rGO) aerogels with macroporous structure [73].

6 Metal-decorated carbon aerogels for chemical and electrochemical catalysis

Due to the diverse porosity and high specific surface area, metal-doped carbon aerogels have been pursued for various catalytic applications, For example, chromium, molybdenum, and tungsten oxide-doped macroporous carbon aerogels were utilized for isomerization reaction of 1-butene.The metal oxides were either found to be amorphous or nanocrystalline. Catalytic performance of these samples towards isomerization of 1-butene was found to be efficient for the tungsten oxide aerogels [74]. The superior catalytic activity of the tungsten oxide-doped carbon aerogels was traced to the high acidic nature of the oxide. The main products of this process were isobutene and trans-2-butene [75]. Keggin-type heteropolyacids were immobilized in the R-F networks and were used for the synthesis of methyl-*tert*-butylether from methyl alcohol and *tert*-butyl alcohol [76]. Cobalt- and nickel-doped carbon aerogels were used for the synthesis of graphic nanoribbons. Gunten et al. investigated the role of Co (II)-, Mn (II)-, and Ti (IV)-doped carbon aerogels for the conversion of ozone to OH radicals [77]. The Mn (II)-doped carbon aerogels were found to be active compared to other aerogels. Fe-, Co-, and Ni-doped carbon aerogels were found to be active for nonbiodegradable azo-dye Orange II degradation using the heterogeneous Fenton-like reaction. Job et al. reported Ni- and Pd-decorated carbon xerogels for ethylene hydrogenation [78]. Pd-Ag alloy catalysts were evaluated for selective hydro-dechlorination of 1,2dichloroethane into ethylene. The activity was found to increase with increasing Ag concentration. Pt- and Pt-Sn-based bimetallic catalysts were found to be active for the hydrogenation of cinnamaldehyde [79]. In another approach, Na-, K-, Mg-, and Zr-doped carbon xerogels were prepared from incorporating respective metal ions with RF aerogels [80]. The activity of these aerogels was tested towards Knoevenagel condensation of benzaldehyde with different methylenic compounds. Recently, Leventis et al. demonstrated the catalytic activities of carbon-supported Au or Pt, towards the oxidation of benzyl alcohol to benzaldehyde, Fe-decorated carbon catalysts for the reduction of nitrobenzene to aniline using hydrazine, and Pd-supported carbon for Heck-coupling reactions of iodobenzene with styrene or butyl acrylate [81].

Transition metal and their derivatives are of interests due to their excellent electroactivity on par with the benchmark precious metal catalysts. Apart from transition metals, the metals and metal oxides from the main groups, like Sn and Pb, also offer good and selective electrocatalytic activity [62,80]. Though the metal oxides offer good catalytic activity, their poor electronic conductivity (semiconducting or insulating) challenges them to be used as an electrocatalyst, as the electronic conductivity is a crucial factor in electrocatalysis. Distribution or decoration of metal/metal oxide active sites over the conducting material supports like conducting carbon frameworks, graphene oxide, and conducting polymers is a common strategy in the development of electrocatalyst. The conducting carbon frameworks stands are attractive due to their higher adsorption properties, superior electrical conductivity, inertness, and stability in the wide potential window and the solvents/electrolytes used. Owing to excellent electrical conductivity provided by the carbon networks and porosity facilitating enhanced diffusion of ions to quench the active centers, metal-decorated carbon aerogels are considered as efficient electrocatalyts for various electrocatalytic reactions such as oxygen evolution, oxygen reduction (ORR), hydrogen evolution, and carbon dioxide reduction reactions. Fe- and Co-doped carbon shows better electrocatalytic activity towards ORR. Li and coworkers prepared single Fe atoms anchored on N-doped porous carbon by the pyrolysis of $Fe(acac)_3$@ZIF-8 as electrocatalyst for oxygen reduction [82]. Carbon frameworks doped with Cu/CuO and heteroatoms can convert CO_2 to C_2 and C_3 products [83]. Single atom-doped carbon matrixes reported unusual electrocatalytic activity unlike from the bulk metal-doped systems. Nickel-based catalysts are well-known for the hydrogen evolution reactions; in contrast, Ni single atom-doped carbon framework shows good selectivity towards electrochemical CO_2 reduction because of the unique active site structure. Xie et al. synthesized single atom-doped Ni catalysts on N-doped carbon (Ni—N4—C) through a topochemical transformation approach. The active site was found to have Ni—N4 structure, exhibiting an excellent selectivity towards CO over a wide potential range; a Faradaic efficiency of 99% was achieved at −0.81 V (vs. RHE) [84]. In situ carbothermal approach to derive metal/metal oxide-doped carbon networks is versatile because of the easiness to tune the composition and percentage of metal/metal oxide active centers. Mono metal/metal oxide and mixed multi-metal/metal oxide catalysts and the ratio of metal to metal oxide can also be attuned through this approach. Mixed metal/metal oxides often exhibit synergistic activity in electrocatalysis. Electrocatalysts with lesser cobalt and larger nickel content are found to be efficient towards electrocatalytic methanol oxidation. Conducting carbon-supported metal/metal oxide sometimes shows better activity than bulk metal/metal oxides. Pt nanoparticles supported on carbon aerogels were tested for proton exchange membrane fuel cells. The catalyst was found to show better performance than the commercial Pt catalysts. Unlike the conducting polymer-supported catalysts, the carbon-supported electrocatalysts exhibit better film stability and retain the catalytic activity for extended hours. Co_3O_4, $Co_{0.85}Se$, and Fe_3C decorated on carbon are found to have an amazing OER activity; apart from aerogel approach, MOF materials were also been used to obtain metal/metal oxide carbon catalysts. Metal doping is also effective in hydrogen evolution reactions; metal dispersed on N-doped carbon has been extensively studied and the synthesis generally involves high-temperature pyrolytic treatments. Metal doping in the single atom level is most appropriate to achieve high catalytic activity. Fei et al. have synthesized atomically dispersed cobalt in carbon by heating a mixture of cobalt salt and graphene oxide in NH_3 atmosphere; the resulting material is found to have Cobalt loading of 0.57%, with an overpotential of 147 mV at 10 mA/cm^{-2} [85]. Various non-noble metal-doped carbon aerogels were also found to be active for electrocatalytic reactions.

7 Conclusions

In summary, carbon aerogels can be prepared from pyrolysis of polymeric aerogels derived from various organic monomers and biomass. Heteroatoms can be doped during pyrolysis of the polymeric aerogels under inert atmosphere. Four main strategies have been employed to introduce metal ion species into the carbon aerogels matrix: (1) dissolving the metal precursors in the organic sol before gelation; (2) Use an organic monomer (resorcinol) derivative containing ion exchange moiety, which can be ion-exchanged with other metal ions; (3) Physical, chemical, or electrodeposition of the metal or its derivatives on the polymer or carbon aerogel; (4) Dipping the organic gels in the metal ion solution. Different metal containing carbon aerogels have been prepared for catalysis and energy storage applications. Earlier attempts of incorporating metal species into the carbon matrix were focused only on RF gels. Additionally, metal ions can be utilized to polymerize organic monomers such as resorcinol, formaldehyde, etc. This chapter summarizes the synthesis of organic gels catalyzed by metal salts, the metal ions-decorated organic gels, and their pyrolyzed derivatives decorated on electrically conducting carbon, which can be explored in chemical and electrically driven catalysis for sustainable energy production.

References

[1] S.S. Kistler, Coherent expanded aerogels and jellies, Nature 127 (1931) 3211.
[2] S.S. Kistler, Coherent expanded-aerogels, J. Phys. Chem. 36 (1932) 52–63.
[3] X. Lu, M.C. Arduini-Schuster, J. Kuhn, O. Nilsson, J. Fricke, R.W. Pekala, Thermal conductivity of monolithic carbon aerogels, Science 255 (1992) 971–972.
[4] R.W. Pekala, Organic aerogels from the polycondensation of resorcinol with formaldehyde, J. Mater. Sci. 24 (1989) 3221–3227.
[5] A.C. Pierre, G.M. Pajonk, Chemistry of aerogels and their applications, Chem. Rev. 102 (2002) 4243–4265.
[6] S.A. Al-Muhtaseb, J.A. Ritter, Preparation and properties of resorcinol-formaldehyde organic and carbon aerogels, Adv. Mater. 2003 (15) (2003) 101–114.
[7] S.T. Mayer, R.W. Pekala, J.L. Kaschmitter, The aerocapacitor: an electrochemical double layer energy-storage device, J. Electrochem. Soc. 140 (1993) 446–451.
[8] R.W. Pekala, C.T. Alviso, X. Lu, J. Gross, J. Fricke, New organic aerogels based upon phenol-furfural reaction, J. Non-Cryst. Solids 188 (1995) 34–40.
[9] R.W. Pekala, Low density, resorcinol-formaldehyde aerogels, 1989. US patent 4873218-A.

[10] R.W. Pekala, C.T. Alviso, F.M. Kong, S.S. Hulsey, Aerogels derived from multifunctional organic monomers, J. Non-Cryst. Solids 145 (1992) 90–98.
[11] R.W. Pekala, J.C. Farmer, C.T. Alviso, T.D. Tran, S.T. Mayer, J.M. Miller, Carbon aerogels for electrochemical applications, J. Non-Cryst. Solids 225 (1998) 74–80.
[12] T. Yamamoto, T. Sugimoto, S. Suzuki, S.R. Mukai, H. Tamon, Preparation and characterization of carbon cryogel microspheres, Carbon 40 (2002) 1345–1351.
[13] K. Chan, J. Ding, J. Ren, S. Cheng, K.J. Tsang, Supported mixed metal nanoparticles as electrocatalysts in low temperature fuel cells, J. Mater. Chem. 14 (2004) 505–516.
[14] S. Mulik, C.S. Leventis, N. Leventis, Time-efficient acid-catalyzed synthesis of resorcinol-formaldehyde aerogels, Chem. Mater. 19 (2007) 6138–6144.
[15] S. Mulik, C.S. Leventis, N. Leventis, Macroporous electrically conducting carbon networks by pyrolysis of isocyanate-crosslinked resorcinol-formaldehyde aerogels, Chem. Mater. 20 (2008) 6985–6997.
[16] N. Leventis, C.S. Leventis, N. Chandrasekaran, S. Mulik, Z.J. Larimore, H. Lu, G. Churu, J.T. Mang, Multifunctional polyurea aerogels from isocyanates and water. A structure-property case study, Chem. Mater. 22 (2010) 6692–6710.
[17] C. Chidambareswarapattar, Z. Larimore, S.C. Leventis, J.T. Mang, N. Leventis, One-step room temperature synthesis of fibrous polyimide aerogels from anhydrides and isocyanates and conversion to isomorphic carbons, J. Mater. Chem. 20 (2010) 9666–9678.
[18] N. Leventis, C.S. Leventis, N. Chandrasekaran, S. Mulik, C. Chidambareswarapattar, A.G. Sadekar, D. Mohite, S. Mahadik, Z.J. Larimore, H. Lu, G. Churu, J. Mang, Isocyanate-derived organic aerogels: polyureas, polyimides, polyamides, MRS Proc. 1306 (2011). mrsf10-1306-bb03-0.
[19] C. Chidambareswarapattar, L. Xu, S.C. Leventis, N. Leventis, Robust monolithic multiscale nanoporous polyimides and conversion to isomorphic carbons, RSC Adv. 3 (2013) 26459–26469.
[20] M.A. Saeed, M.P. Rewatkar, M.H. Far, T. Taghvaee, S. Donthula, C. Mandal, S. C. Leventis, N. Leventis, Selective CO_2 sequestration with monolithic bimodal micro/macroporous carbon aerogels derived from polyamide-polyimide-polyurea random copolymers, ACS Appl. Mater. Interfaces 9 (2017) 13520–13536.
[21] R.W. Pekala, C.T. Alviso, J.D. Lemay, Organic aerogels: microstructural dependence of mechanical properties in compression, J. Non-Cryst. Solids 125 (1990) 67–75.
[22] L. Weigold, D.P. Mohite, S. Mahadik-Khanolkar, N. Leventis, G. Reichenauer, Correlation of microstructure and thermal conductivity in nanoporous solids: the case of polyurea aerogels synthesized from an aliphatic tri-isocyanate and water, J. Non-Cryst. Solids 368 (2013) 105–111.
[23] H.M. Far, M.P. Rewatkar, S. Donthula, T. Taghavee, M.A. Saeed, S.C. Leventis, N. Leventis, Exceptionally high CO_2 adsorption at 273 K by microporous carbons from phenolic aerogels: the role of heteroatoms in comaparison with carbons from polybenzoxazine and other organic aerogels, Macromol. Chem. Phys. 220 (2019) 1800333.
[24] M.S. Khanolkar, S. Donthula, S.C. Leventis, N. Leventis, 1. High-yield room-temperature acid catalyzed synthesis of robust monoliths, oxidative aromatization, and conversion to microporous carbons, Chem. Mater. 26 (3) (2014) 1303–1317.
[25] G. Biesmans, A. Mertens, L. Duffours, T. Woignier, J. Phalippou, Polyurethane based organic aerogels and their transformation into carbon aerogels, J. Non-Cryst. Solids 225 (1998) 64–68.
[26] A.M. Worsley, J.P. Pauzauskie, Y.T. Olson, J. Biener, F.J. Satcher, F.T. Baumann, Synthesis of graphene aerogel with high electrical conductivity, J. Am. Chem. Soc. 132 (2010) 14067–14069.
[27] A.M. Worsley, O.S. Kucheyev, H.E. Mason, D.M. Merrill, P.B. Meyer, J. Lewicki, M. E. Suss, M. Stadermann, J.P. Pauzauskie, H.J. Satcher, J. Biener, F.T. Baumann,

Mechanically robust 3D graphene macroassembly with high surface area, Chem. Commun. 48 (2012) 8428–8430.
[28] X. Zhang, Z. Sui, B. Xu, S. Yue, Y. Luo, W. Zhan, B. Liu, Mechanically strong and highly conducting graphene aerogel and its use as electrodes for electrochemical power sources, J. Mater. Chem. 21 (2011) 6494–6497.
[29] D. Zhang, X. Wen, L. Shi, T. Yan, J. Zhang, Enhanced capacitive deionization of graphene/mesoporous carbon composites, Nanoscale 4 (2012) 5440–5446.
[30] H. Li, L. Pan, C. Nie, Y. Liu, Z. Sun, Reduced graphene oxide and activated carbon composites for capacitive deionization, J. Mater. Chem. 22 (2012) 15556–15561.
[31] N. Chandrasekaran, V. Premkumar, M.S. Senthil Kumar, R. Ram, Single-step rapid synthesis of monolithic mesoporous carbon aerogels with improved double layer capacitance, New J. Chem. 42 (2018) 7371.
[32] T.K. Shruthi, M. Saravana Kumar, M. Arjunan, A. Pratap, N. Chandrasekaran, Graphene oxide aided structural tailoring of 3-D N-doped amorphous carbon network for enhanced energy storage, RSC Adv. 5 (2015) 93423.
[33] N. Leventis, N. Chandrasekaran, C.S. Leventis, A. Mumtaz, Smelting in the age of nano: iron aerogels, J. Mater. Chem. 19 (2009) 63–65.
[34] N. Leventis, N. Chandrasekaran, A.G. Sadekar, S. Mulik, C.S. Leventis, The effect of compactness on the carbothermal conversion of interpenetrating metal oxide/resorcinol-formaldehyde nanoparticle networks to porous metals and carbides, J. Mater. Chem. 20 (2010) 7456–7471.
[35] S. Mahadik-Khanolkar, S. Donthula, A. Bang, C. Wisner, C. Sotiriou-Leventis, N. Leventis, Polybenzoxazine aerogels. 2. Interpenetrating networks with iron oxide and the carbothermal synthesis of highly porous monolithic pure iron (0) aerogels as energetic materials, Chem. Mater. 26 (2014) 1318–1331.
[36] N. Leventis, S. Donthula, C. Mandal, M.S. Ding, C. Sotiriou Leventis, Explosive versus thermite behaviour in iron (0) aerogels infiltrated with perchlorates, Chem. Mater. 27 (2015) 8126–8137.
[37] B.C. Tappan, S.A. Steiner, E.P. Luther, Nanoporous metal foams, Angew. Chem. Int. Ed. 49 (2010) 4544–4565.
[38] (a) L. Lu, A. Eychmüller, Ordered macroporous bimetallic nanostructures: Design, characterization and applications, Acc. Chem. Res. 41 (2008) 244. (b) N. Gaponik, A.K. Herrmann, A. Eychmüller, Colloidal nanocrystal-based gels and aerogels: material aspects and application perspectives, J. Phys. Chem. Lett. 3 (2012) 8–17.
[39] S. Chen, J. Bi, Y. Zhao, L. Yang, C. Zhang, Y. Ma, Q. Wu, X. Wang, Z. Hu, Nitrogen doped carbon nanocages as efficient metal-free electrocatalysts for oxygen reduction reaction, Adv. Mater. 24 (2012) 5593–5597.
[40] Z. Chen, D. Higgins, H. Tao, R.S. Hsu, Z. Chen, Highly active nitrogen doped carbon nanotubes for oxygen reduction reaction in fuel cell applications, J. Phys. Chem. C 113 (2009) 21008–21013.
[41] M. Seredych, K. Laszlo, T.J. Bandosz, Sulfur-doped carbon aerogel as a metal-free oxygen reduction catalyst, ChemCatChem 7 (2015) 2924–2931.
[42] J. Liang, Y. Jiao, M. Jaroniec, S.Z. Qiao, Sulfur and nitrogen dual-doped mesoporous graphene electrocatalyst for oxygen reduction with synergistically enhanced performance, Angew. Chem. Int. Ed. 51 (2012) 11496–11500.
[43] J. Zhang, Z. Zhao, Z. Xia, L.A. Dai, A metal-free bifunctional electrocatalyst for oxygen reduction and oxygen evolution reactions, Nat. Nanotechnol. 10 (2015) 444–452.
[44] D. Geng, Y. Chen, Y. Chen, Y. Li, R. Li, X. Sun, S. Ye, S. Knights, High oxygen-reduction activity and durability of nitrogen-doped graphene, Energy Environ. Sci. 24 (2011) 760–764.
[45] X.L. Wu, T. Wen, H.L. Guo, S. Yang, X. Wang, A.W. Xu, Bio-mass derived sponge like carbonaceous hydrogels and aerogels for supercapacitors, ACS Nano 7 (2013) 3589–3597.

[46] P. Cheng, T. Li, H. Yu, L. Zhi, Z. Liu, Z. Lei, Bio-mass derived carbon fiber aerogel as a binder-free electrode for high-rate supercapacitors, J. Phys. Chem. C 120 (4) (2016) 2079–2086.
[47] Y. Si, X. Wang, C. Yan, L. Yang, L. Yu, B. Ding, Ultralight biomass-derived carbonaceous nanofibrous aerogels with superelasticicty and high-pressure sensitivity, Adv. Mater. 28 (2016) 9512–9518.
[48] S. Gao, X. Li, L. Li, X. Wei, A versatile biomass derived carbon material for oxygen reduction reaction, supercapacitors and oil/water separation, Nano Energy 33 (2017) 334–342.
[49] F. Schipper, A. Vizintin, J. Ren, R. Dominiko, T.P. Fellinger, Biomass derived heteroatom doped carbon aerogels from a salt melt sol-gel synthesis and their performance in Li-S batteries, ChemSusChem 18 (2015) 3077–3083.
[50] R. Jasinski, A new fuel cell cathode catalyst, Nature 201 (1964) 1212–1213.
[51] Z. Yang, Z. Yao, G. Li, G. Fang, H. Nie, Z. Liu, X. Zhou, X. Chen, S. Huang, Sulfur-doped graphene as an efficient metal-free cathode catalyst for oxygen reduction, ACS Nano 6 (2012) 205–211.
[52] S.A. Wohlgemuth, R.J. White, M.-G. Willinger, M.M. Titirici, M.A. Antonietti, A one pot hydrothermal synthesis of sulfur and nitrogen doped carbon aerogels with enhanced electrocatalytic activity in the oxygen reduction reaction, Green Chem. 14 (2012) 1515–1523.
[53] N. Chandrasekaran, K. Selvakumar, V. Premkumar, S. Muthusamy, S.M. Senthil Kumar, R. Thangamuthu, Dual heteroatom doped carbon monoliths derived from catalyst free preparation of porous polyisocyanurate for oxygen reduction reaction, ACS Sustain. Chem. Eng. 6 (2018) 9094–9103.
[54] R. Li, Z. Wei, X. Gou, Nitorgen and phosphorous dual doped graphene/carbon nanosheets as bifunctional electrocatalysts for oxygen reduction and evolution, ACS Catal. 5 (2015) 4133–4142.
[55] Y. Zhao, K. Kamiya, K. Hashimoto, S. Nakanishi, Efficient bifunctional Fe/C/N electrocatalysts for oxygen reduction and evolution reaction, J. Phys. Chem. C 119 (2015) 2583–2588.
[56] P. Chen, T. Zhou, L. Xing, K. Xu, Y. Tong, H. Xie, L. Zhang, W. Yan, W. Chu, C. Wu, Y. Xie, Atomically dispersed iron-nitrogen species as electrocatalysts for bifunctional oxygen evolution and reduction reactions, Angew. Chem. Int. Ed. 56 (2017) 610–614.
[57] X. Huang, Y. Zhao, Z. Ao, G. Wang, Micelle-template synthesis of nitrogen-doped mesoporous graphene as an efficient metal-free electrocatalyst for hydrogen production, Sci. Rep. 4 (2014) 7557.
[58] R.K. Shervedani, A. Amini, Sulfur-doped graphene as a catalyst support: Influences of carbon black and ruthenium nanoparticles on the hydrogen evolution reaction performance, Carbon 93 (2015) 762–773.
[59] B.R. Sathe, X. Zou, T. Asefa, Metal-free B-doped graphene with effiecient electrocatalytic activity for hydrogen evolution reaction, Catal. Sci. Technol. 4 (2014) 2023–2030.
[60] C. Moreno-Castilla, F.J. Maldonado-Hódar, Carbon aerogels for catalysis applications: an overview, Carbon 43 (2005) 455–465.
[61] F.J. Maldonado-Hódar, M.A. Ferro-Garcia, J. Rivera-Utrilla, C. Moreno-Castilla, Synthesis and textural characteristics of organic aerogels, transition-metal-containing organic aerogels and their carbonized derivatives, Carbon 37 (1999) 1199–1205.
[62] C. Moreno-Castilla, F.J. Maldonado-Hódar, J. Rivera-Utrilla, E. Rodriguez-Castellón, Group 6 metal oxide-carbon aerogel. Their synthesis, characterization and catalytic activity in the skeletal isomerisation of 1-butene, Appl. Catal. A 183 (1999) 345–356.

[63] E. Bekyarova, K. Kaneko, Microporous nature of Ce, Zr-doped carbon aerogels, Langmuir 1999 (15) (1999) 7119–7121.
[64] F.J. Maldonado-Hódar, C. Moreno-Castilla, J. Rivera-Utrilla, Y. Hanzawa, Y. Yamada, Catalytic graphitization of carbon aerogels by transition metals, Langmuir 16 (2000) 4367–4373.
[65] E. Bekyarova, K. Kaneko, Structure and physical properties of tailor-made Ce, Zr-doped carbon aerogels, Adv. Mater. 12 (2000) 1625–1628.
[66] S.T. Mayer, J.L. Kaschmitter, R.W. Pekala, Carbon aerogel electrodes for direct energy conversion, 1997. US Patent 5,601,938.
[67] C.D. Saquing, T.T. Cheng, M. Aindow, C. Erkey, Preparation of platinum/carbon aerogel nanocomposites using a supercritical deposition method, J. Phys. Chem. B 108 (2004) 7716–7722.
[68] T.F. Baumann, G.A. Fox, J.H. Satcher, N. Yoshizawa, R. Fu, M.S. Dresselhaus, Synthesis and characterization of copper-doped carbon aerogels, Langmuir 18 (2002) 7073–7076.
[69] J.M. Miller, B. Dunn, Morphology and electrochemistry of ruthenium/carbon aerogel nanostructures, Langmuir 15 (1999) 799–806.
[70] C. Moreno-Castilla, F.J. Maldonado-Hódar, A.F. Pérez-Cadenas, Phyiscochemical surface properties of Fe, Co, Ni, and Cu-doped monolithic carbon aerogels, Langmuir 19 (2003) 5650–5655.
[71] F.J. Maldonado-Hódar, A.F. Pérez-Cadenas, C. Moreno-Castilla, Morphology of heat treated tungsten doped monolithic carbon aerogels, Carbon 41 (2003) 1291–1299.
[72] T.N. Kumar, S. Sivabalan, N. Chandrasekaran, K.L.N. Phani, Synergism between polyurethane and polydopamine in the synthesis of Ni-Fe alloy monoliths, Chem. Commun. 51 (2015) 1922–1925.
[73] N. Chandrasekaran, S. Muthusamy, Binderless, free-standing porous interconnects of Ni-Fe alloy decorated reduced graphene oxide for oxygen evolution reaction, Langmuir 33 (2017) 2–10.
[74] M.A.M. Alvarez-Merino, F. Carrasco-Marín, C. Moreno-Castilla, Tungsten catalysts supported on activated carbon: II. Skeletal Isomerization of 1-butene, J. Catal. 192 (2000) 374–380.
[75] I. Pelech, O. Soares, M.F.R. Pereira, J.L. Figueiredo, Oxidative dehydrogenation of isobutene on carbon xerogel catalysts, Catal. Today 249 (2015) 176–183.
[76] S.R. Mukai, T. Sugiyama, H. Tamon, Immobilization of heteropoly acids in the network structure of carbon aerogels, Appl. Catal. A 256 (2003) 99–105.
[77] M. Sánchez-Polo, J. Rivera-Utrilla, U. von Gunten, Metal-doped carbon aerogels as catalysts during ozonation processes in aqueous solutions, Water Res. 40 (2006) 3375–3384.
[78] N. Job, B. Heinrichs, F. Ferauche, J. Marien, J.P. Pirard, International Symposium of Carbon for Catalysis, Carbocat-2004, Lausanne, Switzerland, July 2004, Book of Abstract, 2004. pp. 55–56.
[79] V.S. Purnakala, J.L. Figeuiredo, International Symposium of Carbon for Catalysis, Carbocat-2004, Lausanne, Switzerland, July 2004, Book of Abstract, 2004. pp. 51–52.
[80] M.L. Rojas-Cervantes, L. Alonso, J. Diaz-Teràna, A.J. López-Peinadoa, R.M. Martin-Aranda, V. Gómez-Serrano, Basic metal-carbons catalysts prepared by sol-gel method, Carbon 2004 (42) (2004) 1575–1582.
[81] A.M. Saeed, C.A. Wisner, S. Donthula, H.M. Far, C.S. Leventis, N. Leventis, Reusable monolithic nanoporous graphite-supported nanocatalysts (Fe, Au, Pt, Pd, Ni and Rh) from pyrolysis and galvanic transmetalation of ferrocene-based polyimide aerogels, Chem. Mater. 28 (13) (2016) 4867–4877.
[82] Y. Chen, S. Ji, Y. Wang, J. Dong, W. Chen, Z. Li, R. Shen, L. Zheng, Z. Zhuang, D. Wang, Y. Li, Isolated single iron atoms anchored on N-doped porous carbon as

an efficient electrocatalyst for the oxygen reduction reaction, Angew. Chem. Int. Ed. 56 (2017) 6937–6941.
[83] Y.X. Duan, F.L. Meng, K.H. Liu, S.S. Yi, S.J. Li, J.M. Yan, Q. Jiang, Amorphizing of cu nanoparticles toward high efficient and robust electrocatalyst for CO_2 reduction to liquid fuels with high faradaic efficiencies, Adv. Mater. 30 (2018) 1706194.
[84] X. Li, W. Bi, M. Chen, Y. Sun, H. Ju, W. Yan, J. Zhu, X. Wu, W. Chu, C. Wu, Y. Xie, Exclusive Ni-N4 sites realize near-unity CO selectivity for electrochemical CO_2 reduction, J. Am. Chem. Soc. 139 (2017) 14889–14892.
[85] H. Fei, J. Dong, M.J. Arellano-Jiménez, G. Ye, N.D. Kim, E.L. Samuel, Z. Peng, Z. Zhu, F. Qin, J. Bao, Atomic cobalt on nitrogen-doped graphene for hydrogen generation, Nat. Commun. 6 (2015) 8668.

CHAPTER 8

Carbon materials functionalized with sulfonic groups as acid catalysts

Enrique García-Bordejé[a], Elisabet Pires[b], and José M. Fraile[b]
[a]Institute of Carbochemistry (ICB-CSIC), Zaragoza, Spain
[b]Faculty of Sciences, Institute of Chemical Synthesis and Homogeneous Catalysis (ISQCH), CSIC-University of Zaragoza, Zaragoza, Spain

1 Introduction

Carbon materials are very interesting precursors for the preparation of acid catalysts bearing sulfonic groups. The variability in surface composition, and hence in the hydrophobic/hydrophilic balance, textural properties (surface area and porosity), and morphology, confers to the carbon materials a versatility that allows the preparation of tailored acid catalysts. In this chapter, we will present a brief overview of the different carbon materials, with their properties and the methods for their modification, the procedures to introduce sulfonic groups and the available techniques for their characterization, as well as some examples of catalytic applications and the deactivation mechanisms detected for this kind of materials.

2 Types of carbon materials functionalized with sulfonic groups

The way in which carbon materials are obtained influences the nature and properties of the final solids. The different types of carbon materials used as acid catalyst after functionalization with sulfonic groups are compiled in Fig. 1. The carbon materials functionalized with sulfonic groups have significant differences in terms of graphitic character, oxygen content, sp^2/sp^3 ratio, surface area, and pore size distribution. These different features affect their functionalization and subsequent catalytic performance in acid-catalyzed reactions.

2.1 Highly graphitic carbon materials

Traditionally, activated carbons have been used as material precursors for carbon catalysts. They possess a high carbonization degree with low heteroatom

Emerging Carbon Materials for Catalysis
https://doi.org/10.1016/B978-0-12-817561-3.00008-1

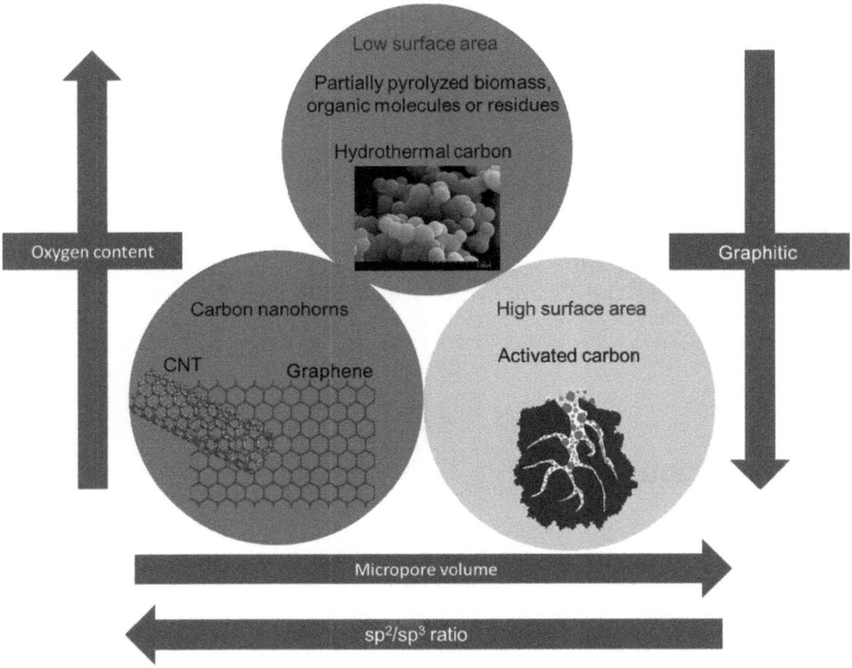

Fig. 1 Types of carbon materials used as precursors for sulfonic acid catalysts.

content and high surface area. These carbon materials are produced by pyrolysis at high temperature of a biomass residue and subsequent activation either by chemical agents (KOH, $ZnCl_2$, H_3PO_4) or physical methods (CO_2, H_2O) [1].

Activated carbons subjected to a treatment in HNO_3 are weakly acidic, due to the formation of carboxylic groups, and can be used directly as acid catalyst. In fact, there are some examples in the literature about the use of these acidic activated carbons as catalysts [2–4]. Thus, the activated carbons have high hydrothermal stability and lead to enhanced selectivity in the dehydration of sugars compared to the autocatalytic process without activated carbon.

The main drawback of activated carbons as acid catalyst is that the major part of pore volume corresponds to micropores (<2 nm) and these small pores are inaccessible for bulky organic molecules. This can be circumvented by using carbon materials without micropores or by the creation of mesoporosity (see Section 3).

Regarding graphitic materials without micropores, different materials, such as graphene, carbon nanotubes [5], carbon nanofibers, or carbon nanohorns [6], have been also functionalized with sulfonic groups. All these materials are based on graphene platelets that adopt different curvature degrees, leading to nanotubes or nanohorns. They are prepared by either top-down approaches (exfoliation of graphite) or decomposition of hydrocarbon at high temperatures in vapor phase. In principle, these materials have low amount of reactive sites to functionalize with sulfonic groups by the usual treatment with H_2SO_4. Nevertheless, other protocols of functionalization of graphitic materials, either non-covalently or through covalent grafting, can be used, for example, by covering the surface of CNT with poly(*p*-styrenesulfonic) acid [7]. On the other hand, the graphitic carbon materials exhibit some advantages such as the hydrophobicity or the absence of microporosity. The presence of hydrophobic domains can be beneficial for preventing the inhibition of the reaction by the water produced in the catalytic reaction. Particularly, sulfonated graphene has been reported to be water-tolerant [8]. Moreover, these materials usually lack microporosity and most of their pores are accessible to the bulky reactant molecules. A material related to graphene is graphene oxide (GO). Structurally, GO has a 2D structure similar to graphene, but it contains domains that are highly functionalized with oxygenated groups. This fact allows an easy functionalization, and hence, its use as acid catalyst (see Section 5.5.1).

2.2 Poorly graphitized carbon materials

As mentioned in Section 2.1, a drawback of highly graphitic materials such as activated carbons is the low amount of reactive sites for functionalization with sulfonic groups. To solve this problem, carbons with low graphitization degree have been prepared by the partial pyrolization of biomass-derived compounds. The partial pyrolization consists in the heat treatment of molecules containing carbon at relatively low temperatures. This is the case of the partial pyrolization of polycyclic aromatic compounds [9]. When heteroatoms containing precursors are used, heteroatoms partially remain on the produced carbonaceous material. For the preparation of these carbons, a wide range of raw materials can be used, ranging from materials with well-defined composition and natural origin to even wastes. Polysaccharides, such as starch or cellulose, or other biomass-derived materials, such as lignin, are examples of raw materials of defined composition and natural origin. Polymeric residues, such as phenolic resins or polyurethane, and biomass residues, such as bagasse or brown bin waste, are examples of wastes used to prepare partially pyrolized carbons. The advantages of using wastes are the low cost of the raw material and the simultaneous transformation of waste into a valuable material.

The temperature of pyrolysis is critical and it should be optimized for each material. At low pyrolysis temperature (<250°C), some reactions take place, such as condensation of CH_2OH groups and glycosidic ring opening, generating complex polymers consisting of aliphatic and alkene/aromatic functions. As the pyrolysis temperature increases, the aliphatic groups convert to small graphene sheets comprising fused aromatic rings. Upon increasing pyrolysis temperature over 450°C, this polycyclic aromatic carbon becomes larger and the carbon material becomes less prone to functionalization. Thus, intermediate temperatures around 400°C are usually found optimum for maximization of the acidity upon sulfonation. Anyway, the pyrolysis temperature should be optimized for each particular precursor and sulfonation method. The carbonization temperature can also affect the porous structure upon the subsequent sulfonation, besides affecting the composition.

Instead of dry pyrolysis, hydrothermal carbonization (HTC) is typically carried out at the temperature range of 130–250°C in subcritical aqueous media [10]. The precursor is usually a dissolved carbohydrate that polymerizes in situ forming a polyfuranic carbon network, but solid polymers are also amenable to HTC carbonization. The carbon material produced, referred as hydrochar, is formed by nonporous microspheres consisting of a hydrophobic core and a hydrophilic shell. The carbon material possesses a high oxygen content (O/C ratio about 0.3) and almost negligible surface area. Compared with dry pyrolysis, this method emits less amount of gases, the yield to carbon is larger, and the final material contains abundant oxygenated functional groups (OFGs), which make it more hydrophilic and reactive for functionalization. There are other variants for the preparation of hydrothermal carbons. One alternative is to carry out the hydrothermal carbonization only with the vapor phase, instead of doing the treatment with an excess of liquid water [11]. The approach avoids the separation of water and provides a carbon material with higher surface area and mesoporosity. Other more recently described variant is the use of an alcohol instead of water [12]. This alcohothermal carbonization is reported to lead to higher amount of sulfonic groups upon sulfonation than hydrothermal carbonization and thermal pyrolysis (Fig. 2). The carbonization of the carbon precursor by any of those methods entails reactions of dehydration, condensation, polymerization, and aromatization.

3 Modification of carbons

The porosity and surface chemistry of carbon materials have profound impacts on acid-catalyzed reactions. Comparing the catalytic performance of synthetic carbons with different porous properties [13], it was demonstrated that the

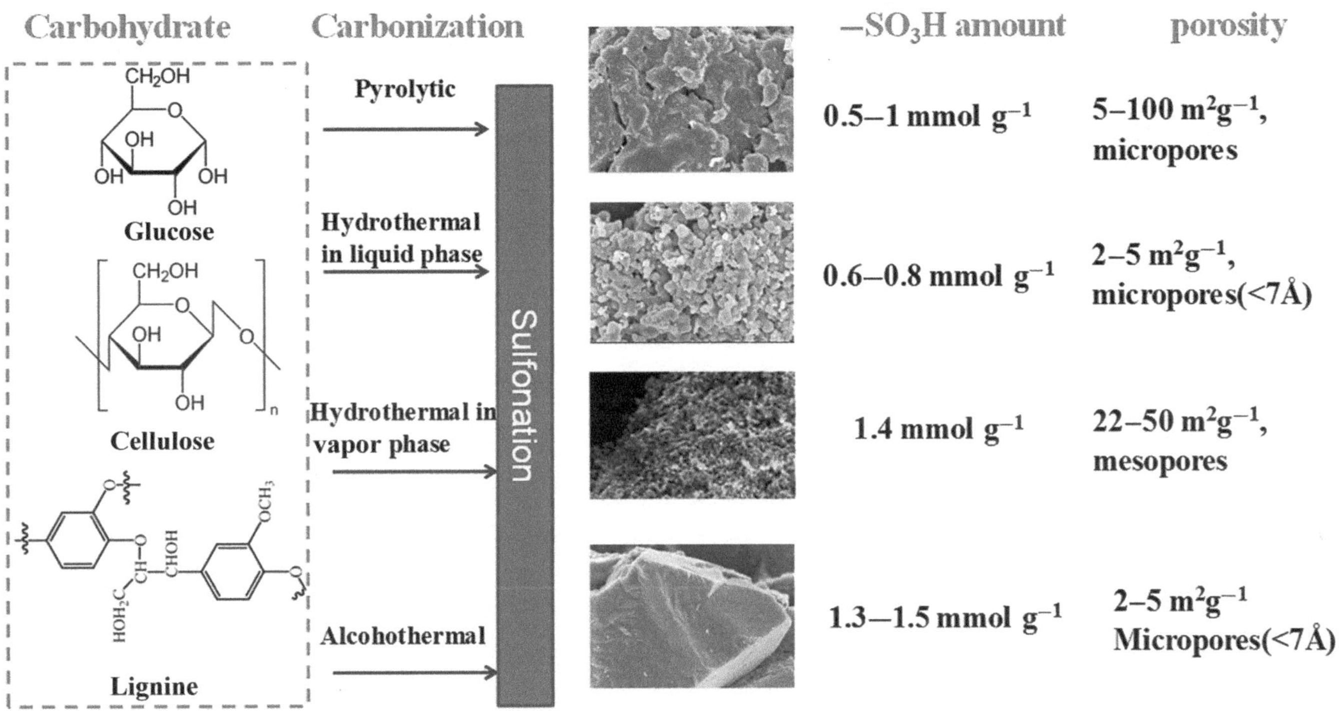

Fig. 2 Effect of the different types of carbonization, applied to biomass precursors, on the subsequent sulfonic acid functionalization. *(Adapted from Y.T. Yang, X.X. Yang, Y.T. Wang, J. Luo, F. Zhang, W.J. Yang, J.H. Chen, Alcohothermal carbonization of biomass to prepare novel solid catalysts for oleic acid esterification, Fuel 219 (2018) 166–175, https://doi.org/10.1016/j.fuel.2018.01.072, with permission of Elsevier.)*

porosity did not affect the esterification activity of small molecules such as acetic acid. However, the esterification of oleic acid with methanol requires a larger mesoporous volume. The esterification reaction with methanol works well with a hydrophilic surface, but other reactions involving hydrophobic substrates, such as the condensation of 2-methylfuran with furfural, are enhanced on a catalyst with hydrophobic surface and large mesopores. Therefore, both porosity and surface chemistry must be tuned to fit the requirements of each specific reaction.

3.1 Porosity of carbons

It has been mentioned that the microporosity of activated carbons hampers the accessibility of the bulky organic reagents to the catalytic sites in some acid-catalyzed reactions. In contrast, the intrinsic mesoporous structure of several graphitic carbon materials, such as carbon nanotubes or graphene, allows the access to these bulky molecules to the active sites. Therefore, the intrinsically inherent porosity of the selected commercial carbon material can be advantageously used for the desired reaction.

On the other hand, poorly graphitized carbon materials derived from partial pyrolization of biomass-derived compounds are readily functionalized with sulfonic groups, but they usually exhibit low surface areas. Therefore, the creation of porosity is essential to fully exploit the possibilities of these sulfonated carbon materials. Fig. 3 displays the different strategies to synthetize carbon materials with tunable porosity.

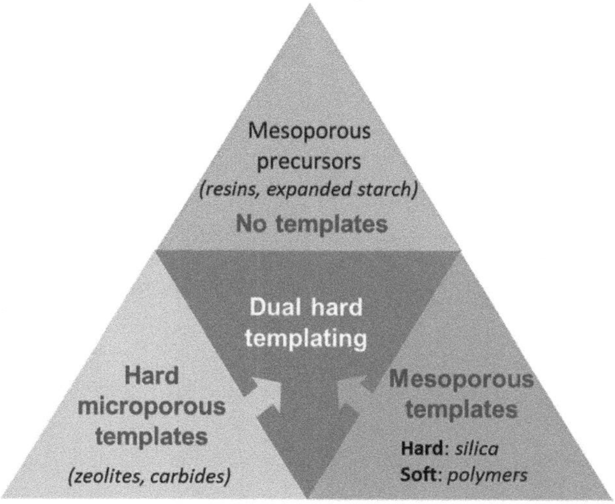

Fig. 3 Different strategies to prepare synthetic carbon materials with the desired micro-mesoporosity.

Mesoporous carbon materials can be prepared by pyrolysis/carbonization of a mesoporous precursor (top triangle in Fig. 3) or via using a sacrificial template, which has been previously filled with a carbon precursor (bottom part of Fig. 3). The hard templating or nanocasting method uses solid templates, whereas the soft templating method employs templates that leave no residues upon thermal treatment. Below, we describe some of the carbon materials prepared using the mentioned strategies that have been sulfonated to deliver acid catalysts.

3.1.1 Mesoporous carbons by pyrolysis of mesoporous precursors

The first strategy does not use a template, but the porosity of the prepared carbon is inherited from that of a parent thermosetting aerogel or xerogel resin. The porosity of the pristine aerogels is tunable from micro to mesopores. Mesoporous carbon aerogels or xerogels are prepared by pyrolysis of phenolic resins or resorcinol-formaldehyde resins [14]. The difference between aerogel and xerogel is that the former is freeze-dried, whereas the latter is dried under atmospheric conditions. Sometimes, freeze-drying is necessary to avoid the collapse of the porous structure. Another mesoporous carbon precursor is expanded starch, which is a natural resource that gives rise to a mesoporous carbon material upon pyrolization. The mesoporous carbon material prepared in this way received the name of "starbon" [15, 16].

3.1.2 .Mesoporous carbons prepared using hard or soft templates

To prepare carbons with an ordered mesoporous structure (bottom-right triangle in Fig. 3), both hard and soft templating can be used. In case a hard sacrificial template is used, such as ordered porous silica, it is impregnated with a carbonizable precursor, such as a resin, a polymer, or an ionic liquid [17]. Subsequently, this organic precursor is pyrolized and the silica template is removed by etching with hydrofluoric acid. Alternative to thermal carbonization, hydrothermal carbonization can be used to prepare silica-carbon nanocomposites. However, hydrothermal carbon is not enough "rigid" and collapses upon silica removal. For this reason, the HTC process is usually followed by further pyrolysis to stabilize the carbon skeleton. The resulting carbon material is a negative replica of the template. In the cases that the porosity can be destroyed upon sulfonation, this process is carried out before template removal. Depending on the wall thickness of the template, the carbon materials can be microporous, mesoporous, or macroporous (Fig. 4). When the silica is not removed, the material is a carbon-silica hybrid or nanocomposite (see Section 3.3). The use of both sulfonated

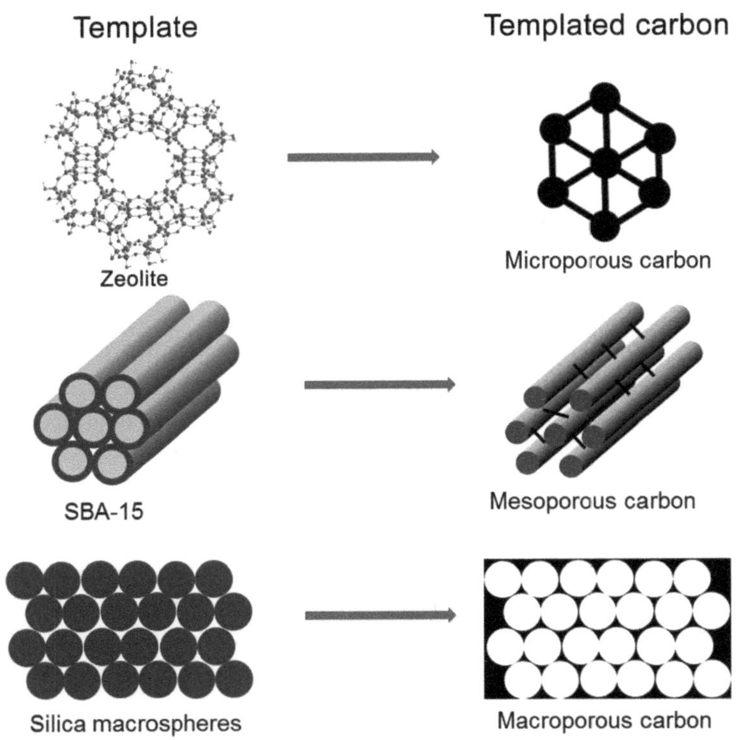

Fig. 4 Templated micro, meso, or macroporous carbons depending on the hard template used.

silica-templated mesoporous carbons and sulfonated carbon-silica nanocomposites as acid catalyst has been recently reviewed [18].

In case of using a soft template, template is easily removed without the use of a corrosive etching agent. Thus, a polymer such as phenol-formaldehyde resin is synthesized in the presence of triblock copolymer Pluronic F127. The soft template is decomposed when the material is pyrolized, leaving behind mesoporosity within a carbon material framework [19]. This approach has some advantages over nanocasting such as lower number of preparation steps, shortest synthesis time, possibility of large-scale production, and avoiding the use of corrosive reagents. The drawback is that the pyrolysis of polymer/template can lead to severe shrinkage of the framework, destroying the well-ordered mesostructure during pyrolysis.

The different carbonization methods shown in Fig. 2 are also applicable in the case of the polymer + template composites. The thermal treatment has two objectives, the transformation of the precursor in a carbon framework and the removal of the template to leave the porosity. Suitable precursors are thermosetting polymers (phenol-formaldehyde polymers or furan derivatives) which do not fully decompose at high temperature, leaving some carbon residue upon pyrolysis.

3.1.3 Microporous and micro-mesoporous template carbons

Sometimes the templating approach is used to prepare microporous carbons with high surface area (bottom-left triangle in Fig. 3). The sulfonated materials can be used in reactions where the diffusion of reactants in micropores is not hindered. For instance, some microporous carbons that have been sulfonated are carbide-derived carbons [20] or zeolite-templated carbons [21, 22]. For the preparation of carbide-derived carbons, silicon is etched from silicon carbide by treating with Cl_2 at high temperatures. For the preparation of zeolite-templated carbons, the porous structure of zeolite is filled with furfuryl alcohol that is polymerized to form a thermosetting resin. Despite its microporosity, these solids have shown good catalytic activity after pyrolysis and sulfonation, demonstrating that both mesopores and micropores are accessible (see Section 5.3).

Carbon materials with bimodal porosity can be prepared using dual templating approaches (bottom-center triangle in Fig. 3). For instance, it is possible to prepare both micro- and mesoporous materials by dual templating approach such as hard template using mesoporous silicon carbide followed by silicon etching to create micropores [20]. Thus, the material possesses a high surface area due to micropores and favorable diffusion of reactants due to mesopores.

3.2 Surface chemistry

Carbon materials usually contain oxygenated functional groups (OFGs). Some of them are acidic, such as carboxylic groups, acting as catalytic sites in some reactions. The OFGs also confer hydrophilicity to carbon material. The more labile acidic groups can be removed by heat treatment at high temperature rendering the surface hydrophobic. The hydrophilicity-hydrophobicity of the material is very important to optimize each specific reaction. Thommes and coworkers defined a hydrophilic-index (H-index) capable of identifying differences in surface chemistry between porous materials, regardless of their chemical composition [23]. This index can be determined by comparison of the adsorption isotherms of two adsorptives, one of them sensitive to surface chemistry (water) and another one that completely wets the adsorbent surface (nitrogen or argon). Other authors have correlated the catalytic behavior of the carbons with this hydrophilic-index. Beyond the optimal value, the highly hydrophilic solids are deactivated by adsorption of water generated in the catalytic reaction (see Section 5.4) [17]. On the contrary, hydrophobic catalysts are favorable for reactions with hydrophobic substrates. Thus, the control of the nature of the carbon surface, either during the synthesis or by thermal posttreatment, is of crucial importance.

3.3 Hybrid and nanocomposite materials

In Section 3.1, the preparation of mesoporous carbons using silica as sacrificial template is described. It was found that the sulfonation of mesoporous carbon could result in a remarkable collapse of the pore structure. For this reason, when the sulfonation is carried out before removing the silica template, the silica provides the composite with higher mechanical strength, preserving the porosity even after harsh sulfonation treatment. Sometimes, the silica template is not sacrificed (removed) resulting in a carbon-silica hybrid after sulfonation that can be used as acid catalyst. Therefore, this approach avoids the step of template removal with hazardous reagents. Compared with pure silica, the carbon material supplies higher hydrothermal stability, hydrophobicity, microporosity, and high density of sites for sulfonic groups' anchorage. It is reported that the sulfonated SBA15-carbon hybrid catalyst has higher surface area than the bulk carbon material, leading to enhanced performance in some hydrophobic reactions [24].

Instead of infiltrating a preformed silica with a carbon precursor, silica-carbon precursor hybrids can be also produced in a single step using the soft templating approach, starting from a mixture of sucrose as carbon source, tetraethyl ortho-silicate (TEOS) as silica source, and triblock copolymer Pluronic 127 as structure directing agent [25]. The porous texture of the nanocomposite is easily tunable by changing the amount of sucrose and it has a high impact on the catalytic performance in acid-catalyzed reactions. Under optimized preparation conditions, the material has a high mesoporous volume and hydrophobicity, which are favorable for reaction. The balance between the mesoporous volume, provided by the silica structure, and the microporosity of carbon gives the nanocomposite its excellent performance. While silica provides the mesoporosity to facilitate entrance and fast transport of reactants, carbon provides microporosity to attain a fast catalysis. It is worth to mention that the soft template itself (Pluronic 123 or alcohol), dehydrated in the acidic medium, can also give some carbon residue upon pyrolysis. This approach avoids the additional steps of infiltration of silica with a carbon precursor and template removal.

3.4 Three-dimensional structuration

For application in flow chemistry, monolithic sulfonated carbons with a hierarchical porous structure are highly desired. This structure comprises not only micropores and mesopores, which are relevant to the molecular length-scale, but also flow-through macropores of diameters sizing hundreds of microns. Sometimes, it is even sought for straight and parallel macropores, reminiscent of channels in honeycombs [26]. Monolith macropores enable fast transfer of substrates by convection and much lower pressure drop than a fixed packed bed. There are different approaches to create 3D macroporous structures. Hierarchical monolithic carbons with mesoporous arrays have been prepared by soft templating approach. When starting from hydrogels, ice-templating is a widely used approach that allows the preparation of straight macrochannels. It consists of the unidirectional growing of ice rods that are subsequently freeze-dried. Ice-templating has been used to create straight macropores in resorcinol-formaldehyde resins. The monolithic macroporous solid prepared from resorcinol-formaldehyde copolymerization has been pyrolized, sulfonated, and tested in esterification in batch and condensation reactions in flow [13, 26]. The hydrothermal carbonization of glucose in autoclave does not led to monolithic material, but when GO is added, the latter acts as structure-directing agent leading to monolithic blocks. In this case, the monoliths are composites of hydrothermal carbon and reduced GO.

Another approach consists in using existing macrostructures as scaffolds to coat the active phase. For instance, graphite felts can be coated with sulfonated hydrothermal carbon. This material has been tested in esterification reaction [27]. This approach has been also used to coat cordierite monoliths with sucrose-derived sulfonated carbon [28]. The sulfonation is usually a harsh treatment (H_2SO_4, $\sim$150°C) and it should be checked that the porosity of the macroporous solid has not changed after the sulfonation treatment. If sulfonation under those conditions is deleterious for the structure, an alternative method should be employed for the introduction of sulfonic sites.

4 Methods to introduce sulfonic groups and their characterization

4.1 Introduction of sulfonic groups

The most typical strong acid sites introduced in the carbon materials are sulfonic ones. Probably, the two most commonly used methods to functionalize these materials are the direct sulfonation with sulfuric acid or alternative

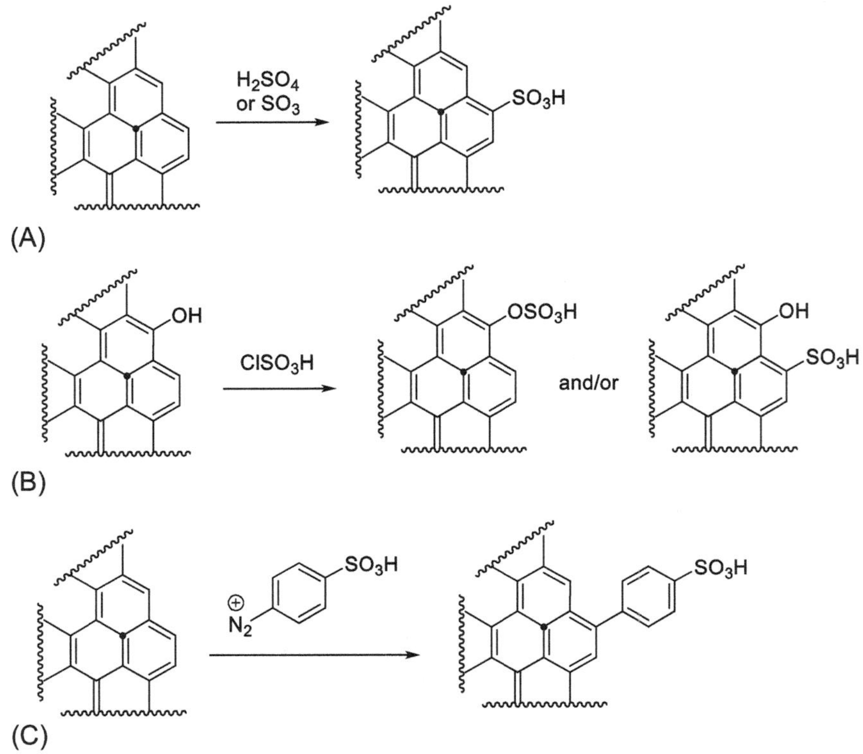

Fig. 5 Functionalization of carbon materials with sulfonic groups: (A) sulfonation with sulfuric acid, (B) sulfonation with chlorosulfonic acid, and (C) grafting of arylsulfonic groups.

reagents and the reaction with a sulfonic containing diazonium salt (Fig. 5). In a third group of methods, the sulfonic groups are present in the starting material for the preparation of the carbon.

4.1.1 Sulfonation with sulfuric acid

In principle, the simplest method to introduce a sulfonic group into an aromatic substrate is the direct sulfonation via an electrophilic aromatic substitution (S_EAr, Fig. 5A). Different electrophilic sulfur species may be involved in the process, although SO_3 is considered the main reactive species. Probably, the reagent of choice for this reaction is concentrated sulfuric acid, used at temperatures in the range of 150–280°C. It has been used to sulfonate different types of carbonaceous materials, such as multiwall carbon nanotubes (MWCNTs) [29], partially reduced graphene oxide (RGO) [30], pyrolized starch carbon [31], or hydrothermal carbon (HTC) microspheres prepared from different starting materials, such as waste *Camellia oleifera* shells and phloroglucinol [32], starch [33], or sucrose [11].

Alternatively, sulfonation can be carried out using SO_3 in the gas phase, by heating fuming H_2SO_4 (a 50% SO_3/H_2SO_4 solution) in an autoclave. This method has been used to sulfonate magnetic Fe_3O_4@carbon nanoparticles [34]. Another method, which precludes the use of highly concentrated (or fuming) sulfuric acid, is the generation of SO_3 via plasma, by an electric discharge into a diluted sulfuric acid solution. This approach has been used for the sulfonation of carbon black [35].

The oxidant properties of the sulfonation medium produce several types of side reactions, mainly oxidation reactions that increase the content of different oxidized groups on the carbon materials, such as carboxylic acids, ketones or quinones.

4.1.2 One-pot carbonization and sulfonation

In some cases, the sulfonation step is accomplished along with carbonization step in a one-pot process. Two examples are the hydrothermal treatment of glucose [36] or glycerol [37] in the presence of concentrated H_2SO_4 at 180°C. A less common starting material is the red liquor solid, a papermaking by-product, composed mainly of magnesium lignosulfonate. This material was used for the preparation of sulfonated carbon by reaction with H_2SO_4 at 200°C [38].

Spray pyrolysis is a straightforward method of carbon materials preparation, and sulfonated carbon was prepared by passing a solution of sucrose in aqueous sulfuric acid through an ultrasonic nebulizer to an oven at temperatures ranging 400–1000°C [39].

4.1.3 Reaction with chlorosulfonic acid

Another classical reagent able to perform aromatic sulfonation reactions is chlorosulfonic acid. However, in many cases, the reaction conditions used to modify carbon materials do not seem to be suitable for the aromatic sulfonation and other types of sulfonic sites are probably generated. In fact, a partially oxidized graphene was made to react with chlorosulfonic acid in an organic solvent at 0 °C, and the formation of Graphene–O–SO_3H groups was proposed (Fig. 5B) [40].

The same method was used for modification of hollow carbon nanospheres, prepared by calcination (650°C) of SiO_2@resorcinol-formaldehyde resin and then removal of silica with HF [41]. The same reaction, but at room temperature, was carried out on monoliths prepared by hydrothermal carbonization of glucose in the presence of graphene oxide [42]. The formed species are not specifically described, but given the mild reaction conditions, they presumably would be –O–SO_3H groups, as in the first example.

4.1.4 Grafting of arylsulfonic groups

The typical method includes the formation of 4-benzenediazoniumsulfonate by diazotization of sulfanilic acid with $NaNO_2$ in acidic medium and then reaction with the carbon material in an aqueous solution of H_3PO_2 to produce the coupling (Fig. 5C). Different carbonaceous materials have been functionalized in this way, including ordered mesoporous carbon CMK-5 (SBA-15 as template) [43] and mesoporous carbons prepared with different types of alumina as template [44], as well as partially reduced GO [8, 45]. The formation of the diazonium salt has been also described with isoamyl nitrite, instead of sodium nitrite, and used in the functionalization of organic mesoporous carbons, prepared from phenol-formaldehyde resins [46].

This reaction is not fully selective and nitrogen is always detected on the final solids. The presence of electron-rich aromatic rings may give rise to diazo coupling.

4.1.5 Preparation of carbon materials from sulfonated precursors

An alternative method to prepare sulfonic carbonaceous materials is the use of sulfonated precursors in the carbonization process. Due to the lability of sulfonic groups, this method is not suitable for thermal carbonization and most of the examples described in the literature deal with the hydrothermal synthesis of sulfonated carbons at 180°C: using only glucose and *p*-toluenesulfonic acid [36], a mixture of glucose, resorcinol, and *p*-toluenesulfonic acid [47], or glucose, *p*-toluenesulfonic acid, and [Ph_2NH_2][OTos] [48]. In the latter, two types of acid sites are generated, strong sulfonic groups from *p*-toluenesulfonic acid and weak diarylammonium tosylate groups taking part in the carbon structure.

It has been also described in the formation of a carbonaceous material only from *p*-toluenesulfonic acid, by carbonization at 160°C on silica. However, this material had to be activated by treatment with sulfuric acid at different concentrations (0.25, 1, and 2.5 M) in order to afford active catalysts [49].

In other cases, the carbonaceous materials are part of composites, where the sulfonic groups are present in a phase different from carbon. Single-walled carbon nanotubes (SWCNTs) [50] and multi-walled carbon nanotubes (MWCNTs) [7] have been covered with poly-styrene sulfonic acid by radical polymerization of sodium *p*-styrenesulfonate. The formation of covalent bonds between the nanotubes and the polymer has been proposed.

In a different approach, Aquivion was entrapped in hydrothermal carbon prepared from cellulose or guar gum, forming a composite material with strong perfluoroalkylsulfonic sites [51].

4.2 Characterization of sulfonic carbon materials

A large variety of techniques have been used to characterize the sulfonic carbon materials, many of them devoted to the morphological, textural, and chemical nature of the material. However, all those techniques are general for the study of carbon materials and this section will only deal with those used for the specific study of the sulfonic sites.

4.2.1 Sulfur content, distribution, and oxidation state

The most obvious method to determine the sulfur content is the chemical (combustion) analysis, which is routinely employed. However, an important question is also the distribution of the sulfur atoms within the catalyst particles, as it may control the accessibility of the reagents to the active sites, mainly in the case of micro- or mesoporous materials. Energy-dispersive X-ray spectroscopy (EDS) elemental mapping allows to determine the distribution of different elements (S and O) along the particles of the catalyst [39, 40].

Another indirect method to determine the presence of sulfur inside the particles is the XPS (X-ray photoelectron spectroscopy) analysis. As it is a surface technique, the variations of the composition, determined by chemical analysis and XPS, give an idea about the presence of sulfur only on the external surface of the particles or also inside them. This is the case of biomass hydrochars prepared from different starting materials, which are sulfonated mainly on the external surface of the particles according to the two types of analysis [52].

Additionally, XPS is also able to determine the oxidation state of sulfur, as the band at 169 eV is attributed to SO_3H groups (S(VI)), whereas the possible SH groups (S(II)) are observed at 164 eV [49] (Fig. 6).

4.2.2 Nature and stability of the sulfur functional groups

Together with the oxidation state, the presence of SO_3H groups can be detected by the FTIR bands corresponding to the different types of vibrations in the group. However, there is no complete agreement about the

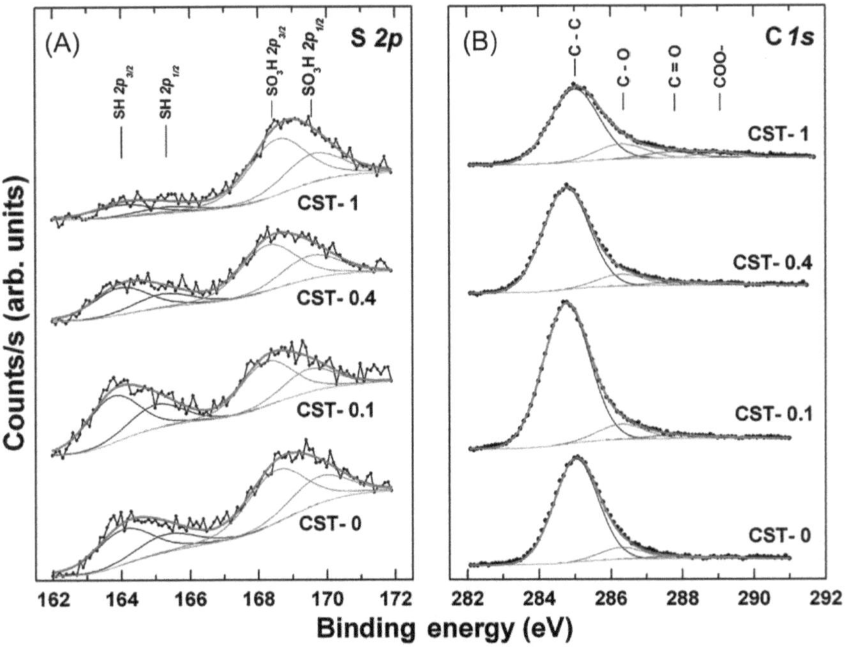

Fig. 6 Deconvoluted (A) S 2p and (B) C 1s XPS of sulfonated carbon-silica composites *(Reprinted from P.A. Russo, M.M. Antunes, P. Neves, P. V. Wiper, E. Fazio, F. Neri, et al., Solid acids with SO_3H groups and tunable surface properties: versatile catalysts for biomass conversion, J. Mater. Chem. A 2 (2014) 11813–11824, https://doi.org/10.1039/C4TA02320J, published by the Royal Society of Chemistry.)*

assignment of the different bands [53]. The asymmetric O=S=O stretching band should appear at 1360–1340 cm^{-1} and bands around that range have been detected, for example, in sulfonated amorphous carbon obtained by cellulose pyrolysis (450°C) [54]. The symmetric O=S=O stretching and the overtone of the SO_2 deformation bands should appear at 1190–1150 and 1100–1000 cm^{-1}, respectively. These bands are commonly observed in many types of sulfonated carbons [30, 39, 55].

An additional proof for the presence of sulfonic groups is the temperature-programmed decomposition coupled with a mass spectrometer. The decomposition of SO_3H produces SO_2 ($m/z=64$) that can be monitored, and it appears at temperatures in the range of 280–300°C [56].

4.2.3 Acidity

The temperature-programmed desorption of ammonia has been used to determine the distribution of acid sites with different strength [56].

However, this method does not monitor the true ammonia desorption, but the decomposition of the $-SO_3NH_4$ and $-SO_3H$ species (sulfonic groups decompose at around 200°C), which is not a reversible process. Thus, its interpretation is not straightforward and nonthermal methods are commonly preferred.

Thus, titration is one of the methods of choice to determine the acidity. The use of several bases of different strengths (Boehm titration) allows the determination of different types of sites [13]. However, the weakest used base ($NaHCO_3$) is not able to distinguish between sulfonic and carboxylic sites, and hence, this method must be used in combination with alternative analytical methods.

The full scale of acidity can be obtained by ^{31}P NMR spectroscopy of adsorbed triethylphosphine oxide (TEPO) using the magic angle spinning (MAS) technique. The chemical shift varies depending on the strength of the acid site, from values around 50 ppm for weak acids (phenols, alcohols) to near 90 ppm for arylsulfonic groups [57]. This method had been described using TEPO solutions in nonpolar solvents, but the flexible porous structure of hydrothermal carbons requires the use of polar solvents, such as methanol, to allow the access of TEPO to the sulfonic sites [57]. In the case of rigid carbon structures with permanent porosity, the adsorption can be carried out from pentane solutions of TEPO [30, 49]. In spite of the interest of this technique, the obtained results are not fully understood, given the presence of bands in the 70–75 ppm range, and other beyond the upper limit (97–100 ppm), (Fig. 7) whose nature is still not explained. Although less frequently used, trimethylphosphine oxide has also been described as probe molecule for this purpose [25].

5 Acid-catalyzed reactions

Sulfonated carbons have been used as catalysts in a plethora of organic reactions mainly in biorefinery processes, that is, in renewable raw materials transformations. The use of such catalysts, normally also obtained from renewable sources, aims to improve the sustainability of these processes.

Sulfonated carbons appear to be potential substitutes to traditional sulfonic solids such as Amberlyst 15, Nafion NR50, Dowex 50 W × 2, or SBA15-SO_3H and they have been applied to acid-catalyzed reactions such as esterification, acetalization, hydrolysis, or dehydration reactions, among others. The use of sulfonated mesoporous carbons and sulfonated carbon-silica nanocomposites as acid catalyst has been recently reviewed [18].

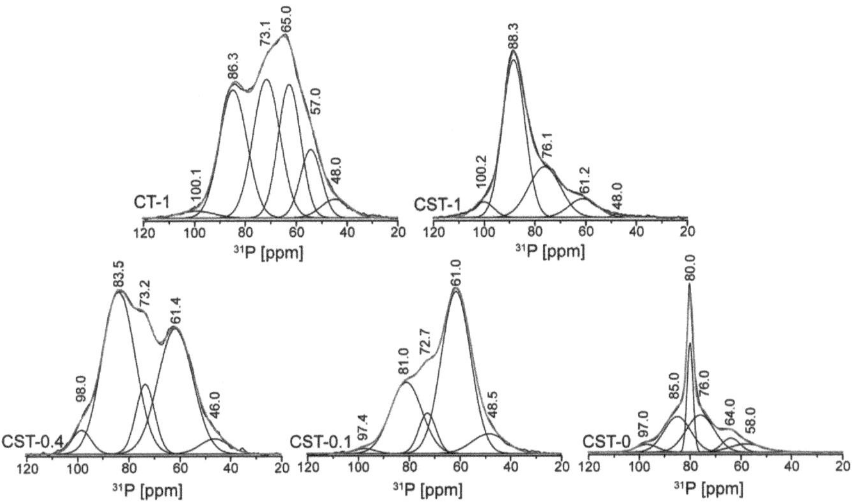

Fig. 7 ^{31}P MAS NMR spectra of TEPO adsorbed onto sulfonated carbon-silica materials prepared from *p*-toluenesulfonic acid. *(Reprinted from P.A. Russo, M.M. Antunes, P. Neves, P. V. Wiper, E. Fazio, F. Neri, etal., Solid acids with SO_3H groups and tunable surface properties: versatile catalysts for biomass conversion, J. Mater. Chem. A 2 (2014) 11813–11824, https://doi.org/10.1039/C4TA02320J, published by the Royal Society of Chemistry.)*

5.1 Esterification

Esterification is normally used as a benchmark reaction for new carbonaceous acid catalysts. In this reaction, the nature of the carbon surface can play an important role in the catalytic performance. Thus, accessibility of the reactants to the catalytic sites is crucial and swelling processes become important in some cases. In addition, the surface hydrophobicity can favor esterification reactions as reactants' adsorption is promoted and the formed water can be repelled from the surface far away from catalytic sites.

These facts have been evidenced by the comparison of esterification reactions of acetic acid and caprylic or oleic acid on different carbon catalysts.

Thus, Mo et al. compared the activity and recycling behavior of sulfonated carbons obtained from D-glucose pyrolysis in esterification reaction of acetic and caprylic acid with methanol [58]. These solids possess very low surface areas ($<1\,m^2\,g^{-1}$) and acid sites density of sulfonic groups between 0.5 and $0.64\,mmol\,g^{-1}$. Comparing activities, a different trend was observed. In the case of acetic acid esterification, high conversions were achieved in the first run, while an induction period was observed in the case of using caprylic acid. This fact is confirmed by an increase in activity upon

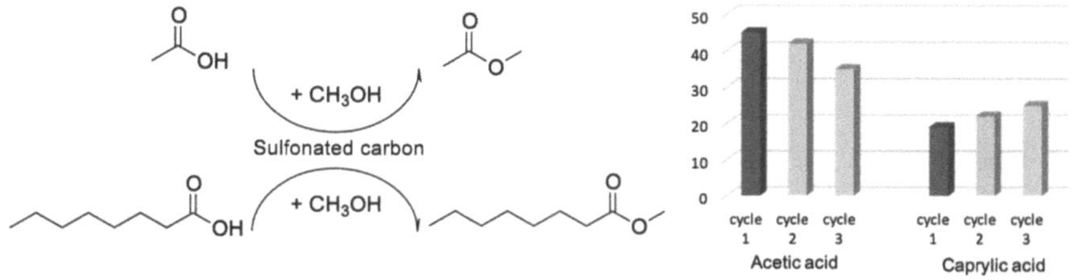

Fig. 8 Esterification reactions of acetic and caprylic acid with methanol catalyzed by sulfonated carbon prepared from pyrolyzed D-glucose. *(Adapted from X. Mo, D.E. López, K. Suwannakarn, Y. Liu, E. Lotero, J.G. Goodwin Jr., et al., Activation and deactivation characteristics of sulfonated carbon catalysts, J. Catal. 254 (2008) 332–338, https://doi.org/10.1016/j.jcat.2008.01.011, with permission of Elsevier.)*

catalyst reuse (Fig. 8). These results demonstrated that catalyst swelling was crucial for the activity of these sulfonated carbons and appeared to greatly affect their initial activity. This conclusion was corroborated by the increase in induction period and the reduced catalyst activity when the reactions were carried out in gas phase.

As previously described (see Section 3.1), a way of fine-tuning the structural properties of the carbonaceous material is to use a soft templating strategy. Thus, sulfonated carbons with different surface areas and pore volume can be achieved by using phenolic resins as precursors. These carbons present different behaviors in esterification reactions, showing how the control of the pore volume is a crucial factor in order to improve mass transfer of reactants and to reach high acid conversion. Thus, Ogino and coworkers prepared several sulfonic-acid-containing carbon catalysts through the pyrolysis of resorcinol-formaldehyde resins and subsequent sulfonation [13]. These solids present surface areas between 570 and $670\,m^2\,g^{-1}$ and mesopore volumes in the range from 0.17 to $1.5\,cm^3\,g^{-1}$. When using these solids as catalysts in esterification reactions of acetic acid with ethanol, similar initial TOF values were obtained ($0.024\,s^{-1}$), while when using oleic acid as reactant TOF values were greatly improved (from 0.015 to $0.040\,s^{-1}$) when using carbon catalyst with a larger mesopore volume.

Hard templating has also been used for the preparation of large mesoporous materials which facilitates esterification reactions of oleic acid with methanol. Thus, mesoporous carbon material with sulfonic acid groups was prepared by sulfonation of a carbon/SBA-15 hybrid composite in liquid phase and subsequent silica removal [59]. Sulfonation previous to silica removal seems crucial in order to keep the ordering in the porous structure. The optimal mesoporous SO_3H–carbon material exhibits a high surface area and a uniform pore size. The large pore size and the hydrophobic surface favor the esterification reaction. Better performance was obtained with this solid compared to amorphous sugar catalyst (96.1 $\mu mol\, min^{-1}$ vs 69.7 $\mu mol\, min^{-1}$).

The design of two-dimensional catalyst using hard templating methods or graphene oxide as precursor can be envisaged as a way to overcome swelling or mass transfer problems. Thus, carbons composed of thin and small carbon sheets, which facilitate catalyst dispersion in polar solvents, can be prepared by sulfonation of thin-wall mesoporous carbons prepared by γ-Al_2O_3 impregnation and subsequent carbonization and sulfonation [60]. These catalysts exhibited twice the activity of Amberlyst 15 or sulfonated activated carbons in the esterification of oleic acid and methanol.

An alternative is starting from a two-dimensional material. Sulfonated graphene (G-SO_3H) synthesized by hydrothermal treatment of graphene oxide and graphene oxide itself prepared by the Hummers method were applied as catalysts to different esterification reactions. In this case, no limitation of mass transfer in the catalytic reactions was observed [61, 62].

Finally, controlling the nature of chemical groups on the surface can also strongly influence their activity as catalyst in esterification reaction. Carbon catalysts obtained either by pyrolysis, hydrothermal, or alcoholthermal carbonization have been compared in the esterification reaction of oleic acid with methanol (Fig. 9) [12]. Better activity and recyclability were observed with the solids prepared by alcoholthermal carbonization. The authors attributed this behavior to the presence of more –SO_3H and C—O groups in these solids.

Apart from esterification reactions, sulfonated carbon catalysts have been successfully applied to triglycerides transesterification reactions, mainly devoted to biodiesel production.

Although acid-catalyzed transesterification reactions normally need more severe conditions as high temperature, the use of this type of catalysts offers a great advantage, as esterification and transesterification can take place simultaneously. This fact is of special interest when using oils with a high free fatty acids content. In this case, basic catalysis needs a previous esterification process in order to avoid soap formation.

Several sulfonated carbons have been prepared by incomplete carbonization of renewables materials such as lignin, cellulose [63], or glycerol [37]. The last two have been successfully tested in the reaction of triolein and ethanol, both being highly stable and reusable.

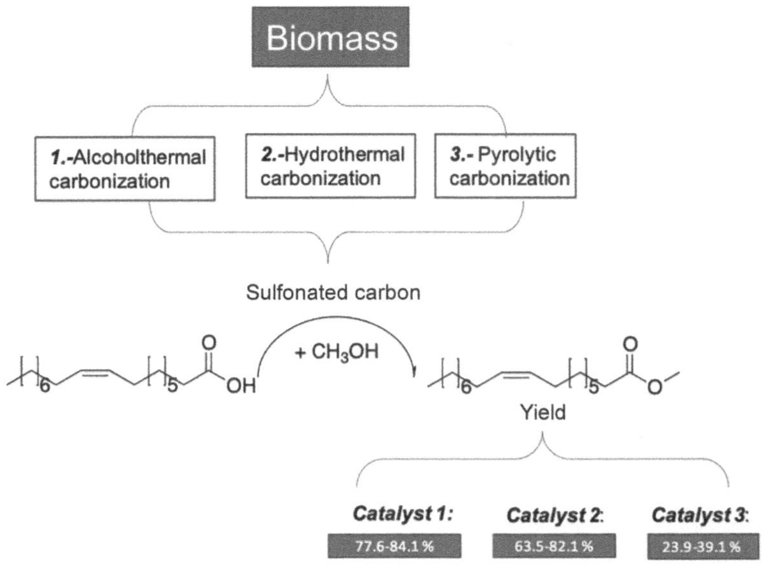

Fig. 9 Influence of the preparation method on the activity of carbonaceous materials in the esterification reaction of oleic acid with methanol. *(Adapted from Y.T. Yang, X.X. Yang, Y.T. Wang, J. Luo, F. Zhang, W.J. Yang, et al., Alcohothermal carbonization of biomass to prepare novel solid catalysts for oleic acid esterification, Fuel 219 (2018) 166–175, https://doi.org/10.1016/j.fuel.2018.01.072, with permission of Elsevier.)*

In order to avoid mass transfer limitations, sulfonated carbon nanofibers (CNF) have been designed and used in triolein transesterification. Functionalized CNFs (Fig. 10) outperformed previously reported sulfonated carbons, due to the high accessibility of active sulfonic acid sites on the CNF surface. These catalysts show also a better stability than catalysts sulfonated with sulfuric acid [64].

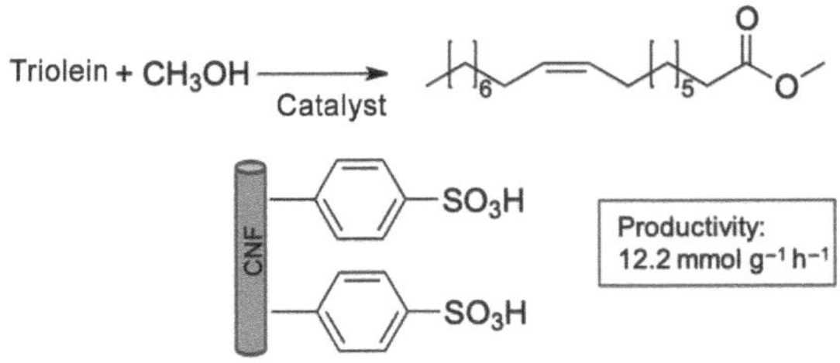

Fig. 10 Triolein transesterification with methanol catalyzed by sulfonated carbon nanofibers (CNF). *(Adapted from D.R. Stellwagen, F. Van Der Klis, D.S. Van Es, K.P. De Jong, J.H. Bitter, Functionalized carbon nanofibers as solid-acid catalysts for transesterification, ChemSusChem 6 (2013) 1668–1672, https://doi.org/10.1002/cssc.201300372, with permission of John Wiley and Sons.)*

It is important to notice that some of the sulfonated carbon catalysts have proven their activity in the simultaneous esterification and transesterification reaction of vegetable oils. Thus, both sulfonated carbon nanohorns dispersed with Fe/Fe_2O_3 nanoparticles (SO_3H-Fe/Fe_2O_3-SWCNH) [6] and sulfonated carbonized D-glucose [65] have shown high activity in the esterification + transesterification reactions. Interestingly, in the case of using SO_3H-Fe/Fe_2O_3-SWCNH, the esterification of palmitic acid takes place first and subsequent tributyrin transesterification seems to be favored by the adsorption of palmitic acid near the Fe/Fe_2O_3 nanoparticles.

A comparison of the main features and performance of the major carbon catalyst used in the biodiesel synthesis can be found in the work of Konwar et al. [66].

Apart from fatty acid esterifications, sulfonated carbons have been tested in the reaction of levulinic acid and succinic acid with ethanol. In the first case, sulfonated hydrothermal carbons provided conversion and ester selectivity over 90% [67] and the reaction of succinic acid with ethanol can be successfully carried out with the use of acid carbon catalysts readily prepared by thermal treatment of glucose in the presence of *p*-toluenesulfonic acid [36].

Some of the sulfonated carbons have been used in the esterification of a broader scope of acids. That is the case of the catalysts prepared by thermal treatment of a mixture of D-glucose, *p*-toluenesulfonic acid, and diphenylammonium tosylate [48] or sulfonated graphene oxides [42].

Finally, glycerol esterification has also been envisaged by using sulfonated carbons as catalysts. These solids exhibited better performances than commercial sulfonic resins such as Amberlyst 15, Dowex NR50, or Deloxan. Thus, sulfonated activated carbons catalyzed glycerol acetylation leading to high glycerol conversions and mixtures of 38% mono, 28% di, and 34% triacetylglycerol at 120–135°C [68].

Mono, di, or triesters can be selectively obtained by carefully selecting the appropriate catalyst and reaction conditions (Fig. 11). Thus, high yields of triacylglycerol have been described when using either sulfonated carbon prepared by sucrose carbonization using silica as template [69] or sulfonated hydrothermal carbon coming from glucose [70]. It is worth to mention that the latter also provided monoacyglycerol as the major reaction product when the reaction was carried out at lower temperatures without the need of glycerol excess.

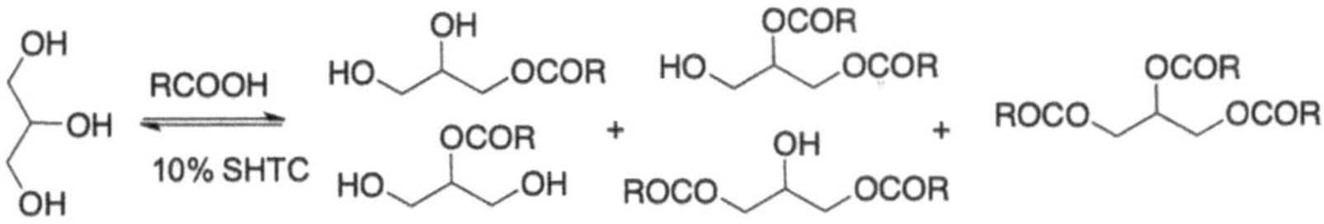

Acid	Reaction temperature	MAG yield	DAG yield	TAG yield
Acetic	40°C	89	11	0
	115°C	5	38	57
Butyric	115°C	1	18	81
Caprylic	115°C	1	19	80

Fig. 11 Glycerol esterification catalyzed by sulfonated hydrothermal carbon: effect of reaction conditions on product selectivity. *(Adapted from C. De La Calle, J.M. Fraile, E. García-Bordejé, E. Pires, L. Roldán, Biobased catalyst in biorefinery processes: sulphonated hydrothermal carbon for glycerol esterification, Catal. Sci. Technol. 5 (2015) 2897–2903, https://doi.org/10.1039/c5cy00059a, with permission of Royal Society of Chemistry.)*

5.2 Acetalization

Solketal has been considered as a green solvent, coming from renewable raw material. It is obtained upon glycerol and acetone condensation, promoted by acid catalysts. In this case, the use of heterogeneous catalysis is crucial in order to prevent the loss of water-soluble solketal upon the aqueous work-up required by the homogeneous acids. Sometimes the use of cosolvents such as ethanol or dioxane is described in order to overcome the lack of miscibility of the reagents. Above the use of traditional heterogeneous acid catalysts, such as sulfonic resins or zeolites, some efforts have been devoted to develop sulfonic carbonaceous materials as catalyst for this reaction.

A sulfonated hydrothermal carbon prepared from crude glycerol coming from biodiesel wastes has been used in solketal synthesis [71]. Although good conversions and high solketal selectivity were described, the hydrophilicity of the surface and its low surface area seemed to limit the recoverability of the catalyst, probably due to site blocking by adsorbed glycerol. The role of both the nature of acid sites and hydrophilicity of the surface, in the synthesis of solketal, was evidenced when using acid-activated carbons [72]. The catalysts were prepared from olive stones by chemical activation and subsequent treatment with HNO_3 or H_2SO_4. The treatment with HNO_3 led to a solid with a high amount of acid sites (6.52 mmol g^{-1}), but lower surface area (1240 $m^2 g^{-1}$) and pore volume than the parent carbon, while treatment with H_2SO_4 caused just a slight decrease in surface area (1495 $m^2 g^{-1}$) and pore volume, but a lower amount of total acid sites (1.88 mmol g^{-1}), half of them being -SO_3H groups. Good solketal yields were obtained, although slight differences were observed depending on the catalyst. The best catalyst was the one obtained by H_2SO_4 treatment. In this case, the authors stated that this behavior is due to the population of sulfonic and carboxylic sites in the solid and to the lower hydrophilicity. Avoiding the presence of water near the catalytic sites prevents the reversal of the solketal formation equilibrium.

In order to overcome catalytic site blocking and improve the recyclability of the carbon catalyst in the solketal synthesis, the design of solids with high surface area seems to be crucial. The use of hard templating with silica provides carbons with higher surface areas that can be easily recovered from reaction media without a loss in activity. Sulfonated carbon-silica-mesoporous composites can be prepared from glucose in a one-step synthetic method involving simultaneous carbonization and sulfonation [73]. The election of a thermal or hydrothermal treatment allows obtaining carbons with tunable acidity and porosity. Thermal treatment produces materials with higher surface areas (779–425 $m^2 g^{-1}$) and lower total acid site density (0.9–1.35 mmol SO_3H g^{-1}) than the ones obtained by hydrothermal treatment (242–176 $m^2 g^{-1}$ and 1.05–2.25 mmol SO_3H g^{-1}). In this case, the best glycerol conversions were achieved with the solid obtained by hydrothermal treatment, these catalysts being recoverable in 4 consecutive runs without loss of activity.

Sulfonated hollow carbon spheres (HSC-SO_3H) can be synthesized from carbonization of a SiO_2-polymer core-shell structure, followed by removal of the SiO_2 core and sulfonation with chlorosulfonic acid. These carbons, which exhibit a high surface area (419 $m^2 g^{-1}$), provided superior performances in acetalization of glycerol (Fig. 12) than traditional heterogeneous acid catalyst [41]. The authors attributed this behavior to the hollow sphere carbon shell with rich microporosity that favors the mass transfer in the reaction. This carbon can also be reused four consecutive runs without observing loss of activity.

Hydrothermal carbonization of furfural and hydroxyethanesulfonic acid leads to a carbon functionalized with sulfonic acid sites. This solid presents

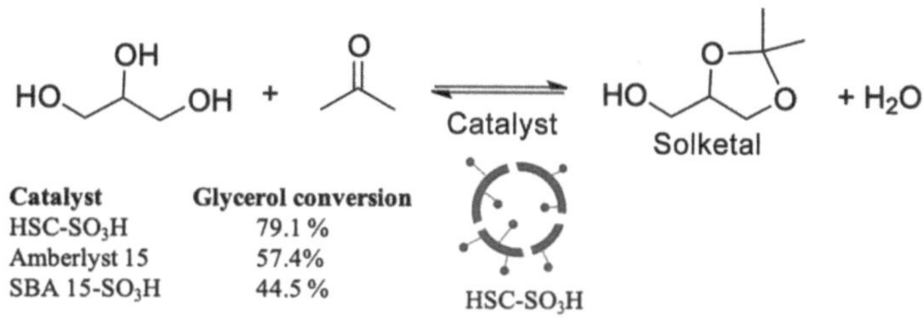

Fig. 12 Solketal synthesis using sulfonated hollow carbon spheres. *(Adapted from L. Wang, J. Zhang, S. Yang, Q. Sun, L. Zhu, Q. Wu, et al., Sulfonated hollow sphere carbon as an efficient catalyst for acetalisation of glycerol, J. Mater. Chem. A 1 (2013) 9422–9426, https://doi.org/10.1039/c3ta10956a, with permission of Royal Society of Chemistry.)*

high surface area (148 $m^2 g^{-1}$) and higher activity and recoverability in the thioacetalization reaction of cyclohexanone with mercaptoethanol (Fig. 13) than HY zeolite or Amberlyst 15 [74].

Finally, a way to overcome the low miscibility of the reagents and the need of the use of cosolvents in ketalization reactions is the design of amphiphilic surfaces. This has been achieved by the preparation of Aquivion/carbon composites by hydrothermal carbonization of guar gum over perflurosulfonic resin Aquivion [75]. These solids have been successfully applied to the reaction of dodecanal and ethylene glycol.

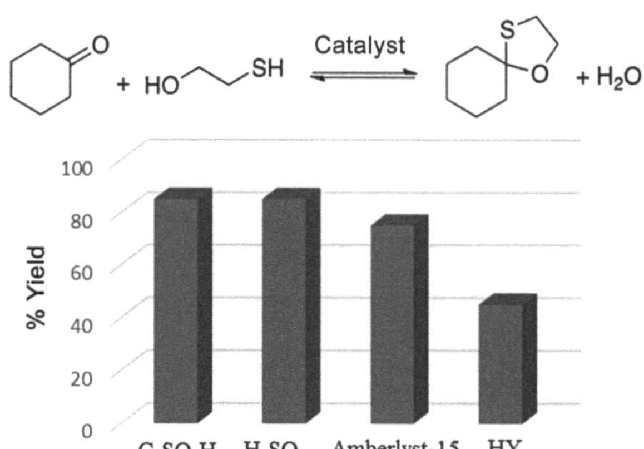

Fig. 13 Comparison of the activity of sulfonated hydrothermal carbon (C-SO_3H) with other acid catalysts. *(Reprinted from X. Liang, M. Zeng, C. Qi, One-step synthesis of carbon functionalized with sulfonic acid groups using hydrothermal carbonization, Carbon N. Y. 48 (2010) 1844–1848, https://doi.org/10.1016/j.carbon.2010.01.030, with permission of Elsevier.)*

5.3 Hydrolysis

Hydrolysis reactions are getting increasing importance, as they are the first step to obtain biofuels and platform molecules from carbohydrates. Among carbohydrates, cellulose is preferred as raw material as it is readily available from agriculture or wood wastes. Cellulose hydrolysis implies the breaking of β-1,4-glycosidic bonds (Fig. 14). This reaction has been tackled using hot compressed water, ionic liquids, enzymatic catalysis, and homogeneous acid catalysis. The development of heterogeneous acid catalyst for this reaction can improve the sustainability of the process as no neutralization should be needed and salt formation would be avoided.

In this area, sulfonated carbons can play a key role. Several authors have described an outstanding activity of different sulfonated carbons in cellulose hydrolysis. As in the case of esterification reactions, the nature of the surface and the acid sites together with acid sites density seem to be crucial factors for the optimization of cellulose conversion and glucose yield.

In the first example, sulfonated activated carbons (AC) exhibited higher activity than zeolites or sulfonated resins such as Amberlyst 15 [76]. Cellulose was selectively hydrolysed into glucose with yields of almost 50%. The authors attributed this result to the presence of strong acid sites (SO_3H groups) and the hydrophobic planes of AC. Partial carbonization of cellulose followed by sulfonation with fuming H_2SO_4 provided a highly active and reusable carbon catalyst for cellulose hydrolysis [54]. Sixty-four percent of water-soluble glucans were obtained, this yield being much higher than the ones observed with niobic acid, sulfuric acid, Nafion, mordenite, or Amberlyst. This behavior can be explained neither by the surface area of the solid ($2\,m^2g^{-1}$) nor by acid site strength, similar to other catalysts. Studies of cellobiose adsorption over the different solids showed a high capacity of adsorption of the sulfonated carbon; this fact is responsible for its activity. Moreover, studies of water vapor adsorption-desorption isotherm of the carbon material allowed estimating an effective area of the carbon material in

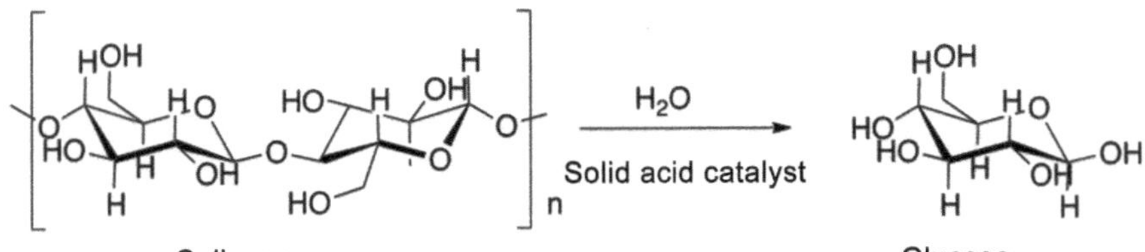

Fig. 14 Cellulose hydrolysis.

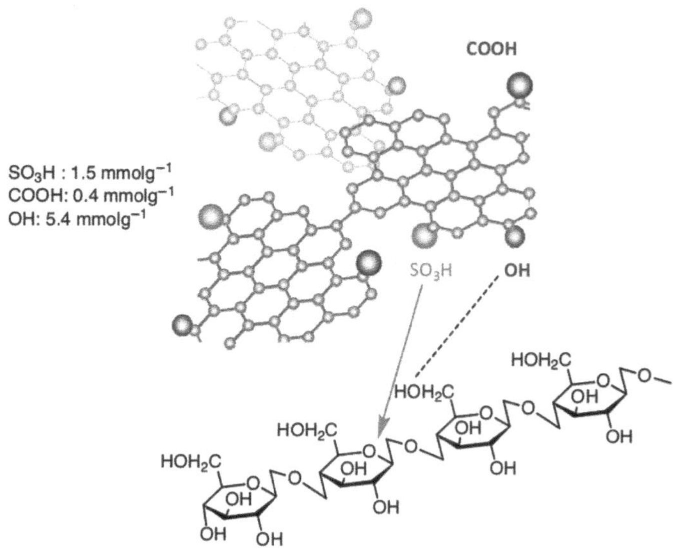

Fig. 15 Proposal of the interactions of β-glucan with the functional groups on the carbon surface. *(Adapted from M. Hara, Biomass conversion by a solid acid catalyst, Energy Environ. Sci. 3 (2010) 601–607, https://doi.org/10.1039/b922917e, with permission of Royal Society of Chemistry.)*

water over $560\,m^2\,g^{-1}$. Thus, the authors attributed its higher activity to this large effective surface area in water and the presence of water-tolerant SO_3H groups. A deeper study of the hydrolysis of β-1,4-glucan showed that the synergetic combination of strong Brønsted acid sites (SO_3H), effective sites for β-1,4-glucan adsorption (OH and COOH groups), and facile access to the material bulk as a result of the incorporation of water are the main driving forces in this case (Fig. 15) [77, 78].

This synergetic effect was also invoked in the case of sulfonated activated carbon (AC). The treatment of AC with H_2O_2 provokes pore enlargement and functionalization of the carbon sheets with –OH, –COOH, and lactone groups. Subsequent sulfonation with H_2SO_4 afforded a carbon catalyst with lower graphitization degree and some defects on the carbon surface. These defects seem to help in the adsorption of the polysaccharides, forcing the glycosidic bonds to interact with the in-plane acidic functional groups, leading to higher cellulose conversion and glucose yields [79].

Again, the high adsorption capacity for cellobiose is the driving force for the higher catalytic performance of sulfonated microporous amorphous carbons in cellobiose hydrolysis [20].

Sulfonated hydrothermal carbons obtained from biomass have also been applied as catalysts for cellulose hydrolysis. Materials obtained from kenaf core [80] or from defatted rice bran [81] provided moderate cellulose conversions and glucose yields. In this case, an effect of sulfonic sites density was observed and the best results were obtained with carbons with higher density of SO_3H groups. Seed waste from biodiesel production has been used to prepared more sophisticated mesoporous materials [82].

Carbons were obtained by both hydrothermal treatment and carbonization, whereas sulfonic groups were introduced either by sulfonation with sulfuric acid or by reaction with 4-benzenediazoniumsulfonate (4-BDS). Comparison of the solids in cellulose hydrolysis showed that catalysts with mesoporous structure obtained by carbonization and subsequent functionalization with 4-BDS provided the best results in comparison with nonporous catalysts with high sulfonic groups' density obtained by hydrothermal treatment, demonstrating the importance of the cellulose accessibility to active sites by means of mesoporosity.

As described in Section 3.1, a way of inducing mesoporosity in carbonaceous materials is the use of templating agents. Mesoporous carbon nanoparticles were prepared using MCM48 as template and they were subsequently sulfonated with sulfuric acid or functionalized with COOH groups via nitric acid treatment to afford highly active catalysts for β-glucans hydrolysis into glucose at pH 2. In this case, reaction of β-glucans with HCl is favored by adsorption on mesoporous carbon nanoparticles functionalized with weak carboxylic acid sites after treatment with nitric acid [83].

Similar conclusions were reached when carrying out the hydrolysis of water-soluble glucan oligomers over three-dimensionally ordered sulfonated mesoporous carbons [84]. The adsorption capacity of glucans on the carbon materials increased with their chain length, which was attributed to a denser packing structure of the glucans compared to small glucan oligomers or glucose and to the layout of glucan molecules with respect to carbon surface, and the enhanced adsorption favored the hydrolysis reaction. The results suggested that the catalytic activity of these 3D ordered materials is strongly affected by several factors such as hydrophobicity, mesoporous structure, and acid site density.

A way of tuning surface properties can be envisaged by the preparation of sulfonated silica/carbon nanocomposites by the evaporation-induced triconstituent co-assembly method in order to ensure high entanglement of carbon and silica [85]. In this case, the authors observed an increase in TOF of cellulose hydrolysis when increasing silica/carbon mass ratio, and

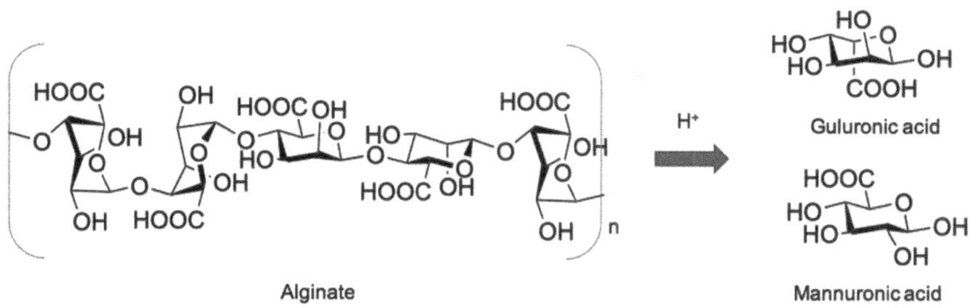

Fig. 16 Hydrolysis of alginate to obtain guluronic and mannuronic acids. *(Adapted from C. Ban, W. Jeon, G. Park, H.C. Woo, D.H. Kim, Hydrothermal conversion of alginate into uronic acids over a sulfonated glucose-derived carbon catalyst, ChemCatChem 9 (2017) 329–337, https://doi.org/10.1002/cctc.201600969, with permission of John Wiley and Sons.)*

hence, the hydrophilicity of the surface, which can be explained by a better adsorption of reagents.

Sulfonated carbon catalysts have also been successfully applied to depolymerization reactions of other carbohydrates, such as the hydrolysis of alginate to obtain uronic acids (Fig. 16) [86]. The activities of sulfonated carbons were compared to the ones of Y-zeolites. Glucose-derived catalysts provided the best conversions compared to sulfonated carbon black or sulfonated activated carbons. Unfortunately, this catalyst suffers from a gradual deactivation upon recovery due to the leaching of sulfur to reaction media.

Finally, several acid catalysts have been tested in cellulose methanolysis combined with transacetalization with long alkyl chain alcohols for nonionic surfactants production (Fig. 17) [87]. In this case, Nafion SAC-13, β-zeolite, Amberlyst-15, sulfonated graphene oxide, sulfonated carbon coming from glucose carbonization, and sulfonated MCM41-carbon composite were tested. Among those acid catalysts, amorphous sulfonated carbon provided

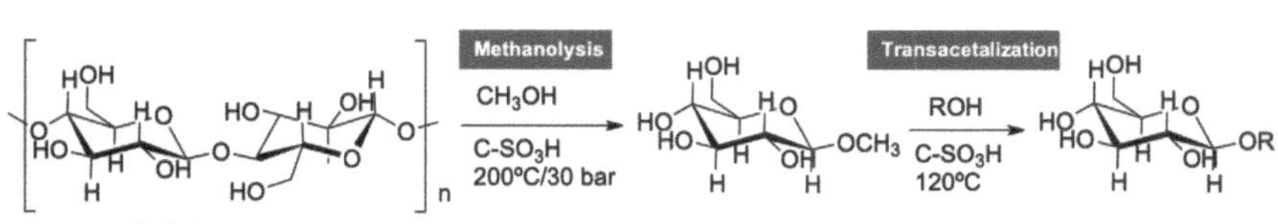

Fig. 17 Nonionic surfactants' production by combined cellulose methanolysis and methyl glucosides transacetalization with fatty alcohols. *(Adapted from M.J. Climent, A. Corma, S. Iborra, S. Martínez-Silvestre, Transformation of cellulose into nonionic surfactants using a one-pot catalytic process, ChemSusChem 9 (2016) 3492–3502, https://doi.org/10.1002/cssc.201600977, with permission of John Wiley and Sons.)*

the best results in both methanolysis and subsequent transacetalization, being reusable without loss of activity. This fact allowed the synthesis of alkyl glucoside surfactants in a cascade process under mild conditions.

5.4 Dehydration

5-Hydroxymethylfurfural (HMF) is addressed as a versatile key precursor for the production of fine chemicals. This platform molecule is readily available by acid-catalyzed dehydration of renewable sugars (Fig. 18). As described in previous sections, carbons are known as water-tolerant materials and are stable under hydrothermal conditions, which make them interesting catalysts for sugar dehydration in aqueous environment.

Thus, the use of Norit-activated carbon in xylose, glucose, and fructose dehydration allowed the obtaining of higher reaction selectivity towards HMF compared to sugars' thermal dehydration processes with similar conversions, in spite of the very weak acidity of this solid [3].

Furthermore, a series of carbon materials with different structures, including carbon nanotubes (CNT), carbon nanofibers (CNF) and structured mesoporous carbon CMK-5, functionalized with 4-benzenediazonium sulfonate have been tested in fructose, inulin, sucrose, and glucose dehydration processes (Fig. 19) [7]. Sulfonated CNT exhibited the highest acid density as well as the best performance for fructose dehydration to HMF. Thus, 100% fructose conversion was achieved in short reaction time with 89% yield of HMF. When using inulin or sucrose, selectivity was slightly reduced to 74% and 68%, respectively, and with glucose as reagent, selectivity decreased down to 46%.

Sulfonated ordered mesoporous carbons were obtained by carbonization of 1-methyl-3-phenethyl-1*H*-imidazolium hydrogen sulfate (MPIHS) ionic liquid using SBA15 as templating agent. These materials were also used in fructose dehydration. In this case, the activity of the carbon-derived catalyst was compared to several materials with different hydrophilicity index (SBA15-SO_3H). An increase in fructose conversions and HMF yields was observed when using more hydrophilic solids, but up to a limit

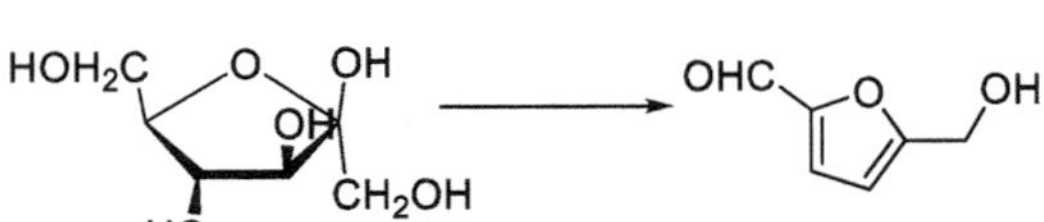

Fig. 18 Synthesis of 5-hydroxymethylfurfural by sugar dehydration.

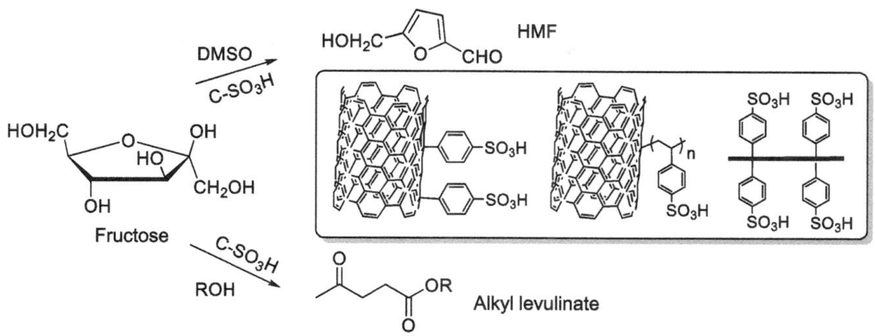

Fig. 19 Fructose transformation catalyzed by sulfonated CNT, CNF, and mesoporous CMK-5. *(Adapted from R. Liu, J. Chen, X. Huang, L. Chen, L. Ma, X. Li, Conversion of fructose into 5-hydroxymethylfurfural and alkyl levulinates catalyzed by sulfonic acid-functionalized carbon materials, Green Chem. 15 (2013) 2895–2903, https://doi.org/10.1039/c3gc41139g, with permission of Royal Society of Chemistry.)*

(HI = 0.8). For solids with higher H-Index, the H_2O adsorption deactivated the catalyst and negatively affected the catalyst recoverability [17].

Sulfonated hydrothermal carbons have also been applied to dehydration reactions of fructose in subcritical water conditions. Although high fructose conversions were obtained (87%), only moderate HMF yields (20%) were achieved [88].

The combination of cellulose-derived sulfonated hydrothermal carbon with [BMIM][Cl] ionic liquid as reaction medium seems to provide better results in fructose dehydration reaction yielding *ca* 80% of HMF [89]. In this case, the dispersion of the carbon particles in the ionic liquid seems to favor the reaction and the recyclability of the catalyst.

Carbons and silica/carbon composites, obtained by carbonization of *p*-toluenesulfonic acid and subsequent sulfonation, have been tested in fructose dehydration in ethanol/water medium to obtain a mixture of HMF, 5-ethoxymethyl furfural, and ethyl levulinate [49]. The use of sulfonated carbon-silica composites provided better results compared to solely sulfonated carbon catalyst prepared by the same procedure. In this case, tuning the surface properties and acid site density by the increase of the carbon-silica ratio allowed getting more active catalysts.

The combination of the hydrolysis of disaccharides or polysaccharides together with the sugar dehydration may open the door to the use of inulin, sucrose, cellobiose, or cellulose as raw materials for HMF synthesis. Thus, Chang and coworkers described the one-pot transformation of inulin to hydroxymethylfurfural using sulfonated hydrochars prepared from biomass as catalysts and 1-allyl-3-methylimidazolium chloride as reaction medium [52]. In this case, HMF yield of 70% was achieved and no significant differences on activity were observed related to the raw materials used for carbon preparation.

A more challenging approach is the use of agricultural wastes as raw material for the one-pot hydrolysis and dehydration reaction. Shotipruk and coworkers envisaged the conversion of Cassava waste into HMF and furfural using sulfonated carbon prepared in one-pot naphthalene carbonization in sulfuric acid [90]. Although rather low HMF and furfural yields were obtained due to the characteristics of the biomass, the carbon catalysts showed comparable activity to homogeneous catalyst such as sulfuric acid.

5.5 Other reactions

5.5.1 Etherification

Ethers can be formed in acid medium by either condensation of two alcohols or addition of an alcohol on an alkene. Both types of reactions have been described with sulfonic carbon materials.

2-Hydroxymethylfurfural reacts with ethanol to give 2-ethoxymethylfurfural and its acetal, together with ethyl levulinate as product of the degradation of the furan ring. Both sulfonated graphene oxide [30] and sulfonated carbon-silica composites [49] are able to promote this reaction, albeit in a nonselective manner.

In this reaction, sulfonated GO exhibited superior catalytic performance in 2-hydroxymethylfurfural etherification (Fig. 20) than several other acid catalysts with much higher sulfonic acid content and stronger acidity [30]. This was attributed to the cooperative effects of the sulfonic acid groups and other types of acid sites (e.g., carboxylic acids) and to the enhanced accessibility to the active sites because of the 2D structure.

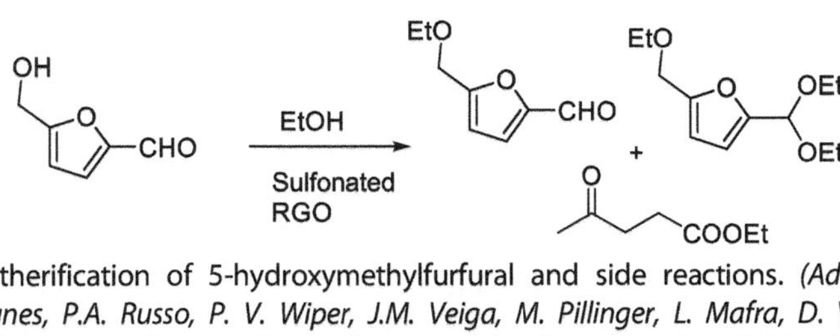

Fig. 20 Etherification of 5-hydroxymethylfurfural and side reactions. *(Adapted from M.M. Antunes, P.A. Russo, P. V. Wiper, J.M. Veiga, M. Pillinger, L. Mafra, D. V. Evtuguin, et al., Sulfonated graphene oxide as effective catalyst for conversion of 5-(hydroxymethyl)-2-furfural into biofuels, ChemSusChem 7 (2014) 804–812, https://doi.org/10.1002/cssc.201301149, with permission of John Wiley and Sons.)*

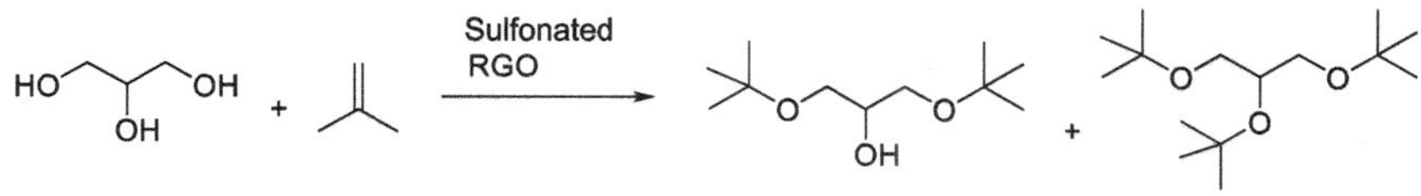

Fig. 21 Etherification of glycerol with isobutylene [91].

Glycerol ethers are interesting compounds from renewable origin with a variety of applications. For example, *tert*-butyl ethers can be used as diesel and biodiesel additives. The etherification of glycerol with isobutene has been described with sulfonic graphene oxide [91]. The selectivity to diether + triether (Fig. 21), the most interesting products, was higher than in the case of *p*-toluenesulfonic acid. The sulfonic graphene oxide was successfully recycled and reused 5 times through a reaction-extraction process, using only glycerol as reagent-solvent able to separate the monophasic reaction crude into a biphasic system.

On the other hand, Starbon-400-SO_3H promotes the reaction of glycerol with primary allyl and benzyl alcohols, as well as secondary benzyl alcohols and *tert*-butanol under microwaves activation to give a mixture of 1-monoether and 2-monoether of glycerol (Fig. 22). The 1-ether/2-ether selectivity depends on the type of alcohol, from 1/1 with allyl alcohol to 10/1 with 1-phenyl-1-propanol [92].

5.5.2 Dimerization

The acid-catalyzed dimerization of α-methylstyrene to unsaturated compounds (Fig. 23) is promoted by a sulfonated mesoporous carbon material

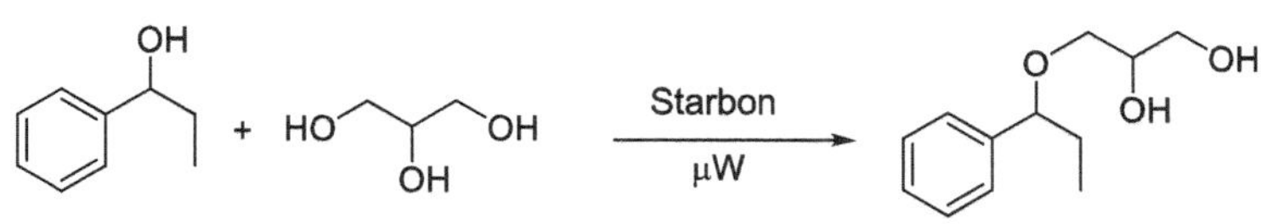

Fig. 22 Etherification of glycerol with 1-phenyl-1-propanol. *(Adapted from R. Luque, V. Budarin, J.H. Clark, D.J. Macquarrie, Glycerol transformations on polysaccharide derived mesoporous materials, Appl. Catal. B Environ. 82 (2008) 157–162, https://doi.org/10.1016/j.apcatb.2008.01.015, with permission of Elsevier.)*

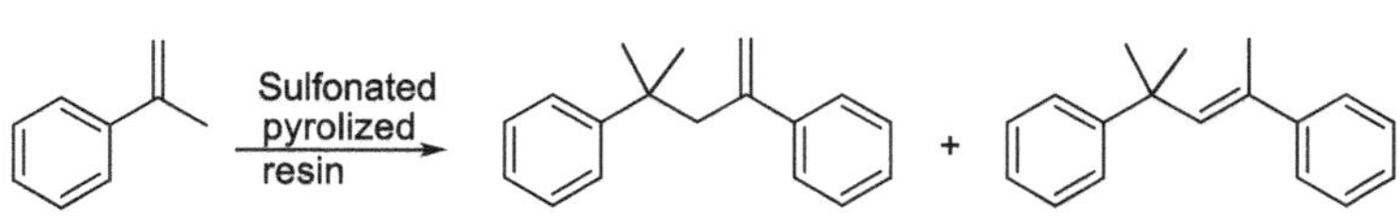

Fig. 23 Dimerization of α-methylstyrene. *(Adapted from S. Suganuma, K. Nakajima, M. Kitano, H. Kato, A. Tamura, H. Kondo, et al., SO_3H-bearing mesoporous carbon with highly selective catalysis, Microporous Mesoporous Mater. 143 (2011) 443–450, https://doi.org/10.1016/j.micromeso.2011.03.028, with permission of Elsevier.)*

prepared by pyrolysis of phenol-formaldehyde resin [93]. The control of the pyrolysis temperature is very important to get high surface area (optimal of $433\,m^2\,g^{-1}$ at 773 K), which seems to be crucial for getting high catalytic activity. Interestingly, the sulfonated carbon material leads to very high selectivity to the linear dimeric alkenes, whereas other acid solids, such as Nafion or Amberlyst 15, produce the cyclic dimer in high amount or even as the major product after an intramolecular Friedel-Crafts alkylation, which is apparently inhibited in the case of the sulfonated carbon material.

A similar trend, both in activity and selectivity, is observed in the dimerization of styrene and α-methylstyrene catalyzed by sulfonated carbon-silica composites prepared from glucose and tetraethoxysilane [25]. The reaction seems to be controlled by the diffusion of the reactant, and hence, the load of carbon is crucial, with the best TOF values at lower carbon content.

5.5.3 Friedel-Crafts

Furans can be considered as platform molecules in the transformation of biomass, given that they are obtained from abundant carbohydrates. Their condensation through Friedel-Crafts reaction is very interesting to get C12-C16 hydrocarbon precursors (Fig. 24). The reaction between 2-methylfuran and furfural is usually used as test reaction. Two types of sulfonic carbon materials have been described as efficient catalysts for this reaction, Lignosulfonate-derived materials [94] and macroporous monoliths prepared by pyrolysis of resorcinol-formaldehyde resins [13]. In both cases, the sulfonic carbon materials were more active than standard sulfonic resins, such as Amberlyst 15, and some of them were as active as the homogeneous *p*-toluenesulfonic acid, not only for the test reaction, but also for other molecules with carbonyl groups, such as butanal, acetone, or cyclohexanone. However, the reason for this behavior is not clear, and the hydrophilic/hydrophobic character of the solid, together with the lack of microporosity, seems to be the origin of the high activity. This reaction has been also tested in a continuous flow system [13].

The synthesis of bisphenol A through reaction of phenol with acetone (Fig. 25) has been described using a sulfonated CMK-5 ordered mesoporous

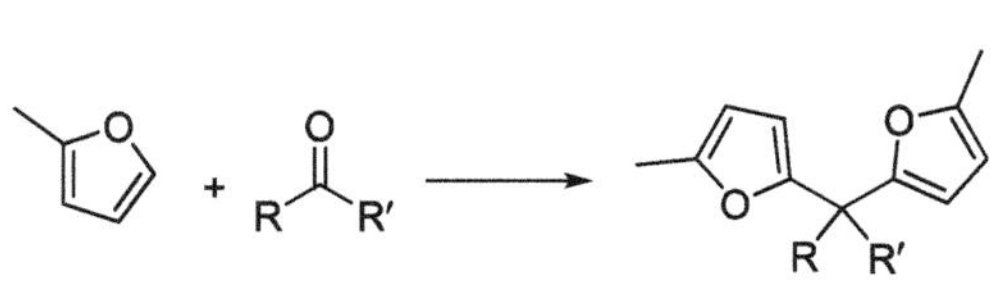

Fig. 24 Friedel-Crafts reaction between 2-methylfuran and a carbonyl compound.

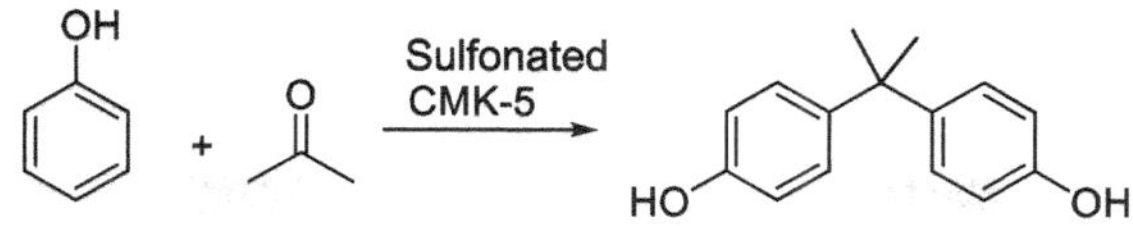

Fig. 25 Synthesis of *p,p′*-bisphenol A [43].

carbon [43]. This solid is significantly more efficient than mesoporous silicas functionalized with alkyl- or arylsulfonic groups, both in TON and selectivity to the *p,p′*-bisphenol A. The hydrophobic character of the carbon material has been proposed as the origin for this improved behavior.

Starbon-400-SO_3H efficiently catalyzes the Friedel-Crafts alkylation of different aromatics (toluene, xylenes, anisole) with benzyl chloride under microwave irradiation [15].

6 Deactivation mechanisms

One of the main problems of this kind of catalysts is the deactivation upon recycling and reuse. Several mechanisms have been detected for deactivation. As it happens in other types of catalysts, one deactivation mechanism is the deposition of side products, mainly those of polymeric nature. This problem has been solved by treatment with sulfuric acid, presumably by transformation of the deposits into carbonaceous material and sulfonation of it to get new active sites [11].

Regarding the specific deactivation mechanisms for sulfonic sites, they can be classified into two types, the leaching of active sites and their chemical modification.

6.1 Leaching of active sites

It has been shown in many cases that the slow drop in catalytic performance parallels a slow loss of sulfonic sites, detected by chemical analysis [33, 95]. In fact, the activity of the leachate in methanol and other solvents demonstrates the leaching of active species in different sulfonated carbons [47, 58].

The nature of the leached active sites is not fully understood. Some evidences point to the leaching of soluble polycyclic arylsulfonic compounds [58], or particles of colloidal size, depending on the preparation method [96, 97]. In any case, this type of leaching is linked to either an incomplete growing of the carbonaceous material, leading to particles of colloidal size, or to the degradation of the carbon structure under reaction conditions. Leaching may be easier in the case of composites, such as

Aquivion-HTC, where the phase containing the sulfonic groups are not covalently bonded to the carbon support [51].

Another different type of leaching is the hydrolysis of the C—S bond. Studies on the hydrothermal stability of sulfonated carbons at 160°C [98] or 200°C [2] under autogenous pressure showed the poor stability against the hydrolysis of sulfonic sites, mainly when the carbons were sulfonated with sulfuric acid. Sulfonic sites anchored using benzenediazoniumsulfonate were more stable under such conditions [98].

The presence of cationic impurities in the feed may produce the deactivation of the sulfonated carbon by proton leaching; in fact, a cation exchange process whose reversibility allows the catalyst to be easily regenerated [99].

6.2 Chemical reaction of the active sites

The acid sites on the sulfonated hydrothermal carbons may react with alcohols to form surface esters, with the loss of the acid sites. This reaction was detected by ^{13}C NMR first with methanol and n-propanol (Fig. 26) [100], but its general character was confirmed after NMR analysis of carbons reacted with glycerol [70] and sorbitol [101]. The regeneration of a catalyst, whose deactivation cannot be ascribed to the leaching of sulfonic sites, with sulfuric acid is proposed to produce the hydrolysis of the esters, given that the sulfur content did not significantly increased [38].

7 Concluding remarks

Sulfonic carbons are very promising acid catalysts with interesting features: they can be prepared from a large variety of precursors using different procedures, and several methods can be applied for functionalization with sulfonic groups. The use of renewable precursors improves the sustainability of the preparation processes for this kind of materials. In this way, the properties of the final catalysts, including morphological (particle size and shape), textural (surface area and porosity), hydrophilic/hydrophobic, and acidic properties, can be tailored. These materials have shown to be efficient catalysts for a variety of organic reactions, with improved performance over well-established commercially available catalysts, such as sulfonic resins.

However, it is still difficult to compare sulfonic carbons prepared from different precursors and using different methods, as the systematic studies with these materials are scarce, and usually they deal with variations on a small number of preparation parameters. Thus, it is still difficult to design

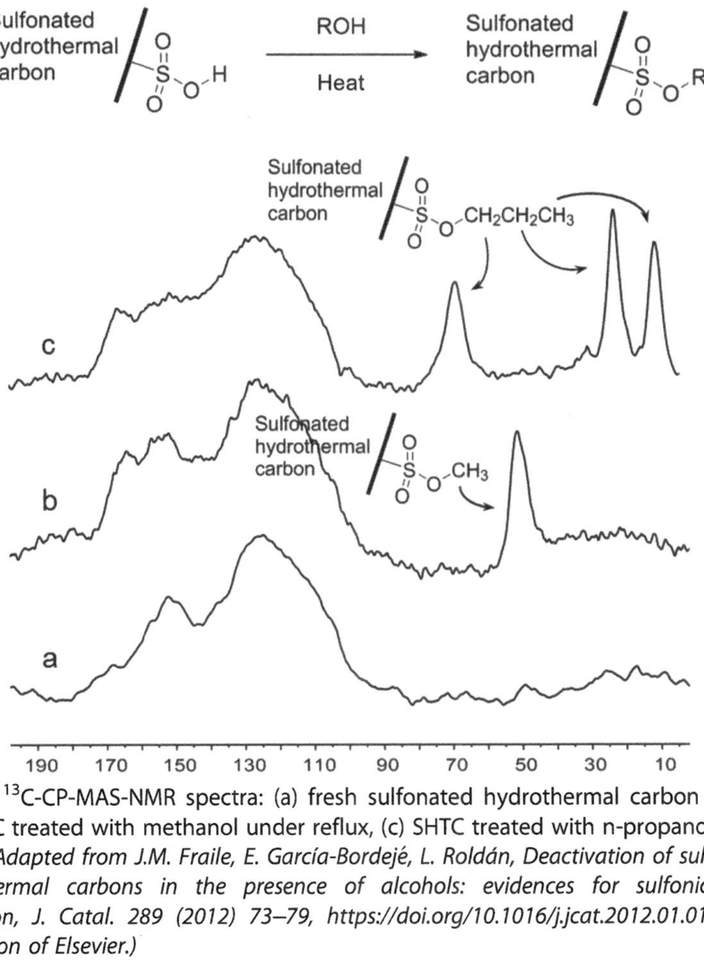

Fig. 26 ^{13}C-CP-MAS-NMR spectra: (a) fresh sulfonated hydrothermal carbon (SHTC), (b) SHTC treated with methanol under reflux, (c) SHTC treated with n-propanol under reflux. *(Adapted from J.M. Fraile, E. García-Bordejé, L. Roldán, Deactivation of sulfonated hydrothermal carbons in the presence of alcohols: evidences for sulfonic esters formation, J. Catal. 289 (2012) 73–79, https://doi.org/10.1016/j.jcat.2012.01.017, with permission of Elsevier.)*

an optimal sulfonic carbon catalyst ab initio for a specific reaction, and comparative studies are necessary in order to establish more general correlations between preparation methods, properties, and catalytic efficiency.

Each type of reaction and reactant requires specific properties of the sulfonated carbon catalyst to maximize its efficiency. The versatility of sulfonated carbon allows their customization to get the desired requirements, but systematic studies in a case-by-case basis are needed.

As future perspectives, the improvement of the recyclability of these materials is very important, and it will come from a deeper knowledge about the deactivation mechanisms. Furthermore, the applicability in continuous flow reactions is a subject of interest still in its initial steps. To this end, more research is needed on the synthesis of sulfonated carbons with a hierarchical porous structure to allow the flowing of the liquid at nearly atmospheric pressure.

References

[1] L.R. Radovic, F. Rodriguez-Reinoso, Carbon materials in catalysis, in: P.A. Thrower (Ed.), Chem. Phys. Carbon, Vol. 25, Marcel Dekker, New York, 1997, , pp. 243–358.

[2] A.H. Van Pelt, O.A. Simakova, S.M. Schimming, J.L. Ewbank, G.S. Foo, E.A. Pidko, E.J.M. Hensen, C. Sievers, Stability of functionalized activated carbon in hot liquid water, Carbon 77 (2014) 143–154, https://doi.org/10.1016/j.carbon.2014.05.015.

[3] E. Sairanen, R. Karinen, J. Lehtonen, Comparison of solid acid-catalyzed and autocatalyzed C5 and C6 sugar dehydration reactions with water as a solvent, Catal. Lett. 144 (2014) 1839–1850, https://doi.org/10.1007/s10562-014-1350-1.

[4] E. Sairanen, K. Vilonen, R. Karinen, J. Lehtonen, Functionalized activated carbon catalysts in xylose dehydration, Top. Catal. 56 (2013) 512–521, https://doi.org/10.1007/s11244-013-0013-6.

[5] S.H. Shuit, K.F. Yee, K.T. Lee, B. Subhash, S.H. Tan, Evolution towards the utilisation of functionalised carbon nanotubes as a new generation catalyst support in biodiesel production: an overview, RSC Adv. 3 (2013) 9070–9094, https://doi.org/10.1039/c3ra22945a.

[6] C. Poonjarernsilp, N. Sano, H. Tamon, Simultaneous esterification and transesterification for biodiesel synthesis by a catalyst consisting of sulfonated single-walled carbon nanohorn dispersed with Fe/Fe_2O_3 nanoparticles, Appl. Catal. A Gen. 497 (2015) 145–152, https://doi.org/10.1016/j.apcata.2015.03.013.

[7] R. Liu, J. Chen, X. Huang, L. Chen, L. Ma, X. Li, Conversion of fructose into 5-hydroxymethylfurfural and alkyl levulinates catalyzed by sulfonic acid-functionalized carbon materials, Green Chem. 15 (2013) 2895–2903, https://doi.org/10.1039/c3gc41139g.

[8] J. Ji, G. Zhang, H. Chen, S. Wang, G. Zhang, F. Zhang, X. Fan, Sulfonated graphene as water-tolerant solid acid catalyst, Chem. Sci. 2 (2011) 484–487, https://doi.org/10.1039/c0sc00484g.

[9] M. Okamura, A. Takagaki, M. Toda, J.N. Kondo, K. Domen, T. Tatsumi, M. Hara, S. Hayashi, Acid-catalyzed reactions on flexible polycyclic aromatic carbon in amorphous carbon, Chem. Mater. 18 (2006) 3039–3045, https://doi.org/10.1021/cm0605623.

[10] M.M. Titirici, M. Antonietti, Chemistry and materials options of sustainable carbon materials made by hydrothermal carbonization, Chem. Soc. Rev. 39 (2010) 103–116, https://doi.org/10.1039/b819318p.

[11] R. Zhong, Y. Liao, R. Shu, L. Ma, B.F. Sels, Vapor-phase assisted hydrothermal carbon from sucrose and its application in acid catalysis, Green Chem. 20 (2018) 1345–1353, https://doi.org/10.1039/c7gc02236k.

[12] Y.T. Yang, X.X. Yang, Y.T. Wang, J. Luo, F. Zhang, W.J. Yang, J.H. Chen, Alcohothermal carbonization of biomass to prepare novel solid catalysts for oleic acid esterification, Fuel 219 (2018) 166–175, https://doi.org/10.1016/j.fuel.2018.01.072.

[13] I. Ogino, Y. Suzuki, S.R. Mukai, Tuning the pore structure and surface properties of carbon-based acid catalysts for liquid-phase reactions, ACS Catal. 5 (2015) 4951–4958, https://doi.org/10.1021/acscatal.5b01022.

[14] R.P. Rocha, M.F.R. Pereira, J.L. Figueiredo, Carbon as a catalyst: esterification of acetic acid with ethanol, Catal. Today 218–219 (2013) 51–56, https://doi.org/10.1016/j.cattod.2013.09.049.
[15] V.L. Budarin, J.H. Clark, R. Luque, D.J. Macquarrie, Versatile mesoporous carbonaceous materials for acid catalysis, Chem. Commun. (2007) 634–636, https://doi.org/10.1039/B614537J.
[16] R. Luque, V. Budarin, J.H. Clark, D.J. MacQuarrie, Microwave-assisted preparation of amides using a stable and reusable mesoporous carbonaceous solid acid, Green Chem. 11 (2009) 459–461, https://doi.org/10.1039/b821392p.
[17] B. Karimi, H.M. Mirzaei, H. Behzadnia, H. Vali, Novel ordered mesoporous carbon based sulfonic acid as an efficient catalyst in the selective dehydration of fructose into 5-HMF: the role of solvent and surface chemistry, ACS Appl. Mater. Interfaces 7 (2015) 19050–19059, https://doi.org/10.1021/acsami.5b03985.
[18] R. Zhong, B.F. Sels, Sulfonated mesoporous carbon and silica-carbon nanocomposites for biomass conversion, Appl. Catal. B Environ. 236 (2018) 518–545, https://doi.org/10.1016/j.apcatb.2018.05.012.
[19] B. Chang, J. Fu, Y. Tian, X. Dong, Multifunctionalized ordered mesoporous carbon as an efficient and stable solid acid catalyst for biodiesel preparation, J. Phys. Chem. C 117 (2013) 6252–6258, https://doi.org/10.1021/jp312820g.
[20] P. Krawiec, E. Kockrick, L. Borchardt, D. Geiger, A. Corma, S. Kaskel, Ordered mesoporous carbide derived carbons: novel materials for catalysis and adsorption, J. Phys. Chem. C 113 (2009) 7755–7761, https://doi.org/10.1021/jp808470s.
[21] K. Fukuhara, K. Nakajima, M. Kitano, S. Hayashi, M. Hara, Synthesis and acid catalysis of zeolite-templated microporous carbons with SO3H groups, Phys. Chem. Chem. Phys. 15 (2013) 9343–9350, https://doi.org/10.1039/c3cp43853h.
[22] K. Fukuhara, K. Nakajima, M. Kitano, S. Hayashi, M. Hara, Transesterification of Triolein over hydrophobic microporous carbon with SO3H groups, ChemCatChem 7 (2015) 3945–3950, https://doi.org/10.1002/cctc.201500525.
[23] M. Thommes, S. Mitchell, J. Pérez-Ramírez, Surface and pore structure assessment of hierarchical MFI zeolites by advanced water and argon sorption studies, J. Phys. Chem. C 116 (2012) 18816–18823, https://doi.org/10.1021/jp3051214.
[24] K. Nakajima, M. Okamura, J.N. Kondo, K. Domen, T. Tatsumi, S. Hayashi, M. Hara, Amorphous carbon bearing sulfonic acid groups in mesoporous silica as a selective catalyst, Chem. Mater. 21 (2009) 186–193, https://doi.org/10.1021/cm801441c.
[25] R. Zhong, L. Peng, F. De Clippel, C. Gommes, B. Goderis, X. Ke, G. Van Tendeloo, P.A. Jacobs, B.F. Sels, An eco-friendly soft template synthesis of mesostructured silica-carbon nanocomposites for acid catalysis, ChemCatChem 7 (2015) 3047–3058, https://doi.org/10.1002/cctc.201500728.
[26] K. Murakami, Y. Satoh, I. Ogino, S.R. Mukai, Synthesis of a monolithic carbon-based acid catalyst with a honeycomb structure for flow reaction systems, Ind. Eng. Chem. Res. 52 (2013) 15372–15376, https://doi.org/10.1021/ie400656x.
[27] L. Roldán, I. Santos, S. Armenise, J.M. Fraile, E. García-Bordejé, The formation of a hydrothermal carbon coating on graphite microfiber felts for using as structured acid catalyst, Carbon 50 (2012) 1363–1372, https://doi.org/10.1016/j.carbon.2011.11.008.
[28] S. Hosseini, J. Janaun, T.S.Y. Choong, Feasibility of honeycomb monolith supported sugar catalyst to produce biodiesel from palm fatty acid distillate (PFAD), Process. Saf. Environ. Prot. 98 (2015) 285–295, https://doi.org/10.1016/j.psep.2015.08.011.
[29] F. Peng, L. Zhang, H. Wang, P. Lv, H. Yu, Sulfonated carbon nanotubes as a strong protonic acid catalyst, Carbon 43 (2005) 2405–2408, https://doi.org/10.1016/j.carbon.2005.04.004.

[30] M.M. Antunes, P.A. Russo, P.V. Wiper, J.M. Veiga, M. Pillinger, L. Mafra, D.V. Evtuguin, N. Pinna, A.A. Valente, Sulfonated graphene oxide as effective catalyst for conversion of 5-(hydroxymethyl)-2-furfural into biofuels, ChemSusChem 7 (2014) 804–812, https://doi.org/10.1002/cssc.201301149.
[31] L. Zhou, B. Dong, S. Tang, H. Ma, C. Chen, X. Yang, J. Xu, Sulfonated carbon catalyzed oxidation of aldehydes to carboxylic acids by hydrogen peroxide, J. Energy Chem. 22 (2013) 659–664, https://doi.org/10.1016/S2095-4956(13)60087-X.
[32] K. Zhao, S. Liu, K. Li, Z. Hu, Y. Yuan, L. Yan, H. Guo, X. Luo, Fabrication of −SO3H functionalized aromatic carbon microspheres directly from waste Camellia oleifera shells and their application on heterogeneous acid catalysis, Mol. Catal. 433 (2017) 193–201, https://doi.org/10.1016/j.mcat.2017.02.032.
[33] H. Song, C. Xing, B. Li, W. Shen, Spherical carbon with SO_3H groups as an efficient solid acid catalyst for 2,4,5-triphenyl-imidazole synthesis, ChemistrySelect 1 (2016) 301–308, https://doi.org/10.1002/slct.201500013.
[34] F.C. Zheng, Q.W. Chen, L. Hu, N. Yan, X.K. Kong, Synthesis of sulfonic acid-functionalized Fe_3O_4@C nanoparticles as magnetically recyclable solid acid catalysts for acetalization reaction, Dalton Trans. 43 (2014) 1220–1227, https://doi.org/10.1039/c3dt52098f.
[35] O.L. Li, R. Ikura, T. Ishizaki, Hydrolysis of cellulose to glucose over carbon catalysts sulfonated via a plasma process in dilute acids, Green Chem. 19 (2017) 4774–4777, https://doi.org/10.1039/c7gc02143g.
[36] B. Zhang, J. Ren, X. Liu, Y. Guo, Y. Guo, G. Lu, Y. Wang, Novel sulfonated carbonaceous materials from p-toluenesulfonic acid/glucose as a high-performance solid-acid catalyst, Catal. Commun. 11 (2010) 629–632, https://doi.org/10.1016/j.catcom.2010.01.010.
[37] X. Song, X. Fu, C. Zhang, W. Huang, Y. Zhu, J. Yang, Y. Zhang, Preparation of a novel carbon based solid acid catalyst for biodiesel production via a sustainable route, Catal. Lett. 142 (2012) 869–874, https://doi.org/10.1007/s10562-012-0840-2.
[38] S. Kang, J. Chang, J. Fan, One step preparation of sulfonated solid catalyst and its effect in esterification reaction, Chin. J. Chem. Eng. 22 (2014) 392–397, https://doi.org/10.1016/S1004-9541(14)60058-6.
[39] N. Duyckaerts, I.-T. Trotuş, V. Nese, A.-C. Swertz, S. Auris, H. Wiggers, F. Schüth, Mesoporous sulfonated carbon materials prepared by spray pyrolysis, ChemCatChem 7 (2015) 2891–2896, https://doi.org/10.1002/cctc.201500483.
[40] L. Wang, D. Wang, S. Zhang, H. Tian, Synthesis and characterization of sulfonated graphene as a highly active solid acid catalyst for the ester-exchange reaction, Cat. Sci. Technol. 3 (2013) 1194–1197, https://doi.org/10.1039/c3cy20646g.
[41] L. Wang, J. Zhang, S. Yang, Q. Sun, L. Zhu, Q. Wu, H. Zhang, X. Meng, F.S. Xiao, Sulfonated hollow sphere carbon as an efficient catalyst for acetalisation of glycerol, J. Mater. Chem. A 1 (2013) 9422–9426, https://doi.org/10.1039/c3ta10956a.
[42] A.V. Nakhate, G.D. Yadav, Synthesis and characterization of sulfonated carbon-based graphene oxide monolith by solvothermal carbonization for esterification and unsymmetrical ether formation, ACS Sustain. Chem. Eng. 4 (2016) 1963–1973, https://doi.org/10.1021/acssuschemeng.5b01205.
[43] X. Wang, R. Liu, M.M. Waje, Z. Chen, Y. Yan, K.N. Bozhilov, P. Feng, Sulfonated ordered mesoporous carbon as a stable and highly active protonic acid catalyst, Chem. Mater. 19 (2007) 2395–2397, https://doi.org/10.1021/cm070278r.
[44] L. Geng, G. Yu, Y. Wang, Y. Zhu, Ph-SO_3H-modified mesoporous carbon as an efficient catalyst for the esterification of oleic acid, Appl. Catal. A Gen. 427–428 (2012) 137–144, https://doi.org/10.1016/j.apcata.2012.03.044.

[45] M.C. Nongbe, T. Ekou, L. Ekou, K.B. Yao, E. Le Grognec, F.X. Felpin, Biodiesel production from palm oil using sulfonated graphene catalyst, Renew. Energy 106 (2017) 135–141, https://doi.org/10.1016/j.renene.2017.01.024.
[46] Z. Gao, S. Tang, X. Cui, S. Tian, M. Zhang, Efficient mesoporous carbon-based solid catalyst for the esterification of oleic acid, Fuel 140 (2015) 669–676, https://doi.org/10.1016/j.fuel.2014.10.012.
[47] W. Zhang, H. Tao, B. Zhang, J. Ren, G. Lu, Y. Wang, One-pot synthesis of carbonaceous monolith with surface sulfonic groups and its carbonization/activation, Carbon 49 (2011) 1811–1820, https://doi.org/10.1016/j.carbon.2010.12.050.
[48] S. Zhan, X. Tao, L. Cai, X. Liu, T. Liu, The carbon material functionalized with NH_2^+ and SO_3H groups catalyzed esterification with high activity and selectivity, Green Chem. 16 (2014) 4649–4653, https://doi.org/10.1039/C4GC01395F.
[49] P.A. Russo, M.M. Antunes, P. Neves, P.V. Wiper, E. Fazio, F. Neri, F. Barreca, L. Mafra, M. Pillinger, N. Pinna, A.A. Valente, Solid acids with SO_3H groups and tunable surface properties: versatile catalysts for biomass conversion, J. Mater. Chem. A 2 (2014) 11813–11824, https://doi.org/10.1039/C4TA02320J.
[50] S. Qin, D. Qin, W.T. Ford, J.E. Herrera, D.E. Resasco, S.M. Bachilo, R.B. Weisman, Solubilization and purification of single-wall carbon nanotubes in water by in situ radical polymerization of sodium 4-styrenesulfonate, Macromolecules 37 (2004) 3965–3967, https://doi.org/10.1021/ma049681z.
[51] W. Fang, Z. Fan, H. Shi, S. Wang, W. Shen, H. Xu, J.M. Clacens, F. De Campo, A. Liebens, M. Pera-Titus, Aquivion®-carbon composites via hydrothermal carbonization: Amphiphilic catalysts for solvent-free biphasic acetalization, J. Mater. Chem. A 4 (2016) 4380–4385, https://doi.org/10.1039/c5ta09705c.
[52] S. Kang, J. Ye, Y. Zhang, J. Chang, Preparation of biomass hydrochar derived sulfonated catalysts and their catalytic effects for 5-hydroxymethylfurfural production, RSC Adv. 3 (2013) 7360–7366, https://doi.org/10.1039/C3RA23314F.
[53] C.N.R. Rao, Chemical Applications of Infrared Spectroscopy, Academic Press, New York, 1963.
[54] S. Suganuma, K. Nakajima, M. Kitano, D. Yamaguchi, H. Kato, S. Hayashi, M. Hara, Hydrolysis of cellulose by amorphous carbon bearing SO_3H, COOH, and OH groups, J. Am. Chem. Soc. 130 (2008) 12787–12793, https://doi.org/10.1021/ja803983h.
[55] Q. Fu, B. Gao, H. Dou, L. Hao, X. Lu, K. Sun, J. Jiang, X. Zhang, Novel non-covalent sulfonated multiwalled carbon nanotubes from p-toluenesulfonic acid/glucose doped polypyrrole for electrochemical capacitors, Synth. Met. 161 (2011) 373–378, https://doi.org/10.1016/j.synthmet.2010.12.009.
[56] B.L. Oliveira, V. Teixeira Da Silva, Sulfonated carbon nanotubes as catalysts for the conversion of levulinic acid into ethyl levulinate, Catal. Today 234 (2014) 257–263, https://doi.org/10.1016/j.cattod.2013.11.028.
[57] J.M. Fraile, E. García-Bordejé, E. Pires, L. Roldán, Catalytic performance and deactivation of sulfonated hydrothermal carbon in the esterification of fatty acids: comparison with sulfonic solids of different nature, J. Catal. 324 (2015) 107–118, https://doi.org/10.1016/j.jcat.2014.12.032.
[58] X. Mo, D.E. López, K. Suwannakarn, Y. Liu, E. Lotero, J.G. Goodwin Jr., C. Lu, Activation and deactivation characteristics of sulfonated carbon catalysts, J. Catal. 254 (2008) 332–338, https://doi.org/10.1016/j.jcat.2008.01.011.
[59] L. Peng, A. Philippaerts, X. Ke, J. Van Noyen, F. De Clippel, G. Van Tendeloo, P.A. Jacobs, B.F. Sels, Preparation of sulfonated ordered mesoporous carbon and its use for the esterification of fatty acids, Catal. Today 150 (2010) 140–146, https://doi.org/10.1016/j.cattod.2009.07.066.

[60] L. Geng, Y. Wang, G. Yu, Y. Zhu, Efficient carbon-based solid acid catalysts for the esterification of oleic acid, Catal. Commun. 13 (2011) 26–30, https://doi.org/10.1016/j.catcom.2011.06.014.
[61] F. Liu, J. Sun, L. Zhu, X. Meng, C. Qi, F.S. Xiao, Sulfated graphene as an efficient solid catalyst for acid-catalyzed liquid reactions, J. Mater. Chem. 22 (2012) 5495–5502, https://doi.org/10.1039/c2jm16608a.
[62] H.P. Mungse, N. Bhakuni, D. Tripathi, O.P. Sharma, B. Sain, O.P. Khatri, Fractional distribution of graphene oxide and its potential as an efficient and reusable solid catalyst for esterification reactions, J. Phys. Org. Chem. 27 (2014) 944–951, https://doi.org/10.1002/poc.3375.
[63] K. Nakajima, M. Hara, Amorphous carbon with SO_3H groups as a solid brønsted acid catalyst, ACS Catal. 2 (2012) 1296–1304, https://doi.org/10.1021/cs300103k.
[64] D.R. Stellwagen, F. Van Der Klis, D.S. Van Es, K.P. De Jong, J.H. Bitter, Functionalized carbon nanofibers as solid-acid catalysts for transesterification, ChemSusChem 6 (2013) 1668–1672, https://doi.org/10.1002/cssc.201300372.
[65] M.-H. Zong, Z.-Q. Duan, W.-Y. Lou, T.J. Smith, H. Wu, Preparation of a sugar catalyst and its use for highly efficient production of biodiesel, Green Chem. 9 (2007) 434–437, https://doi.org/10.1039/B615447F.
[66] L.J. Konwar, J. Boro, D. Deka, Review on latest developments in biodiesel production using carbon-based catalysts, Renew. Sust. Energ. Rev. 29 (2014) 546–564, https://doi.org/10.1016/j.rser.2013.09.003.
[67] F.D. Pileidis, M. Tabassum, S. Coutts, M.-M. Titirici, Esterification of levulinic acid into ethyl levulinate catalysed by sulfonated hydrothermal carbons, Chin. J. Catal. 35 (2014) 929–936, https://doi.org/10.1016/S1872-2067(14)60125-X.
[68] M.S. Khayoon, B.H. Hameed, Acetylation of glycerol to biofuel additives over sulfated activated carbon catalyst, Bioresour. Technol. 102 (2011) 9229–9235, https://doi.org/10.1016/j.biortech.2011.07.035.
[69] J.A. Sánchez, D.L. Hernández, J.A. Moreno, F. Mondragón, J.J. Fernández, Alternative carbon based acid catalyst for selective esterification of glycerol to acetylglycerols, Appl. Catal. A Gen. 405 (2011) 55–60, https://doi.org/10.1016/j.apcata.2011.07.027.
[70] C. De La Calle, J.M. Fraile, E. García-Bordejé, E. Pires, L. Roldán, Biobased catalyst in biorefinery processes: Sulphonated hydrothermal carbon for glycerol esterification, Cat. Sci. Technol. 5 (2015) 2897–2903, https://doi.org/10.1039/c5cy00059a.
[71] M. Gonçalves, R. Rodrigues, T.S. Galhardo, W.A. Carvalho, Highly selective acetalization of glycerol with acetone to solketal over acidic carbon-based catalysts from biodiesel waste, Fuel 181 (2016) 46–54, https://doi.org/10.1016/j.fuel.2016.04.083.
[72] R. Rodrigues, M. Gonçalves, D. Mandelli, P.P. Pescarmona, W.A. Carvalho, Solvent-free conversion of glycerol to solketal catalysed by activated carbons functionalised with acid groups, Cat. Sci. Technol. 4 (2014) 2293–2301, https://doi.org/10.1039/C4CY00181H.
[73] D. Nandan, P. Sreenivasulu, L.N. Sivakumar Konathala, M. Kumar, N. Viswanadham, Acid functionalized carbon–silica composite and its application for solketal production, Microporous Mesoporous Mater. 179 (2013) 182–190, https://doi.org/10.1016/j.micromeso.2013.06.004.
[74] X. Liang, M. Zeng, C. Qi, One-step synthesis of carbon functionalized with sulfonic acid groups using hydrothermal carbonization, Carbon 48 (2010) 1844–1848, https://doi.org/10.1016/j.carbon.2010.01.030.
[75] S. Zhang, B. Hong, Z. Fan, J. Lu, Y. Xu, M. Pera-Titus, Aquivion–carbon composites with tunable amphiphilicity for Pickering interfacial catalysis, ACS Appl. Mater. Interfaces 10 (2018) 26795–26804, https://doi.org/10.1021/acsami.8b08649.

[76] A. Onda, T. Ochi, K. Yanagisawa, Selective hydrolysis of cellulose into glucose over solid acid catalysts, Green Chem. 10 (2008) 1033–1037, https://doi.org/10.1039/b808471h.
[77] M. Kitano, D. Yamaguchi, S. Suganuma, K. Nakajima, H. Kato, S. Hayashi, M. Hara, Adsorption-enhanced hydrolysis of β-1,4-glucan on graphene-based amorphous carbon bearing SO_3H, COOH, and OH groups, Langmuir 25 (2009) 5068–5075, https://doi.org/10.1021/la8040506.
[78] M. Hara, Biomass conversion by a solid acid catalyst, Energy Environ. Sci. 3 (2010) 601–607, https://doi.org/10.1039/b922917e.
[79] G.S. Foo, C. Sievers, Synergistic effect between defect sites and functional groups on the hydrolysis of cellulose over activated carbon, ChemSusChem 8 (2014) 534–543, https://doi.org/10.1002/cssc.201402928.
[80] Y. Chen, X. Ai, B. Huang, M. Huang, Y. Huang, Y. Lu, Consecutive preparation of hydrochar catalyst functionalized in situ with sulfonic groups for efficient cellulose hydrolysis, Cellulose 24 (2017) 2743–2752, https://doi.org/10.1007/s10570-017-1306-x.
[81] P. Wataniyakul, P. Boonnoun, A.T. Quitain, T. Kida, N. Laosiripojana, A. Shotipruk, Preparation of hydrothermal carbon acid catalyst from defatted rice bran, Ind. Crop. Prod. 117 (2018) 286–294, https://doi.org/10.1016/j.indcrop.2018.03.002.
[82] L.J. Konwar, P. Mäki-Arvela, E. Salminen, N. Kumar, A.J. Thakur, J.-P. Mikkola, D. Deka, Towards carbon efficient biorefining: multifunctional mesoporous solid acids obtained from biodiesel production wastes for biomass conversion, Appl. Catal. B Environ. 176–177 (2015) 20–35, https://doi.org/10.1016/j.apcatb.2015.03.005.
[83] A. Charmot, P.W. Chung, A. Katz, Catalytic hydrolysis of cellulose to glucose using weak-acid surface sites on postsynthetically modified carbon, ACS Sustain. Chem. Eng. 2 (2014) 2866–2872, https://doi.org/10.1021/sc500669q.
[84] P. Dornath, S. Ruzycky, S. Pang, L. He, P. Dauenhauer, W. Fan, Adsorption-enhanced hydrolysis of glucan oligomers into glucose over sulfonated three-dimensionally ordered mesoporous carbon catalysts, Green Chem. 18 (2016) 6637–6647, https://doi.org/10.1039/c6gc02221a.
[85] S. Van De Vyver, L. Peng, J. Geboers, H. Schepers, F. De Clippel, C.J. Gommes, B. Goderis, P.A. Jacobs, B.F. Sels, Sulfonated silica/carbon nanocomposites as novel catalysts for hydrolysis of cellulose to glucose, Green Chem. 12 (2010) 1560–1563, https://doi.org/10.1039/c0gc00235f.
[86] C. Ban, W. Jeon, G. Park, H.C. Woo, D.H. Kim, Hydrothermal conversion of alginate into uronic acids over a sulfonated glucose-derived carbon catalyst, ChemCatChem 9 (2017) 329–337, https://doi.org/10.1002/cctc.201600969.
[87] M.J. Climent, A. Corma, S. Iborra, S. Martínez-Silvestre, Transformation of cellulose into nonionic surfactants using a one-pot catalytic process, ChemSusChem 9 (2016) 3492–3502, https://doi.org/10.1002/cssc.201600977.
[88] P. Wataniyakul, P. Boonnoun, A.T. Quitain, M. Sasaki, T. Kida, N. Laosiripojana, A. Shotipruk, Preparation of hydrothermal carbon as catalyst support for conversion of biomass to 5-hydroxymethylfurfural, Catal. Commun. 104 (2018) 41–47, https://doi.org/10.1016/j.catcom.2017.10.014.
[89] X. Qi, H. Guo, L. Li, R.L. Smith, Acid-catalyzed dehydration of fructose into 5-hydroxymethylfurfural by cellulose-derived amorphous carbon, ChemSusChem 5 (2012) 2215–2220, https://doi.org/10.1002/cssc.201200363.
[90] W. Daengprasert, P. Boonnoun, N. Laosiripojana, M. Goto, A. Shotipruk, Application of sulfonated carbon-based catalyst for solvothermal conversion of cassava waste to hydroxymethylfurfural and furfural, Ind. Eng. Chem. Res. 50 (2011) 7903–7910, https://doi.org/10.1021/ie102487w.

[91] J. Zhou, Y. Wang, X. Guo, J. Mao, S. Zhang, Etherification of glycerol with isobutene on sulfonated graphene: reaction and separation, Green Chem. 16 (2014) 4669–4679, https://doi.org/10.1039/C4GC01044B.
[92] R. Luque, V. Budarin, J.H. Clark, D.J. Macquarrie, Glycerol transformations on polysaccharide derived mesoporous materials, Appl. Catal. B Environ. 82 (2008) 157–162, https://doi.org/10.1016/j.apcatb.2008.01.015.
[93] S. Suganuma, K. Nakajima, M. Kitano, H. Kato, A. Tamura, H. Kondo, S. Yanagawa, S. Hayashi, M. Hara, SO_3H-bearing mesoporous carbon with highly selective catalysis, Microporous Mesoporous Mater. 143 (2011) 443–450, https://doi.org/10.1016/j.micromeso.2011.03.028.
[94] L.J. Konwar, A. Samikannu, P. Mäki-Arvela, J.P. Mikkola, Efficient C-C coupling of bio-based furanics and carbonyl compounds to liquid hydrocarbon precursors over lignosulfonate derived acidic carbocatalysts, Cat. Sci. Technol. 8 (2018) 2449–2459, https://doi.org/10.1039/c7cy02601c.
[95] Y. Zhu, W. Li, Y. Lu, T. Zhang, H. Jameel, H.M. Chang, L. Ma, Production of furfural from xylose and corn Stover catalyzed by a novel porous carbon solid acid in γ-valerolactone, RSC Adv. 7 (2017) 29916–29924, https://doi.org/10.1039/c7ra03995f.
[96] C.A. Deshmane, M.W. Wright, A. Lachgar, M. Rohlfing, Z. Liu, J. Le, B.E. Hanson, A comparative study of solid carbon acid catalysts for the esterification of free fatty acids for biodiesel production. Evidence for the leaching of colloidal carbon, Bioresour. Technol. 147 (2013) 597–604, https://doi.org/10.1016/J.BIORTECH.2013.08.073.
[97] G. Chen, X. Wang, Y. Jiang, X. Mu, H. Liu, Insights into deactivation mechanism of sulfonated carbonaceous solid acids probed by cellulose hydrolysis, Catal. Today 319 (2018) 25–30, https://doi.org/10.1016/j.cattod.2018.03.069.
[98] J.M. Anderson, R.L. Johnson, K. Schmidt-Rohr, B.H. Shanks, Solid state NMR study of chemical structure and hydrothermal deactivation of moderate-temperature carbon materials with acidic SO_3H sites, Carbon 74 (2014) 333–345, https://doi.org/10.1016/j.carbon.2014.03.041.
[99] D. Scholz, O. Kröcher, F. Vogel, Deactivation and regeneration of sulfonated carbon catalysts in hydrothermal reaction environments, ChemSusChem 11 (2018) 2189–2201, https://doi.org/10.1002/cssc.201800678.
[100] J.M. Fraile, E. García-Bordejé, L. Roldán, Deactivation of sulfonated hydrothermal carbons in the presence of alcohols: Evidences for sulfonic esters formation, J. Catal. 289 (2012) 73–79, https://doi.org/10.1016/j.jcat.2012.01.017.
[101] J.M. Fraile, C.J. Saavedra, Synthesis of isosorbide esters from sorbitol with heterogeneous catalysts, ChemistrySelect 2 (2017) 1013–1018, https://doi.org/10.1002/slct.201601866.

CHAPTER 9

Functional porous carbons: Synthetic strategies and catalytic application in fine chemical synthesis

E. Pérez Mayoral[a], I. Matos[b], M. Bernardo[b], C. Durán-Valle[c], and I. Fonseca[b]

[a]Department of Inorganic Chemistry and Technical Chemistry, Faculty of Science, National University of Distance Education, UNED, Madrid, Spain
[b]LAQV/REQUIMTE, Department of Chemistry, Faculty of Science and Technology, NOVA University of Lisbon, Caparica, Portugal
[c]Department of Organic and Inorganic Chemistry, University of Extremadura, Badajoz, Spain

1 Introduction

Nowadays, environmental pollution is intimately related to the contemporary lifestyles, and sooner than later, sustainable solutions should be found. Among the major problems are the air and water pollution, as consequence of fossil fuel-based society, originating adverse environmental impacts which strongly affect the ecosystems and, therefore, with implications for the animal and human health.

Carbon-based materials are considered as extremely versatile materials applied in energy generation and storage [1], water treatment for removal of dyes, heavy metal, organic pollutants including pharmaceuticals but also products for personal care and others [2], NOx removal originated from industrial processes, production and distribution of energy, transport and traffic, etc. [3], biomedical engineering [4], and catalysis [5, 6], among other research fields, limiting or reducing the environmental impact.

Chemistry is present in everything that surrounds us and makes life much easier; however, the chemical industry is highly polluting. Green chemistry is then the key to face the major challenges toward sustainability [7]. In fact, many pharmaceutical companies (i.e., Amgen, Merck Group, Abbott, Johnson & Johnson, and Roche) practice green chemistry for drug discovery, development, and manufacture [8]. Catalysis in all its variants is one of the great pillars for sustainable chemistry and carbon-based materials, in the past and at the present, have been extensively studied for this application.

Emerging Carbon Materials for Catalysis
https://doi.org/10.1016/B978-0-12-817561-3.00009-3

1.1 Carbon materials as catalyst supports and catalysts

In general, inherent properties of carbon-based materials such as large specific surface area, tunable porosity, extraordinarily high thermal and chemical stability, and hydrophobic character are determinants for catalytic performance. All these characteristics make them excellent catalytic materials acting as both supports or catalysts [9], advantageously providing the adequate chemical environment to produce fine chemicals [10, 11].

These materials can be prepared from diverse sources including biomass residues [12] and even pure chemicals [13–16]. Carbon materials also include advanced nanomaterials as Metal-Organic frameworks (MOFs) and Covalent-Polymer Frameworks (COFs) [17–19] whose surface can be modified by introduction of different functional groups or metal nanoparticles [20] or even by immobilization of enzymes [21, 22].

Carbon-based materials have been used in catalysis for a long time [5]. In more recent times, the development of more sustainable variants of this type of materials has undergone a spectacular progress being an important hot topic in the modern material science, paying attention on designing carbon nanomaterials with improved properties.

Metal-supported carbon-based materials for catalytic applications have particularly impressed much interest [23] with activated carbons (ACs) being the most explored carbonaceous materials due to their large surface areas strongly favoring the high dispersion and stability of the metallic phases [24]. It is well-known that metal nanoparticles' sizes and dispersion influence the catalytic performance. In this sense, special attention is referred to the development of single atom carbon-based catalysts (SAC), in which the metal atoms can be easily visualized [25]. In this regard, SACs can be prepared not only by single atom insertion as primary it could be thought, but also using direct or even postsynthesis methods mainly depending on the metal loading. All the synthetic approaches to obtain SACs seek to increase the interaction between metal and support in order to prevent the formation of aggregates during the synthesis or postsynthetic subsequent modifications. In this context, the presence of defects in a carbonaceous material is then key for the synthesis of this type of carbons which can be originated by physical or chemical activation as detailed in subsequent sections.

The use of carbon nanotubes (CNTs) and carbon nanofibers (CNFs) as catalysts or supports is an active hot topic [26, 27]. Enhanced catalytic performances are reported for both carbon forms, specifically for CNTs, compared to other traditional supports. Especially interesting is the possibility of eliminating the diffusional and mass transfer limitations when using

CNT-based catalysts. It is not only possible due to the high accessibility of the active phase and the absence of microporosity, but also the confinement effects related with the location of active sites, at the most external surface, or even inside the channels when opened, making them useful materials as nanoreactors.

Because graphene and single-wall carbon nanotubes (SWCNTs) exhibit a great tendency to grouping and considering their high economic cost, graphene with a few layers of that and particularly multiwalled carbon nanotubes (MWCNT) constitute the preferred carbon-based material involved in the SAC synthesis. Basal plane surface of graphene or CNTs, both constituted by C atoms with pure sp^2 hybridization, is often a homogeneous surface, whereas majority of the structural defects, mainly as oxygenated functions, are contained at the edges, showing different reactivity as function of the surface and porosity of the material and, hence, distinct adsorptive properties. The nature and concentration of defects as well as the surface curvature, helicity, the inner cavity, and electrical conductivity of CNTs are all crucial factors influencing the metal-supports interactions. At this regard, different adsorption sites are possible for basal plane surface of graphene: (i) on top of a carbon atom; (ii) bridging a C—C bond; and (iii) at the center of hexagons [28].

On the other side, nitrogen-doped carbons are also particularly interesting materials for the preparation of SACs [29, 30]. Obviously, the introduction of nitrogen atoms changes both the acid-base properties and the adsorption properties. At this respect, MOFs are considered emerging precursors of highly porous, strongly basic, and highly graphitized *N*-doped carbon materials with considerably high *N* loadings, even superior to those synthetized by traditional methods [31]. Direct pyrolysis of the appropriate MOF often results in the formation of metal-supported porous carbons or even metal-free carbons, as function of the carbonization temperature, suitable for modification with other metallic species.

The goal of this chapter is then to provide an overview concerning the development and use of porous carbon materials showing a very different structure, morphology, and composition applied to fine chemical synthesis.

2 Synthetic strategies for the preparation of porous carbons

2.1 Activation methods

Activation relates to the processes in which a carbonaceous precursor gains porosity (space, voids, sites, pores, holes, etc.) or it suffers the widening of its incipient porosity. Porous carbons can be considered as segments of graphene sheets of different sizes and degree of perfection linked together in an infinite number of ways. These segments are polycyclic aromatic compounds of various complexities that are bonding together to create a three-dimensional network of carbon atoms. The random bonding of these polycyclic groups of carbon atoms, with linear carbon atoms, creates spaces (absent of carbon atoms and heteroatoms) called porosity [32]. The porosity is classified according to the pore sizes into: microporosity (<2 nm), mesoporosity (2–50 nm), and macroporosity (>50 nm) [33].

Activation methods to prepare porous carbon materials typically involve carbonization (pyrolysis) and activation itself. Activation can be a physical process in which gasifying agents remove carbon atoms from the carbonaceous matrix, or a chemical process involving the co-carbonization of the precursor with chemical agents such as acids, bases, or alkali salts. Also, physical and chemical activation can be used simultaneously.

Physical activation is usually a two-step process: the raw material is first carbonized in the absence of oxygen (pyrolysis/carbonization step), at temperatures between 400°C and 850°C, to remove volatiles (mainly with noncarbon atoms) and increase the fixed carbon; the second step consists in the activation of the resulting char with oxidant/gasifying gases such as steam, air, N_2, O_2, NH_3, CO_2, or a mixture of these gases, which react with the carbon atoms from the carbonaceous matrix, at temperatures around 800–1000°C [32, 34–37]. The fact that carbon atoms are eliminated in this type of activation leads to quite low production yields. The development of porosity is strongly dependent on the type, amount, and time of the gas supply and temperature and structure/composition of the precursor [38].

The fundamental reactions with the most used gasifying agents, CO_2 and water steam, are represented in Eqs. (1) and (2) [32, 37, 38]

$$C + CO_2 \leftrightarrow 2CO \text{ (Boudouard reaction)} \tag{1}$$

$$C + H_2O \leftrightarrow CO + H_2 \text{ (Water steam gasification)} \tag{2}$$

Both reactions are endothermic and relatively easy to control since the reaction rate of carbon with carbon dioxide and water steam is slowed down due to an inhibitory effect of the reaction products (CO, H_2) [38].

The activation with CO_2 will produce ACs with narrow pore size distribution, while water steam activation will generate ACs with wider pore size distribution and smaller micropore volume [32, 38].

The gasification reactions lead to the formation of chemisorbed oxygen, C(O), which in turn results in the formation of surface oxygen functional groups [32, 39]. Whenever an activation is performed with NH_3, usually coupled with CO_2, it results also in the introduction of *N*-containing groups on carbon's surface [39–41].

Some of the inorganic matter present in the precursors may act as catalysts of the gasification reactions during activation. That is the case of salts, oxides, and metals of the series of alkali metals, alkali earth metals, and transition metals. Basically, these species catalyze the dissociation rate of CO_2 and H_2O, providing chemisorbed oxygen that reacts with the carbon atoms to produce oxidized species [32, 38]. These catalyst particles can be mobile and induce the production of pores by themselves in the carbon matrix with the size of the particle.

As said above, chemical activation consists in the co-carbonization of a carbonaceous precursor with a chemical agent that sometimes can provide simultaneously the porosity development and the introduction of surface functional groups at carbon's surface.

The first step of the chemical activation process is the impregnation of the precursor with a chemical agent (dehydrating agents and/or oxidants) and the second step is the heating under inert atmosphere at temperatures between 400°C and 900°C to promote simultaneously carbonization (removal of noncarbon elements) and activation (porosity development) [32, 34, 37, 42, 43]. The last step of the process is the washing of the resulting porous carbons to remove the residual impregnating agent and expose the porosity developed. Acidic or basic solutions, or simply water, can be used in the washing step. Thus, the success of a chemical activation depends on the nature of the precursor, the type and amount of activating agent, method of impregnation (dry, wet, mixing), temperature and time of activation, and finally the postwashing [38].

The most common chemical agents used in chemical activations are H_3PO_4, H_2SO_4, $ZnCl_2$, K_2CO_3, NaOH, and KOH. Other chemical agents that can be used, although less common, are HNO_3, HCl, H_2O_2, $FeCl_3$, and $KMnO_4$ [32, 34, 36–38, 42, 43]. H_3PO_4 and $ZnCl_2$ are the most common dehydrating and degrading agents being active at low temperatures, around 450–500°C, and are particularly effective for lignocellulosic precursors. $ZnCl_2$ acts as template for the creation of microporosity which means that the volume of micropores developed is similar to the volume of $ZnCl_2$ introduced, creating an uniform microporosity [32, 42–44]. On the other hand, the mechanisms of action for H_3PO_4

are different since the acid promotes dehydration, depolymerization, condensation, and cross-linking reactions in the lignocellulosic precursors, promoting at the same time the creation of phosphate ester linkage structures within the carbon matrix. The insertion of these phosphate linkages creates a dilatation that, after removal of the acid, leaves the matrix in an expanded state with an accessible pore structure [32, 38, 42–45]. At temperature above 450–500°C, these phosphate linkages become unstable and contract which causes a reduction in the porosity. $ZnCl_2$ develops both wide micropores and small mesopores and H_3PO_4 develops essentially microporosity with some larger mesopores and even macropores [32, 43, 45].

The use of alkali metal salts (oxides/hydroxides/carbonates) to prepare ACs requires higher temperatures, between 700°C and 900°C. The reaction mechanism for the creation of porosity with alkali salts usually involves a first step in which occurs a redox reaction where the carbon atoms oxidized to CO or CO_2 and this gasification being catalyzed by the alkali metal; then, the reduction of the alkali metal salt by carbon occurs to produce the metallic atom which will be intercalated between the graphene layers expanding the carbon structure [32, 36, 38, 42–44]. The activation with an alkali (M) hydroxide, for example, proceeds through a series of simultaneous/consecutive reactions (Eqs. 3–9) represented below [38, 39, 43, 44]:

$$2MOH + CO_2 \rightarrow M_2CO_3 + H_2O \tag{3}$$

$$2C + 2MOH \rightarrow 2CO + 2M + H_2 \tag{4}$$

$$CO + H_2O \rightarrow H_2 + CO_2 \tag{5}$$

$$C + H_2O \rightarrow CO + H_2 \tag{6}$$

$$M_2CO_3 + C \rightarrow M_2O + 2CO \tag{7}$$

$$M_2O + C \rightarrow 2M + CO \tag{8}$$

$$M_2O + H_2 \rightarrow 2M + H_2O \tag{9}$$

There are advantages and disadvantages associated with the two types of activations. Chemical activation allows to produce porous carbons with high surface areas in one single step that combines carbonization and activation which saves energy. However, the use of chemical agents increases the costs of the process and the washing of the resulting carbons makes the process time-consuming and environmentally nonfriendly. On the other hand, physical activation proceeds at higher temperatures and there is a poorer control of the porosity [32, 37, 45, 46].

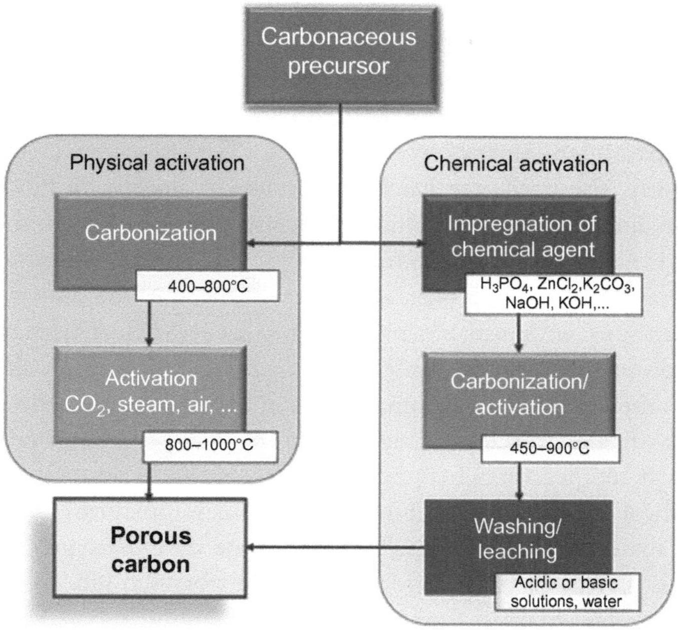

Fig. 1 General flowsheet for the preparation of porous carbons by using the different activation methods.

Fig. 1 presents a flow diagram for the preparation of porous carbons with the different activation methods.

2.2 Biomass-derived porous carbon materials

The use of biomass, particularly biowaste, is very attractive to prepare porous carbon materials. These are abundant, renewable, sustainable, and cost-effective precursors; therefore, in recent years, the number of studies dealing with the preparation of porous carbons from biomass, especially lignocellulosic biomass, has significantly increased [37, 39, 42–46].

Traditionally, commercial ACs have been prepared not only from bituminous coals, lignite and anthracite, but also from biomasses such as wood, coconut shells, peat, fruit stones/shells, and animal bones [35, 42, 47, 48]. However, the potential of biowastes from agroforestry industries to be converted into porous carbons has been academic and scientifically demonstrated. Some examples of the reported biowastes are as follow: rice husk [49], rice straw [50], corn cob [51], olive stones [52], fruit stones [53–57], sugarcane bagasse [58], manure [59], sewage sludge [60, 61], nut shells [62–66], wood wastes [45, 67–70], among many others [12, 39, 42–46].

The physical and mechanical properties of the biomasses will be reflected in the resulting carbons: soft, compressible, and low density biowastes such as husks, peanut shells, soybean shells, etc. will generate low-density and friable carbons; hard and dense biowastes such as nut shells and fruit stones will result in hard, high-density carbons [46].

Recently, the challenge has been the use of alternative biowastes as precursors and innovative methodologies to convert them into porous carbons. Isolated carbohydrate materials such as glucose, sucrose, starch, amylose, and so on have been also successfully used to prepare highly porous carbons with tunable properties [12, 71, 72]. This new range of bio-porous carbons have been mainly synthesized through alternative routes to the traditional carbonization/activation process. One of those routes is the hydrothermal carbonization (HTC). HTC commonly proceeds at temperatures between 180 and 260°C in aqueous medium, at subcritical conditions, to convert the biomass precursor mainly into a carbon material, along with an aqueous phase rich in oxygenated compounds, as well as a small fraction of gases are obtained [48, 71–73]. The carbon material obtained from HTC may present an incipient porosity, being designated as hydrochar, and has to be further activated through physical and/or chemical methods. However, different chemical agents can be used in situ to catalyze or template-assist the process, so the resulting carbon from HTC already presents a high porous structure and well-defined morphology [4, 74–79]. The main advantages of HTC compared to dry carbonization are the lower temperature used, the higher yields of carbon product, and the possibility of process wet biomasses.

The use of biomaterials as precursors of porous carbons presents additional advantages to those before mentioned, the accumulation of heteroatoms (O, N, S, P, B) in the form of functional groups at carbons surface. Depending on the biomass used and the synthesis method, an in situ functionalization or a heteroatom self-doping of the produced carbon may occur, since the starting material can offer these surface functional groups [80]. For example, human hair, gelatin from animal wastes, egg wastes, kraft lignin, rape pollen, fish waste, among others [81–86] were used as precursors of porous carbons self-doped with O, N, S, B, and P. Also, the produced porous carbon can be enriched with minerals typically found in biomasses, composed of elements such as Ca, Mg, K, Na, P, Si, Al, Fe, etc., which can represent an advantage for specific applications [87–90].

2.3 Pure carbon gels: Xerogels, cryogels, and aerogels

Pure carbon gels are prepared by the sol-gel method. The sol-gel process is a very versatile technique used in the preparation of mass catalysts and supports. Due to the number of parameters that can be manipulated throughout the various stages of preparation, it allows the control of the texture, composition, homogeneity, and structural properties of the solids. As the name implies, this method first involves the formation of colloidal solution of monomers, a sol, followed by the formation of a gel (an interconnected network of particles) [91] .

The organic gels used in the preparation of porous carbons are formed by the polymerization reaction either in aqueous phase or in organic solvents in the presence of a catalyst. The reagents are commonly hydroxybenzenes such as resorcinol, phenol, phloroglucinol, and cresol, which react with aldehydes such as formaldehyde and furfural, in a catalyzed reaction using agents like Na_2CO_3, NaOH, HCl, acetic acid, among others [13–15].

However, the use of suitable biomass-derived precursors as structure-directing agents of carbon aerogels has received some attention in the last years. Starch, alginic acid, pectin, and amylose have been applied in the so-called Starbon® approach [92–94]. Also, chitosan/chitin, tannins, glucose, fructose, and xylose have been used as precursors for carbon gels [13, 94].

The gel results from the cross-linking of micelles, forming three-dimensional networks whose solvent is encapsulated in the solid. The point at which the gel is formed is called the gelation point and depends on different factors, such as temperature or pH, for example. Normally, the gelation step occurs at temperatures below 100°C. The aging stage is the time that elapses from the formation of the gel to the removal of the solvent entrapped in its structure. The aging of the formed gel is critical to promote the cross-linking and it strongly affects the porosity of the final carbon gel. Since there is a solvent in the pores, the gel is not static and can undergo different transformations, the main factors affecting this stage being the temperature, the aging time, and the pH of the liquid in the pores. The aging step may take 3–7 days at temperatures around 80–90°C.

The final step is the removal of solvent from the pores of the solid gel, the drying step. This step can be critical because, in the case of small pores, the capillary pressure associated with the liquid/vapor interface is high, which may lead to the collapse of the porous structure.

Different materials will be obtained depending on the drying method applied:

(1) Ambient drying—it simply consists in the exchange of the water used as solvent trapped in the gel, by an organic solvent with lower surface tension such as acetone. This is intended to prevent the collapse of the structure. It follows the solvent evaporation at room temperature and at atmospheric pressure. This drying is usually carried out with a gas flow rate (air or nitrogen being the most common). Some gradual warming may be done, although a careful choice of operating conditions is required because an internal vapor pressure can develop which can lead to the collapse of the solid structure, with consequent reduction of surface area and pore volume. This can be avoided by lowering the temperature gradient or the gas flow rate. A xerogel is thus obtained.

(2) Drying in supercritical conditions—after the solvent exchanged (acetone), it is further exchanged by carbon dioxide (CO_2) which is then removed out at high pressure, obtaining an aerogel.

(3) Freeze drying—allows to eliminate the liquid/vapor interface, freezing, and subsequent sublimation of the solvent inside the pores, furnishing a cryogel [13–15].

The final and very important step is pyrolysis, as it directly influences the textural properties (specific area, pore volume, pore size distribution) and indirectly the activity, selectivity, and stability of the carbon porous catalyst. It is usually carried out at high temperature under nitrogen atmosphere, in order to generate porosity and give mechanical resistance to the material. Usually, temperatures between 600 and 900°C are used, with very low heating rates, to widen the narrow pores formed during the previous steps.

Although the carbonization step influences the final porous structure of the carbon gel, the gelation and aging stages are the most critical stages to tune the carbon.

2.4 Template methods and ordered porous carbon materials

Ordered porous carbon materials with uniform and regular pore size and structure are of great interest for different applications, and synthetic methodologies have been developed to obtain such feature. Additionally, tunable pore sizes are advantageous as they allow for tailored materials for specific applications.

The use of templates to enforce structure in the carbon materials is a popular synthesis approach [95]. Two main approaches are used in this method: one consists in the introduction of the template into the carbon source, followed by carbonization and final removal of the template, the porosity is formed by the removal of the template; in the other approach, the porous template is filled with the carbon precursors followed by carbonization; the posterior removal of the template results in a carbon with a porous structure shaped by the internal structure of the template presenting an interconnected pore system [96]. These template-assisted synthesis can be further classified as hard-templating and soft-templating [95].

Hard templates are inorganic, or sometimes organic, materials with well-known and developed structure that do not interact with the carbon precursors. The obtained structure is determined by the template; the internal framework of the different used inorganic substances is replicated into the porous carbon structure. In general, the obtained structure is a negative replica of the template [97].

Soft templates are organic species and block copolymers that self-assemble in specific oriented manner and may interact with the carbon precursors. In this case, the obtained porous structure also depends on the synthesis conditions such as temperature, pH, reactant ratios, and solvent polarity, as they will influence the self-orientation of the template molecules.

Fig. 2 demonstrates the main differences between hard and soft templating.

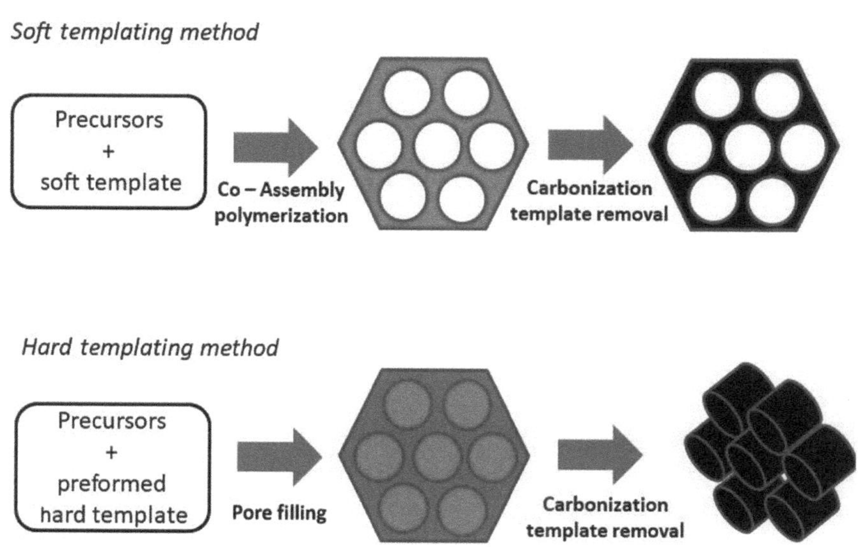

Fig. 2 Main steps for soft and hard templating methods of porous carbon materials.

2.4.1 Hard template

One of the most straightforward approaches to obtain ordered porous carbon with controlled porous structure is hard templating. Using inorganic materials such as zeolites or silicas and applying a casting technique, it is possible to produce carbon materials with different textural properties [98]. The morphology of the carbon materials is controlled by the morphology of the template materials. It is possible to obtain spherical nanoparticles starting from nanospherical silicas as hard templates and using a phenol formaldehyde mixture as carbon precursor [99].

It is also possible to form monoliths using this technique. Using hierarchical porous silicas with tunable morphologies as a hard template, it is possible to obtain a porous carbon presenting meso- and macroporosity.

The carbon source can also play an important role, as does the experimental technique used to incorporate it, since besides the solution-based impregnation, chemical vapor deposition (CVD) has also been reported as technique for the pore volume filling with the precursor [100].

Even though furfuryl alcohol has been commonly used as carbon precursor, bulky molecular precursors are not the most suitable for some zeolite structures. Some authors have suggested the use of other precursors such as propylene or acetylene gas, and Ryoo et al. [101] concluded that using ethanol vapor as a carbon source and hierarchically *meso*−/microporous zeolite beta as a template, hierarchically porous carbons exhibiting high BET surface area could be generated.

Joo et al. [102] also reported highly ordered mesoporous carbon using MCM-48 silica as template and sucrose as a carbon source. The authors claimed that the synthesis method was suitable for large-scale production of ordered mesoporous carbons.

Using a double filling approach, Masika et al. [103] developed ultrahigh surface area porous carbons through a two-step process combining liquid impregnation and chemical vapor deposition. The impregnation of zeolite13X with furfuryl alcohol followed by chemical vapor deposition of ethylene gas resulted in a suite of porous carbons with zeolite-like structural ordering presenting microporosity with a unimodal pore size distribution.

A pulsed chemical vapor-deposition technique was reported by Nueangnoraj et al. [104], ensuring that large amount of carbon was introduced into the confined nanochannels of the inorganic structure. The authors claim that the results suggest the presence of a cross-linked-fullerene-like framework in the obtained carbon.

Although valuable and promising results are obtained using this technique, there are great disadvantages in the use of sacrificial templates and the complex synthetic procedure.

2.4.2 Soft template

Soft-templating technique allows for the direct synthesis of ordered porous carbon via the self-assembly of organic molecules and the carbon precursors. This approach includes the polymerization of carbon precursors, commonly a phenolic resin—RESOL, around the organized domains of the soft template, normally a block copolymer, followed by the carbonization during which the controlled decomposition of the template occurs.

Instead of using a sacrificial template that is later removed, soft templates generate porous structures due to the interaction between the carbon precursor and the surfactant, which together with the synthesis conditions drive the manner in which self-assembly occurs, thus determining the porous structure [105]. This work was first developed by Liang et al. [106, 107], Tanaka et al. [108], Meng et al., and Zhang et al. [109, 110].

Compared with hard template method, the number of steps in soft templating syntheses is significantly reduced; the step for template removal is eliminated, which, besides being an advantage in terms of procedure simplification, removes the need for what can be a structure destructive step. However, the desired and predictable final porous structure is not so straightforward as with hard templating.

Two main processes can be used in this approach, the evaporation-induced self-assembly (EISA) process or a hydrothermal process. The EISA method is a simple procedure suitable for a variety of synthetic conditions. The evaporation of the solvent can be simply performed in an open-mouthed dish, and the ordered structure is formed on the solvent evaporating surface, thus obtaining films. A spin-coating or dip-coating technique can sometimes be applied [111]. Using this method, ordered mesoporous carbon with a variety of structures has been prepared; 2D hexagonal, lamellar, 3D bicontinuous, body-centered cubic, and other structures are reported in the literature [111]. The drawback of this technique is that it is currently limited to block copolymers as templates and phenolic resins as carbon precursors. The development of new amphiphilic molecules template is of great research interest [100].

Hydrothermal process preparation method consists in an autoclaving treatment of the solution containing the carbon precursors and templates. The autoclaving of the mixture at a certain temperature results in an autogenous pressure that promotes the interaction between the molecules resulting in faster polymerization rate and in the periodic assembly between precursor and template molecules. This is a faster and efficient process, but more energetically demanding in order to work at high temperature and pressure.

2.5 Emerging porous carbons

In the last decades, Metal-Organic Frameworks (MOFs) and Covalent Organic Frameworks (COFs) emerged as new source materials for the preparation of porous carbons with improved and tailored physico-chemical properties. COFs and MOFs have been subjected to exponentially increased number of studies and developments due to their wide variety of application fields.

MOFs are known as crystalline porous coordination polymers, consisting of poly-dentated organic ligands and coordinating metallic centers [112, 113]. In these porous materials, the metals are uniformly placed and large surface areas and high pore volumes are reported. Because a great number of metals can be applied for the MOF synthesis along with a large selection of organic components as coordinating ligands, there are already many different proposed structures and an enormous potential for new materials development, offering great molecular diversity [114]. MOFs are particularly attractive because, in principle, the composition, functionalities, and texture of this type of materials can be tuned [18] by careful selection of the organic and metal components targeting specific applications such as gases sorption and separation [115], energy storage and conversion [116], chemical sensing [117], catalysis [114], degradation of organic pollutants [118], and even drug delivery [119], among others.

COFs are class of porous crystalline polymers formed by covalent bonding of organic building blocks forming periodic architectures and are highly porous materials with controllable pore geometries and remarkably high pore volumes [120]. Most of these structures do not have metals and the geometry and composition are determined by the organic molecules chosen for building blocks and the way they react. In this manner, it is possible to obtain materials with tailored heteroatom species, high surface areas, and controllable structures at molecular levels [121, 122]. The different tailored functionalities impart to the material the possibility of use in multiple applications, such as gas capture and separation [123], heterogeneous catalysis [124, 125], and energy storage [126].

The preparation of new materials using MOFs and COFs as precursors or templates has gained much attention. Aiming at the replication of the characteristics of the parent materials in the products or in the search for taking advantage of the great features like metal dispersion, possibility of heteroatom incorporation and structure tuning, new multifunctional nanomaterials including porous carbons, nanostructured metal oxides, and carbon-metal/metal oxide hybrids have been developed based on the transformation of COFs and MOFs [18].

In this sense, highly dispersed heteroatom-doped porous carbons and new porous carbons comprising the presence of single atoms of metallic species can be synthesized by direct pyrolysis of COFs or MOFs. These materials are reported to present a great development of the porosity, good metal dispersion, and tailored heteroatoms incorporation, all of which are of supreme importance for catalytic applications.

2.5.1 Porous carbon from COFs

COFs have been considered as outstanding templates or precursors to fabricate highly tailored porous carbons [18, 126] due to their higher tolerance to harsh treatments; a closer structural relation between the COFs precursors and the porous carbon obtained is expected.

By simple direct carbonization of the COF precursor under controlled conditions, a wide variety of new porous carbon material has been prepared and normally a microporous carbon is obtained [122]. Tailored carbons with specific morphologies and efficient heteroatoms incorporation on the porous carbon structure, like nitrogen, sulfur, and boron, have been reported. Considering the potential of new COF structures development, it is foreseen that many new and improved materials are yet to be developed by careful selection of the precursor framework.

Microporous CNTs and CNFs with a surface area up to 900 m^2/g were obtained by Feng et al. [127] through a simple carbonization of the as-formed carbon-rich tubular and fiber-like polyphenylene precursors. Xiang et al. [128] obtained a *N*-doped graphitic porous carbon, reporting on the possibility of controlling the exact location of the heteroatom. Hollow, spherical nitrogen-rich porous carbon shells [129] composed of amorphous carbon with micropores on the shell frame and with specific surface areas as high as 525 m^2/g were prepared by direct carbonization of structured porous organic frameworks at high temperature.

However, the tailored design of COFs precursors remains a challenge as it is not always easy to accurately predict the association between the sacrificial COFs and the final porous carbon [130].

Other complementary techniques have been used for the preparation of these COF-derived carbons such as chemical and physical activation [131], microwave irradiation [132], and the molten-salt method [133], aiming at improving the final characteristics of the material by enhanced textural properties or new heteroatoms introduction.

2.5.2 Porous carbon from MOFs

New and exciting materials are being developed based on MOFs. Metal carbon hybrids and even metal-free porous carbons can be produced from MOFs under strictly controlled experimental thermal treatment conditions. The direct carbonization under an inert gas flow allows the formation of metal/carbon hybrids, or in some materials using higher temperatures, metal-free porous carbon materials can be obtained. The use of an aerobic atmosphere during the thermal treatment is used targeting the formation of nanostructured metal oxides.

Porous carbons from MOFs can present a more rational relation between their morphology and hierarchical porosity and that of the parent MOF. These inherited characteristics from the MOF patterns suggest a more reasoning design of the porous structures aimed at improved catalytic performance.

By thermal decomposition of MOF-5, Zhang et al. [134] reported the synthesis of metal-free mesoporous carbon with up to 1800 m^2/g of surface area. Hu et al. [135], by direct thermolysis of MOF-5, at 900°C, reported the preparation of a metal-free carbon material; the Zn cations of the parent material were reduced and vaporized. Hierarchically, porous carbons were prepared using an additional carbon source such as glucose [136] or glycerol [137, 138]. Srinivas et al.'s [139] work showed how selection of the MOF precursor and changes in the synthesis conditions strongly influence the texture and porosity of the resulting materials. Many other examples can already be found in recent literature [140].

2.6 Functionalization of porous carbons

As mentioned, carbon materials consist of a basic structure of graphene layers interacting by van der Waals forces. The edge of these layers can present heteroatoms that give rise to various functional groups. These groups can come from the raw material, the treatment received, and the functionalization of porous carbon; many of them are indicated in Fig. 3. The methods used to create functional groups are very numerous [141], although not all of them are used with the same frequency. Some of the best known are described below.

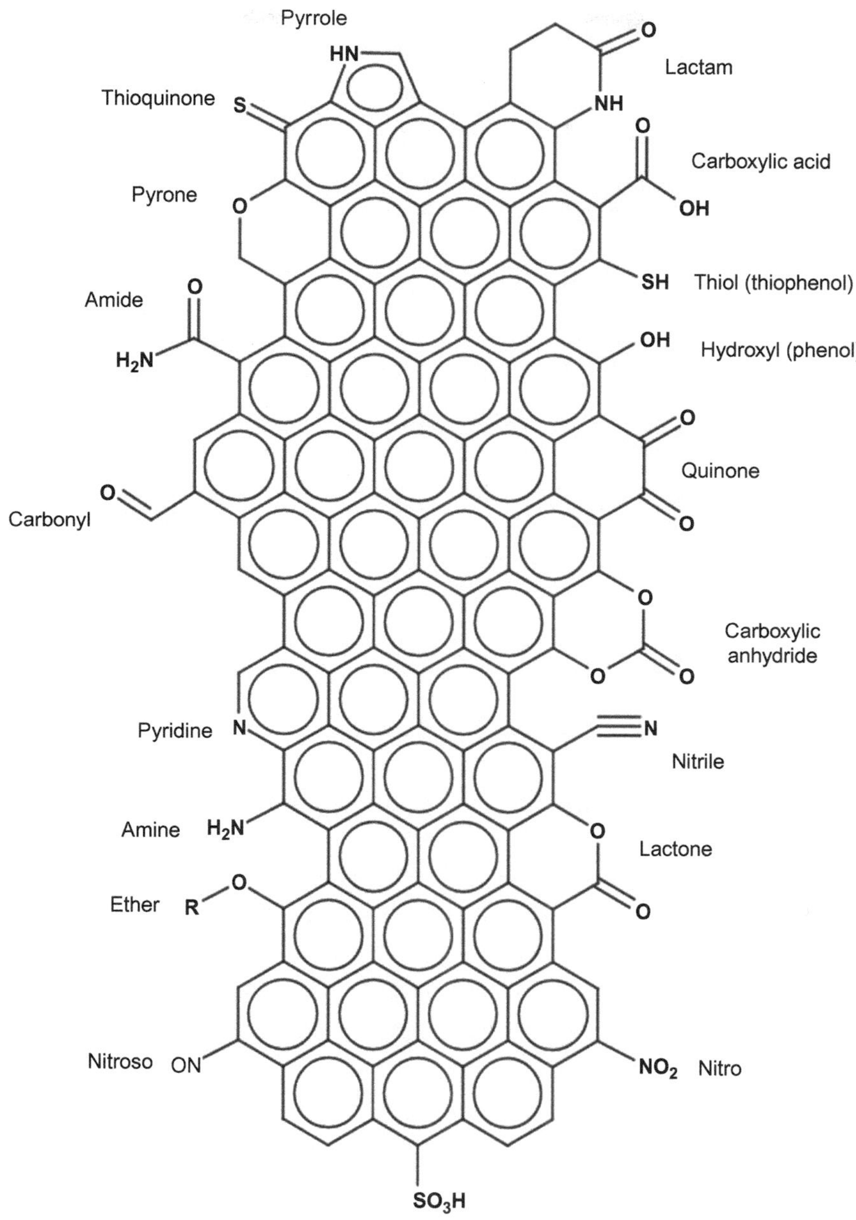

Fig. 3 Types of surface functionalities at porous carbon materials. *(Adapted from E. Pérez-mayoral, I. Matos, M. Bernardo, I.M. Fonseca, D.Q. Inorgánica, P. Senda, New and Advanced Porous Carbon Materials in Fine Chemical Synthesis. Emerging Precursors of Porous Carbons,* Catalysts *(2019) 1–35, https://doi.org/10.3390/catal9020133; T.J. Bandosz, Surface Chemistry of Carbon Materials, in: Carbon Materials for Catalysis, John Wiley & Sons, Inc., Hoboken, NJ, USA, 2008: pp. 45–92. https://doi.org/10.1002/9780470403709.ch2; Adapted from E. Pérez-Mayoral, V. Calvino-Casilda, E. Soriano, Metal-supported carbon-based materials: Opportunities and challenges in the synthesis of valuable products, Catal. Sci. Technol. 6 (2016) 1265–1291. https://doi.org/10.1039/c5cy01437a with permission of Royal Society of Chemistry.)*

2.6.1 Oxygen-containing functional groups

Oxidation is the most commonly used method for chemical modification of the surface of carbon materials. It can be carried out both in gas or liquid phase, and the number of oxidants used is very high. In gas phase, the most used oxidant is air, at temperatures between 473 and 673 K; at lower temperatures, any oxidation effect is barely observed, whereas at higher temperatures a degradation of the functional groups can surpass their formation. Among other oxidants are pure oxygen, ozone, or plasma treatment.

A greater variety of oxidizing agents have been used in the liquid phase and the effect is usually more intense than in the gas phase treatment. Among the most frequently used are nitric acid, hydrogen peroxide, and potassium permanganate. The oxidation degree depends on the oxidant, its concentration, temperature, and reaction time. A low oxidation degree usually produces hydroxyl groups such as phenols, while a high degree of oxidation favors the formation of carboxyl groups; in this last case, a higher polarity surface and a more acidic solid are obtained.

2.6.2 Nitrogen-containing functional groups

In the gaseous phase, ammonia is usually employed at higher temperatures than when air is used. In the liquid phase, porous carbon is often treated with a nitrogen precursor—urea, amines, melamine, etc.—and then treated at elevated temperatures; a better result is usually obtained if the carbon is previously oxidized. Nitrogen functional groups such as amines, pyrroles, pyridines, amides, etc. can be created, most of them exhibiting basic character resulting of a carbon material with increased pH at point of zero charge.

2.6.3 Sulfur-containing functional groups

As in the previous cases, the functionalization not only can be carried out in the liquid or gaseous, but also in solid phase. In gas phase, the carbon material is often heated in the presence of a gas containing sulfur, generally H_2S, while in solid phase, carbon is usually heated in the presence of elemental sulfur. Another method of choice is sulfonation consisting of the treatment of the carbon with commercial sulfuric acid at room temperature [142]; sulfonic groups are interesting because their acidity exceeds that of carboxylic acids. Remarkably, if an aqueous solution of sulfuric acid is used, no sulfonic groups are formed and only a mild surface oxidation is produced.

2.6.4 Supported catalysts

Some catalysts present a better activity if they are supported on the surface of another material mainly due to the chemical activity of the support, its porous structure, or the fact that the dispersion on a solid of high specific surface allows the existence of smaller particles; the latter is most frequent and provides a greater active surface for the catalyst. One of the most frequently used methods is the impregnation within an active phase or precursor of that via (i) impregnation excess-solution or (ii) incipient-wetness impregnation. In the first case, the solution volume is much greater than the pore volume of the support, and after a period of contact, the solid is filtered or the excess solvent is removed by evaporation. Although this method is simpler to carry out, it is more difficult to quantify the exact amount of the active phase supported on the porous carbon. In the incipient-wetness impregnation method, the solution volume used is equal to the volume of the pores of the support, and it is added drop by drop allowing a better control of the active phase loading.

The electrostatic interactions between the catalyst precursor and the carbonaceous support are important in impregnation. Therefore, the conditions that can influence these interactions—polarity of the solvent, functional groups in the carbon, cationic or anionic precursors and, above all, the pH—are important to achieve a catalyst supported by the desired qualities. In order to obtain a good dispersion of the catalyst, a large number of anchorage sites are needed, so the presence of abundant functional groups is usually beneficial.

2.6.5 Thermal treatment

The main chemical effect of thermal treatment is the elimination of labile functional groups, usually formed by heteroatoms. A more hydrophobic surface is obtained, as well as an increase in basicity in most cases.

3 Use of porous carbons in the synthesis of fine chemicals

This section seeks to summarize the most representative catalytic methodologies which use porous carbon-based materials for the preparation of fine chemicals. It includes the use of ACs and other advanced porous carbon catalysts based on CNT, graphene, and highly porous carbons synthetized from MOFs and COFs.

3.1 Activated carbon materials

The use of carbonaceous materials in catalysis has been the subject of several reviews in recent years, which shows its emerging interest. In some cases, the review is more general [143, 144], and in others, it focuses specifically on the application of catalysis in the synthesis of fine chemicals [23, 145]. It has been shown that, although the surface area and the shape of carbon porosity may be very important in the preparation and properties of the corresponding catalysts, the role of carbon surface chemistry is also extremely important.

Some reactions catalyzed by ACs are described below. These reactions have been grouped according to the type of catalyst activity. However, there are some reactions that can be catalyzed by different types of catalysts. The Friedländer reaction is an example of this, which is employed in the preparation of nitrogen heterocycles, such as quinolines or quinolones, naphthyridines, phenanthrolines, and others. When the catalyst is an acid-AC, its activity and selectivity depend on the acid strength and the porosity of the solid [146], although it should be noted that the most strong acidic active sites are not always the best catalysts [147]. Moreover, the Friedländer reaction can be catalyzed by alkaline catalysts, such as MgO supported on AC [148]. A third option to catalyze Friedländer reaction apart from acid or basic catalysis is the use of transition metals in the form of nanoparticles of Co(0) or Cu(0) doping an aerogel of carbon [149]. This material presents better activity if activated with water vapor.

3.1.1 Acid-catalyzed reactions

Heterogeneous acid catalysts have traditionally been applied in petrochemistry. However, they can also be applied in the synthesis of fine chemicals, with the advantage of greater easy use, less corrosive power, and shorter reaction time compared to conventional homogeneous catalysts.

An AC can exhibit acidic properties on its own if properly prepared. Thus, a sulfonated carbon has been used as a catalyst [142] in the alkylation of imidazole. *N*-alkyl-imidazoles and ionic liquids can be obtained by single alkylation and double alkylation, respectively. A similar carbon has been used for the acetylation (esterification) of alcohols [150], although this reaction can also be catalyzed with a basic activated carbon.

Acidic ACs have been used as catalyst in the epoxy opening reaction, which can be catalyzed by various types of catalyst, including Brönsted and Lewis acids. Using an acid carbon gel, alcoholysis and aminolysis of epoxides have been performed under mild conditions with excellent yields and selectivity [151, 152].

An important group of fine chemicals are the substances used as flavor and fragrance. An example of this is α-terpinyl methyl ether, which can be obtained by methoxylation of α-pinene using ACs as acid catalysts [153].

3.1.2 Alkaline-catalyzed reactions

Research on the use of basic catalysts began later, since they are not as useful in oil processing as acid catalysts. In recent decades, this difference has decreased, due to the large number of reactions of interest which can be catalyzed by bases.

Alkylation of *N*-heterocycles can be carried out with basic catalysts as well. In this case, only monosubstituted products can be obtained [154]. Other similar reactions are Michael addition [155] and the preparation of *N*-substituted-gamma-lactams [156].

Chalcones are other important intermediates for the synthesis of many pharmaceuticals. They are obtained generally by Claisen-Schmidt condensation, and this reaction can be catalyzed by acid or alkaline catalysts. The use of alternative energy sources such as microwave or ultrasound combined with the use of catalysts improves the activity performance [157].

From methanol and acetonitrile, propionitrile can be obtained in a reaction catalyzed by a sodium compound supported on AC. The activity depends on the used sodium salt [158].

3.1.3 Catalyzed reductions

In the most reduction reactions catalyzed by ACs (for example, in hydrogenation, hydrogenolysis, or hydrodechlorination), dihydrogen is used as reducer agent. Platinum group metals, cobalt and nickel, are frequently used as catalysts supported on ACs. Among other reactions, nitro to amino groups can be reduced with dihydrogen using nickel supported on AC as a catalyst [159], many of them commercially available. However, other reducing agents can be used. Such is the case of the transformation of nitroarenes to arylamines using a Co-promoted Mo carbide catalyst on ACs in the presence of hydrazine as a reducing agent. [160].

3.1.4 Catalyzed oxidations

The most studied catalyzed oxidation reaction with ACs is probably the degradation of volatile organic compounds to carbon dioxide and water. However, oxidative catalysis has also been used in the synthesis of fine chemicals. Most of the reactions for the synthesis of fine chemicals and pharmaceuticals involve one or more steps of oxidation of organic functional groups.

In recent years, great efforts have been made to achieve selective oxidation processes with low environmental impact.

Although in oxidation processes the catalysts supported on ACs are frequently considered as active phase, it has been demonstrated, by theoretical calculations, that the structure of the carbon material can play an important role in the oxidation processes, for example, activating H_2O_2 in the oxidation ofbenzene to phenol [161]. An example of this reaction is the use of a Fe catalyst supported on AC [162]. Phenol can be transformed, by oxidation, into highly oxidized species such as catechol and hydroquinone, using also a Fe catalyst supported on AC by impregnation [163], although the use of supported catalysts is not necessary. For example, an AC treated with nitric acid affords similar results in obtaining phenol from benzene and H_2O_2 [164].

3.1.5 Other catalyzed reactions

Some other reactions which cannot be framed within the above groups give rise to products of high added value. Such is the case of the synthesis of benzodiazepines, an important type of nitrogen heterocycles with relevant pharmacological activities. Although they can be synthesized using zirconia (sulfated or not), better yields are obtained [165] with greater selectivity if these catalysts are supported on ACs.

A widely used catalyst is palladium supported on AC. It has been reported in Heck's coupling [166] and Suzuki cross-coupling [167] reactions.

The use of a $PdCl_2$-$CuCl_2$ catalyst supported on ACs allows diethyl carbonate to be obtained from ethanol, NO and CO; better results are also achieved compared with other catalyst supports [168]. This substance is interesting, since it can be used as an additive in fuels with a low environmental impact.

3.2 Advanced porous carbon catalysts

In recent time, nanostructured carbons are being explored in many research fields, finding also a relevant place in the catalytic synthesis of high valuable compounds. CNTs and graphene are particularly in great expansion related to their special properties—electronic, optical, mechanical and chemical properties—as well as a type of highly porous carbons from emerging precursors—MOFs and COFs—with enhanced textural properties. In this context, this section summarizes some examples of advanced porous carbon-based materials involved in a great variety of catalytic transformations, particularly in fine chemical synthesis.

3.2.1 Carbon nanotubes

During the last years, CNTs have been proposed as catalytic supports of metal nanoparticles, mainly comprising transition metals, and applied in organic synthesis, especially in oxidation-reduction reactions, but also acting as acid-base catalysts.

Particularly, palladium catalysts have been widely used in a great variety of organic transformations, in which oxidation-reduction steps are involved. In this regard, for instance, Cano et al. [169] reported the rapid and easy microwave-induced synthesis of Pd/MWCNT hybrids, presenting high Pd loadings, in which the Pd(0) nanoparticles are homogenously distributed on the surface. Decomposition of Pd source by using microwave irradiation produces the formation of smaller metal nanoparticles compared to the conventional heating. These catalysts are not only active and recyclable in the C—C coupling reactions—Heck and Suzuki—upon microwave irradiation, but also in hydrogenation reactions of aryl ketones, α,β-unsaturated ketones, alkenes, cyclic di-, tri- and tetraenes, aromatic compounds, terpenes, and terpenoids, at atmospheric pressure, in high conversions and selectivities [170]. Pd-CNT nanohybrid also successfully catalyzes the Tsuji-Wacker oxidation of terminal olefins, in the presence of CuCl acting as cocatalyst, selectively giving the corresponding ketones in good-to-excellent yields (Scheme 1) [171]. Interestingly, new SACs consisting of thiolated MWCNTs, in which single atoms of Pd or Pt were detected, have been developed for efficiently catalyzing Suzuki coupling reaction [172] and intramolecular hydroalkoxylation reaction [173]. The authors demonstrated that the Pt-S-MWCNT was more active than other traditional Pt-catalysts, especially for the Suzuki coupling reaction between 4-iodoanisol and 4-methylbenzene boronic acid when using other halides such as -Br or -Cl. The observed reactivity could be related to the metal oxidation state, close to Pt(0) for Pt-S-MWCNT catalyst, the high electron density of the metal species favoring the cleavage of the leaving groups.

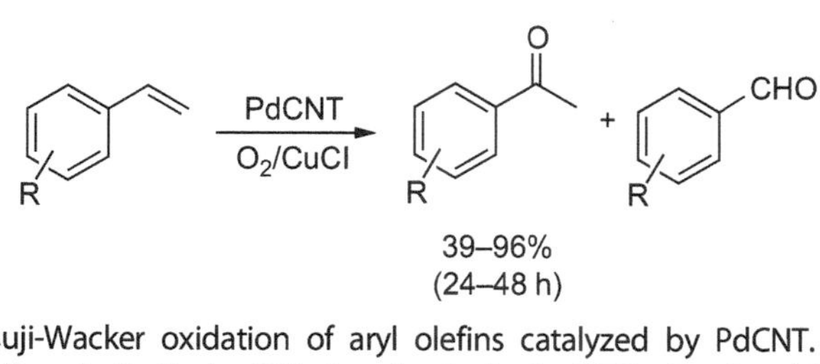

Scheme 1 Tsuji-Wacker oxidation of aryl olefins catalyzed by PdCNT. *(Adapted from S. Donck, E. Gravel, N. Shah, D.V. Jawale, E. Doris, I.N.N. Namboothiri, Tsuji-Wackier oxidation of terminal olefins using a palladium-carbon nanotube nanohybrid, ChemCatChem 7 (2015) 2318–2322, https://doi.org/10.1002/cctc.201500241 with permission of John Wiley and Sons.)*

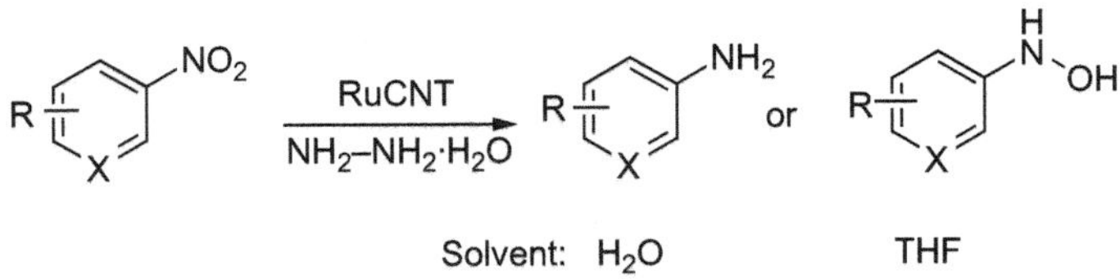

Scheme 2 Reduction of nitroarenes to aryl hydroxylamines or aryl amines catalyzed by Ru CNT. *(Adapted from D. V. Jawale, E. Gravel, C. Boudet, N. Shah, V. Geertsen, H. Li, I.N.N. Namboothiri, E. Doris, Selective conversion of nitroarenes using a carbon nanotube-ruthenium nanohybrid, Chem. Commun. 51 (2015) 1739–1742, https://doi.org/10.1039/c4cc09192b with permission of Royal Society of Chemistry.)*

In another approach reported by Jawale et al., Ru nanoparticles stabilized on CNTs were applied as active catalysts for the reduction of nitroarenes to arylhydroxylamines or arylamines, using hydrazine monohydrate as H_2 donor [174] (Scheme 2). Both products that are valuable compounds in the pharmaceutical and agrochemical industry were selectively obtained in excellent yields (up to 95%) as function of the used solvent. RuCNTs were synthetized through a layer-by-layer approach comprising the covering of MWCNTs with diacetylene nitrilotriacetic amphiphile (DANTA) yielding half-cylinders with nanoring-like structures, which are stabilized by photo-polymerization, followed by the formation of an additional layer by adding poly(diallyldimethylammonium chloride) (PDADMAC) and, finally, the anchoring of preformed Ru nanoparticles into the PDADMAC network.

CNT-based catalysts have also been reported for the synthesis of heterocyclic systems of interest. Such is the case of RuCNT nanohybrid involved in the selective catalytic reduction of nitrochalcones to quinoline *N*-oxides (Scheme 3), which were prepared in moderate to good yields following a reduction-condensation sequence [175]; quinoline *N*-oxides are part of many naturally occurring compounds and relevant synthetic drugs.

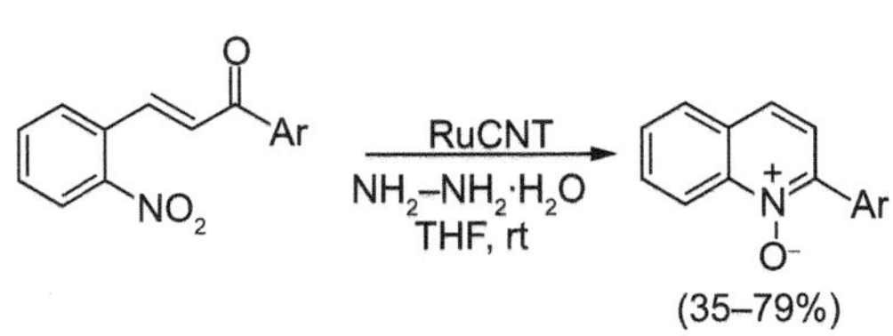

Scheme 3 Reduction of nitrochalcones to quinoline *N*-oxides catalyzed by RuCNT. *(Adapted from P. Basu, P. Prakash, E. Gravel, N. Shah, K. Bera, E. Doris, I.N.N. Namboothiri, Carbon nanotube-ruthenium hybrids for the partial reduction of 2-nitrochalcones: easy access to quinoline N-oxides, ChemCatChem 8 (2016) 1298–1302, https://doi.org/10.1002/cctc.201600042 with permission of John Wiley and Sons.)*

In the same context, the catalytic system formed by RhCNTs hybrid materials, constructed also by using a layer-by-layer strategy, in the presence of *tert*-butylcatechol as cocatalyst, has been reported in the dehydrogenation of *N*-heterocycles [176].

Much more recently, Godino-Ojer et al. [177] have developed new families of eco-sustainable catalysts comprising CoO nanoparticles on different carbon supports—AC, CNTs, or carbon aerogels—involved in the synthesis of nitrogen heterocyclic systems, through the Friedländer condensation (Scheme 4). The nature of the carbonaceous support and its surface chemistry together with the Co-salt precursor strongly affects the texture, dispersion, and location of the nanoparticles in the catalysts. In this case, Cobalt-supported $MWCNT_{ox}$ showed an intermediate catalytic behavior compared with mesoporous carbon aerogels and microporous ACs catalysts.

Gold chemistry has received a great interest in the last decades. AuCNTs have also been found to be interesting catalysts in diverse chemical transformations such as oxidation reaction of silanes [178] and alcohols [179] and in the synthesis of quinoxalines [180]. This type of catalysts has been also reported for selective hydrogenation of nitroaromatics, a great variety of functional groups being tolerated [181]. Au nanoparticles consisting of Au(0) and Au(I) species are deposited and firmly anchored over the pristine CNTs surface which confers a great stability to the hybrid material. The authors attributed the high chemoselectivity of the catalyst in the hydrogenation of nitroaromatics to a preferential orientation of nitro compound on the CNT surface by π-π stacking, whereas nitro functions and reducing agent, $NaBH_4$, interacting with the gold cationic species, the selection of the support being a crucial factor (Fig. 4).

Metal complexes covalently bound to CNTs have been also reported as useful heterogeneous catalysts [182]. For instance, iridium–*N*-heterocyclic carbene complexes anchored to oxidized surface of CNTs have been described as efficient catalysts for the transfer of hydrogenation reaction of

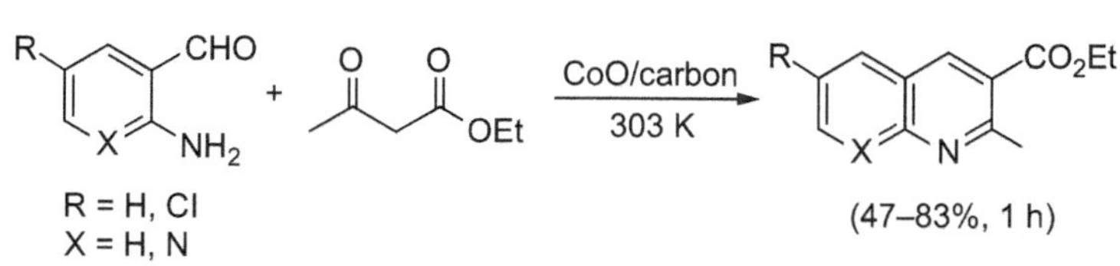

Scheme 4 Friedländer reaction catalyzed by CoO/CNT. *(Adapted from M. Godino-Ojer, A.J. López-Peinado, F.J. Maldonado-Hódar, E. Bailón-García, E. Pérez-Mayoral, Cobalt oxide-carbon nanocatalysts with highly enhanced catalytic performance for the green synthesis of nitrogen heterocycles through the Friedländer condensation, Dalt. Trans. 48 (2019) 5637–5648, https://doi.org/10.1039/c8dt04403a with permission of Royal Society of Chemistry.)*

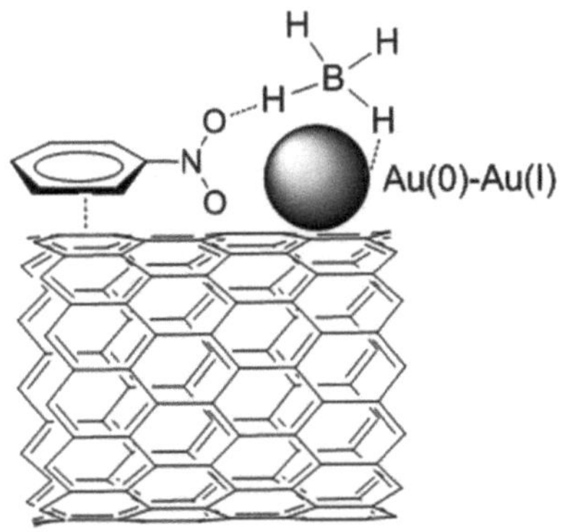

Fig. 4 Interactions between reactants and catalyst in the hydrogenation of nitroarenes catalyzed by AuCNTs. *(Adapted from M. Cano, P. Villuendas, A.M. Benito, E.P. Urriolabeitia, W.K. Maser, Carbon nanotube-supported gold nanoparticles as efficient catalyst for the selective hydrogenation of nitroaromatic derivatives to anilines, Mater. Today Commun. 3 (2015) 104–113. https://doi.org/10.1016/j.mtcomm.2015.02.002 with permission of Elsevier.)*

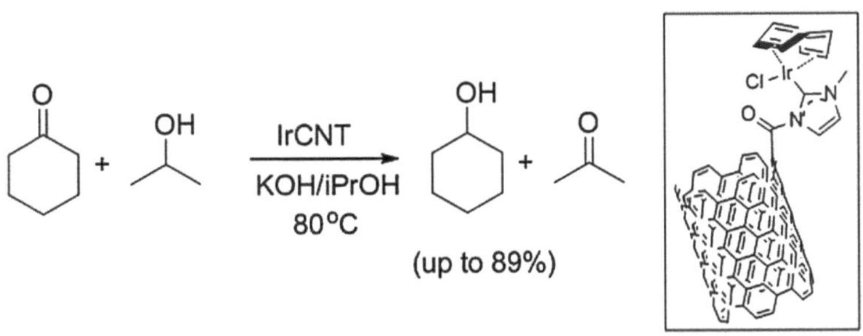

Scheme 5 Transfer hydrogenation reaction of cyclohexanone to cyclohexanol catalyzed by iridium–*N*-heterocyclic carbene complexes anchored to CNTs. *(Adapted from M. Blanco, P. Álvarez, C. Blanco, M.V. Jiménez, J.J. Pérez-Torrente, L.A. Oro, J. Blasco, V. Cuartero, R. Menéndez, Enhancing the hydrogen transfer catalytic activity of hybrid carbon nanotube-based NHC-iridium catalysts by increasing the oxidation degree of the nanosupport, Catal. Sci. Technol. 6 (2016) 5504–5514, https://doi.org/10.1039/c5cy01998b, with permission of Royal Society of Chemistry.)*

cyclohexanone in the presence of 2-propanol/KOH as hydrogen donor (Scheme 5). The authors claim that the catalytic activity is enhanced when increasing the concentration of oxygenated functions in surroundings of active centers probably because both are acting in cooperation. In other context, CNTs just modified with oxygenated functional groups have been also used for catalytic applications. As one example, poly(*p*-styrenesulfonic acid)-grafted MWCNTs (CNT-PSSA) and analogous carbon nanofibers (CNF-PSSA) and

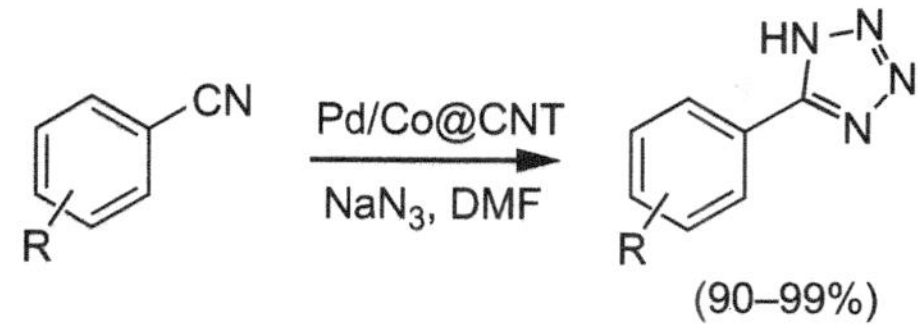

Scheme 6 Synthesis of substituted tetrazoles catalyzed by Pd/Co@CNT. *(Adapted from Y. Yıldız, L. Esirden, E. Erken, E. Demir, M. Kaya, F. Şen, Microwave (Mw)-assisted synthesis of 5-substituted 1H-tetrazoles via [3+2] cycloaddition catalyzed by Mw-Pd/Co nanoparticles decorated on multi-walled carbon nanotubes, ChemistrySelect 1 (2016) 1695–1701. https://doi.org/10.1002/slct.201600265 with permission of John Wiley and Sons.)*

the corresponding benzenesulfonic acid-grafted ones are active in the fructose transformation to hydroxymethyl furfural or ethyl levulinate [183].

In continuation with the synthesis of heterocyclic systems, bimetallic catalysts consisting of Pd/Co nanoparticles supported on MWCNTs have recently been reported in the microwave-assisted efficient synthesis of substituted tetrazoles through a [3+2] cycloaddition [184] (Scheme 6); a type of heterocyclic system with a huge number of applications such as stabilizers in photo imaging, chelating agents, herbicides, fungicides, and also as part of drugs with diverse biological activity. Metal nanoparticles in Pd/Co@CNT catalyst are homogeneously distributed forming Pd/Co alloy, in which the metallic species are as zero-valent transition metals.

3.2.2 Graphene

In the last decades, the interest for the development of new synthesis methods, characterization, and applicability of graphene is notably growing. Particularly in catalysis, covalently functionalized nano-graphene oxide [185] and graphene-supported metal/metal oxide nanohybrids have been also explored in fine chemical synthesis [186]. In this section, it is shown that some relevant interesting transformations have been recently reported.

Selective oxidation of alcohols and hydrogenation reactions are important transformations with a great utility in organic synthesis. There are huge number of methodologies reported for both types, most of them in the presence of metallic species. At this regard, Pd nanoparticles supported on graphene have been reported for oxidation of alcohol, the graphene support having an important role not only on the dispersion and size of metallic nanoparticles, but also in adsorption capability of both benzyl alcohols and molecular oxygen [187]. The authors demonstrated that the adsorption of the aromatic alcohols, via π,π-stacking, is notably superior for modified graphene catalyst than for other carbon supports such as CNTs or ACs.

This circumstance could be behind the considerably improved catalytic activity and selectivity. Much more recently, Li et al. [188] reported that atomically cobalt dispersed over nitrogen-doped graphene as highly active and selective (conversion and selectivity up to 87%) in the oxidation of benzyl alcohols. The catalyst was prepared from graphene oxide by impregnation with cobalt nitrate and subsequent calcination under NH_3, in which Co atoms were uniformly dispersed on the *N*-doped graphene where Co was found to be as Co(II) and Co(III). In other sense, and motivated by the searching of more sustainable methodologies, metal-free catalysts fabricated by grafting of $C_3N_3S_3$ polymer over graphene oxide, in different ratios, have been reported in the photocatalytic totally selective oxidation of benzyl alcohols under aerobic conditions [189]. The catalysts present a sandwich-like structure, both components interacting by π,π-stacking. An optimal amount of graphene oxide (0.3 wt%) is required for the best catalytic performance; remarkably, visible light is adsorbed by the polymer in the presence of superior graphene contents decreasing the active catalytic sites and diminishing the photocatalytic activity.

Another strategy, widely investigated for the functionalization of graphene, consists of the stabilization of catalytic active complexes over the surface through noncovalent interactions, using metal complexes containing polyaromatic functions which can interact with graphene via π,π-stacking interactions. In this regard, carboxylic acids have been selectively prepared by dehydrogenation of alcohols, in aqueous medium, in the presence of a recyclable catalytic system comprising reduced graphene oxide with homogenously distributed Ru molecular complexes (Scheme 7) [190].

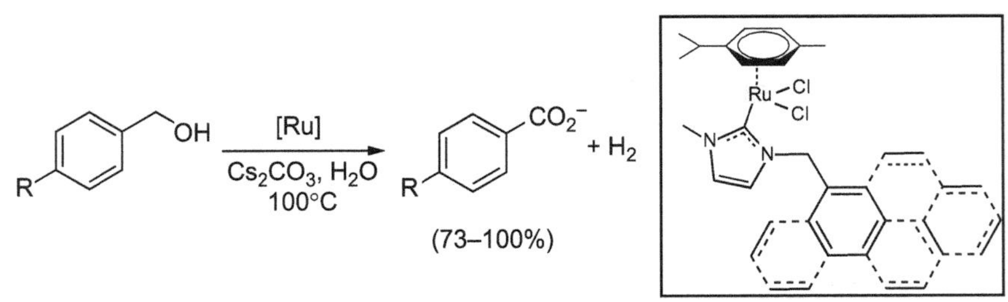

Scheme 7 Dehydrogenation of benzyl alcohols to benzoic acids, in water medium, catalyzed by Ru complexes supported over graphene oxide by noncovalent interactions. *(Adapted from D. Ventura-Espinosa, C. Vicent, M. Baya, J.A. Mata, Ruthenium molecular complexes immobilized on graphene as active catalysts for the synthesis of carboxylic acids from alcohol dehydrogenation, Catal. Sci. Technol. 6 (2016) 8024–8035, https://doi.org/10.1039/c6cy01455k with permission of Royal Society of Chemistry.)*

In another context, a well-dispersed Co_3O_4 nanoparticles supported on magnetic *N*-doped graphene catalyst, RGO/Fe_3O_4@$C_{-N}Co_3O_4$, has been developed for the epoxidations of styrene and ciclooctene, under aerobic conditions [191]. This catalyst represents a more sustainable alternative presenting the advantage of being separated of the reaction mixture by magnetic methods yielding the desired products in high conversions (up to 83%) and selectivity (up to 82%).

Selective reduction of C—C or C—O multiple bonds is traditionally catalyzed by precious metals such as Pd, Pt, and Ru. At this respect, Pd [192] and Ru nanoparticles supported on graphene oxide [193], among others, have been reported for the catalytic hydrogenation of a variety of alkenes and also ketones; graphene supporting Pt nanoparticles even involved in the selective hydrogenation of cinnamaldehyde to cinnamal alcohol [194]. Interestingly, nanodiamond-graphene (ND@G) has recently been used for the synthesis of Pd SACs, Pd1/ND@G, as catalyst for the selective acetylene hydrogenation to ethylene with enhanced catalytic performance (100% of conversion with high selectivity, 90%) [195, 196]. Pd species in Pd1/ND@G are atomically dispersed and strongly anchored to the support through Pd—C bonds in the graphene defects. The developed catalyst is able to selectively transform traces of acetylene contained acetylene/ethylene mixtures. In the same context, Xi et al. [197] developed a strategy to prepare a Pd SACs, in which Pd single atoms are located at the confined interface between graphene oxide and AC. This material presents an excellent stability and notably enhanced catalytic activity for both *p*-nitrophenol catalytic hydrogenation and Suzuki coupling reaction.

Other metal nanoparticles, among them Au nanoparticles, [198, 199] and bimetallic catalysts supported on graphene [200, 201] have been found to be efficient catalysts for the reduction of nitroarenes.

In other approach, Jagadeesh et al. [202] prepared different and highly stable *N*-doped graphene-activated Fe_2O_3-based nanocatalysts, by pyrolysis of different iron-amine complexes—1,10-phenanthroline, bipyridine, terpyridine, and pyridinebisbenzimidazole as ligands, highly efficient in the transfer hydrogenation reaction of nitro aromatics to anilines. The methodology implies the use of HCO_2H-Et_3N as H_2 donor and tolerates a great variety of functional groups, giving the corresponding anilines and related compounds with excellent yields.

As mentioned above, supported metal nanoparticles on graphene catalyzing coupling reactions have been reported. Among them, Pd [203], Au [204], and Ni hybrids [205] were applied in the Ullmann homocoupling

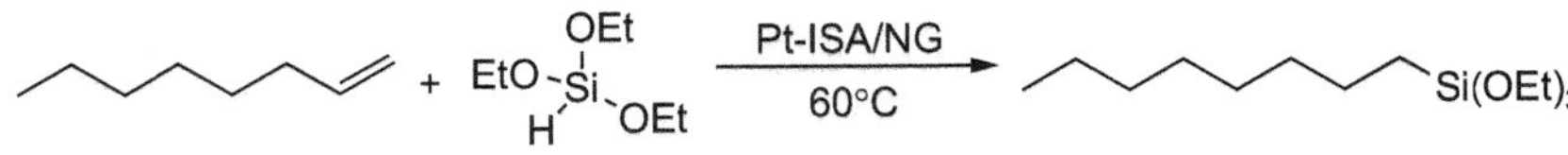

Scheme 8 Hydrosilylation of terminal alkenes catalyzed by Pt-ISA/NG. *(Adapted from Y. Zhu, T. Cao, C. Cao, J. Luo, W. Chen, L. Zheng, J. Dong, J. Zhang, Y. Han, Z. Li, C. Chen, Q. Peng, D. Wang, Y. Li, One-pot pyrolysis to N-doped graphene with high-density Pt single atomic sites as heterogeneous catalyst for alkene hydrosilylation, ACS Catal. 8 (2018) 10004–10011, https://doi.org/10.1021/acscatal.8b02624 with permission of American Chemical Society.)*

of aryl halides and Kumada–Corriu cross-coupling of aryl halides with phenylmagnesium chloride, respectively.

In other context, alkylsilanes are very useful substrates for a wide range of industrial applications. Very recently, Zhu et al. [206] have reported a new SAC, presenting high Pt loading and consisting of single atoms of Pt tethered on *N*-doped graphene, Pt-ISA/NG, highly efficient in the hydrosilylation of numerous terminal alkenes showing excellent catalytic performance and selectivity (Scheme 8). This material was prepared by pyrolysis of EDTA-Pt complex and Na_2CO_3, where Pt species were reduced by carbon matrix and bound and stabilized by nitrogen atoms in Pt–N4 structure.

3.2.3 Carbon material from emerging precursors

Direct pyrolysis of MOFs or COFs represents an easy and practical strategy which can be applied for the synthesis of highly dispersed heteroatom-doped porous carbons and new SACs, showing a highly developed porosity and very recently applied to the fine chemical synthesis. Catalytic behavior of porous carbons from both emerging precursors is an interesting research field still at the infancy. Only a small and reduced number of organic transformations have been explored such as oxidation and hydrogenation reactions and some condensation reactions [140].

3.2.3.1 Carbon material from MOFs

Oxidation reactions Highly porous carbon from MOFs containing or not metallic species has been reported in oxidation reactions such as oxidation of alkanes and olefins, alcohols to aldehydes or ketones, carboxylic acids, and oxidation of carbonyl compounds. In this context, highly graphitized and recyclable metal-free *N*-doped nanoporous carbons prepared from ZIF-67, after Co removal with acids, have been reported for the aerobic oxidation of cyclohexane, toluene, and oxidative coupling of benzyl amines [207] (Scheme 9). In this case, the graphitic nitrogen species are key in

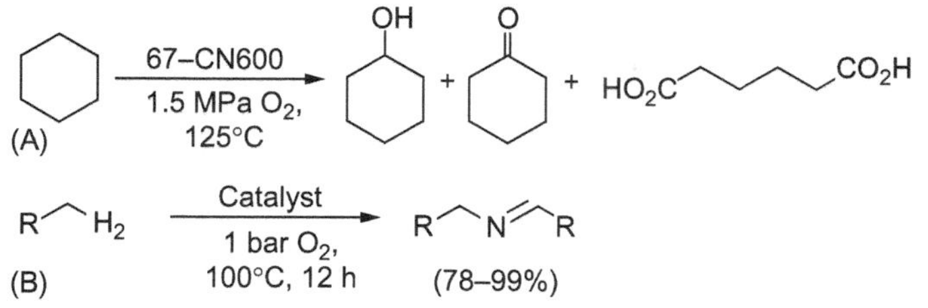

Scheme 9 Metal-free *N*-doped nanoporous carbons in the aerobic oxidation of a) cyclohexane and b) synthesis of benzyl imines. *(Adapted from X. Wang, Y. Li, Nanoporous carbons derived from MOFs as metal-free catalysts for selective aerobic oxidations, J. Mater. Chem. A 4 (2016) 5247–5257, https://doi.org/10.1039/c6ta00324a with permission of Royal Society of Chemistry.)*

the C-H oxidation reaction observing an enhancement of alkane conversion attributed to the homogeneous distribution of those and the developed mesoporosity. Nitrogen-cobalt catalysts also from ZIF-67 containing Co(0) nanoparticles, which favor the magnetically separation of the catalyst, have been used in the selective oxidation of styrene, under aerobic conditions, yielding styrene oxide in high conversions (76–91%) and excellent selectivity (up to 81%) [208].

Recently, Pendem et al. [209] reported Co_3O_4 nanocrystals embedded into *N*-doped graphitic carbon hybrid materials, Co_3O_4@GNC, involved in the selective aerobic oxidation of aryl alkanes showing high catalytic performance. These hybrids were prepared from ZIF-9, Co-based zeolitic imidazolate framework, and graphene oxide by pyrolysis. Enhanced catalytic activity is attributed not only to the combined effect of the presence of Co−N−C active sites and $\pi-\pi$ interactions between the graphene and ethylbenzene, but also to porosity of the hybrids.

Facile synthetized magnetic Cobalt-composites from $[Co_9(btc)_6(tpt)_2(H_2O)_{15}]$, where btc is 1,3,5-benzenetricarboxylate and tpt is 2,4,6-tris (4-pyridyl)-1,3,5-triazine), with Co contents ranging 30–53%, resulted in high selectivity in the aerobic oxidation of alcohols to ketones, in aqueous medium and in the absence of any base, for great variety of alcohols [210]. The strong basicity of the sample is probably behind the enhanced catalytic performance. Interestingly, Co@C-N catalyst was also reported in the preparation of amides, by oxidative amidation of different aldehydes with formamides, through a radical route (Scheme 10) [211].

In the same context, Cu@C, a mesoporous carbon showing Cu(0)-supported nanoparticles, obtained by direct pyrolysis of microporous $Cu_3(BTC)_2$, efficiently catalyzed the aerobic benzyl alcohol oxidation to

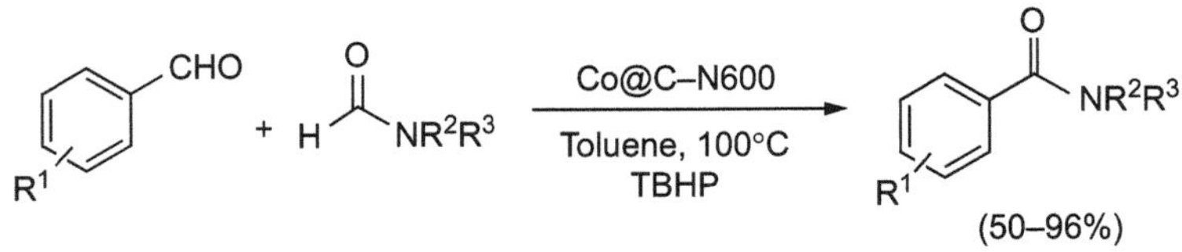

Scheme 10 Oxidative amidation of benzaldehydes with formamides catalyzed by Co@C-N. *(Adapted from C. Bai, X. Yao, Y. Li, Easy access to amides through aldehydic C–H bond functionalization catalyzed by heterogeneous Co-based catalysts, ACS Catal. 5 (2015) 884–891, https://doi.org/10.1021/cs501822r with permission of American Chemical Society.)*

benzaldehyde, in CH_3CN as solvent [212]. The catalytic system, Cu@C / TEMPO/NMI, is composed of the carbon catalyst, 2,2,6,6-tetramethylpiperidine-*N*-oxyl (TEMPO), in the presence of an organic base; in this case, *N*-methylimidazole (NMI).

Co-based porous carbon catalysts, Co@C-N, prepared by pyrolysis of ZIF-67, involved in the alcohol oxidation to esters have been also reported [213] (Scheme 11). It seems that the gradual decomposition of methyl imidazole as organic ligands in MOF is key in the formation of Co nanoparticles embedded in carbon-nitrogen matrix. The catalytic performance of Co@C-N was checked in the oxidative cross-esterification of *p*-nitrobenzyl alcohol and methanol, under aerobic ambient conditions, and in the absence of any base or additive, observing a decrease of catalytic activity for the samples pyrolyzed at higher temperatures. The reported methodology was also applied for not only the oxidation of other aromatic and aliphatic alcohols, but also for the preparation of lactones when using the appropriate diols.

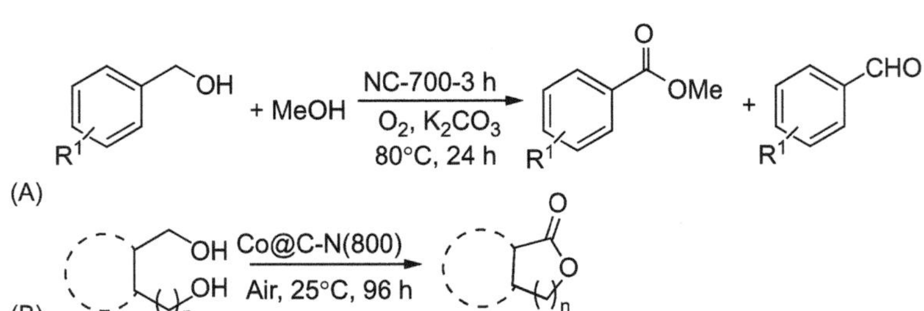

Scheme 11 (A) Oxidation of benzyl alcohols to esters and (B) Aerobic diol oxidation to lactones catalyzed by porous carbons catalysts prepared from ZIF-67. *(Adapted from W. Zhong, H. Liu, C. Bai, S. Liao, Y. Li, Base-free oxidation of alcohols to esters at room temperature and atmospheric conditions using nanoscale Co-based catalysts, ACS Catal. 5 (2015) 1850–1856, https://doi.org/10.1021/cs502101c with permission of American Chemical Society.)*

Hydrogenation reactions This type of porous carbons has been also explored in hydrogenation reactions such as hydrogenation of unsaturated compounds, heterocyclic systems, nitroaromatics, and nitriles, among others. In this sense, different porous carbons from structurally distinct MOFs have recently been reported for the chemoselective hydrogenation of nitroarenes. Thus, magnetically separable catalyst containing γ-Fe_2O_3 nanoparticles dispersed on porous carbon was prepared from Fe-MIL-88A, $Fe_3O(FA)_3(H_2O)_2(NO_3)$ (FA = fumaric acid), by direct pyrolysis at 500°C [214] and successfully applied for the efficient and selective transfer hydrogenation of nitro compounds to anilines, using hydrazine as H_2 donor. Niu et al. [215] developed a Cu/Cu_2O porous carbon from HKUST-1 ($Cu_3(BTC)_2$) and additional carbon source such as phenol-formaldehyde involved in the selective synthesis of 4-aminophenol in the presence of $NaBH_4$. Tang et al. [216] developed microporous Ni-based nanocarbons, Ni@C, by thermal treatment of a Ni-based MOF, in which Ni(II) is coordinated to *p*-benzenedicarboxylic acid ligand, with monodispersed metallic nanoparticles highly crystallized and with demonstrated catalytic activity in the synthesis of amines by hydrogenation of nitro compounds, with excellent yields (up to 93%) and selectivity superior to 90% with a great variety of functional groups' tolerance.

As mentioned above, Co porous carbons have been widely studied not only in the oxidation reactions, but also in hydrogenation reactions. Thus, Co@C-N catalyst with basic properties, prepared by pyrolyzing [Co(bdc)(ted)0.5]·2DMF·0.2H_2O, where bdc is 1,4-benzenedicarboxylic acid and ted is triethylenediamine, resulted in high activity and selectivity in the transfer hydrogenation of unsaturated compounds—hydrocarbons, carbonyl components, and nitro and cyano compounds—using isopropanol as hydrogen donor, at 150°C, without the presence of any base [217]. While the MOF precursor is a microporous sample, Co@C-N catalysts present a developed mesoporosity, with Co nanoparticles homogeneously dispersed (6–17 nm) mainly depending on the pyrolysis temperature. In the same context, Liu et al. [218] used a basic mesoporous nanocarbon with supported Co(0) nanoparticles, from [Co(TPA)(ted)0.5] (where TPA is terephthalic acid and ted is 1,4-diazabicyclooctane), for the selective hydrogenation of a variety of substituted cinnamaldehydes, in alcohols as solvents as well as the hydrogenation of acetophenones, benzaldehydes, benzylimines, nitroaromatics, and benzonitriles. Well-dispersed coexisted Co(0)-Co(II) species on *N*-doped graphitic structures have been applied in the synthesis of amines from a great variety of benzonitriles in the presence of isopropanol

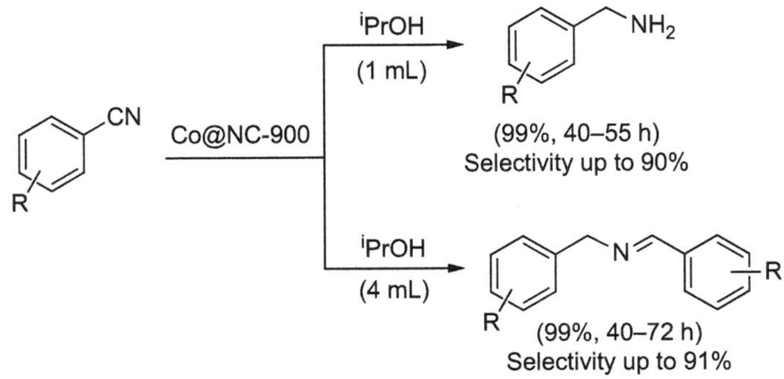

Scheme 12 Transfer hydrogenation of benzonitriles in the presence of isopropanol. *(Adapted from J. Long, K. Shen, Y. Li, Bifunctional N-doped Co@C catalysts for base-free transfer hydrogenations of nitriles: controllable selectivity to primary amines vs imines, ACS Catal. 7 (2017) 275–284, https://doi.org/10.1021/acscatal.6b02327 with permission of American Chemical Society.)*

as hydrogen donor and solvent, without any base, at 80°C [219] (Scheme 12). It is noteworthy that the isopropanol amount has an important effect on the observed selectivity.

Bimetallic *N*-doped porous carbons, Co@Pd/NC, prepared from ZIF-67 and $Pd(NO_3)_2$ as Co-MOF and Pd precursors, respectively, were reported by Shen et al. [220]. During the thermal treatment, Co and Pd were transformed into zero-valent metal nanoparticles leading to porous carbons with different texture and nitrogen loading, as function of the carbonization temperature, in which Pd atoms are highly exposed on Co nanoparticles. The investigated catalyst presents an enhanced catalytic behavior in hydrogenation of nitrobenzene compared to the classical Pd-catalysts. Yang et al. [221] synthetized monometallic or bimetallic nanoparticles confined within hollow nitrogen-doped porous carbon capsules by using a more sophisticated strategy, highly efficient in the nitroarenes hydrogenation, under mild conditions. In the case of the monometallic porous carbons, the synthetic approach consisted in the pyrolysis of ZIF-8@M-TA (where M is Co or Ni), prepared from ZIF-8@K-TA material in which potassium tannic acid (K-TA) coated the ZIF-8 surface and Co(II) or Ni(II) are introduced by replacement of K^+ ions. For the bimetallic catalysts, the ZIF surface was first decorated with platinum nanoparticles and subsequently coated with the tannic acid layer able to capture the cobalt or nickel species. The thermal heating implies the reduction of metal ions, as Co(0) or Ni(0), able to migrate to the Pt species, producing bimetallic alloyed nanoparticles hosted on the carbon walls of the capsules.

Different transition metal alloy nanoparticles embedded in *N*-doped carbon mesoporous carbons—Co-Ni@C-N, Co-Cu@C-N, Ni-Cu@C-N, and Co-Ni-Cu@C-N—were synthetized by carbonization of symmetric heterometallic M-M'-MOF, M-M'(1,4-BDC)$_2$(DABCO)·4DMF·1/2H$_2$O (where M/M' = Co, Ni, Cu) [222]. These catalysts resulted in high activity in the transfer hydrogenation of different nitriles, in the presence of isopropanol, leading to the formation of the corresponding benzyl imines with excellent conversions and selectivities (up to 98% and 80%, respectively). Much more recently, Murugesan et al. [223] have developed a new methodology for the selective hydrogenation of benzonitriles, cyano-heterocycles, and nitro compounds to primary amines in the presence of nanostructured catalysts, consisting of Co_3O_4 and Co(0) nanoparticles encapsulated within graphene shells, prepared from Cobalt-terephthalic acid MOF.

CuNi bimetallic nanoparticles embedded in mesoporous porous carbons, CuNi@C, synthetized by carbonization of $Cu_3(BTC)_2$ impregnated with nickel, in which Cu(0) and Ni(0) nanoparticles are homogeneously dispersed (15 nm) on the carbon surface, showed also high catalytic activity in the transformation of furfural to cyclopentanone as major reaction product [224].

Interestingly, Wang et al. [225] developed a new SAC from Zr-based MOF, which is highly active and selective in the hydrogenation of quinolines. The MOF precursors were UiO-66 and its analog UiO-66 functionalized with ammonium, UiO-66–NH$_2$, which by treatment with $RuCl_3$, subsequent pyrolysis and, finally, Zr etching in HF solutions led to Ru SAs/N–C and Ru NCs/C samples, respectively, both exhibiting high surface areas. While Ru(III) single atoms were visualized in Ru SAs/N–C sample, favored by the strong coordination of Ru to amine functions in UiO-66–NH$_2$, the absence of such amine groups UiO-66 produces the formation of Ru aggregates. High dispersion of Ru in Ru SAs/N–C is probably behind both excellent conversion and selectivity (99%).

Other interesting reactions Although poorly, some interesting transformations such as alcohol dehydration [226], aerobic transformation of benzaldehydes to benzonitriles [227], acetylene hydrochlorination [228, 229], Knoevenagel condensation [230], and multicomponent synthesis of *N*-Sulfonyl amidines [231] have been also reported.

It deserves to be highlighted that the Knoevenagel condensation, one of the most useful condensation reactions in organic synthesis, is promoted by a

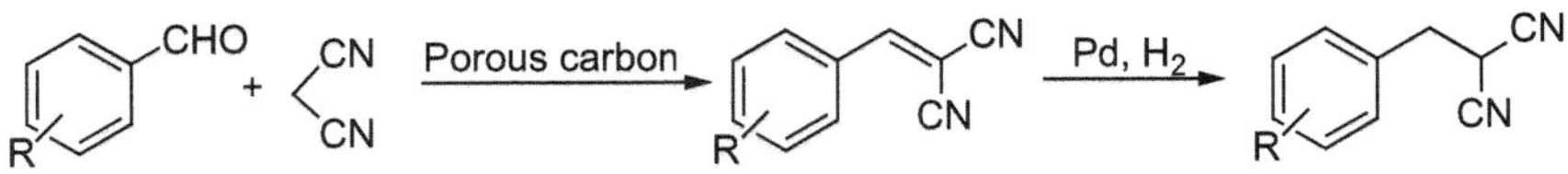

Scheme 13 Knoevenagel condensation of benzaldehyde and malononitrile followed by hydrogenation in the presence of porous carbons. *(Adapted from X. Li, B. Zhang, Y. Fang, W. Sun, Z. Qi, Y. Pei, S. Qi, P. Yuan, X. Luan, T.W. Goh, W. Huang, Metal-organic-framework-derived carbons: applications as solid-base catalyst and support for Pd nanoparticles in tandem catalysis, Chem. A Eur. J. 23 (2017) 4266–4270, https://doi.org/10.1002/chem.201605852 with permission of John Wiley and Sons.)*

basic *N*-doped porous carbon prepared by carbonization of microporous aluminum-based MOF-253 and subsequent Al(III) removal by acid treatment. These *N*-doped porous carbons within highly dispersed ultrafine Pd(0) nanoparticles were assayed for tandem reactions through Knoevenagel condensation and hydrogenation of C—C double bond sequence, yielding to the corresponding nitriles in excellent yields with total selectivity (99%) (Scheme 13) [230].

3.2.3.2 Carbon material from COFs

Porous carbons from COFs applied in the synthesis of high valuable compounds have been barely investigated. At this regard, recyclable metal-free B,*N*-doped hierarchical hollow carbon spheres involved in the synthesis of nitro aromatics, in the presence of hydrazine in ethanol as solvent, have been investigated [232]. In the same context, Hu et al. [233] synthesized mesoporous carbon materials with high N loadings active in the reduction of nitroarenes reaction as well as condensation reactions such as Knoevenagel condensation. N species seem to be the predominant catalytic active sites being responsible for the high catalytic performance.

Finally, COFs doped with Pd have been used as precursor to prepare Pd nanoparticles encapsulated into *N*-doped hollow carbon spheres (NHCS), exhibiting an enhanced activity and selectivity in the selective reduction of nitrobenzene or cinnamyl alcohol oxidation [234].

4 Concluding remarks

Development of new advanced materials with great potential is an extremely important issue for environmental sustainability in terms of energy storage and generation, water treatment, NOx removal, and chemical production, among other disciplines. At this respect, a large variety of porous solids have been investigated—zeolites, clays, MOFs, carbonaceous materials, etc.

Among all of them, carbon materials, particularly ACs, are highlighted because of their industrial applications. In this context, carbon-based catalysts emerge as interesting economic and environmental alternatives to other porous solid catalysts for the preparation of highly valuable compounds [23, 235].

In the last decades, the accumulated knowledge in material science and engineering has allowed the development of new synthetic strategies focused on the improvement of the physico-chemical properties of modern and advanced functional nanostructured materials, among which the carbon materials occupy a preferential place. For instance, CNTs and graphene have been extensively explored mainly due to their multiple application fields and utility for the development of catalytic materials in fine chemical synthesis.

Much more recently, a new generation of highly porous carbons with ultra-high surface areas, large pore volumes and enhanced porosity, plenty metal/organic species or doping heteroatoms, and functional surface, from direct pyrolysis of MOFs and COFs, has been reported. Both carbon precursors are found to be green porous materials for a more sustainable future, as long as its availability and manufacture in large-scale production are feasible. Sooner, it will be possible, since in the last years a great molecular diversity of MOFs, due to the large selection of organic chelating ligands and metal ions, has been developed. However, there are still many efforts to be made in the use of biocompatible building blocks, benign reaction media including water or strategies under solvent-less conditions [236], saving energy costs, and other aspects which allow the manufacture continuously, all of them considered future environmental challenges. In this context, these advanced porous carbons emerge as versatile and more robust catalytic systems useful as catalysts and supports with demonstrated efficiency for the green catalytic synthesis of valuable compounds and alternative to their MOFs precursors [140, 237].

Dispersion and location of metal nanoparticles and metal loadings onto the catalyst surface are key factors conditioning the accessibility and diffusion of reactants and products toward the catalytic sites, therefore playing an important role in the catalytic performance. Particularly, the small-size limit of the metal particles represents an important issue to be considered. This concept is already implemented in the development of the atomically dispersed carbon catalysts, in which highly dispersed metal atoms greatly influence the adsorptive properties, providing a marked distinguished reactivity compared to the conventional metallic nanoparticles. The use of SACs then represents a new horizon in heterogeneous catalysis probably to be still discovered [238]. In this sense, catalytic applications of SACs are interestingly in rapid expansion. In fact, SACs derived from CNTs and graphene with demonstrated efficiency have been developed for the synthesis of fine chemicals. Much more recently, MOFs and metal-doped COFs, in which the active sites are precisely located, emerge as interesting opportunities for development of SACs and applied in the synthesis of valuable compounds.

Although the type and number of catalytic reactions, in which these advanced porous carbons are involved, are still notably reduced, mostly limited to oxidation-reduction reactions, predictably in the nearest future, this class of porous carbon-based catalysts are found to be excellent candidates catalyzing the synthesis of relevant heterocyclic scaffolds and others interesting substrates through cascade reactions.

Acknowledgments

This work has been supported by Spanish Ministry (CTM2014-56668-R project) and the Associate Laboratory for Green Chemistry-LAQV which is financed by national funds from FCT/MCTES (UID/QUI/50006/2019). Ines Matos thanks FCT for the Investigator FCT contract IF/01242/2014/CP1224/CT0008.

References

[1] M.M. Titirici, R.J. White, N. Brun, V.L. Budarin, D.S. Su, F. Del Monte, J.H. Clark, M.J. MacLachlan, Sustainable carbon materials, Chem. Soc. Rev. 44 (2015) 250–290, https://doi.org/10.1039/c4cs00232f.

[2] S. Wong, N. Ngadi, I.M. Inuwa, O. Hassan, Recent advances in applications of activated carbon from biowaste for wastewater treatment: a short review, J. Clean. Prod. 175 (2018) 361–375, https://doi.org/10.1016/j.jclepro.2017.12.059.

[3] S.B.I. Anthonysamy, S.B. Afandi, M. Khavarian, A.R. Bin Mohamed, A review of carbon-based and non-carbon-based catalyst supports for the selective catalytic reduction of nitric oxide, Beilstein J. Nanotechnol. 9 (2018) 740–761, https://doi.org/10.3762/bjnano.9.68.

[4] M.M. Titirici, A. Thomas, S.H. Yu, J.O. Müller, M. Antonietti, A direct synthesis of mesoporous carbons with bicontinuous pore morphology from crude plant material by hydrothermal carbonization, Chem. Mater. 19 (2007) 4205–4212, https://doi.org/10.1021/cm0707408.

[5] P. Serp, J.L. Figueiredo, Wiley InterScience (Online Service), Carbon Materials for Catalysis, John Wiley & Sons, 2009.

[6] J.L. Figueiredo, Application of nanocarbon materials to catalysis, in: M.H. van de Voorde, B. Sels (Eds.), Nanotechnology in Catalysis: Application in the Chemical Industry, Energy Development, and Environment Protection, Wiley-VCH Verlag GmbH & Co. KGaA, 2017, pp. 37–56. https://www.wiley.com/en-us/Nanotechnology+in+Catalysis%3A+Applications+in+the+Chemical+Industry%2C+Energy+Development%2C+and+Environment+Protection%2C+3+Volumes-p-9783527339143 (accessed June 5, 2019).

[7] P.T. Anastas, Meeting the challenges to sustainability through green chemistry, in: Green Chemistry, The Royal Society of Chemistry, 2003, p. G29, https://doi.org/10.1039/b211620k.
[8] J.L. Tucker, M.M. Faul, Industrial research: drug companies must adopt green chemistry, Nature 534 (2016) 27–29, https://doi.org/10.1038/534027a.
[9] E. Lam, J.H.T. Luong, Carbon materials as catalyst supports and catalysts in the transformation of biomass to fuels and chemicals, ACS Catal. 4 (2014) 3393–3410, https://doi.org/10.1021/cs5008393.
[10] E. Pérez-Mayoral, E. Soriano, F.J. Martín-Aranda, R.M. Maldonado-Hódar, Mesoporous catalytic materials and fine chemistry, in: M. Aliofkhazrae (Ed.), Comprehensive Guide for Mesoporous Materials, Synthesis and Characterisation, vol. 1, Nova Science Publishers, Inc., 2015, pp. 83–118.
[11] E. Pérez-Mayoral, V. Calvino-Casilda, M. Godino, A.J. López-Peinado, R.M. Martín-Aranda, Porous catalytic systems in the synthesis of bioactive heterocycles and related compounds, in: G. Brahmachari (Ed.), Green Synthetic Approaches for Biologically Relevant Heterocycles, Elsevier, 2015, pp. 378–403, https://doi.org/10.1016/B978-0-12-800070-0.00001-3.
[12] S. De, A.M. Balu, J.C. van der Waal, R. Luque, Biomass-derived porous carbon materials: synthesis and catalytic applications, ChemCatChem 7 (2015) 1608–1629, https://doi.org/10.1002/cctc.201500081.
[13] M. Enterría, J.L. Figueiredo, Nanostructured mesoporous carbons: tuning texture and surface chemistry, Carbon 108 (2016) 79–102, https://doi.org/10.1016/j.carbon.2016.06.108.
[14] N. Rey-Raap, A. Arenillas, J.A. Menéndez, Carbon gels and their applications: a review of patents, in: P. Bettotti (Ed.), Submicron Porous Materials, Springer International Publishing, Cham, 2017, pp. 25–52, https://doi.org/10.1007/978-3-319-53035-2_2.
[15] M.L. Rojas-Cervantes, Some strategies to lower the production cost of carbon gels, J. Mater. Sci. 50 (2015) 1017–1040, https://doi.org/10.1007/s10853-014-8617-1.
[16] F.J. Maldonado-Hódar, Advances in the development of nanostructured catalysts based on carbon gels, Catal. Today 218–219 (2013) 43–50, https://doi.org/10.1016/j.cattod.2013.06.005.
[17] B. Liu, H. Shioyama, T. Akita, Q. Xu, Metal-organic framework as a template for porous carbon synthesis, J. Am. Chem. Soc. 130 (2008) 5390–5391, https://doi.org/10.1021/ja7106146.
[18] J.K. Sun, Q. Xu, Functional materials derived from open framework templates/precursors: synthesis and applications, Energy Environ. Sci. 7 (2014) 2071–2100, https://doi.org/10.1039/c4ee00517a.
[19] S. Dang, Q.L. Zhu, Q. Xu, Nanomaterials derived from metal-organic frameworks, Nat. Rev. Mater. 3 (2017) 17075, https://doi.org/10.1038/natrevmats.2017.75.
[20] V.K. Thakur, M.K. Thakur, Chemical functionalization of carbon nanomaterials: properties and applications, in: Chemical Functionalization of Carbon Nanomaterials, 2015, p. 571, https://doi.org/10.1201/b18724-25.
[21] M. Bilal, H.M.N. Iqbal, Chemical, physical, and biological coordination: an interplay between materials and enzymes as potential platforms for immobilization, Coord. Chem. Rev. 388 (2019) 1–23, https://doi.org/10.1016/j.ccr.2019.02.024.
[22] W. Feng, P. Ji, Enzymes immobilized on carbon nanotubes, Biotechnol. Adv. 29 (2011) 889–895, https://doi.org/10.1016/j.biotechadv.2011.07.007.
[23] E. Pérez-Mayoral, V. Calvino-Casilda, E. Soriano, Metal-supported carbon-based materials: opportunities and challenges in the synthesis of valuable products, Catal. Sci. Technol. 6 (2016) 1265–1291, https://doi.org/10.1039/c5cy01437a.
[24] J.L. Figueiredo, Functionalization of porous carbons for catalytic applications, J. Mater. Chem. A 1 (2013) 9351, https://doi.org/10.1039/c3ta10876g.

[25] C. Rivera-Cárcamo, P. Serp, Single atom catalysts on carbon-based materials, ChemCatChem 10 (2018) 5058–5091, https://doi.org/10.1002/cctc.201801174.
[26] K.P. De Jong, J.W. Geus, Carbon nanofibers: catalytic synthesis and applications, Catal. Rev. Sci. Eng. 42 (2000) 481–510, https://doi.org/10.1081/CR-100101954.
[27] P. Serp, E. Castillejos, Catalysis in carbon nanotubes, ChemCatChem 2 (2010) 41–47, https://doi.org/10.1002/cctc.200900283.
[28] K. Nakada, A. Ishii, DFT calculation for adatom adsorption on graphene, in: J.R. Gong (Ed.), Graphene Simulation, InTech, 2011, p. 376, https://doi.org/10.5772/854.
[29] B. Li, X.Y. Sun, D. Su, Calibration of the basic strength of the nitrogen groups on the nanostructured carbon materials, Phys. Chem. Chem. Phys. 17 (2015) 6691–6694, https://doi.org/10.1039/c4cp05765a.
[30] Q. Wei, X. Tong, G. Zhang, J. Qiao, Q. Gong, S. Sun, Nitrogen-doped carbon nanotube and graphene materials for oxygen reduction reactions, Catalysts 5 (2015) 1574–1602, https://doi.org/10.3390/catal5031574.
[31] T. Sun, L. Xu, D. Wang, Y. Li, Metal organic frameworks derived single atom catalysts for electrocatalytic energy conversion, Nano Res. 12 (2019) 2067–2080, https://doi.org/10.1007/s12274-019-2345-4.
[32] H. Marsh, F.R. Reinoso, Activated Carbon, Elsevier, 2006.
[33] K.S.W. Sing, D.H. Everett, R.A.W. Haul, L. Moscou, R.A. Pieroti, J. Rouquerol, T. Siemieniewska, Reporting physisorption data for gas/solid systems with special reference to the determination of surface area and porosity (Recommendations 1984), Pure Appl. Chem. 57 (1985) 603–619, https://doi.org/10.1351/pac198557040603.
[34] M.I. Din, S. Ashraf, A. Intisar, Comparative study of different activation treatments for the preparation of activated carbon: a mini-review, Sci. Prog. 100 (2017) 299–312, https://doi.org/10.3184/003685017X14967570531606.
[35] R. Ansari, A. Mohammad-Khah, Activated charcoal: preparation, characterization and applications: a review article, Int. J. ChemTech Res. 1 (2009) 859–864.
[36] E. Pérez-mayoral, I. Matos, M. Bernardo, I.M. Fonseca, D.Q. Inorgánica, P. Senda, New and advanced porous carbon materials in fine chemical synthesis. Emerging precursors of porous carbons, Catalysts (2019) 1–35, https://doi.org/10.3390/catal9020133.
[37] D. Bergna, H. Romar, S. Tuomikoski, H. Runtti, T. Kangas, P. Tynjälä, U. Lassi, Activated carbon from renewable sources: thermochemical conversion and activation of biomass and carbon residues from biomass gasification, in: Waste Biomass Management—A Holistic Approach, Springer International Publishing, Cham, 2017, pp. 187–213, https://doi.org/10.1007/978-3-319-49595-8_9.
[38] Z. Yue, J. Economy, Carbonization and activation for production of activated carbon fibers, in: J.Y. Chen (Ed.), Activated Carbon Fiber and Textiles, Woodhead Publishing Series in Textiles, 2016, pp. 61–139. https://doi.org/10.1016/B978-0-08-100660-3.00004-3.
[39] X. Tan, S. Liu, Y. Liu, Y. Gu, G. Zeng, X. Hu, X. Wang, S. Liu, L. Jiang, Biochar as potential sustainable precursors for activated carbon production: multiple applications in environmental protection and energy storage, Bioresour. Technol. 227 (2017) 359–372, https://doi.org/10.1016/j.biortech.2016.12.083.
[40] H.S. Kim, M.S. Kang, W.C. Yoo, Highly enhanced gas sorption capacities of N-doped porous carbon spheres by hot NH_3 and CO_2 treatments, J. Phys. Chem. C 119 (2015) 28512–28522, https://doi.org/10.1021/acs.jpcc.5b10552.
[41] X. Zhang, S. Zhang, H. Yang, Y. Feng, Y. Chen, X. Wang, H. Chen, Nitrogen enriched biochar modified by high temperature CO_2–ammonia treatment: characterization and adsorption of CO_2, Chem. Eng. J. 257 (2014) 20–27, https://doi.org/10.1016/j.cej.2014.07.024.

[42] M.A. Yahya, Z. Al-Qodah, C.W.Z. Ngah, Agricultural bio-waste materials as potential sustainable precursors used for activated carbon production: a review, Renew. Sustain. Energy Rev. 46 (2015) 218–235, https://doi.org/10.1016/j.rser.2015.02.051.
[43] P. González-García, Activated carbon from lignocellulosics precursors: a review of the synthesis methods, characterization techniques and applications, Renew. Sustain. Energy Rev. 82 (2018) 1393–1414, https://doi.org/10.1016/j.rser.2017.04.117.
[44] W.-J. Liu, H. Jiang, H.-Q. Yu, Development of biochar-based functional materials: toward a sustainable platform carbon material, Chem. Rev. 115 (2015) 12251–12285, https://doi.org/10.1021/acs.chemrev.5b00195.
[45] M. Danish, T. Ahmad, A review on utilization of wood biomass as a sustainable precursor for activated carbon production and application, Renew. Sustain. Energy Rev. 87 (2018) 1–21, https://doi.org/10.1016/j.rser.2018.02.003.
[46] P. Paraskeva, D. Kalderis, E. Diamadopoulos, Production of activated carbon from agricultural by-products, J. Chem. Technol. Biotechnol. 83 (2008) 581–592, https://doi.org/10.1002/jctb.1847.
[47] M.A. Tadda, A. Ahsan, A. Shitu, M. ElSergany, T. Arunkumar, B. Jose, M.D. A. Razzaque, N. Nik, A review on activated carbon: process, application and prospects, J. Adv. Civil Eng. Pract. Res. 2 (2016) 7–13.
[48] C. Correa, A. Kruse, Biobased functional carbon materials: production, characterization, and applications—a review, Materials 11 (2018) 1568, https://doi.org/10.3390/ma11091568.
[49] Y. Shen, N. Zhang, Y. Fu, Synthesis of high-performance hierarchically porous carbons from rice husk for sorption of phenol in the gas phase, J. Environ. Manage. 241 (2019) 53–58, https://doi.org/10.1016/j.jenvman.2019.04.012.
[50] S. Sangon, A.J. Hunt, T.M. Attard, P. Mengchang, Y. Ngernyen, N. Supanchaiyamat, Valorisation of waste rice straw for the production of highly effective carbon based adsorbents for dyes removal, J. Clean. Prod. 172 (2018) 1128–1139, https://doi.org/10.1016/j.jclepro.2017.10.210.
[51] H. Cheng, Y. Bian, F. Wang, X. Jiang, R. Ji, C. Gu, X. Yang, Y. Song, Green conversion of crop residues into porous carbons and their application to efficiently remove polycyclic aromatic hydrocarbons from water: sorption kinetics, isotherms and mechanism, Bioresour. Technol. 284 (2019) 1–8, https://doi.org/10.1016/j.biortech.2019.03.104.
[52] M. Calzado, M.J. Valero-Romero, P. Garriga, A. Chica, M.O. Guerrero-Pérez, J. Rodríguez-Mirasol, T. Cordero, Lignocellulosic waste-derived basic solids and their catalytic applications for the transformation of biomass waste, Catal. Today 257 (2015) 229–236, https://doi.org/10.1016/j.cattod.2014.06.038.
[53] A.M. Puziy, O.I. Poddubnaya, A. Martínez-Alonso, F. Suárez-García, A. Castro-Muñiz, J.M.D. Tascón, Impact of the carbonization atmosphere on the properties of phosphoric acid-activated carbons from fruit stones, Adsorpt. Sci. Technol. 26 (2008) 843–851, https://doi.org/10.1260/026361708788708261.
[54] O.F. Cruz, J. Silvestre-Albero, M.E. Casco, D. Hotza, C.R. Rambo, Activated nanocarbons produced by microwave-assisted hydrothermal carbonization of Amazonian fruit waste for methane storage, Mater. Chem. Phys. 216 (2018) 42–46, https://doi.org/10.1016/j.matchemphys.2018.05.079.
[55] X. Su, S. Jiang, X. Zheng, X. Guan, P. Liu, Z. Peng, Hierarchical porous carbon materials from bio waste-mango stone for high-performance supercapacitor electrodes, Mater. Lett. 230 (2018) 123–127, https://doi.org/10.1016/j.matlet.2018.07.096.
[56] S. Pap, T. Šolević Knudsen, J. Radonić, S. Maletić, S.M. Igić, M. Turk Sekulić, Utilization of fruit processing industry waste as green activated carbon for the treatment of heavy metals and chlorophenols contaminated water, J. Clean. Prod. 162 (2017) 958–972, https://doi.org/10.1016/j.jclepro.2017.06.083.

[57] P. Nowicki, J. Kazmierczak, R. Pietrzak, Comparison of physicochemical and sorption properties of activated carbons prepared by physical and chemical activation of cherry stones, Powder Technol. 269 (2015) 312–319, https://doi.org/10.1016/j.powtec.2014.09.023.
[58] R.A. Reza, M. Ahmaruzzaman, Removal of naproxen from aqueous environment using porous sugarcane bagasse: impact of ionic strength, hardness and surfactant, Res. Chem. Intermed. 42 (2016) 1463–1485, https://doi.org/10.1007/s11164-015-2097-z.
[59] W. Kong, J. Liu, Nitrogen-decorated, porous carbons derived from waste cow manure as efficient catalysts for the selective capture and conversion of CO_2, RSC Adv. 9 (2019) 4925–4931, https://doi.org/10.1039/C8RA10497B.
[60] A.F.M. Streit, L.N. Côrtes, S.P. Druzian, M. Godinho, G.C. Collazzo, D. Perondi, G. L. Dotto, Development of high quality activated carbon from biological sludge and its application for dyes removal from aqueous solutions, Sci. Total Environ. 660 (2019) 277–287, https://doi.org/10.1016/j.scitotenv.2019.01.027.
[61] K.-W. Jung, S.Y. Lee, Y.J. Lee, Optimized preparation of activated carbon from municipal sewage sludge and application to the adsorption of azo dye from aqueous solutions, Environ. Prog. Sustain. Energy. 38 (2019) S267–S276, https://doi.org/10.1002/ep.13008.
[62] L. Rao, S. Liu, L. Wang, C. Ma, J. Wu, L. An, X. Hu, N-doped porous carbons from low-temperature and single-step sodium amide activation of carbonized water chestnut shell with excellent CO_2 capture performance, Chem. Eng. J. 359 (2019) 428–435, https://doi.org/10.1016/j.cej.2018.11.065.
[63] L. Guan, L. Pan, T. Peng, C. Gao, W. Zhao, Z. Yang, H. Hu, M. Wu, Synthesis of biomass-derived nitrogen-doped porous carbon nanosheests for high-performance supercapacitors, ACS Sustain. Chem. Eng. 7 (2019) 8405–8412, https://doi.org/10.1021/acssuschemeng.9b00050.
[64] X. Lu, K. Xiang, W. Zhou, Y. Zhu, H. Chen, Biomass carbon materials derived from macadamia nut shells for high-performance supercapacitors, Bull. Mater. Sci. 41 (2018) 138, https://doi.org/10.1007/s12034-018-1666-3.
[65] M. Zbair, K. Ainassaari, Z. El Assal, S. Ojala, N. El Ouahedy, R.L. Keiski, M. Bensitel, R. Brahmi, Steam activation of waste biomass: highly microporous carbon, optimization of bisphenol A, and diuron adsorption by response surface methodology, Environ. Sci. Pollut. Res. 25 (2018) 35657–35671, https://doi.org/10.1007/s11356-018-3455-3.
[66] A. Bazan-Wozniak, P. Nowicki, R. Pietrzak, The influence of activation procedure on the physicochemical and sorption properties of activated carbons prepared from pistachio nutshells for removal of NO_2/H_2S gases and dyes, J. Clean. Prod. 152 (2017) 211–222, https://doi.org/10.1016/j.jclepro.2017.03.114.
[67] S.-A. Sajjadi, A. Meknati, E.C. Lima, G.L. Dotto, D.I. Mendoza-Castillo, I. Anastopoulos, F. Alakhras, E.I. Unuabonah, P. Singh, A. Hosseini-Bandegharaei, A novel route for preparation of chemically activated carbon from pistachio wood for highly efficient Pb(II) sorption, J. Environ. Manage. 236 (2019) 34–44, https://doi.org/10.1016/j.jenvman.2019.01.087.
[68] U. Thubsuang, S. Laebang, N. Manmuanpom, S. Wongkasemjit, T. Chaisuwan, Tuning pore characteristics of porous carbon monoliths prepared from rubber wood waste treated with H_3PO_4 or NaOH and their potential as supercapacitor electrode materials, J. Mater. Sci. 52 (2017) 6837–6855, https://doi.org/10.1007/s10853-017-0922-z.
[69] G.V. Nunell, M.E. Fernandez, P.R. Bonelli, A.L. Cukierman, Nitrate uptake from water by means of tailored adsorbents, Water Air Soil Pollut. 226 (2015) 278, https://doi.org/10.1007/s11270-015-2546-8.

[70] K. Sun, Q. Huang, X. Meng, Y. Chi, J. Yan, Catalytic pyrolysis of waste polyethylene into aromatics by H_3PO_4-activated carbon, Energy Fuels 32 (2018) 9772–9781, https://doi.org/10.1021/acs.energyfuels.8b02091.
[71] M.-M. Titirici, M. Antonietti, Chemistry and materials options of sustainable carbon materials made by hydrothermal carbonization, Chem. Soc. Rev. 39 (2010) 103–116, https://doi.org/10.1039/B819318P.
[72] A. Jain, R. Balasubramanian, M.P. Srinivasan, Hydrothermal conversion of biomass waste to activated carbon with high porosity: a review, Chem. Eng. J. 283 (2016) 789–805, https://doi.org/10.1016/j.cej.2015.08.014.
[73] B. Hu, K. Wang, L. W.u, S.-H. Yu, M. Antonietti, M.-M. Titirici, Engineering carbon materials from the hydrothermal carbonization process of biomass, Adv. Mater. 22 (2010) 813–828, https://doi.org/10.1002/adma.200902812.
[74] A. Jain, R. Balasubramanian, M.P. Srinivasan, Tuning hydrochar properties for enhanced mesopore development in activated carbon by hydrothermal carbonization, Microporous Mesoporous Mater. 203 (2015) 178–185, https://doi.org/10.1016/j.micromeso.2014.10.036.
[75] N. Fechler, S.-A. Wohlgemuth, P. Jäker, M. Antonietti, Salt and sugar: direct synthesis of high surface area carbon materials at low temperatures via hydrothermal carbonization of glucose under hypersaline conditions, J. Mater. Chem. A. 1 (2013) 9418, https://doi.org/10.1039/c3ta10674h.
[76] H.-S. Qian, S.-H. Yu, L.-B. Luo, J.-Y. Gong, L.-F. Fei, X.-M. Liu, Synthesis of uniform Te@carbon-rich composite nanocables with photoluminescence properties and carbonaceous nanofibers by the hydrothermal carbonization of glucose, Chem. Mater. 18 (2006) 2102–2108, https://doi.org/10.1021/cm052848y.
[77] X. Cui, M. Antonietti, S.-H. Yu, Structural effects of iron oxide nanoparticles and iron ions on the hydrothermal carbonization of starch and rice carbohydrates, Small 2 (2006) 756–759, https://doi.org/10.1002/smll.200600047.
[78] X. Lu, C. Jiang, Y. Hu, H. Zhong, Y. Zhao, X. Xu, H. Liu, Preparation of hierarchically porous carbon spheres by hydrothermal carbonization process for high-performance electrochemical capacitors, J. Appl. Electrochem. 48 (2018) 233–241, https://doi.org/10.1007/s10800-018-1146-x.
[79] C.-X. Bai, F. Shen, X.-H. Qi, Preparation of porous carbon directly from hydrothermal carbonization of fructose and phloroglucinol for adsorption of tetracycline, Chinese Chem. Lett. 28 (2017) 960–962, https://doi.org/10.1016/j.cclet.2016.12.026.
[80] D. Chen, L. Yang, J. Li, Q. Wu, Effect of self-doped heteroatoms in biomass-derived activated carbon for supercapacitor applications, ChemistrySelect 4 (2019) 1586–1595, https://doi.org/10.1002/slct.201803413.
[81] W. Qian, F. Sun, Y. Xu, L. Qiu, C. Liu, S. Wang, F. Yan, Human hair-derived carbon flakes for electrochemical supercapacitors, Energy Environ. Sci. 7 (2014) 379–386, https://doi.org/10.1039/C3EE43111H.
[82] B. Xu, S. Hou, G. Cao, F. Wu, Y. Yang, Sustainable nitrogen-doped porous carbon with high surface areas prepared from gelatin for supercapacitors, J. Mater. Chem. 22 (2012) 19088, https://doi.org/10.1039/c2jm32759g.
[83] Y. Lu, N. Zhu, F. Yin, T. Yang, P. Wu, Z. Dang, M. Liu, X. Wei, Biomass-derived heteroatoms-doped mesoporous carbon for efficient oxygen reduction in microbial fuel cells, Biosens. Bioelectron. 98 (2017) 350–356, https://doi.org/10.1016/j.bios.2017.07.006.
[84] F. Liu, Z. Wang, H. Zhang, L. Jin, X. Chu, B. Gu, H. Huang, W. Yang, Nitrogen, oxygen and sulfur co-doped hierarchical porous carbons toward high-performance supercapacitors by direct pyrolysis of kraft lignin, Carbon 149 (2019) 105–116, https://doi.org/10.1016/j.carbon.2019.04.023.

[85] L. Wan, W. Wei, M. Xie, Y. Zhang, X. Li, R. Xiao, J. Chen, C. Du, Nitrogen, sulfur co-doped hierarchically porous carbon from rape pollen as high-performance supercapacitor electrode, Electrochim. Acta 311 (2019) 72–82, https://doi.org/10.1016/j.electacta.2019.04.106.

[86] Z. Liu, F. Zhang, T. Liu, N. Peng, C. Gai, Removal of azo dye by a highly graphitized and heteroatom doped carbon derived from fish waste: adsorption equilibrium and kinetics, J. Environ. Manage. 182 (2016) 446–454, https://doi.org/10.1016/j.jenvman.2016.08.008.

[87] T.I. Bhuiyan, J.K. Tak, S. Sessarego, D. Harfield, J.M. Hill, Adsorption of acid-extractable organics from oil sands process-affected water onto biomass-based biochar: metal content matters, Chemosphere 168 (2017) 1337–1344, https://doi.org/10.1016/j.chemosphere.2016.11.126.

[88] Q. Wei, X. Yang, G. Zhang, D. Wang, L. Zuin, D. Banham, L. Yang, S. Ye, Y. Wang, M. Mohamedi, S. Sun, An active and robust Si-Fe/N/C catalyst derived from waste reed for oxygen reduction, Appl. Catal. B Environ. 237 (2018) 85–93, https://doi.org/10.1016/j.apcatb.2018.05.046.

[89] M. Uchimiya, I.M. Lima, K. Thomas Klasson, S. Chang, L.H. Wartelle, J.E. Rodgers, Immobilization of heavy metal ions (Cu II , Cd II , Ni II , and Pb II) by broiler litter-derived biochars in water and soil, J. Agric. Food Chem. 58 (2010) 5538–5544, https://doi.org/10.1021/jf9044217.

[90] T. Shen, Y. Tang, X.-Y. Lu, Z. Meng, Mechanisms of copper stabilization by mineral constituents in sewage sludge biochar, J. Clean. Prod. 193 (2018) 185–193, https://doi.org/10.1016/j.jclepro.2018.05.071.

[91] M. Kakunuri, C.S. Sharma, Resorcinol-formaldehyde derived carbon xerogels: a promising anode material for lithium-ion battery, J. Mater. Res. 33 (2018) 1074–1087, https://doi.org/10.1557/jmr.2017.461.

[92] R.J. White, V.L. Budarin, J.H. Clark, Tuneable mesoporous materials from α-D-polysaccharides, ChemSusChem 1 (2008) 408–411, https://doi.org/10.1002/cssc.200800012.

[93] R.J. White, V. Budarin, R. Luque, J.H. Clark, D.J. Macquarrie, Tuneable porous carbonaceous materials from renewable resources, Chem. Soc. Rev. 38 (2009) 3401, https://doi.org/10.1039/b822668g.

[94] R.J. White, N. Brun, V.L. Budarin, J.H. Clark, M.-M. Titirici, Always look on the "Light" side of life: sustainable carbon aerogels, ChemSusChem 7 (2014) 670–689, https://doi.org/10.1002/cssc.201300961.

[95] V. Malgras, J. Tang, J. Wang, J. Kim, N.L. Torad, S. Dutta, K. Ariga, M.S.A. Hossain, Y. Yamauchi, K.C.W. Wu, Fabrication of nanoporous carbon materials with hard- and soft-templating approaches: a review, J. Nanosci. Nanotechnol. 19 (2019) 3673–3685, https://doi.org/10.1166/jnn.2019.16745.

[96] H. Nishihara, T. Kyotani, Zeolite-templated carbons-three-dimensional microporous graphene frameworks, Chem. Commun. 54 (2018) 5648–5673, https://doi.org/10.1039/c8cc01932k.

[97] M. Zhang, L. He, T. Shi, R. Zha, Nanocasting and direct synthesis strategies for mesoporous carbons as supercapacitor electrodes, Chem. Mater. 30 (2018) 7391–7412, https://doi.org/10.1021/acs.chemmater.8b03345.

[98] M.R. Benzigar, S.N. Talapaneni, S. Joseph, K. Ramadass, G. Singh, J. Scaranto, U. Ravon, K. Al-Bahily, A. Vinu, Recent advances in functionalized micro and mesoporous carbon materials: synthesis and applications, Chem. Soc. Rev. 47 (2018) 2680–2721, https://doi.org/10.1039/c7cs00787f.

[99] J. Schuster, G. He, B. Mandlmeier, T. Yim, K.T. Lee, T. Bein, L.F. Nazar, Spherical ordered mesoporous carbon nanoparticles with high porosity for lithium-sulfur

batteries, Angew. Chem. Int. Ed. 51 (2012) 3591–3595, https://doi.org/10.1002/anie.201107817.
[100] A. Eftekhari, Z. Fan, Ordered mesoporous carbon and its applications for electrochemical energy storage and conversion, Mater. Chem. Front. 1 (2017) 1001–1027, https://doi.org/10.1039/c6qm00298f.
[101] K. Kim, M. Choi, R. Ryoo, Ethanol-based synthesis of hierarchically porous carbon using nanocrystalline beta zeolite template for high-rate electrical double layer capacitor, Carbon 60 (2013) 175–185, https://doi.org/10.1016/j.carbon.2013.04.011.
[102] S.H. Joo, S. Jun, R. Ryoo, Synthesis of ordered mesoporous carbon molecular sieves CMK-1, Microporous Mesoporous Mater. 44–45 (2001) 153–158, https://doi.org/10.1016/S1387-1811(01)00179-2.
[103] E. Masika, R. Mokaya, Preparation of ultrahigh surface area porous carbons templated using zeolite 13X for enhanced hydrogen storage, Prog. Nat. Sci. Mater. Int. 23 (2013) 308–316, https://doi.org/10.1016/J.PNSC.2013.04.007.
[104] K. Nueangnoraj, H. Nishihara, K. Imai, H. Itoi, T. Ishii, M. Kiguchi, Y. Sato, M. Terauchi, T. Kyotani, Formation of crosslinked-fullerene-like framework as negative replica of zeolite Y, Carbon 62 (2013) 455–464, https://doi.org/10.1016/J.CARBON.2013.06.033.
[105] C. Young, T. Park, J.W. Yi, J. Kim, M.S.A. Hossain, Y.V. Kaneti, Y. Yamauchi, Advanced functional carbons and their hybrid nanoarchitectures towards supercapacitor applications, ChemSusChem 11 (2018) 3546–3558, https://doi.org/10.1002/cssc.201801525.
[106] C. Liang, S. Dai, Synthesis of mesoporous carbon materials via enhanced hydrogen-bonding interaction, J. Am. Chem. Soc. 128 (2006) 5316–5317, https://doi.org/10.1021/ja060242k.
[107] C. Liang, K. Hong, G.A. Guiochon, J.W. Mays, S. Dai, Synthesis of a large-scale highly ordered porous carbon film by self-assembly of block copolymers, Angew. Chem. Int. Ed. 43 (2004) 5785–5789, https://doi.org/10.1002/anie.200461051.
[108] S. Tanaka, N. Nishiyama, Y. Egashira, K. Ueyama, Synthesis of ordered mesoporous carbons with channel structure from an organic–organic nanocomposite, Chem. Commun. (2005) 2125–2127, https://doi.org/10.1039/B501259G.
[109] Y. Meng, D. Gu, F. Zhang, Y. Shi, H. Yang, Z. Li, C. Yu, B. Tu, D. Zhao, Ordered mesoporous polymers and homologous carbon frameworks: amphiphilic surfactant templating and direct transformation, Angew. Chem. Int. Ed. 44 (2005) 7053–7059, https://doi.org/10.1002/anie.200501561.
[110] F. Zhang, Y. Meng, D. Gu, Y. Yan, C. Yu, B. Tu, D. Zhao, A facile aqueous route to synthesize highly ordered mesoporous polymers and carbon frameworks with Ia3d bicontinuous cubic structure, J. Am. Chem. Soc. 127 (2005) 13508–13509, https://doi.org/10.1021/ja0545721.
[111] T.Y. Ma, L. Liu, Z.Y. Yuan, Direct synthesis of ordered mesoporous carbons, Chem. Soc. Rev. 42 (2013) 3977–4003, https://doi.org/10.1039/c2cs35301f.
[112] N. Stock, S. Biswas, Synthesis of metal-organic frameworks (MOFs): routes to various MOF topologies, morphologies, and composites, Chem. Rev. 112 (2012) 933–969, https://doi.org/10.1021/cr200304e.
[113] H. Furukawa, K.E. Cordova, M. O'Keeffe, O.M. Yaghi, The chemistry and applications of metal-organic frameworks, Science 341 (2013) 1230444.
[114] Y. Wen, J. Zhang, Q. Xu, X.-T. Wu, Q.-L. Zhu, Pore surface engineering of metal–organic frameworks for heterogeneous catalysis, Coord. Chem. Rev. 376 (2018) 248–276, https://doi.org/10.1016/j.ccr.2018.08.012.
[115] J.-R. Li, Y. Ma, M.C. McCarthy, J. Sculley, J. Yu, H.-K. Jeong, P.B. Balbuena, H.-C. Zhou, Carbon dioxide capture-related gas adsorption and separation in

metal-organic frameworks, Coord. Chem. Rev. 255 (2011) 1791–1823, https://doi.org/10.1016/j.ccr.2011.02.012.
[116] Y. Fang, Y. Ma, M. Zheng, P. Yang, A.M. Asiri, X. Wang, Metal–organic frameworks for solar energy conversion by photoredox catalysis, Coord. Chem. Rev. 373 (2018) 83–115, https://doi.org/10.1016/j.ccr.2017.09.013.
[117] P. Yin, T. Yao, Y. Wu, L. Zheng, Y. Lin, W. Liu, H. Ju, J. Zhu, X. Hong, Z. Deng, G. Zhou, S. Wei, Y. Li, Single cobalt atoms with precise N-coordination as superior oxygen reduction reaction catalysts, Angew. Chem. Int. Ed. 55 (2016) 10800–10805, https://doi.org/10.1002/anie.201604802.
[118] D. Jiang, P. Xu, H. Wang, G. Zeng, D. Huang, M. Chen, C. Lai, C. Zhang, J. Wan, W. Xue, Strategies to improve metal organic frameworks photocatalyst's performance for degradation of organic pollutants, Coord. Chem. Rev. 376 (2018) 449–466, https://doi.org/10.1016/j.ccr.2018.08.005.
[119] N.A. Khan, Z. Hasan, S.H. Jhung, Beyond pristine metal-organic frameworks: preparation and application of nanostructured, nanosized, and analogous MOFs, Coord. Chem. Rev. 376 (2018) 20–45, https://doi.org/10.1016/J.CCR.2018.07.016.
[120] D. Liu, L. Dai, X. Lin, J.F. Chen, J. Zhang, X. Feng, K. Müllen, X. Zhu, S. Dai, Chemical approaches to carbon-based metal-free catalysts, Adv. Mater. 31 (2019) 1804863, https://doi.org/10.1002/adma.201804863.
[121] K. Matsuyama, Supercritical fluid processing for metal–organic frameworks, porous coordination polymers, and covalent organic frameworks, J. Supercrit. Fluids. 134 (2018) 197–203, https://doi.org/10.1016/j.supflu.2017.12.004.
[122] F. Xu, D. Wu, R. Fu, B. Wei, Design and preparation of porous carbons from conjugated polymer precursors, Mater. Today 20 (2017) 629–656, https://doi.org/10.1016/j.mattod.2017.04.026.
[123] P.J. Waller, F. Gándara, O.M. Yaghi, Chemistry of covalent organic frameworks, Acc. Chem. Res. 48 (2015) 3053–3063, https://doi.org/10.1021/acs.accounts.5b00369.
[124] S.Y. Ding, J. Gao, Q. Wang, Y. Zhang, W.G. Song, C.Y. Su, W. Wang, Construction of covalent organic framework for catalysis: Pd/COF-LZU1 in Suzuki-Miyaura coupling reaction, J. Am. Chem. Soc. 133 (2011) 19816–19822, https://doi.org/10.1021/ja206846p.
[125] H. Hu, Q. Yan, R. Ge, Y. Gao, Covalent organic frameworks as heterogeneous catalysts, Chin. J. Catal. 39 (2018) 1167–1179, https://doi.org/10.1016/S1872-2067(18)63057-8.
[126] S.Y. Ding, W. Wang, Covalent organic frameworks (COFs): from design to applications, Chem. Soc. Rev. 42 (2013) 548–568, https://doi.org/10.1039/c2cs35072f.
[127] X. Feng, Y. Liang, L. Zhi, A. Thomas, D. Wu, I. Lieberwirth, U. Kolb, K. Müllen, Synthesis of microporous carbon nanofibers and nanotubes from conjugated polymer network and evaluation in electrochemical capacitor, Adv. Funct. Mater. 19 (2009) 2125–2129, https://doi.org/10.1002/adfm.200900264.
[128] Z. Xiang, D. Cao, L. Huang, J. Shui, M. Wang, L. Dai, Nitrogen-doped holey graphitic carbon from 2D covalent organic polymers for oxygen reduction, Adv. Mater. 26 (2014) 3315–3320, https://doi.org/10.1002/adma.201306328.
[129] X. Liu, L. Zhou, Y. Zhao, L. Bian, X. Feng, Q. Pu, Hollow, spherical nitrogen-rich porous carbon shells obtained from a porous organic framework for the supercapacitor, ACS Appl. Mater. Interfaces 5 (2013) 10280–10287, https://doi.org/10.1021/am403175q.
[130] Y. Zhu, X. Chen, J. Liu, J. Zhang, D.Y. Xu, W. Peng, Y. Li, G. Zhang, F. Zhang, X. Fan, Rational design of Fe/N/S-doped nanoporous carbon catalysts from covalent triazine frameworks for efficient oxygen reduction, ChemSusChem 11 (2018) 2402–2409, https://doi.org/10.1002/cssc.201800855.

[131] Y.J. Lee, S.N. Talapaneni, A. Coskun, Chemically activated covalent triazine frameworks with enhanced textural properties for high capacity gas storage, ACS Appl. Mater. Interfaces 9 (2017) 30679–30685, https://doi.org/10.1021/acsami.7b08930.

[132] C. Bai, J. Li, S. Liu, X. Yang, X. Yang, Y. Tian, K. Cao, Y. Huang, L. Ma, S. Li, In situ preparation of nitrogen-rich and functional ultramicroporous carbonaceous COFs by "segregated" microwave irradiation, Microporous Mesoporous Mater. 197 (2014) 148–155, https://doi.org/10.1016/j.micromeso.2014.06.004.

[133] Y.-B. Huang, P. Pachfule, J.-K. Sun, Q. Xu, From covalent–organic frameworks to hierarchically porous B-doped carbons: a molten-salt approach, J. Mater. Chem. A 4 (2016) 4273–4279, https://doi.org/10.1039/C5TA10170K.

[134] L. Zhang, Y.H. Hu, A systematic investigation of decomposition of nano Zn_4O(C $8H_4O_4)_3$ metal-organic framework, J. Phys. Chem. C 114 (2010) 2566–2572, https://doi.org/10.1021/jp911043r.

[135] J. Hu, H. Wang, Q. Gao, H. Guo, Porous carbons prepared by using metal-organic framework as the precursor for supercapacitors, Carbon 48 (2010) 3599–3606, https://doi.org/10.1016/j.carbon.2010.06.008.

[136] S.L. Jin, H.G. Deng, L. Zhan, W.M. Qiao, L.C. Ling, Synthesis of 3D hierarchical porous carbon as an electrode material for electric double layer capacitors, New Carbon Mater. 27 (2012) 87–92, https://doi.org/10.1016/S1872-5805(12)60005-5.

[137] S. Mo, Z. Sun, X. Huang, W. Zou, J. Chen, D. Yuan, Synthesis, characterization and supercapacitive properties of hierarchical porous carbons, Synth. Met. 162 (2012) 85–88, https://doi.org/10.1016/j.synthmet.2011.11.015.

[138] D. Yuan, J. Chen, S. Tan, N. Xia, Y. Liu, Worm-like mesoporous carbon synthesized from metal-organic coordination polymers for supercapacitors, Electrochem. Commun. 11 (2009) 1191–1194, https://doi.org/10.1016/j.elecom.2009.03.045.

[139] G. Srinivas, V. Krungleviciute, Z.X. Guo, T. Yildirim, Exceptional CO_2 capture in a hierarchically porous carbon with simultaneous high surface area and pore volume, Energy Environ. Sci. 7 (2014) 335–342, https://doi.org/10.1039/c3ee42918k.

[140] E. Pérez-Mayoral, I. Matos, M. Bernardo, I.M. Fonseca, E. Pérez-Mayoral, I. Matos, M. Bernardo, I.M. Fonseca, New and advanced porous carbon materials in fine chemical synthesis. Emerging precursors of porous carbons, Catalysts 9 (2019) 133, https://doi.org/10.3390/catal9020133.

[141] T.J. Bandosz, Surface chemistry of carbon materials, in: P. Serp, J.L. Figueiredo (Eds.), Carbon Materials for Catalysis, John Wiley & Sons, Inc., Hoboken, NJ, USA, 2008, pp. 45–92, https://doi.org/10.1002/9780470403709.ch2

[142] C.J. Durán-Valle, M. Madrigal-Martínez, M. Martínez-Gallego, I.M. Fonseca, I. Matos, A.M.B. Do Rego, Activated carbon as a catalyst for the synthesis of N-alkylimidazoles and imidazolium ionic liquids, Catal. Today 187 (2012) 108–114, https://doi.org/10.1016/j.cattod.2011.12.021.

[143] F. Rodríguez-Reinoso, The role of carbon materials in heterogeneous catalysis, Carbon 36 (1998) 159–175, https://doi.org/10.1016/S0008-6223(97)00173-5.

[144] P. Serp, J.L. Figueiredo (Eds.), Carbon Materials for Catalysis, John Wiley & Sons, Inc., Hoboken, NJ, USA, 2008https://doi.org/10.1002/9780470403709

[145] V. Calvino-Casilda, A.J. Lopez-Peinado, C.J. Duran-Valle, R.M. Martin-Aranda, Last decade of research on activated carbons as catalytic support in chemical processes, Catal. Rev. Sci. Eng. 52 (2010) 325–380, https://doi.org/10.1080/01614940.2010.498748.

[146] J. Lõpez-Sanz, E. Pérez-Mayoral, E. Soriano, D. Omenat-Morán, C.J. Durán, R. M. Martín-Aranda, I. Matos, I. Fonseca, Acid-activated carbon materials: cheaper alternative catalysts for the synthesis of substituted quinolines, ChemCatChem 5 (2013) 3736–3742, https://doi.org/10.1002/cctc.201300626.

[147] M. Godino-Ojer, E. Soriano, V. Calvino-Casilda, F.J. Maldonado-Hódar, E. Pérez-Mayoral, Metal-free synthesis of quinolines catalyzed by carbon aerogels: influence of the porous texture and surface chemistry, Chem. Eng. J. 314 (2017) 488–497, https://doi.org/10.1016/j.cej.2016.12.006.
[148] M. Godino-Ojer, A.J. Lõpez-Peinado, R.M. Martín-Aranda, J. Przepiõrski, E. Pérez-Mayoral, E. Soriano, Eco-friendly catalytic systems based on carbon-supported magnesium oxide materials for the friedländer condensation, ChemCatChem 6 (2014) 3440–3447, https://doi.org/10.1002/cctc.201402602.
[149] M. Godino-Ojer, A.J. López-Peinado, F.J. Maldonado-Hódar, E. Pérez-Mayoral, Highly efficient and selective catalytic synthesis of quinolines involving transition-metal-doped carbon aerogels, ChemCatChem 9 (2017) 1422–1428, https://doi.org/10.1002/cctc.201601657.
[150] I.M. López-Coca, S. Izquierdo, G. Silvero, C.J. Durán-Valle, Sustainable carbon-based materials as heterogeneous catalysts in solvent-free acetylation reactions, Proceedings 9 (2018) 40, https://doi.org/10.3390/ecsoc-22-05685.
[151] I. Matos, P.D. Neves, J.E. Castanheiro, E. Perez-Mayoral, R. Martin-Aranda, C. Duran-Valle, J. Vital, A.M.B. Do Rego, I.M. Fonseca, Mesoporous carbon as an efficient catalyst for alcoholysis and aminolysis of epoxides, Appl. Catal. A Gen. 439–440 (2012) 24–30, https://doi.org/10.1016/j.apcata.2012.06.036.
[152] C.J. Durán-Valle, J.A. García-Vidal, Acidic activated carbons: an efficient catalyst for the epoxide ring-opening reaction with ethanol, Catal. Lett. 130 (2009) 37–41, https://doi.org/10.1007/s10562-009-9860-y.
[153] I. Matos, M.F. Silva, R. Ruiz-Rosas, J. Vital, J. Rodríguez-Mirasol, T. Cordero, J. E. Castanheiro, I.M. Fonseca, Methoxylation of α-pinene over mesoporous carbons and microporous carbons: a comparative study, Microporous Mesoporous Mater. 199 (2014) 66–73, https://doi.org/10.1016/j.micromeso.2014.08.006.
[154] S. Ferrera-Escudero, E. Perozo-Rondón, V. Calvino-Casilda, B. Casal, R.M. Martín-Aranda, A.J. López-Peinado, C.J. Durán-Valle, The effect of ultrasound on the N-alkylation of imidazole over alkaline carbons: kinetic aspects, Appl. Catal. A Gen. 378 (2010) 26–32, https://doi.org/10.1016/j.apcata.2010.01.041.
[155] V. Calvino Casilda, E. Pérez-Mayoral, M.A. Bañares, E. Lozano Diz, Real-time Raman monitoring of dry media heterogeneous alkylation of imidazole with acidic and basic catalysts, Chem. Eng. J. 161 (2010) 371–376, https://doi.org/10.1016/j.cej.2010.01.028.
[156] V. Calvino-Casilda, R.M. Martín-Aranda, A.J. López-Peinado, Alkaline carbons as effective catalysts for the microwave-assisted synthesis of N-substituted-gamma-lactams, Appl. Catal. A Gen. 398 (2011) 73–81, https://doi.org/10.1016/j.apcata.2011.03.017.
[157] V. Calvino, M. Picallo, A.J. López-Peinado, R.M. Martín-Aranda, C.J. Durán-Valle, Ultrasound accelerated Claisen-Schmidt condensation: a green route to chalcones, Appl. Surf. Sci. 252 (2006) 6071–6074, https://doi.org/10.1016/j.apsusc.2005.11.006.
[158] F.L. Wang, T.F. Tsai, W. Bin Lin, Catalytic synthesis of propionitrile from methanol and acetonitrile over activated carbon-supported sodium catalysts, Catal. Lett. 73 (2001) 215–219, https://doi.org/10.1023/A:1016643813217.
[159] F. Cárdenas-Lizana, S. Gómez-Quero, M.A. Keane, Clean production of chloroanilines by selective gas phase hydrogenation over supported Ni catalysts, Appl. Catal. A Gen. 334 (2008) 199–206, https://doi.org/10.1016/j.apcata.2007.10.007.
[160] Z. Zhao, H. Yang, Y. Li, X. Guo, Cobalt-modified molybdenum carbide as an efficient catalyst for chemoselective reduction of aromatic nitro compounds, Green Chem. 16 (2014) 1274–1281, https://doi.org/10.1039/c3gc42049c.

[161] Y.J. Lyu, T. Qi, H.Q. Yang, C.W. Hu, Performance of edges on carbon for the catalytic hydroxylation of benzene to phenol, Catal. Sci. Technol. 8 (2018) 176–186, https://doi.org/10.1039/c7cy01648d.
[162] Y. Zhong, G. Li, L. Zhu, Y. Yan, G. Wu, C. Hu, Low temperature hydroxylation of benzene to phenol by hydrogen peroxide over Fe/activated carbon catalyst, J. Mol. Catal. A Chem. 272 (2007) 169–173, https://doi.org/10.1016/j.molcata.2007.03.015.
[163] M. Jin, R. Yang, M. Zhao, G. Li, C. Hu, Application of Fe/activated carbon catalysts in the hydroxylation of phenol to dihydroxybenzenes, Ind. Eng. Chem. Res. 53 (2014) 2932–2939, https://doi.org/10.1021/ie404010u.
[164] C.H. Chen, J.Q. Xu, M.M. Jin, G.Y. Li, C.W. Hu, Direct synthesis of phenol from benzene on an activated carbon catalyst treated with nitric acid, Chin. J. Chem. Phys. 24 (2011) 358–364, https://doi.org/10.1088/1674-0068/24/03/358-364.
[165] M. Godino-Ojer, L. Milla-Diez, I. Matos, C.J. Durán-Valle, M. Bernardo, I. M. Fonseca, E. Pérez Mayoral, Enhanced catalytic properties of carbon supported zirconia and sulfated zirconia for the green synthesis of benzodiazepines, ChemCatChem 10 (2018) 5215–5223, https://doi.org/10.1002/cctc.201801274.
[166] K.F. Heck, J.P. Nolley, Palladium-catalyzed vinylic hydrogen substitution reactions with aryl, benzyl, and styryl halides, J. Org. Chem. 37 (1972) 2320–2322, https://doi.org/10.1021/jo00979a024.
[167] K. Köhler, R.G. Heidenreich, S.S. Soomro, S.S. Pröckl, Supported palladium catalysts for suzuki reactions: structure-property relationships, optimized reaction protocol and control of palladium leaching, Adv. Synth. Catal. 350 (2008) 2930–2936, https://doi.org/10.1002/adsc.200800575.
[168] M. Fan, P. Zhang, X. Ma, Study on Wacker-type catalysts for catalytic synthesis of diethyl carbonate from ethyl nitrite route, Fuel 86 (2007) 902–905, https://doi.org/10.1016/j.fuel.2006.09.005.
[169] M. Cano, A. Benito, W.K. Maser, E.P. Urriolabeitia, One-step microwave synthesis of palladium-carbon nanotube hybrids with improved catalytic performance, Carbon 49 (2011) 652–658, https://doi.org/10.1016/j.carbon.2010.10.012.
[170] M. Cano, A.M. Benito, W.K. Maser, E.P. Urriolabeitia, High catalytic performance of palladium nanoparticles supported on multiwalled carbon nanotubes in alkene hydrogenation reactions, New J. Chem. 37 (2013) 1968–1972, https://doi.org/10.1039/c3nj00183k.
[171] S. Donck, E. Gravel, N. Shah, D.V. Jawale, E. Doris, I.N.N. Namboothiri, Tsuji-Wacker oxidation of terminal olefins using a palladium-carbon nanotube nanohybrid, ChemCatChem 7 (2015) 2318–2322, https://doi.org/10.1002/cctc.201500241.
[172] E.K. Lee, S.A. Park, H. Woo, K. Hyun Park, D.W. Kang, H. Lim, Y.T. Kim, Platinum single atoms dispersed on carbon nanotubes as reusable catalyst for Suzuki coupling reaction, J. Catal. 352 (2017) 388–393, https://doi.org/10.1016/j.jcat.2017.05.005.
[173] H. Woo, E.K. Lee, S.W. Yun, S.A. Park, K.H. Park, Y.T. Kim, Platinum single atoms on carbon nanotubes as efficient catalyst for hydroalkoxylation, Bull. Korean Chem. Soc. 38 (2017) 1221–1225, https://doi.org/10.1002/bkcs.11252.
[174] D.V. Jawale, E. Gravel, C. Boudet, N. Shah, V. Geertsen, H. Li, I.N.N. Namboothiri, E. Doris, Selective conversion of nitroarenes using a carbon nanotube-ruthenium nanohybrid, Chem. Commun. 51 (2015) 1739–1742, https://doi.org/10.1039/c4cc09192b.
[175] P. Basu, P. Prakash, E. Gravel, N. Shah, K. Bera, E. Doris, I.N.N. Namboothiri, Carbon nanotube-ruthenium hybrids for the partial reduction of 2-nitrochalcones: easy access to quinoline N-oxides, ChemCatChem 8 (2016) 1298–1302, https://doi.org/10.1002/cctc.201600042.

[176] D.V. Jawale, E. Gravel, N. Shah, V. Dauvois, H. Li, I.N.N. Namboothiri, E. Doris, Cooperative dehydrogenation of N-heterocycles using a carbon nanotube-rhodium nanohybrid, Chem. A Eur. J. 21 (2015) 7039–7042, https://doi.org/10.1002/chem.201500148.
[177] M. Godino-Ojer, A.J. López-Peinado, F.J. Maldonado-Hódar, E. Bailón-García, E. Pérez-Mayoral, Cobalt oxide-carbon nanocatalysts with highly enhanced catalytic performance for the green synthesis of nitrogen heterocycles through the Friedländer condensation, Dalt. Trans. 48 (2019) 5637–5648, https://doi.org/10.1039/c8dt04403a.
[178] J. John, E. Gravel, A. Hagège, H. Li, T. Gacoin, E. Doris, Catalytic oxidation of silanes by carbon nanotube-gold nanohybrids, Angew. Chem. Int. Ed. 50 (2011) 7533–7536, https://doi.org/10.1002/anie.201101993.
[179] R. Kumar, E. Gravel, A. Hagège, H. Li, D.V. Jawale, D. Verma, I.N.N. Namboothiri, E. Doris, Carbon nanotube-gold nanohybrids for selective catalytic oxidation of alcohols, Nanoscale 5 (2013) 6491–6497, https://doi.org/10.1039/c3nr01432k.
[180] N. Shah, E. Gravel, D.V. Jawale, E. Doris, I.N.N. Namboothiri, Synthesis of quinoxalines by a carbon nanotube-gold nanohybrid-catalyzed cascade reaction of vicinal diols and keto alcohols with diamines, ChemCatChem 7 (2015) 57–61, https://doi.org/10.1002/cctc.201402782.
[181] M. Cano, P. Villuendas, A.M. Benito, E.P. Urriolabeitia, W.K. Maser, Carbon nanotube-supported gold nanoparticles as efficient catalyst for the selective hydrogenation of nitroaromatic derivatives to anilines, Mater. Today Commun. 3 (2015) 104–113, https://doi.org/10.1016/j.mtcomm.2015.02.002.
[182] M. Blanco, P. Álvarez, C. Blanco, M.V. Jiménez, J.J. Pérez-Torrente, L.A. Oro, J. Blasco, V. Cuartero, R. Menéndez, Enhancing the hydrogen transfer catalytic activity of hybrid carbon nanotube-based NHC-iridium catalysts by increasing the oxidation degree of the nanosupport, Catal. Sci. Technol. 6 (2016) 5504–5514, https://doi.org/10.1039/c5cy01998b.
[183] R. Liu, J. Chen, X. Huang, L. Chen, L. Ma, X. Li, Conversion of fructose into 5-hydroxymethylfurfural and alkyl levulinates catalyzed by sulfonic acid-functionalized carbon materials, Green Chem. 15 (2013) 2895–2903, https://doi.org/10.1039/c3gc41139g.
[184] Y. Yıldız, İ. Esirden, E. Erken, E. Demir, M. Kaya, F. Şen, Microwave (Mw)-assisted synthesis of 5-substituted 1H-tetrazoles via [3+2] cycloaddition catalyzed by Mw-Pd/Co nanoparticles decorated on multi-walled carbon nanotubes, ChemistrySelect 1 (2016) 1695–1701, https://doi.org/10.1002/slct.201600265.
[185] S. Rana, S.B. Jonnalagadda, Covalently functionalized nano-graphene oxide for fine chemical synthesis, in: Functionalized Nanomaterials, InTech, 2016https://doi.org/10.5772/65941.
[186] Y. Cheng, Y. Fan, Y. Pei, M. Qiao, Graphene-supported metal/metal oxide nanohybrids: synthesis and applications in heterogeneous catalysis, Catal. Sci. Technol. 5 (2015) 3903–3916, https://doi.org/10.1039/c5cy00630a.
[187] G. Wu, X. Wang, N. Guan, L. Li, Palladium on graphene as efficient catalyst for solvent-free aerobic oxidation of aromatic alcohols: role of graphene support, Appl. Catal. B Environ. 136–137 (2013) 177–185, https://doi.org/10.1016/j.apcatb.2013.01.067.
[188] M. Li, S. Wu, X. Yang, J. Hu, L. Peng, L. Bai, Q. Huo, J. Guan, Highly efficient single atom cobalt catalyst for selective oxidation of alcohols, Appl. Catal. A Gen. 543 (2017) 61–66, https://doi.org/10.1016/j.apcata.2017.06.018.
[189] J. Xu, L. Luo, G. Xiao, Z. Zhang, H. Lin, X. Wang, J. Long, Layered $C_3N_3S_3$ polymer/graphene hybrids as metal-free catalysts for selective photocatalytic oxidation of

benzylic alcohols under visible light, ACS Catal. 4 (2014) 3302–3306, https://doi.org/10.1021/cs5006597.
[190] D. Ventura-Espinosa, C. Vicent, M. Baya, J.A. Mata, Ruthenium molecular complexes immobilized on graphene as active catalysts for the synthesis of carboxylic acids from alcohol dehydrogenation, Catal. Sci. Technol. 6 (2016) 8024–8035, https://doi.org/10.1039/c6cy01455k.
[191] Z. Li, S. Wu, C. Yang, Y. Ma, X. Fu, L. Peng, J. Guan, Q. Kan, Nano-Co_3O_4 supported on magnetic N-doped graphene as highly efficient catalyst for epoxidation of alkenes, Mol. Catal. 432 (2017) 267–273, https://doi.org/10.1016/j.mcat.2017.02.001.
[192] M. Cano, A.M. Benito, E.P. Urriolabeitia, R. Arenal, W.K. Maser, Reduced graphene oxide: firm support for catalytically active palladium nanoparticles and game changer in selective hydrogenation reactions, Nanoscale 5 (2013) 10189–10193, https://doi.org/10.1039/c3nr02822d.
[193] J. Zhao, W. Hu, H. Li, M. Ji, C. Zhao, Z. Wang, H. Hu, One-step green synthesis of a ruthenium/graphene composite as a highly efficient catalyst, RSC Adv. 5 (2015) 7679–7686, https://doi.org/10.1039/c4ra11397g.
[194] X. Ji, X. Niu, B. Li, Q. Han, F. Yuan, F. Zaera, Y. Zhu, H. Fu, Selective hydrogenation of cinnamaldehyde to cinnamal alcohol over platinum/graphene catalysts, ChemCatChem 6 (2014) 3246–3253, https://doi.org/10.1002/cctc.201402573.
[195] F. Huang, Y. Deng, Y. Chen, X. Cai, M. Peng, Z. Jia, P. Ren, D. Xiao, X. Wen, N. Wang, H. Liu, D. Ma, Atomically dispersed Pd on nanodiamond/graphene hybrid for selective hydrogenation of acetylene, J. Am. Chem. Soc. 140 (2018) 13142–13146, https://doi.org/10.1021/jacs.8b07476.
[196] H. Yan, H. Lv, H. Yi, W. Liu, Y. Xia, X. Huang, W. Huang, S. Wei, X. Wu, J. Lu, Understanding the underlying mechanism of improved selectivity in pd 1 single-atom catalyzed hydrogenation reaction, J. Catal. 366 (2018) 70–79, https://doi.org/10.1016/j.jcat.2018.07.033.
[197] J. Xi, H. Sun, D. Wang, Z. Zhang, X. Duan, J. Xiao, F. Xiao, L. Liu, S. Wang, Confined-interface-directed synthesis of palladium single-atom catalysts on graphene/amorphous carbon, Appl. Catal. B Environ. 225 (2018) 291–297, https://doi.org/10.1016/j.apcatb.2017.11.057.
[198] W. Wang, J. Gu, W. Hua, X. Jia, K. Xi, A novel high efficiency composite catalyst: single crystal triangular Au nanoplates supported by functional reduced graphene oxide, Chem. Commun. 50 (2014) 8889–8891, https://doi.org/10.1039/c4cc03306j.
[199] J. Huang, L. Zhang, B. Chen, N. Ji, F. Chen, Y. Zhang, Z. Zhang, Nanocomposites of size-controlled gold nanoparticles and graphene oxide: formation and applications in SERS and catalysis, Nanoscale 2 (2010) 2733–2738, https://doi.org/10.1039/c0nr00473a.
[200] S. Bai, X. Shen, G. Zhu, M. Li, H. Xi, K. Chen, In situ growth of Ni_xCo_{100-x} nanoparticles on reduced graphene oxide nanosheets and their magnetic and catalytic properties, ACS Appl. Mater. Interfaces 4 (2012) 2378–2386, https://doi.org/10.1021/am300310d.
[201] Y.S. Feng, J.J. Ma, Y.M. Kang, H.J. Xu, PdCu nanoparticles supported on graphene: an efficient and recyclable catalyst for reduction of nitroarenes, Tetrahedron 70 (2014) 6100–6105, https://doi.org/10.1016/j.tet.2014.04.034.
[202] R.V. Jagadeesh, K. Natte, H. Junge, M. Beller, Nitrogen-doped graphene-activated iron-oxide-based nanocatalysts for selective transfer hydrogenation of nitroarenes, ACS Catal. 5 (2015) 1526–1529, https://doi.org/10.1021/cs501916p.
[203] K.H. Lee, S.W. Han, K.Y. Kwon, J.B. Park, Systematic analysis of palladium-graphene nanocomposites and their catalytic applications in Sonogashira reaction,

J. Colloid Interface Sci. 403 (2013) 127–133, https://doi.org/10.1016/j.jcis.2013.04.006.
[204] S.K. Movahed, M. Fakharian, M. Dabiri, A. Bazgir, Gold nanoparticle decorated reduced graphene oxide sheets with high catalytic activity for Ullmann homocoupling, RSC Adv. 4 (2014) 5243–5247, https://doi.org/10.1039/c3ra45518a.
[205] K. Bhowmik, D. Sengupta, B. Basu, G. De, Reduced graphene oxide supported Ni nanoparticles: a high performance reusable heterogeneous catalyst for Kumada-Corriu cross-coupling reactions, RSC Adv. 4 (2014) 35442–35448, https://doi.org/10.1039/c4ra04834b.
[206] Y. Zhu, T. Cao, C. Cao, J. Luo, W. Chen, L. Zheng, J. Dong, J. Zhang, Y. Han, Z. Li, C. Chen, Q. Peng, D. Wang, Y. Li, One-pot pyrolysis to N-doped graphene with high-density Pt single atomic sites as heterogeneous catalyst for alkene hydrosilylation, ACS Catal. 8 (2018) 10004–10011, https://doi.org/10.1021/acscatal.8b02624.
[207] X. Wang, Y. Li, Nanoporous carbons derived from MOFs as metal-free catalysts for selective aerobic oxidations, J. Mater. Chem. A 4 (2016) 5247–5257, https://doi.org/10.1039/c6ta00324a.
[208] G. Yu, J. Sun, F. Muhammad, P. Wang, G. Zhu, Cobalt-based metal organic framework as precursor to achieve superior catalytic activity for aerobic epoxidation of styrene, RSC Adv. 4 (2014) 38804–38811, https://doi.org/10.1039/c4ra03746d.
[209] S. Pendem, R. Singuru, C. Sarkar, B. Joseph, J.-F. Lee, D.B. Shinde, Z. Lai, J. Mondal, Zeolitic imidazolate framework-mediated synthesis of Co_3O_4 nanoparticles encapsulated in N-doped graphitic carbon as an efficient catalyst for selective oxidation of hydrocarbons, ACS Appl. Nano Mater. 1 (2018) 4836–4851, https://doi.org/10.1021/acsanm.8b01027.
[210] C. Bai, A. Li, X. Yao, H. Liu, Y. Li, Efficient and selective aerobic oxidation of alcohols catalysed by MOF-derived Co catalysts, in: Green Chemistry, The Royal Society of Chemistry, 2016, pp. 1061–1069, https://doi.org/10.1039/c5gc02082d.
[211] C. Bai, X. Yao, Y. Li, Easy access to amides through aldehydic C–H bond functionalization catalyzed by heterogeneous Co-based catalysts, ACS Catal. 5 (2015) 884–891, https://doi.org/10.1021/cs501822r.
[212] B.R. Kim, J.S. Oh, J. Kim, C.Y. Lee, Robust aerobic alcohol oxidation catalyst derived from metal–organic frameworks, Catal. Lett. 146 (2016) 734–743, https://doi.org/10.1007/s10562-016-1700-2.
[213] W. Zhong, H. Liu, C. Bai, S. Liao, Y. Li, Base-free oxidation of alcohols to esters at room temperature and atmospheric conditions using nanoscale Co-based catalysts, ACS Catal. 5 (2015) 1850–1856, https://doi.org/10.1021/cs502101c.
[214] Y. Li, Y.X. Zhou, X. Ma, H.L. Jiang, A metal-organic framework-templated synthesis of γ-Fe_2O_3 nanoparticles encapsulated in porous carbon for efficient and chemoselective hydrogenation of nitro compounds, Chem. Commun. 52 (2016) 4199–4202, https://doi.org/10.1039/c6cc00011h.
[215] H. Niu, S. Liu, Y. Cai, F. Wu, X. Zhao, MOF derived porous carbon supported Cu/Cu_2O composite as high performance non-noble catalyst, Microporous Mesoporous Mater. 219 (2016) 48–53, https://doi.org/10.1016/j.micromeso.2015.07.027.
[216] B. Tang, W.C. Song, E.C. Yang, X.J. Zhao, MOF-derived Ni-based nanocomposites as robust catalysts for chemoselective hydrogenation of functionalized nitro compounds, RSC Adv. 7 (2017) 1531–1539, https://doi.org/10.1039/c6ra26699a.
[217] J. Long, Y. Zhou, Y. Li, Transfer hydrogenation of unsaturated bonds in the absence of base additives catalyzed by a cobalt-based heterogeneous catalyst, Chem. Commun. 51 (2015) 2331–2334, https://doi.org/10.1039/c4cc08946d.
[218] X. Liu, S. Cheng, J. Long, W. Zhang, X. Liu, D. Wei, MOFs-derived Co@CN bi-functional catalysts for selective transfer hydrogenation of α,β-unsaturated

aldehydes without use of base additives, Mater. Chem. Front. 1 (2017) 2005–2012, https://doi.org/10.1039/c7qm00189d.
[219] J. Long, K. Shen, Y. Li, Bifunctional N-doped Co@C catalysts for base-free transfer hydrogenations of nitriles: controllable selectivity to primary amines vs imines, ACS Catal. 7 (2017) 275–284, https://doi.org/10.1021/acscatal.6b02327.
[220] K. Shen, L. Chen, J. Long, W. Zhong, Y. Li, MOFs-templated Co@Pd core-shell NPs embedded in N-doped carbon matrix with superior hydrogenation activities, ACS Catal. 5 (2015) 5264–5271, https://doi.org/10.1021/acscatal.5b00998.
[221] H. Yang, S.J. Bradley, A. Chan, G.I.N. Waterhouse, T. Nann, P.E. Kruger, S. G. Telfer, Catalytically active bimetallic nanoparticles supported on porous carbon capsules derived from metal-organic framework composites, J. Am. Chem. Soc. 138 (2016) 11872–11881, https://doi.org/10.1021/jacs.6b06736.
[222] J. Long, K. Shen, L. Chen, Y. Li, Multimetal-MOF-derived transition metal alloy NPs embedded in an N-doped carbon matrix: highly active catalysts for hydrogenation reactions, J. Mater. Chem. A 4 (2016) 10254–10262, https://doi.org/10.1039/c6ta00157b.
[223] K. Murugesan, T. Senthamarai, M. Sohail, A.S. Alshammari, M.M. Pohl, M. Beller, R.V. Jagadeesh, Cobalt-based nanoparticles prepared from MOF-carbon templates as efficient hydrogenation catalysts, Chem. Sci. 9 (2018) 8553–8560, https://doi.org/10.1039/c8sc02807a.
[224] Y. Wang, S. Sang, W. Zhu, L. Gao, G. Xiao, CuNi@C catalysts with high activity derived from metal-organic frameworks precursor for conversion of furfural to cyclopentanone, Chem. Eng. J. 299 (2016) 104–111, https://doi.org/10.1016/j.cej.2016.04.068.
[225] X. Wang, W. Chen, L. Zhang, T. Yao, W. Liu, Y. Lin, H. Ju, J. Dong, L. Zheng, W. Yan, X. Zheng, Z. Li, X. Wang, J. Yang, D. He, Y. Wang, Z. Deng, Y. Wu, Y. Li, Uncoordinated amine groups of metal-organic frameworks to anchor single Ru sites as chemoselective catalysts toward the hydrogenation of quinoline, J. Am. Chem. Soc. 139 (2017) 9419–9422, https://doi.org/10.1021/jacs.7b01686.
[226] S. Mutyala, R. Reddy Kanchamreddy, M. Jonnalagadda, D.R. Burri, K.S. Rama Rao, M. Lakshmi Kantam, P. Srinivasu, Catalytic dehydration of 1-phenylethanol over chromia-carbon composite derived from metal organic framework, CrMIL-101, Indian J. Chem. A Inorg. Phys. Theor. Anal. Chem. 53 (2014) 572–575.
[227] Y. Yoon, B.R. Kim, C.Y. Lee, J. Kim, Heterogeneous copper-catalyzed aerobic oxidative conversions of benzaldehydes with aqueous ammonia to give benzonitriles, Asian J. Org. Chem. 5 (2016) 746–749, https://doi.org/10.1002/ajoc.201600121.
[228] X. Li, J. Zhang, W. Li, MOF-derived nitrogen-doped porous carbon as metal-free catalysts for acetylene hydrochlorination, J. Ind. Eng. Chem. 44 (2016) 146–154, https://doi.org/10.1016/j.jiec.2016.08.024.
[229] S. Chao, F. Zou, F. Wan, X. Dong, Y. Wang, Y. Wang, Q. Guan, G. Wang, W. Li, Nitrogen-doped carbon derived from ZIF-8 as a high-performance metal-free catalyst for acetylene hydrochlorination, Sci. Rep. 7 (2017) 39789, https://doi.org/10.1038/srep39789.
[230] X. Li, B. Zhang, Y. Fang, W. Sun, Z. Qi, Y. Pei, S. Qi, P. Yuan, X. Luan, T.W. Goh, W. Huang, Metal-organic-framework-derived carbons: applications as solid-base catalyst and support for Pd nanoparticles in tandem catalysis, Chem. A Eur. J. 23 (2017) 4266–4270, https://doi.org/10.1002/chem.201605852.
[231] M.J. Kim, B.R. Kim, C.Y. Lee, J. Kim, N-sulfonyl amidine synthesis via three-component coupling reaction using heterogeneous copper catalyst derived from metal-organic frameworks, Tetrahedron Lett. 57 (2016) 4070–4073, https://doi.org/10.1016/j.tetlet.2016.07.085.

[232] L. Li, L. Li, C. Cui, H. Fan, R. Wang, Heteroatom-doped carbon spheres from hierarchical hollow covalent organic framework precursors for metal-free catalysis, ChemSusChem 10 (2017) 4921–4926, https://doi.org/10.1002/cssc.201700979.
[233] X. Hu, Y. Long, M. Fan, M. Yuan, H. Zhao, J. Ma, Z. Dong, Two-dimensional covalent organic frameworks as self-template derived nitrogen-doped carbon nanosheets for eco-friendly metal-free catalysis, Appl. Catal. B Environ. 244 (2019) 25–35, https://doi.org/10.1016/j.apcatb.2018.11.028.
[234] L. Chen, L. Zhang, Z. Chen, H. Liu, R. Luque, Y. Li, A covalent organic framework-based route to the: in situ encapsulation of metal nanoparticles in N-rich hollow carbon spheres, Chem. Sci. 7 (2016) 6015–6020, https://doi.org/10.1039/c6sc01659f.
[235] I. Matos, M. Bernardo, I. Fonseca, Porous carbon: a versatile material for catalysis, Catal. Today 285 (2017) 194–203, https://doi.org/10.1016/j.cattod.2017.01.039.
[236] T. Friščić, P.A. Julien, C. Mottillo, Environmentally-friendly designs and syntheses of metal-organic frameworks (MOFs), in: ACS Symposium Series, 2014, pp. 161–183, https://doi.org/10.1021/bk-2014-1186.ch009.
[237] Y.Z. Chen, R. Zhang, L. Jiao, H.L. Jiang, Metal–organic framework-derived porous materials for catalysis, Coord. Chem. Rev. 362 (2018) 1–23, https://doi.org/10.1016/j.ccr.2018.02.008.
[238] X.F. Yang, A. Wang, B. Qiao, J. Li, J. Liu, T. Zhang, Single-atom catalysts: a new frontier in heterogeneous catalysis, Acc. Chem. Res. 46 (2013) 1740–1748, https://doi.org/10.1021/ar300361m.

CHAPTER 10

Emerging carbon nanostructures in electrochemical processes

S. Pérez-Rodríguez, C. Alegre, D. Sebastián, and M.J. Lázaro
Instituto de Carboquímica, CSIC, Zaragoza, Spain

1 Carbon nanostructures in electrochemistry

Carbon is well-known to form a large diversity of materials with a vast variety of properties. This is a result of the combination of several particular characteristics, such as the three different hybridization of atoms (sp, sp^2, sp^3), the spatial arrangement of carbon structures at different levels (nano-, micro-, and macro-), or bulk and surface properties of particles, just to name a few. One carbon atom has four electrons at its outermost orbitals (2s, 2p), which can present three different hybridizations depending on the number of σ and π electrons: sp (2 and 2), sp^2 (3 and 1), and sp^3 (4 σ electrons). The repetition of C—C bonds using sp^3 hybrid orbitals gives place to well-known diamond. However, the combination of carbon atoms into complex structures with sp, sp^2, and sp^3 hybrid bonds gives place to organic hydrocarbons from aliphatic to aromatic types. In an upper level of organization, carbon can form one-dimensional structures, like nanofilaments, two-dimensional structures like graphene, or three-dimensional structures like those constituted by spheroidal agglomerates. There also exist the so-called zero-dimensional carbons like fullerenes, with the smallest one consisting of 12 pentagons and 20 hexagons of carbon atoms (C_{60} or buckminsterfullerene).

In history, carbon has always had a relevant importance for the development of humanity; for instance, the use of coal for heating, soot for the production of inks and the consequent spread of knowledge by means of written documents, graphite for steel production, or carbon electrodes for the primary batteries. In our days, the development of carbon nanostructures is promoting nanoscience and technology in various fields. Nanomaterials are characterized by a characteristic dimension not higher than 100 nm. They have attracted great interest since the very beginning of nanoscience and nanotechnology due to their unique and on-demand tunable properties.

Emerging Carbon Materials for Catalysis
https://doi.org/10.1016/B978-0-12-817561-3.00010-X

The main peculiarity about nanostructures is based on the appearance of a specific behavior of some properties (conductivity, mechanical, etc.) only when these materials are in nanometric scale.

Electrochemistry, that is, the discipline that studies the combination of chemical reactions together with the exchange of electrons, plays a very important role in the investigation and development of energy storage and conversion devices like batteries, fuel cells, supercapacitors, or electrolyzers.

In this book chapter, a series of emerging carbon nanostructures like carbon nanofibers (CNFs), carbon nanotubes (CNTs), graphene, carbon gels, and mesoporous carbons are reviewed together with potential applications in electrochemistry. The chapter sections are organized attending to the characteristic dimension of the different structures, that is, one-dimensional for filaments, two-dimensional for sheets, and three-dimensional for the rest of nanostructures.

2 One-dimensional carbon nanostructures

2.1 Carbon nanofibers

CNFs are filaments of nanometer size with diameters from 3 to 100 nm, formed by the stack of graphene layers in a certain orientation. The core of the filament can be either: continuously hollow, giving place to tubes (or nanotubes); discontinuously hollow, with a structure that resembles a bamboo; and completely full or rod-like. Moreover, the angle between the graphene layers and the fiber growth axis can take any value between 0 degree (parallel nanofibers) and 90 degree (platelet nanofibers), with most common values in between those two to result in fishbone or herringbone nanofibers [1–3].

CNF can be grown by mainly two methodologies, the precipitation of carbon on certain metal surfaces (catalytic approach at high temperature) or the electrospinning of a carbon containing liquid precursor [4]. The choice of the methodology to obtain CNFs relies on the specific characteristics for the desired product. Carbon has a turbostratic nature in CNF, with an average spacing between graphene layers of 0.34 nm, positioned in between graphite (0.335 nm) and nanotubes (about 0.36 nm) [5].

CNFs are good catalyst supports in specific applications due to their resistance to acid/alkaline media, good electrical and thermal conductivity, and external surface area easily accessible to gases and liquids [6, 7]. Unlike in parallel nanofibers (or CNT), the surface of CNF is present as graphite edges which facilitate a strong interaction between the support and the active phase [8]. The porosity of CNF results from the void space among filaments, with surface areas not higher than $300\,m^2/g$ and with a negligible contribution of micropores [9].

The advantageous behavior of CNFs as catalyst support has been reported in the case of electrocatalysts for electrochemical applications like fuel cells [10–13]. CNFs are more resistant to degradation due to the lower electrooxidation rate, compared to conventional carbon supports like carbon blacks [14]. In Fig. 1, the evolution of the oxidation rate with time of a CNF is reported and compared to a typically used carbon black (Vulcan XC 72R). The charge associated to carbon oxidation for CNF is half the value obtained for carbon black, which indicates a larger resistance to harsh conditions.

The main difficulty with CNF as support for fuel cell electrocatalysts arises from their relatively low surface area. A high metal (Pt-based) concentration is required to minimize the catalytic layer thickness and thus minimize mass and electron transfer constrains, whereas a high dispersion of noble metal nanoparticles is required to maximize catalyst utilization. This drawback has been overcome by using more advanced synthesis procedures

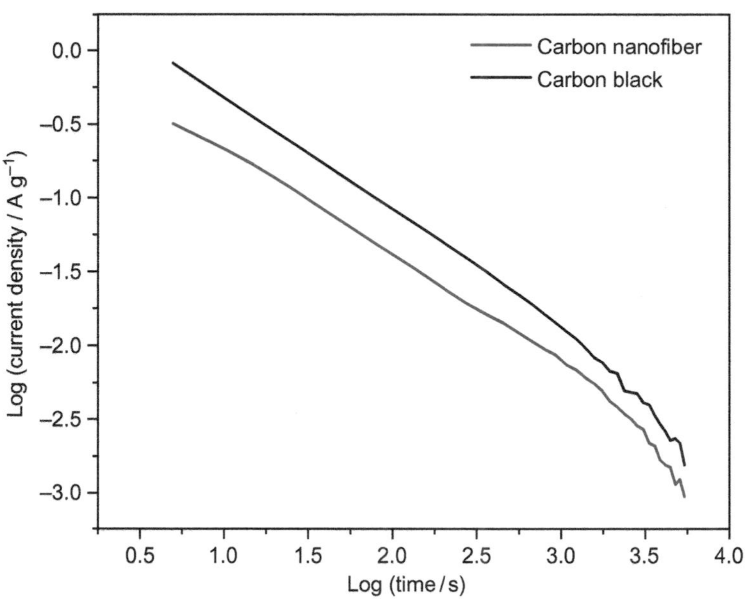

Fig. 1 Comparison of current-time curves for electrooxidation of carbon at 1.2 V versus reversible hydrogen electrode (RHE) in 0.5 M H_2SO_4 for a carbon nanofiber and a commercial carbon black (Vulcan) [14].

that guarantee the production of well-dispersed catalyst nanoparticles regardless of the nature and characteristics of the carbon support.

Initial works to investigate CNF as electrocatalyst support date to 2001, when Bessel et al. observed an improvement in the oxidation of methanol and a better behavior to the presence of carbon monoxide when compared to conventional supports [10]. They attributed this to the preferential formation of certain crystallographic orientations of metal particles on the surface of CNF. Steigerwalt investigated different CNF structures as PtRu support [11, 15, 16]. They concluded that the best activity was achieved with herringbone type CNF. The following works in the early 2000s extended the knowledge on the use of CNFs as catalyst supports for fuel cells [12, 17, 18]. For the catalyst at the anode, both for the oxidation of methanol or hydrogen, the utilization of CNF as support can be beneficial as the combination of the effects of a high dispersion and lower mass transfer constrains. Yang and coworkers observed also a better activity of CNF-supported catalyst for the oxygen reduction reaction (ORR), occurring at the cathode side of fuel cells [19].

In 2006, Lee and collaborators published a paper where they reviewed the state of the art of fuel cell catalysts supported on either CNT or nanofibers [20]. The part dedicated to nanofibers was not so exhaustive, indicating that it was an emerging field of research. The stability was an important aspect for CNF-supported catalyst. In this regard, Zheng et al. and Chou et al. investigated the influence of the spatial configuration of graphene planes in CNF [21, 22]. They concluded that the stability is favored when supporting nanoparticles on parallel type CNF, while the activity was better for the platelet-based CNF. Interestingly, herringbone type CNF presented intermediate characteristics in between the two previous structures. The methodology to incorporate metal nanoparticles is very relevant for the behavior of the catalyst in terms of activity and stability [13]. The relatively low surface area of CNF makes colloidal methodologies the most appropriate to control metal particle size in optimum ranges, like for Pt in the interval 2–3 nm.

CNF can be chemically treated to introduce certain functionalities of interest. Most common procedures involve the oxidation in strong acidic environment to increase surface oxygen concentration. The oxidation of the surface can lead to the decrease of apparent electrical conductivity of CNF [23]. According to several publications [24–27], oxidation appears to favor the catalytic activity as a function of the type of CNF and the kind and degree of oxidation, despite the decrease of surface area compared to untreated carbon support. It is required to balance three main characteristics to maximize the activity of oxidized CNF as support: a sufficient electrochemical surface area of the catalyst, a good metal-support interaction that can be favored upon oxidation, and a good diffusion of reactants like methanol, result of enhanced wettability of oxidized supports [28, 29].

CNFs have also been explored for other electrochemical applications for the storage of energy, where carbon materials are of paramount importance, like batteries or supercapacitors. Zhang and coworkers reviewed in 2016 the recent advances in this field for electrospun CNF [30]. The electrospinning process involves the application of an electrostatic field to a capillary connected to a reservoir with the polymer precursor solution [31]. The good electrical conductivity together with the accessible surface area of CNFs make them ideal candidates for energy storage in applications such as lithium-ion batteries, redox flow batteries, metal-air batteries, or supercapacitors. Some important factors are identified to determine the electrochemical performance: electrical conductivity, nanofiber diameter, porosity, and chemical composition. All of them can be modified by the adequate selection of synthesis parameters.

2.2 Carbon nanotubes

CNTs consist of carbon nanofilaments, which are based on tubular one-dimensional nanostructures, formed by coaxial layers of graphene sheets. Structurally, CNT can be divided into two basic forms: (i) single wall carbon nanotubes (SWCNT), which consist of a single tube of graphene and (ii) multi wall carbon nanotubes (MWCNT), formed by several concentric graphene layers. The diameter of CNT is around 0.4–3 nm and 1.4–100 nm for SWCNT and MWCNT, respectively, while the length ranges from few microns to several mm [32].

Although CNTs were first observed in 1952 by Radushkevich and Lukyanovich [33] or 1976 by Oberlin et al. [34] in the case of SWCNT, these nanostructures did not attract special attention until their discovery by Iijima in 1991 for MWCNT [35] or 1993 for SWCNT [36]. Since 1991, several synthesis routes of CNT have been reported, such as arch discharge, laser ablation, chemical vapor deposition (CVD), electrolysis approaches, or sono-chemical/hydrothermal routes [37–39]. Although high temperature methodologies, such as arch discharge and laser ablation, were first used to obtain CNT, CVD is the most popular method for producing CNT, nowadays. As compared to arc-discharge and laser-ablation methods,

CVD is economic and requires milder synthesis conditions (temperature <800°C, ambient pressure). Additionally, a higher CNT yield and purity can be obtained with a precise control on the orientation, structure, or architecture of CNT [37, 40]. The CVD methodology usually involves the catalytic decomposition of a carbon source (usually a hydrocarbon, such as methane, ethane, ethylene, etc.) at the surface of a catalyst. The most frequently used catalysts are transition metals, primarily Fe, Co, or Ni. Different carbon nanofilaments have been obtained by our research group by the catalytic decomposition of methane or biogas using different metal active sites (Fe, Ni, Mo, Cu) and textural promoters (Al_2O_3, MgO) [41–43].

CNTs have been used in many applications due to their unique electrical, mechanical, optical, and thermal properties. The potential applications of CNT include, but are not limited to, conductive films [44], solar cells [45], catalysis [46, 47], fuel cells [48], supercapacitors [49], electronic devices [50], separation membranes [51], sensors [52], etc. The high conductivity, superior surface area, and tubular structure, which allow both the nanoconfinement of metal nanoparticles in the inner cavity and the deposition on the outer walls, have driven the use of CNT in electrocatalysis [46, 48, 49, 53–55].

CNTs have been extensively used as catalyst support in proton exchange membrane fuel cells (PEMFC) and direct alcohol fuel cells (DAFC) obtaining a high performance and catalyst durability [56–59]. Compared to carbon blacks usually used in electrochemical energy conversion devices, CNT presents a superior electrical conductivity and carbon crystallinity. The latter results in a higher resistance to carbon corrosion, as well as a greater stability and catalyst durability [56]. CNT presents a developed and accessible porosity, which facilitates the dispersion of metal nanoparticles [20, 57, 60]. Additionally, metal nanoparticles may be deposited on the inner and outer walls of CNTs, enhancing the electrochemical surface area [54, 61]. In contrast, CNT-surface is very inert with few defects/binding sites and presents a poor solubility in aqueous solutions, which may hinder the effective anchoring of metal nanoparticles in pristine CNT. In order to overcome these issues, covalent [62] and noncovalent [63] functionalization strategies are often required to modify the CNT surface, and hence, to improve the dispersion of metal nanoparticles. In this context, Jha et al. [59] studied the PEMFC performance using Pt catalysts supported on chemically modified SWCNT with varying degrees of functionalization. The optimized SWCNT-COOH functionality allowed the construction of a PEMFC prototype with a Pt loading of $0.06\,mg\,cm^{-2}$, well below the value of $0.125\,mg\,cm^{-2}$ set as the US DOE 2017 technical target for total Pt group metals (PGM) loading.

A better fuel cell performance has been reported using CNT as catalyst support than that reached with a carbon black [57, 58, 60]. In hydrogen fuel cells (PEMFC), a 10% higher fuel cell voltage was obtained using Pt/CNT-based electrodes in both anode and cathode of a PEMFC, in comparison to a commercial Pt/C electrode, and reducing the Pt usage by 60% [57]. Similarly, an improved performance in a direct methanol fuel cell (DMFC) has been obtained using CNTs as catalyst supports, instead of a commercial carbon black [58, 60]. For example, Li et al. reported a maximum power density of a single cell of $103\,mW\,cm^{-2}$ using a Pt/CNT cathode, while a value of $70\,mW\,cm^{-2}$ was obtained with a commercial Pt/C [58]. On the other hand, Kamat and coworkers [60] showed a higher activity for the methanol oxidation or the oxygen reduction at Pt deposited on SWCNT film than unsupported Pt.

PtRu alloys supported on CNT have been prepared to solve the CO poisoning problem in DAFC, obtaining an improved condition for CO and alcohol oxidation [64–67]. In this context, Prabhuram et al. [66] found a higher mass activity for the methanol oxidation reaction at a PtRu/CNT catalyst than that obtained on PtRu catalyst supported on a carbon black. Other binary systems (such as PtCo [68], PtNi [69], $PtWO_3$ [70], PdFe [71], etc.) or ternary catalytic systems (e.g., PtRuNi [72]) have been also used to enhance CO and alcohol oxidation obtaining promising results.

Surface chemistry of CNT plays also an important role in the activity and stability of the resultant electrode as the interaction between the support and active sites can modify the electronic properties of the catalyst [73, 74]. In low temperature fuel cells, nitrogen-doped CNTs have demonstrated an enhancement of the metal nanoparticles dispersion and a catalyst durability [75–78]. Consequently, a higher performance for the ORR in the cathode or the oxidation of organic molecules, such as methanol or ethanol, which takes place in the anode, has been obtained. Recently, Begum et al. [78] designed a strategy to fabricate CNT-based PdM (M=Mn, Co and Cu) bimetallic catalysts through four-nitrogen system onto CNT (denoted as PdM-N_4/CNTs). The material was used as an efficient bifunctional electrocatalyst for electrochemical ethanol oxidation reaction and hydrogen evolution reaction in 1 M KOH electrolyte. PdMn-N_4/CNTs exhibited superior activity, better reaction kinetics, and higher durability compared to the benchmark Pt/C.

Nanoparticles of nonnoble metals, such as Ni, supported on N-doped CNT have also evidenced a great performance for hydrogen oxidation in alkaline media [79]. Indeed, Yan and coworkers found an increase of the mass catalytic activity for hydrogen oxidation by factor of 33 in comparison to unsupported Ni nanoparticles [79]. Transition metal catalysts encapsulated into nitrogen-doped CNT (M-N-CNTs, M = Ni, Co, Fe) have also obtained an outstanding performance for overall water splitting [80].

Apart from the use of CNTs as catalyst supports, metal-free catalysts based on CNT doped with different heteroatoms (such as N, B, etc.) have attracted special attention in the last years [81–83]. A high activity for the ORR, good stability and immunity towards methanol crossover, and CO poisoning have been showed at metal-free doped-CNT in alkaline media. For example, these metal-free CNT catalysts have been used in other electrochemical reactions, for example, the electrochemical reduction of carbon dioxide [84, 85].

Recently, CNTs have also been used to fabricate new electrolytes for fuel cells [86]. An enhanced proton conductivity and mechanical stability were obtained by the incorporation of SWCNT prefunctionalized with sulfonic acid groups into a Nafion matrix.

3 Two-dimensional carbon nanostructures

The most relevant 2-D carbon nanostructure of our times is undoubtedly graphene. Graphene is essentially a one atom thick layer of sp^2 carbon atoms. Since the groundbreaking experiments on a single layer graphene sheet in 2004 by Geim and Novoselov [87], which was worth of a Nobel Prize in Physics in 2010, graphene materials have become of paramount relevance in different research fields due to the promising and amazing properties that can potentially arise from its application. Graphene possesses high thermal and chemical stability, high mechanical strength, high electrical conductivity, and large surface area [88].

The number of publications related to graphene has rapidly increased since the famous publication of Geim and Novoselov, as can be seen in Fig. 2. According to Scopus, in 2010, the number of publications related to graphene barely achieved 5000, while in 2017 the number overcame 20,000. Interestingly, the number of publications related to graphene and electrochemistry has greatly increased. Before 2010, the percentage of publications related to electrochemistry within the graphene-related ones was below 5%, increasing up to almost 20% in the last 5 years. This is a clear indication of the relevant role that electrochemistry and electrochemical research works are playing in graphene materials. This section does not pretend to be an exhaustive review of such a large number of research works,

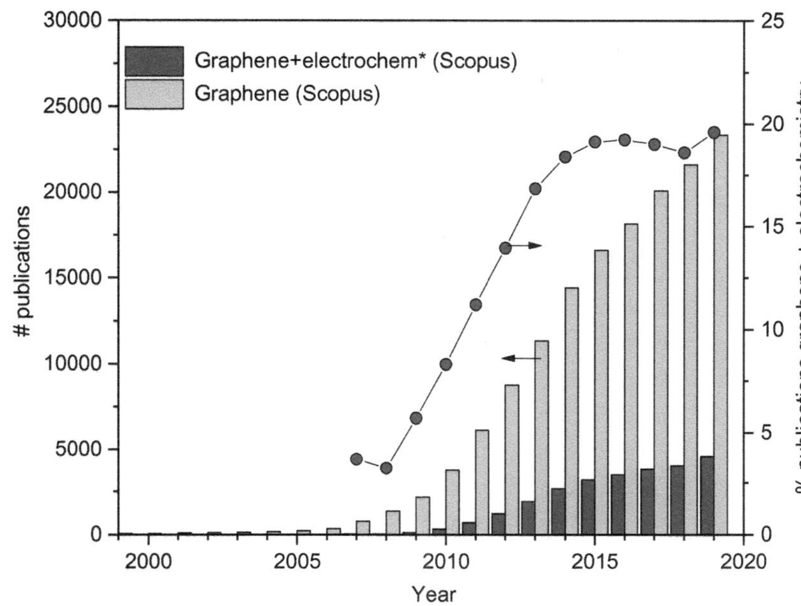

Fig. 2 Evolution of the number of publications related to graphene and to graphene + electrochemistry (electrochem*) according to Scopus, 2020. The *red points* (dark *gray points* in print version) indicate the percentage of graphene + electrochem* search results with respect to graphene results.

but rather a brief summary of the most remarkable findings and general conclusions.

The most common methodology to obtain graphene oxide (GO), the precursor for graphene materials, is the oxidative treatment of graphite in concentrated acid with a strong oxidizing agent. There are several methods to perform such an oxidation, like those described by Brodie [89] or Staudenmaier [90], but the most common ones refer to Hummers and Offeman method [91], more recently modified by Tour and coworkers [92]. The latter consists basically of the oxidation of graphite with potassium permanganate in concentrated sulfuric acid. GO can then be reduced by thermal, chemical, or electrochemical treatment to obtain the so-called reduced graphene oxide (rGO). This is the common top-down approach to produce rGO, which can be also performed using alternative carbon sources like CNT or CNF [93, 94]. There are also bottom-up strategies to synthesize graphene from quantum dots, in turn obtained by chemical assembly methods or template-assisted deposition to build up controllable three-dimensional structures.

Few layer graphene nanosheets have been investigated as electrocatalyst support for fuel cell application [95]. In this particular case, the high surface area of graphene, together with its high electrical conductivity, make it a very promising material to act as a support in electrocatalysis [96, 97]. The high surface area of graphene allows a high metal loading maintaining a high dispersion, with particle sizes in the interval 2–3 nm and metal concentration over 60 wt% [98–100]. Moreover, the surface groups on graphene nanosheets act as anchoring sites for Pt and prevent the agglomeration of nanoparticles [101].

The main drawback arising from the use of two-dimensional materials is the stacking of the layers, which results in a much lower exposed surface for the active sites than expected. One reported strategy to prevent stacking is to use a nanoscale material as spacer between nanosheets. CNT [102] and carbon black [103] have been explored in this field with an enhancement of the behavior of the catalyst as derived from a larger accessible surface area, enabled by the presence of the spacer.

A topic of great and recent interest is the investigation of noble-metal-free catalysts using graphene as a backbone nanostructure. An important number of investigations has focused on the modification of the chemical composition at the surface of rGO by introducing heteroatoms like nitrogen, boron, sulfur, or phosphorus [104–108]. Undoubtedly, nitrogen offers the best perspectives to increase the activity of metal-free graphene-based catalysts from both theoretical point of view and confirmed experimental evidences [109]. The improvement of activity of doped graphene was ascribed to the density of nitrogen-carbon active sites [104]. Nitrogen being more electronegative than carbon, neighboring C—N atoms make carbon deficient in electrons, therefore promoting oxygen adsorption. Still, there is no consensus on the correlation between nitrogen speciation, that is, the different forms on nitrogen bonded to carbon, and ORR electroactivity. Some authors observed that increasing quaternary nitrogen content on carbon resulted in an improvement of ORR activity [110, 111]; on the other hand, some other works led to conclude that it is the pyridinic moiety, the one that favors ORR activity [112, 113]. In this open discussion, some works point out that both pyridinic and quaternary nitrogen species contribute to the electroactivity [114].

The combination of graphene and metal nanoparticles in nanocomposites represents another interesting field of research towards the enhancement of activity and the reduction of costs [115, 116]. It is the synergic effects derived from the combination of graphene and metallic nanoparticles, like cobalt or iron, that result in a higher activity than the single phases considered separately [117, 118]. The two-dimensional structure of graphene offers the opportunity to finely disperse active sites on the surface of carbon sheets, promoting a high density of active sites. The role of transition metals in these nanocomposites has been proposed to rely on the coordination of metal cations with nitrogen-carbon species. In this regard, the efforts are now being directed to the elucidation of the role of carbon nanostructures, and graphene with special interest, at the atomic level and the local bonding environment of carbons as they appear fundamental to design highly active and cost-effective electrocatalysts.

4 Three-dimensional carbon nanostructures

4.1 Carbon gels

Carbon gels are a kind of porous materials with very interesting features given by their reticular structure formed of interconnected spherical carbon nanoparticles [119, 120]. Carbon gels are usually obtained by a sol-gel reaction that involves the use of hydroxylated benzenes and aldehydes, resorcinol and formaldehyde being the most employed ones [120–123]. The particularity of these materials is that the sol-gel synthesis allows the fine-tuning of their structural and textural properties so that carbon gels can be tuned to fit the requirements of a particular application (supercapacitors, insulators, catalysts supports, water decontamination, etc.) [119, 124, 125]. Besides, carbon gels can be synthesized in different forms, from monoliths to powders, films, or microspheres, which extends their potential application in several fields [126–128].

The synthesis of carbon gels includes four main steps: (i) aqueous sol-gel synthesis, known as gelation process, in which the hydroxylated benzene (usually resorcinol, R) reacts with an aldehyde (usually formaldehyde, F) normally in stoichiometric molar ratio, R/F=0.5; (ii) curing process, in which the mixture of R and F is heated (usually 85°C), allowing the cross-linking of the reticular network; (iii) drying; and (iv) pyrolysis (between 600 and 1200°C). Sometimes, prior to the drying step, there is an extra step consisting of the exchange of solvent (usually water) by an organic solvent with a lower surface tension (acetone, ethanol, CO_2, etc.). This way, the shrinkage of the organic gel upon drying is minimized, allowing to better preserve the porous structure [120, 126].

Carbon gels can be classified into three types depending on the drying method employed: freeze drying, produces carbon cryogels (CCGs);

supercritical drying creates carbon aerogels (CAGs); and ambient drying (also called subcritical drying) produces carbon xerogels (CXGs) [120].

The final properties of the carbon gel depend on the accurate control of the synthesis parameters of the organic gel: (i) pH of sol-gel solution; (ii) the concentration and type of reactants (either resorcinol/formaldehyde or other hydroxylated benzenes and aldehydes, like phenol, furfural, etc.); (iii) drying conditions; (iv) temperature and time of gelation and curing; (v) pyrolysis temperature; and (vi) type of solvent [119].

Since their discovery by Pekala et al. in 1989 [123], carbon gels have obtained increasing popularity due to their unique features including high purity, low density, both low thermal conductivity and dielectric constant, specific surface areas ranging from 400 to 1200 m^2/g, high pore volumes (up to 2 cm^3/g), and electrical conductivities in the order of hundreds of S/m.

There is a great number of reviews and book chapters dealing with the synthesis of this type of carbon materials and their features. The focus of this chapter is on the application of these materials in electrochemical devices. For further information on the synthesis of carbon gels and their properties, the reader is addressed to the following references [126–128]. In particular, in the last 15 years, there has been a special interest on the use of these materials as photocatalysts and electrocatalysts.

Carbon materials have been employed as electrocatalysts supports for decades due to several features, such as their good electrical conductivity, chemical stability, good mechanical properties, adjustable porous texture, and modifiable surface [127, 129–131]. Carbon-supported electrocatalysts consist of nanoparticles of the active phase, usually a metal, dispersed in a porous matrix, which increases the exposed surface of the active phase particles, and hence, its reactivity. It is now solidly proved that the nature of the support as well as the interaction between the former and the metal determines important properties of catalysts such as dispersion, stability, and morphology of metallic crystallites. Besides, it has also been demonstrated that the features of the support alter the mass transport and the electronic properties of metal and their stability. Hence, the optimization of carbon supports is of paramount relevance for the proper design of a highly performing electrocatalyst [132–135].

Traditionally, carbon blacks are the most common materials employed as electrocatalyst support [129, 136, 137]. However, in the last 15 years, the research on the use of carbon gels as electrocatalyst support has risen almost exponentially. Over the past 10 years, Alegre and coworkers have studied the use of CXGs as support for Pt and PtRu electrocatalysts [124, 138–143]. Authors studied the influence of the synthesis conditions on the properties of CXGs, analyzing the effect of the composition of the mixture of carbon gel precursors and the pH on the porous structure, the arrangement of carbon, the surface chemistry, and the electrical conductivity [124]. Certain conditions allowed obtaining CXGs with high porosity, high ordering, and polymerization degree. CXGs with higher ordering and polymerization degree resulted in Pt-catalysts with low crystal sizes and good metallic distribution. Such catalysts showed the highest activities towards methanol oxidation, outperforming a Vulcan-supported Pt catalyst [124]. On other hand, the specific activity towards the ORR, that is, the limiting catalytic process in low temperature fuel cells, is significantly favored by highly ordered CXGs due to an enhanced metal-support interaction [119]. Nevertheless, surface defects favor the distribution of the metallic particles on the surface of carbon, which in the end influences the effectiveness of the catalyst. Accelerated degradation tests were conducted to evaluate catalyst stability under potential cycling conditions. The observed decay of performance was considerably lower for the catalysts based on ordered CXGs stabilizing Pt particles in a higher extent than xerogels with a lower ordering degree or Pt-catalysts supported on commercial carbon black.

The same authors also paid attention to the synthesis method of the electrocatalyst, as the success for obtaining a highly dispersed catalyst will depend both on the synthesis method and the support [138, 140]. Different synthesis routes led to catalysts having different physicochemical properties, crystal sizes, and metallic distribution of particles on the surface. These parameters lead to differences in the electrocatalytic activity in a fuel cell. Lower crystal sizes and higher amounts of reduced Pt provided the highest active catalysts for the ORR. On the other hand, methanol oxidation was favored in the case of catalysts supported on highly mesoporous CXGs, as the diffusion of a molecule like methanol determines the catalytic activity. The synthesis method of the catalyst with high Pt loadings (>40%) was also thoroughly studied by Job et al. in several works [144–148]. This group studied the strong electrostatic adsorption (SEA) method, which consists in enhancing electrostatic interactions between the support and the precursor and the charge-enhanced dry impregnation (CEDI).

Alegre and coworkers have also studied the effect of the modification of the surface of CXGs with heteroatoms (O, N, and S) on the activity of Pt and PtRu catalysts for fuel cell-related reactions, that is, methanol oxidation and ORR [139, 142, 143, 149]. Alegre et al. determined that functionalizing

CXGs with oxygen was beneficial for the catalytic activity of Pt and PtRu electrocatalysts towards methanol oxidation [138, 139, 150]. The presence of surface oxygen groups on the surface of the carbon material resulted in enhanced dispersion, lower Pt crystal size, and improved catalytic performance compared to nontreated CXGs. However, when very stringent oxidation treatments were employed (e.g., concentrated nitric acid), Pt particles agglomerated on the surface of the carbon material. Despite Pt agglomeration, in some cases, catalysts presented higher activities (compared to nonoxidized materials), due to the positive influence of the presence of oxygen surface groups in Pt-carbon interaction. For PtRu electrocatalysts supported on functionalized xerogels, authors determined that functionalization of the carbon support played a definitive role in the interaction between carbon material and the active phase, determining metallic dispersion, alloying degree, and reduction extent [150, 151]. The activity of the catalysts in methanol electrooxidation was directly related to the metal oxidation state and alloying degree. However, no positive influence of functionalization was observed for PtRu-electrocatalysts supported on functionalized xerogels, leading to less active catalysts.

Sulfur doping of carbon materials has been reported to enhance the anchoring of the metal particles on the support [152, 153]. Several authors have proved that the chemical bond between the metal particle and sulfur functionalities reduces the effect of dissolution and recrystallization (Ostwald ripening) of the catalyst particles, thus enhancing the resistance to degradation [149, 152, 153]. Alegre et al. obtained highly mesoporous CXGs treated with elemental sulfur [149]. Pt catalysts supported on sulfurized CXGs showed enhanced catalytic activity towards both the methanol electrooxidation and the oxygen electro-reduction reactions, exceeding not only the performance of the catalyst supported on the bare xerogel, but also of the catalyst supported on a commercial carbon black. The sulfurization treatment was also effective in improving the resistance of Pt/CXG catalysts towards corrosion phenomena occurring at the fuel cell cathode.

Another way to improve the catalytic activity and durability of carbon supports is by doping with N [122, 154, 155]. Alegre et al. introduced melamine as a CXG precursor together with resorcinol for an effective in situ N doping (3–4 wt% N) [143]. Pt nanoparticles were supported on nitrogen-doped CXGs and their activities for ORR were evaluated in acid media along with their stability. Results provided new evidences of the type of N groups aiding the activity of Pt for the ORR and of a remarkable stability for N-doped carbon-supported Pt catalysts, providing appropriate physicochemical features. The presence of N in the carbon matrix favored the complete reduction mechanism of oxygen to water through a $4e^-$ pathway. Pt/N-CXGs presented a higher activity towards the ORR than the undoped carbon-supported counterparts. Besides providing adequate physicochemical features, that is, low surface area and high N groups density, doping with N allows obtaining highly durable Pt catalysts.

In summary, the use of carbon gels presents great advantages for electrocatalysis-related processes, and for that reason, research in this field is continuously increasing.

4.2 Ordered mesoporous carbons

Ordered mesoporous carbon materials (OMC) are carbonaceous materials with high range-order and pore sizes in the range of mesoporosity (average pore size in the range 2–10 nm). OMCs have attracted special attention due to their unique nanostructures with controllable pore size, developed surface (800–1500 m^2/g) area, and large pore volumes, which provide them a wide versatility for a broad range of applications, such as biosensors, adsorption, catalysis, or electrochemical energy conversion and storage [131, 156–175].

OMCs are usually prepared by the template carbonization pathway. Both hard- (sacrificial method) and soft-templating (direct route) strategies have been used [176–180]. Fig. 3 shows a scheme of the synthesis methods of OMCs.

The hard-templating route usually involves the following steps: (i) the preparation of an inorganic template with controlled porosity (usually a

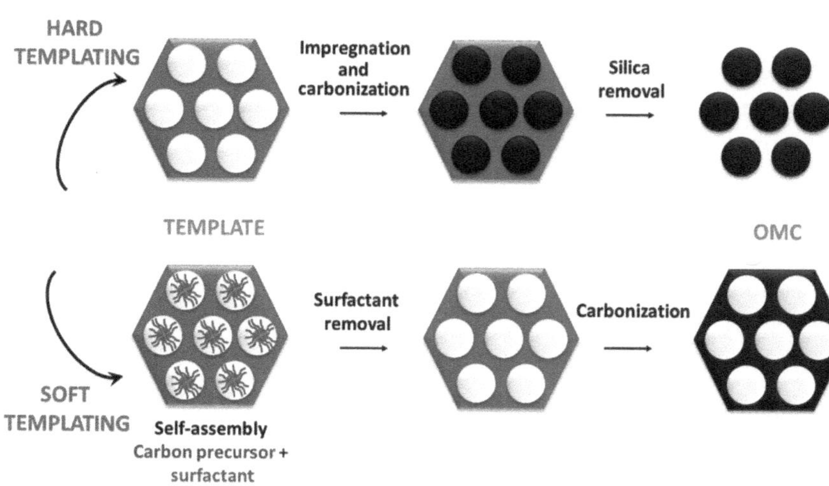

Fig. 3 Scheme of the hard- and soft-templating routes for the synthesis of OMC [176–180].

mesoporous silica template); (ii) the infiltration of a carbon precursor into the pores of the inorganic framework; (iii) the polymerization of the resulting organic-inorganic composite and its carbonization; and (iv) the removal of the inorganic matrix to obtain the desired carbon material [160–162, 167]. In this way, the structure of the resulting carbon is a replica of the inorganic material used as template, although some shrinkage of the inorganic framework may occur during carbonization step as a result of the high temperatures [162, 163]. The porous structure and the range-order of the OMC are influenced by the inorganic matrix. The template needs to display a structure with interconnected pores to obtain an ordered carbon [131, 162, 167]. MCM-48, SBA-1, and SBA-15 silicas have been successfully used to synthesize different carbons, namely CMK (Carbon Mesostructure from Korea [181]), with ordered cubic or hexagonal structures and narrow mesopore-size distributions [163, 164, 168, 182]. Among them, the large pore SBA-15 silica has attracted considerable attention due to its 2D hexagonal mesostructure (P6mm symmetry [183]) with cylindrical-mesopore channels connected by micropores, which provides the cross-linking necessary for obtaining a stable replica of carbon [163]. Furthermore, SBA-15 silica may be synthesized with tuneable pore sizes and surface areas, which allows to obtain carbon materials with desired textural properties [165, 183]. On the other hand, CMK with different structures may be prepared depending on the pore-filling degree of the hexagonal pore structure of SBA-15 silica with the carbon precursor: (i) CMK-3 if the pore structure of the silica is completely filled [162, 163]; and (ii) CMK-5 if the pores are partially coated [162, 164].

OMC have been also successfully obtained by the soft-templating route (direct route) using supramolecular aggregates as templates, generally block copolymers [176, 179]. The main advantage of this approach is the absence of the template removal step, which may lead to the collapse of the mesoporous porous structure. The soft-templating method involves: (i) self-assembly and aggregation of the block copolymer surfactant and the carbon precursor into three-dimensional ordered mesostructures, (ii) removal of the surfactants generating mesoporous polymers with open pores, and (iii) carbonization of the polymers resulting in the mesoporous carbon with the ordered structure [176, 179]. The size and structure of the supramolecular aggregates and the cross-linking ability of the carbon precursors define the textural structure of the resultant OMC. Different organic carbon precursors have been used for the synthesis of OMC, from the conventional phenol [184], resorcinol [185] or phloroglucinol [186] resines, to biosourced molecules, such as tannins [187]. Consequently, a variety of OMC have been obtained by the soft-templating route with different structure and morphologies, for example, spheres [188], wires [189], or films [185].

The tuneable nanostructure and developed mesoporosity of OMC make them promising materials for being used in electrocatalysis, for example in the energy conversion field, where OMCs have been extensively used as catalyst support in low temperature fuel cells [131, 157–159, 167, 173–175, 190]. Noble metal nanoparticles, such as Pt or Pt-group metals, deposited on a carbon material (generally a carbon black) are usually used as electrocatalysts in both sides of these devices, anode and cathode. The large surface area and ordered structure of OMC enhance the dispersion of metal nanoparticles, increasing the electrochemical active area [191–193]. Consequently, an improved performance of PEMFC was found by our group using a Pt/OMC catalyst at the anode side (13–27 $mW\,cm^{-2}$), in comparison to a conventional electrode based on a carbon black (9.5 $mW\,cm^{-2}$) [194].

A well-known strategy to favor the anchoring of metal nanoparticles is the generation of oxygenated species on the carbon surface by chemical oxidation using an acid (such as sulfuric acid, nitric acid, etc.) [195]. Platinum nanoparticles supported on OMC subjected to oxidation treatments with nitric acid showed an improved electrochemical activity for CO oxidation [158]. In this context, CO poisoning of the electrocatalyst occurs in the anode side of fuel cells, which leads to a decrease of the active surface sites. A lower overpotential for CO oxidation was reported at Pt catalysts supported on oxidized OMC than that obtained using Vulcan XC-72R [158, 196]. Similarly, a high electrochemical activity for methanol oxidation has been also achieved by functionalization of OMC [158, 197]. For example, a Pt catalyst supported on OMC treated with diluted nitric acid exhibited a stationary current density around 0.75 $mA\,cm^{-2}$ in a 0.5 M methanol +0.5 M H_2SO_4 solution at 0.70 V, while a value of around 0.50 $mA\,cm^{-2}$ was obtained at Pt/Vulcan [158]. Similarly, a faster oxidation of methanol has been found on catalysts supported on OMC doped with nitrogen due to the enrichment of the Pt at the near surface [155].

OMC-supported alloys of Pt with other noble (such as Ru [159, 198]) and nonnoble metals (e.g., Ni [199], Fe [200]) have been also proposed to enhance CO and methanol oxidation since the presence of foreign atoms in the Pt structure facilitates the formation of Pt-OH. Binary catalysts based on Pd and NiO have also showed an outstanding activity for ethanol oxidation and a better tolerance to CO poisoning [201].

Regarding the cathode side of fuel cells, a 1.5 times higher mass activity was reported on Pt/OMC for the ORR than that obtained using a commercial carbon as support in 0.1 M $HClO_4$ [193]. A similar trend has been found on a Pd-based catalyst in alkaline media [202]. However, Pt and Pd-based catalysts present strong alcohol poisoning, which results from the crossover of alcohol fuel from the anode to cathode. Surprisingly, a high activity for the ORR and good alcohol poisoning tolerance have been obtained using metal-free OMC doped with different heteroatoms (N, S, B or P) [203–212]. N and S co-doped OMC exhibited an electrocatalytic activity for the ORR similar to that obtained on a Pt-based catalyst in alkaline media [203, 209]. Superior performances have been obtained on OMC decorated with N and a transition metal, such as Fe or Co [213, 214].

However, OMCs usually present a low conductivity (especially when they are obtained by the hard-templating route) due to their amorphous character, which may result in a high ohmic resistance and a poor tolerance to the electrochemical oxidation of carbon to CO_2. The latter is one of the main issues facing the lifetime of low temperature fuel cells. In order to address these limitations, graphitization procedures of OMC have been carried out by heat treatments [215–219], catalytic strategies [220, 221], or doping with N, P, or B [222, 223]. Recently, a significantly improved electrical conductivity was obtained on a graphitized OMC (with an intrinsic value of 9523 $S\,m^{-1}$) than that obtained on the as-synthesized material (0.033 $S\,m^{-1}$) [216]. The latter resulted in a better tolerance to carbon degradation by the irreversible oxidation to CO_2 [224, 225]. Fig. 4 shows an example of the current-times curves obtained at 1.4 V versus RHE in 0.5 M H_2SO_4 aqueous solution for a OMC, before and after graphitization. An important decrease of the oxidation current was obtained after graphitization.

Pt, PtRu, and Pd nanoparticles supported on graphitized OMC showed an enhancement of the electrocatalytic activity for CO and the oxidation of organic molecules, such as formic acid or methanol [198, 215]. The improved behavior of graphitized OMC is not limited to the anodic electrode, as a high intrinsic activity for ORR and methanol tolerance have been obtained in ternary catalytic systems of Pt_2CrCo [226].

OMCs have been also used in other electrocatalytic reactions, such as the electrochemical reduction of CO_2 furnishing promising results [227–229]. For example, OMC-supported Pd nanoparticles with an average size around 3–4 nm exhibited a strong inhibition of the hydrogen evolution reaction at −1 V versus Ag/AgCl in aqueous 0.1 M $NaHCO_3$ in the presence of

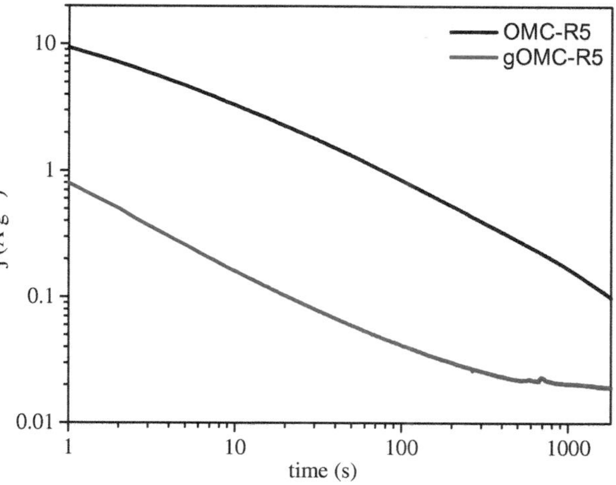

Fig. 4 Current-time curves for carbon electrooxidation at 1.4V versus RHE in 0.5M H_2SO_4 of an ordered mesoporous carbon prepared with silica to surfactant ratio $R=5$ (OMC-R5), before and after graphitization (g prefix indicates graphitization) [224, 225].

CO_2 (Fig. 5) [227]. The latter is associated to the formation of adsorbed species (CO and formats) derived from CO_2 reduction.

High faradaic efficiencies for the formation of different compounds, such as carbon monoxide [84], formic acid [84, 230] or ethanol [84, 231], have been obtained on metal-free OMC doped with N, P, or B.

The well-tailored structure of OMC allows to adapt the pore size to that of solvated ions, which has boosted the use of OMC in electrochemical

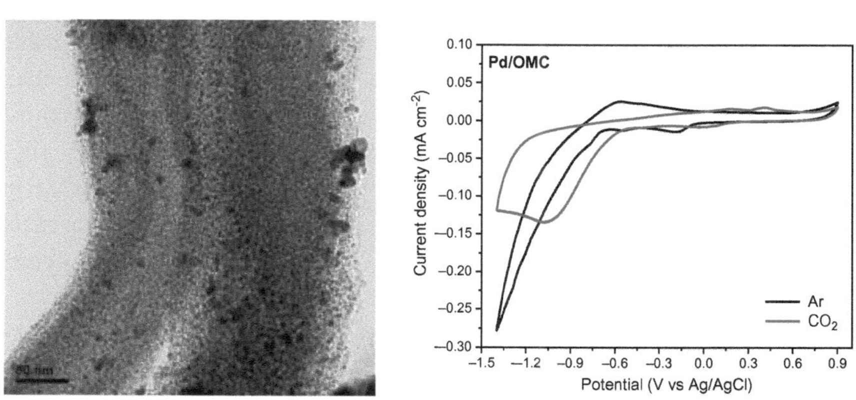

Fig. 5 TEM image of Pd/OMC (left) and cyclic voltammograms of Pd/OMC in 0.1M $NaHCO_3$ saturated with Ar or CO_2, $\upsilon=10\,mV\,s^{-1}$ (right) [227].

energy storage devices, such as supercapacitors, batteries, or hydrogen storage [167, 169–171, 177, 232–234]. Recent works have reviewed the design of OMC for energy storage applications [177, 178].

5 Conclusions

Carbon has played an important role in the development of fundamental electrochemistry, where the use of glassy carbon as a substrate has been essential in electroanalysis because of its lower price in comparison to noble metals. With the beginning of nanotechnology, carbon materials have attracted special attention due to the ability of carbon to generate materials with controllable dimensions in nanometric scale and consequently with specific and unique properties.

Emerging classes of carbon nanostructures attending to their characteristic dimension are: (i) one-dimensional nanofilaments (CNT or CNF); (ii) two-dimensional sheets (graphene); and (iii) three-dimensional for the rest of nanostructures (e.g., carbon gels or ordered mesoporous carbons). The design of carbon-based electrodes carbon electrocatalyst-supports from these emerging nanostructures has been essential in the progress of electrochemical technologies for energy conversion and storage devices, for example, batteries, supercapacitors, fuel cells, or electrolyzers.

Concerning the use of emerging carbon materials as catalyst support for fuel cells, an ideal support must provide: (i) a high surface area with a developed and accessible porosity in order to enhance the dispersion of metal nanoparticles; (ii) an adequate electrical conductivity to facilitate electron transfer; (iii) long-term stability to harsh degradation conditions. Carbon gels and ordered mesoporous carbons present a large specific area with tunable pore size, making them promising candidates for energy conversion devices. On the other hand, the graphitic structure of nanofibers and nanotubes minimizes carbon-support electrooxidation. Similarly, graphene has a large theoretical surface area and an ultrahigh electrical conductivity resulting in enhanced activities for electrochemical reactions. Accordingly, the exact operating conditions and final application of the fuel cell electrodes determine the best choice of carbon by balancing the described characteristics.

A promising route to improve the catalytic activity and durability of carbon supports is by doping the carbon structure with different heteroatoms (such as nitrogen, sulfur, boron, phosphorous). Interestingly, a high activity for different electrocatalytic reactions (such as the ORR, which takes place in the cathode side of a PEMFC or the reduction of CO_2) has been obtained using metal-free carbon materials doped with N, S, B, or P.

These carbon nanostructures have also been explored for the electrochemical storage of energy, such as lithium-ion batteries, redox flow batteries, metal-air batteries, or supercapacitors. The good electrical conductivity together with the accessible surface area of one-dimensional materials make them ideal candidates for energy storage. On the other hand, the well-tailored structure of mesoporous carbons and carbon gels allows to adapt the pore size to that of solvated ions.

Acknowledgments

Authors acknowledge financial support given by Spanish Ministry of Science, Innovation and University (MICINN) through project ENE2017-83976-C2-1-R (cofounded by FEDER). C. Alegre and D. Sebastián acknowledge also MICINN for their Juan de la Cierva (IJCI-2017-32354) and Ramón y Cajal (RyC-2016-20944) research contracts, respectively.

References

[1] N.M. Rodriguez, A review of catalytically grown carbon nanofibers, J. Mater. Res. 8 (1993) 3233–3250, https://doi.org/10.1557/JMR.1993.3233.

[2] E. Boellaard, P.K. de Bokx, A.J.H.M. Kock, J.W. Geus, The formation of filamentous carbon on iron and nickel catalysts. III. Morphology, J. Catal. 96 (1985) 481–490, https://doi.org/10.1016/0021-9517(85)90316-1.

[3] I. Suarez-Martinez, N. Grobert, C.P. Ewels, Nomenclature of sp2 carbon nanoforms, Carbon N. Y. 50 (2012) 741–747, https://doi.org/10.1016/j.carbon.2011.11.002.

[4] J.C. Ruiz-Cornejo, D. Sebastián, M.J. Lázaro, Synthesis and applications of carbon nanofibers: a review, Rev. Chem. Eng. (2018)https://doi.org/10.1515/revce-2018-0021.

[5] M. Endo, K. Takeuchi, T. Hiraoka, T. Furuta, T. Kasai, X. Sun, C.H. Kiang, M. S. Dresselhaus, Stacking nature of graphene layers in carbon nanotubes and nanofibres, J. Phys. Chem. Solid 58 (1997) 1707–1712. http://www.scopus.com/inward/record.url?eid=2-s2.0-0031273886&partnerID=tZOtx3y1.

[6] D. Sebastián, I. Suelves, M.J. Lázaro, R. Moliner, Carbon nanofibers as electrocatalyst support for fuel cells: effect of hydrogen on their properties in CH4 decomposition, J. Power Sources 192 (2009) 51–56, https://doi.org/10.1016/j.jpowsour.2008.11.092.

[7] D. Sebastián, A.G. Ruiz, I. Suelves, R. Moliner, M.J. Lázaro, On the importance of the structure in the electrical conductivity of fishbone carbon nanofibers, J. Mater. Sci. 48 (2013) 1423–1435, https://doi.org/10.1007/s10853-012-6893-1.

[8] A. Romero, A. Garrido, A. Nieto-Márquez, A.R. de la Osa, A. de Lucas, J.L. Valverde, The influence of operating conditions on the growth of carbon nanofibers on carbon nanofiber-supported nickel catalysts, Appl. Catal. A. Gen. 319 (2007) 246–258, https://doi.org/10.1016/j.apcata.2006.12.009.

[9] D. Sebastián, I. Suelves, R. Moliner, M.J. Lázaro, Carbon nanofibers, in: Nanofibers Synthesis Properties Applications, Nova Science Publishers, Inc., 2012, pp. 1–40.

[10] C.A. Bessel, K. Laubernds, N.M. Rodriguez, R.T.K. Baker, Graphite nanofibers as an electrode for fuel cell applications, J. Phys. Chem. B 105 (2001) 1121–1122, https://doi.org/10.1021/jp003280d.

[11] E.S. Steigerwalt, G.A. Deluga, D.E. Cliffel, C.M. Lukehart, A Pt-Ru/graphitic carbon nanofiber nanocomposite exhibiting high relative performance as a direct-methanol fuel cell anode catalyst, J. Phys. Chem. B 105 (2001) 8097–8101, https://doi.org/10.1021/jp011633i.

[12] H. Tang, J. Chen, L. Nie, D. Liu, W. Deng, Y. Kuang, S. Yao, High dispersion and electrocatalytic properties of platinum nanoparticles on graphitic carbon nanofibers (GCNFs), J. Colloid Interface Sci. 269 (2004) 26–31, https://doi.org/10.1016/S0021-9797(03)00608-8.

[13] D. Sebastián, I. Suelves, R. Moliner, M.J. Lázaro, A. Stassi, V. Baglio, A.S. Aricò, Optimizing the synthesis of carbon nanofiber based electrocatalysts for fuel cells, Appl. Catal. Environ. 132–133 (2013) 22–27, https://doi.org/10.1016/j.apcatb.2012.11.023.

[14] D. Sebastián, A.G.A.G. Ruíz, I. Suelves, R. Moliner, M.J. Lázaro, V. Baglio, A. Stassi, A.S. Aricò, Enhanced oxygen reduction activity and durability of Pt catalysts supported on carbon nanofibers, Appl. Catal. Environ. 115–116 (2012) 269–275, https://doi.org/10.1016/j.apcatb.2011.12.041.

[15] E.S. Steigerwalt, G.A. Deluga, C.M. Lukehart, Pt-Ru/carbon fiber nanocomposites: synthesis, characterization, and performance as anode catalysts of direct methanol fuel cells. A search for exceptional performance, J. Phys. Chem. B 106 (2002) 760–766, https://doi.org/10.1021/jp012707t.

[16] E.S. Steigerwalt, G.A. Deluga, C.M. Lukeharta, Rapid preparation of Pt-Ru/graphitic carbon nanofiber nanocomposites as DMFC anode catalysts using microwave processing, J. Nanosci. Nanotechnol. 3 (2003) 247–251, https://doi.org/10.1166/jnn.2003.195.

[17] F. Yuan, H.K. Yu, H. Ryu, Preparation and characterization of carbon nanofibers as catalyst support material for PEMFC, Electrochim. Acta 50 (2004) 685–691, https://doi.org/10.1016/j.electacta.2004.01.106.

[18] Z.R. Ismagilov, M.A. Kerzhentsev, N.V. Shikina, A.S. Lisitsyn, L.B. Okhlopkova, C. N. Barnakov, M. Sakashita, T. Iijima, K. Tadokoro, Development of active catalysts for low Pt loading cathodes of PEMFC by surface tailoring of nanocarbon materials, Catal. Today 102–103 (2005) 58–66, https://doi.org/10.1016/j.cattod.2005.02.007.

[19] W. Yang, S. Yang, J. Guo, G. Sun, Q. Xin, Comparison of CNF and XC-72 carbon supported palladium electrocatalysts for magnesium air fuel cell, Carbon N. Y. 45 (2007) 397–401, https://doi.org/10.1016/j.carbon.2006.09.003.

[20] K. Lee, J. Zhang, H. Wang, D.P. Wilkinson, Progress in the synthesis of carbon nanotube- and nanofiber-supported Pt electrocatalysts for PEM fuel cell catalysis, J. Appl. Electrochem. 36 (2006) 507–522, https://doi.org/10.1007/s10800-006-9120-4.

[21] J.-S. Zheng, X.-S. Zhang, P. Li, J. Zhu, X.-G. Zhou, W.-K. Yuan, Effect of carbon nanofiber microstructure on oxygen reduction activity of supported palladium electrocatalyst, Electrochem. Commun. 9 (2007) 895–900, https://doi.org/10.1016/j.elecom.2006.12.006.

[22] J.-T. Chou, K. Sasaki, Microstructural analysis of catalyst/catalyst support interfaces of PEFC electrocatalysts, Solid State Ion. 179 (2008) 1405–1408, https://doi.org/10.1016/j.ssi.2008.04.011.

[23] D. Sebastián, I. Suelves, R. Moliner, M.J.J. Lázaro, The effect of the functionalization of carbon nanofibers on their electronic conductivity, Carbon N. Y. 48 (2010) 4421–4431, https://doi.org/10.1016/j.carbon.2010.07.059.

[24] J. Guo, G. Sun, Q. Wang, G. Wang, Z. Zhou, S. Tang, L. Jiang, B. Zhou, Q. Xin, Carbon nanofibers supported Pt-Ru electrocatalysts for direct methanol fuel cells, Carbon N. Y. 44 (2006) 152–157, https://doi.org/10.1016/j.carbon.2005.06.047.

[25] A. Guha, T.A. Zawodzinski, D.A. Schiraldi, Evaluation of electrochemical performance for surface-modified carbons as catalyst support in polymer electrolyte membrane (PEM) fuel cells, J. Power Sources 172 (2007) 530–541, https://doi.org/10.1016/j.jpowsour.2007.07.035.

[26] L. Calvillo, M. Gangeri, S. Perathoner, G. Centi, R. Moliner, M.J. Lázaro, Effect of the support properties on the preparation and performance of platinum catalysts supported on carbon nanofibers, J. Power Sources 192 (2009) 144–150, https://doi.org/10.1016/j.jpowsour.2009.01.005.

[27] H.-S. Oh, K. Kim, Y.-J. Ko, H. Kim, Effect of chemical oxidation of CNFs on the electrochemical carbon corrosion in polymer electrolyte membrane fuel cells, Int. J. Hydrogen Energy 35 (2010) 701–708, https://doi.org/10.1016/j.ijhydene.2009.10.105.

[28] D. Sebastián, J.C.C. Calderón, J.A. González-Expósito, E. Pastor, M.V. Martínez-Huerta, I. Suelves, R. Moliner, M.J. Lázaro, Influence of carbon nanofiber properties as electrocatalyst support on the electrochemical performance for PEM fuel cells, Int. J. Hydrogen Energy 35 (2010) 9934–9942, https://doi.org/10.1016/j.ijhydene.2009.12.004.

[29] D. Sebastián, M.J.J. Lázaro, R. Moliner, I. Suelves, A.S.S. Aricò, V. Baglio, Oxidized carbon nanofibers supporting PtRu nanoparticles for direct methanol fuel cells, Int. J. Hydrogen Energy 39 (2014) 5414–5423, https://doi.org/10.1016/j.ijhydene.2013.12.005.

[30] B. Zhang, F. Kang, J.M. Tarascon, J.K. Kim, Recent advances in electrospun carbon nanofibers and their application in electrochemical energy storage, Prog. Mater. Sci. (2016)https://doi.org/10.1016/j.pmatsci.2015.08.002.

[31] M. Bognitzki, W. Czado, T. Frese, A. Schaper, M. Hellwig, M. Steinhart, A. Greiner, J. H. Wendorff, Nanostructured fibers via electrospinning, Adv. Mater. 13 (2001) 70–72. https://doi.org/10.1002/1521-4095(200101) 13:1<70::AID-ADMA70>3.0.CO;2-H.

[32] R.H. Baughman, A.A. Zakhidov, W.A. de Heer, Carbon nanotubes—the route toward applications, Science 297 (2002) 787–792, https://doi.org/10.1126/science.1060928.

[33] L.V. Radushkevich, V.M. Lukyanovich, O strukture ugleroda, obrazujucegosja pri termiceskom razlozenii okisi ugleroda na zeleznom kontakte, Zh. Fiz. Khim. 26 (1952) 88–95.

[34] A. Oberlin, M. Endo, T. Koyama, Filamentous growth of carbon through benzene decomposition, J. Cryst. Growth 32 (1976) 335–349, https://doi.org/10.1016/0022-0248(76)90115-9.

[35] S. Iijima, Helical microtubules of graphitic carbon, Nature 354 (1991) 56–58. http://www.scopus.com/inward/record.url?eid=2-s2.0-0342819025&partnerID=tZOtx3y1.

[36] S. Iijima, T. Ichihashi, Single-shell carbon nanotubes of 1-nm diameter, Nature 363 (1993) 603–605, https://doi.org/10.1038/363603a0.

[37] J. Prasek, J. Drbohlavova, J. Chomoucka, J. Hubalek, O. Jasek, V. Adam, R. Kizek, Methods for carbon nanotubes synthesis—review, J. Mater. Chem. 21 (2011) 15872, https://doi.org/10.1039/c1jm12254a.

[38] L.S. Ying, M.A. Bin Mohd Salleh, H.B. Mohamed Yusoff, S.B. Abdul Rashid, J.b. Abd Razak, Continuous production of carbon nanotubes—a review, J. Ind. Eng. Chem. 17 (2011) 367–376, https://doi.org/10.1016/J.JIEC.2011.05.007.

[39] N. Hamzah, M.F.M. Yasin, M.Z.M. Yusop, A. Saat, N.A.M. Subha, Rapid production of carbon nanotubes: a review on advancement in growth control and morphology manipulations of flame synthesis, J. Mater. Chem. A 5 (2017) 25144–25170, https://doi.org/10.1039/C7TA08668G.

[40] M. Kumar, Y. Ando, Chemical vapor deposition of carbon nanotubes: a review on growth mechanism and mass production, J. Nanosci. Nanotechnol. 10 (2010) 3739–3758, https://doi.org/10.1166/jnn.2010.2939.

[41] D. Torres, J.L. Pinilla, I. Suelves, Co-, Cu- and Fe-doped Ni/Al_2O_3 catalysts for the catalytic decomposition of methane into hydrogen and carbon nanofibers, Catalysts (2018), https://doi.org/10.20944/PREPRINTS201807.0213.V1.
[42] D. Torres, J.L. Pinilla, I. Suelves, Screening of Ni-Cu bimetallic catalysts for hydrogen and carbon nanofilaments production via catalytic decomposition of methane, Appl. Catal. A. Gen. 559 (2018) 10–19, https://doi.org/10.1016/J.APCATA.2018.04.011.
[43] D. Torres, S. De Llobet, J.L. Pinilla, M.J. Lázaro, I. Suelves, R. Moliner, Hydrogen production by catalytic decomposition of methane using a Fe-based catalyst in a fluidized bed reactor, J. Nat. Gas Chem. 21 (2012) 367–373, https://doi.org/10.1016/S1003-9953(11)60378-2.
[44] Y. Zhou, R. Azumi, Carbon nanotube based transparent conductive films: progress, challenges, and perspectives, Sci. Technol. Adv. Mater. 17 (2016) 493–516, https://doi.org/10.1080/14686996.2016.1214526.
[45] F. Wang, D. Kozawa, Y. Miyauchi, K. Hiraoka, S. Mouri, Y. Ohno, K. Matsuda, Considerably improved photovoltaic performance of carbon nanotube-based solar cells using metal oxide layers, Nat. Commun. 6 (2015) 6305, https://doi.org/10.1038/ncomms7305.
[46] Y. Yan, J. Miao, Z. Yang, F.-X. Xiao, H. Bin Yang, B. Liu, Y. Yang, Carbon nanotube catalysts: recent advances in synthesis, characterization and applications, Chem. Soc. Rev. 44 (2015) 3295–3346, https://doi.org/10.1039/C4CS00492B.
[47] M.D. De Los Bernardos, S. Pérez-Rodríguez, A. Gual, C. Claver, C. Godard, Facile synthesis of NHC-stabilized Ni nanoparticles and their catalytic application in the: Z-selective hydrogenation of alkynes, Chem. Commun. 53 (2017), https://doi.org/10.1039/c7cc01779k.
[48] E. Akbari, Z. Buntat, Benefits of using carbon nanotubes in fuel cells: a review, Int. J. Energy Res. 41 (2017) 92–102, https://doi.org/10.1002/er.3600.
[49] L. Dai, D.W. Chang, J.-B. Baek, W. Lu, Carbon nanomaterials for advanced energy conversion and storage, Small 8 (2012) 1130–1166, https://doi.org/10.1002/smll.201101594.
[50] L.-M. Peng, Z. Zhang, C. Qiu, Carbon nanotube digital electronics, Nat. Electron. 2 (2019) 499–505, https://doi.org/10.1038/s41928-019-0330-2.
[51] J. Saththasivam, W. Yiming, K. Wang, J. Jin, Z. Liu, A novel architecture for carbon nanotube membranes towards fast and efficient oil/water separation, Sci. Rep. 8 (2018) 7418, https://doi.org/10.1038/s41598-018-25788-9.
[52] J.J. Gooding, Nanostructuring electrodes with carbon nanotubes: a review on electrochemistry and applications for sensing, Electrochim. Acta 50 (2005) 3049–3060, https://doi.org/10.1016/J.ELECTACTA.2004.08.052.
[53] L. Xue, Y. Li, X. Liu, Q. Liu, J. Shang, H. Duan, L. Dai, J. Shui, Zigzag carbon as efficient and stable oxygen reduction electrocatalyst for proton exchange membrane fuel cells, Nat. Commun. 9 (2018) 3819, https://doi.org/10.1038/s41467-018-06279-x.
[54] H. Tabassum, A. Mahmood, B. Zhu, Z. Liang, R. Zhong, S. Guo, R. Zou, Recent advances in confining metal-based nanoparticles into carbon nanotubes for electrochemical energy conversion and storage devices, Energ. Environ. Sci. 12 (2019) 2924–2956, https://doi.org/10.1039/C9EE00315K.
[55] W. Zhang, P. Sherrell, A.I. Minett, J.M. Razal, J. Chen, Carbon nanotube architectures as catalyst supports for proton exchange membrane fuel cells, Energ. Environ. Sci. 3 (2010) 1286, https://doi.org/10.1039/c0ee00139b.
[56] X. Wang, W. Li, Z. Chen, M. Waje, Y. Yan, Durability investigation of carbon nanotube as catalyst support for proton exchange membrane fuel cell, J. Power Sources 158 (2006) 154–159, https://doi.org/10.1016/J.JPOWSOUR.2005.09.039.

[57] T. Matsumoto, T. Komatsu, K. Arai, T. Yamazaki, M. Kijima, H. Shimizu, Y. Takasawa, J. Nakamura, Reduction of Pt usage in fuel cell electrocatalysts with carbon nanotube electrodes, Chem. Commun. 840 (2004), https://doi.org/10.1039/b400607k.
[58] W. Li, C. Liang, J. Qiu, W. Zhou, H. Han, Z. Wei, G. Sun, Q. Xin, Carbon nanotubes as support for cathode catalyst of a direct methanol fuel cell, Carbon N. Y. 40 (2002) 791–794, https://doi.org/10.1016/S0008-6223(02)00039-8.
[59] N. Jha, P. Ramesh, E. Bekyarova, X. Tian, F. Wang, M.E. Itkis, R.C. Haddon, Functionalized single-walled carbon nanotube-based fuel cell benchmarked against US DOE 2017 technical targets, Sci. Rep. 3 (2013) 2257, https://doi.org/10.1038/srep02257.
[60] G. Girishkumar, K. Vinodgopal, P.V. Kamat, Carbon nanostructures in portable fuel cells: single-walled carbon nanotube electrodes for methanol oxidation and oxygen reduction, J. Phys. Chem. B (2004), https://doi.org/10.1021/JP046872V.
[61] Z. Wen, Q. Wang, J. Li, Template synthesis of aligned carbon nanotube arrays using glucose as a carbon source: Pt decoration of inner and outer nanotube surfaces for fuel-cell catalysts, Adv. Funct. Mater. 18 (2008) 959–964, https://doi.org/10.1002/adfm.200700707.
[62] C.A. Dyke, J.M. Tour, Covalent functionalization of single-walled carbon nanotubes for materials applications, J. Phys. Chem. A 108 (2004) 11151–11159, https://doi.org/10.1021/jp046274g.
[63] W. Yuan, S. Lu, Y. Xiang, S.P. Jiang, Pt-based nanoparticles on non-covalent functionalized carbon nanotubes as effective electrocatalysts for proton exchange membrane fuel cells, RSC Adv. 4 (2014) 46265–46284, https://doi.org/10.1039/C4RA05120C.
[64] H.-J. Ahn, W.J. Moon, T.-Y. Seong, D. Wang, Three-dimensional nanostructured carbon nanotube array/PtRu nanoparticle electrodes for micro-fuel cells, Electrochem. Commun. 11 (2009) 635–638, https://doi.org/10.1016/J.ELECOM.2008.12.056.
[65] L. Liang, Y. Xing, Pt-Ru nanoparticles supported on carbon nanotubes as methanol fuel cell catalysts, J. Phys. Chem. C (2007), https://doi.org/10.1021/JP0655470.
[66] J. Prabhuram, T.S. Zhao, Z.K. Tang, R. Chen, Z.X. Liang, Multiwalled carbon nanotube supported ptru for the anode of direct methanol fuel cells, J. Phys. Chem. B (2006), https://doi.org/10.1021/JP0567063.
[67] M. Borghei, G. Scotti, P. Kanninen, T. Weckman, I.V. Anoshkin, A.G. Nasibulin, S. Franssila, E.I. Kauppinen, T. Kallio, V. Ruiz, Enhanced performance of a silicon microfabricated direct methanol fuel cell with PtRu catalysts supported on few-walled carbon nanotubes, Energy 65 (2014) 612–620, https://doi.org/10.1016/J.ENERGY.2013.11.067.
[68] H. Huang, X. Hu, J. Zhang, N. Su, J. Cheng, Facile fabrication of platinum-cobalt alloy nanoparticles with enhanced electrocatalytic activity for a methanol oxidation reaction, Sci. Rep. 7 (2017) 45555, https://doi.org/10.1038/srep45555.
[69] Y.-Y. Zhou, C.-H. Liu, J. Liu, X.-L. Cai, Y. Lu, H. Zhang, X.-H. Sun, S.-D. Wang, Self-decoration of PtNi alloy nanoparticles on multiwalled carbon nanotubes for highly efficient methanol electro-oxidation, Nanomicro Lett. 8 (2016) 371–380, https://doi.org/10.1007/s40820-016-0096-2.
[70] B. Rajesh, K. Ravindranathan Thampi, J.-M. Bonard, N. Xanthopoulos, H.J. Mathieu, B. Viswanathan, Carbon nanotubes generated from template carbonization of polyphenyl acetylene as the support for electrooxidation of methanol, J. Phys. Chem. B (2003), https://doi.org/10.1021/JP0219350.
[71] L. Juárez-Marmolejo, M.G.M. de Oca-Yemha, M. Palomar-Pardavé, M.A. Romero-Romo, A. Ezeta-Mejía, E. Arce-Estrada, S. Pérez-Rodríguez, M.J. Lázaro, Electrocatalytic performance of palladium-based electrocatalysts supported on carbon

nanotubes for formic acid oxidation, ECS Trans. 92 (2019) 317–324, https://doi.org/10.1149/09208.0317ecst.

[72] Y. Zhao, L. Fan, J. Ren, B. Hong, Electrodeposition of Pt–Ru and Pt–Ru–Ni nanoclusters on multi-walled carbon nanotubes for direct methanol fuel cell, Int. J. Hydrogen Energy 39 (2014) 4544–4557, https://doi.org/10.1016/J.IJHYDENE.2013.12.202.

[73] Y. Shao, G. Yin, J. Wang, Y. Gao, P. Shi, Multi-walled carbon nanotubes based Pt electrodes prepared with in situ ion exchange method for oxygen reduction, J. Power Sources 161 (2006) 47–53, https://doi.org/10.1016/J.JPOWSOUR.2006.03.064.

[74] K. Prehn, R. Adelung, M. Heinen, S.P. Nunes, K. Schulte, Catalytically active CNT–polymer-membrane assemblies: from synthesis to application, J. Membr. Sci. 321 (2008) 123–130, https://doi.org/10.1016/J.MEMSCI.2008.04.041.

[75] Y. Shao, J. Sui, G. Yin, Y. Gao, Nitrogen-doped carbon nanostructures and their composites as catalytic materials for proton exchange membrane fuel cell, Appl. Catal. Environ. 79 (2008) 89–99, https://doi.org/10.1016/J.APCATB.2007.09.047.

[76] G. Vijayaraghavan, K.J. Stevenson, Synergistic assembly of dendrimer-templated platinum catalysts on nitrogen-doped carbon nanotube electrodes for oxygen reduction, Langmuir (2007), https://doi.org/10.1021/LA0637263.

[77] B. Viswanathan, Architecture of carbon support for Pt anodes in direct methanol fuel cells, Catal. Today 141 (2009) 52–55, https://doi.org/10.1016/J.CATTOD.2008.08.025.

[78] H. Begum, M.S. Ahmed, D.-W. Lee, Y.-B. Kim, Carbon nanotubes-based PdM bimetallic catalysts through N4-system for efficient ethanol oxidation and hydrogen evolution reaction, Sci. Rep. 9 (2019) 11051, https://doi.org/10.1038/s41598-019-47575-w.

[79] Z. Zhuang, S.A. Giles, J. Zheng, G.R. Jenness, S. Caratzoulas, D.G. Vlachos, Y. Yan, Nickel supported on nitrogen-doped carbon nanotubes as hydrogen oxidation reaction catalyst in alkaline electrolyte, Nat. Commun. 7 (2016) 10141, https://doi.org/10.1038/ncomms10141.

[80] W. Wan, S. Wei, J. Li, C.A. Triana, Y. Zhou, G.R. Patzke, Transition metal electrocatalysts encapsulated into N-doped carbon nanotubes on reduced graphene oxide nanosheets: efficient water splitting through synergistic effects, J. Mater. Chem. A 7 (2019) 15145–15155, https://doi.org/10.1039/C9TA03213D.

[81] L. Yang, S. Jiang, Y. Zhao, L. Zhu, S. Chen, X. Wang, Q. Wu, J. Ma, Y. Ma, Z. Hu, Boron-doped carbon nanotubes as metal-free electrocatalysts for the oxygen reduction reaction, Angew. Chem. Int. Ed. 50 (2011) 7132–7135, https://doi.org/10.1002/anie.201101287.

[82] B. Winther-Jensen, O. Winther-Jensen, M. Forsyth, D.R. Macfarlane, High rates of oxygen reduction over a vapor phase-polymerized PEDOT electrode, Science 321 (2008) 671–674, https://doi.org/10.1126/science.1159267.

[83] Y. Tang, B.L. Allen, D.R. Kauffman, A. Star, Electrocatalytic activity of nitrogen-doped carbon nanotube cups, J. Am. Chem. Soc. 131 (2009) 13200–13201, https://doi.org/10.1021/ja904595t.

[84] D.M. Fernandes, A.F. Peixoto, C. Freire, Nitrogen-doped metal-free carbon catalysts for (electro)chemical CO_2 conversion and valorisation, Dalton Trans. 48 (2019) 13508–13528, https://doi.org/10.1039/C9DT01691K.

[85] G. Tuci, A. Rossin, L. Luconi, C. Pham-Huu, S. Cicchi, H. Ba, G. Giambastiani, Pyridine-decorated carbon nanotubes as a metal-free heterogeneous catalyst for mild CO_2 reduction to methanol with hydroboranes, Cat. Sci. Technol. 7 (2017) 5833–5837, https://doi.org/10.1039/C7CY01772C.

[86] R. Kannan, B.A. Kakade, V.K. Pillai, Polymer electrolyte fuel cells using nafion-based composite membranes with functionalized carbon nanotubes, Angew. Chem. Int. Ed. 47 (2008) 2653–2656, https://doi.org/10.1002/anie.200704343.

[87] K.S. Novoselov, A.K. Geim, S.V. Morozov, D. Jiang, Y. Zhang, S.V. Dubonos, I. V. Grigorieva, A.A. Firsov, Electric field effect in atomically thin carbon films, Science 306 (2004) 666–669, https://doi.org/10.1126/science.1102896.
[88] A.K. Geim, K.S. Novoselov, The rise of graphene, Nat. Mater. 6 (2007) 183–191, https://doi.org/10.1038/nmat1849.
[89] B.C. Brodie, Sur le poids atomique du graphite, Ann. Chim. Phys. 59 (1960) 466–472. http://www.scopus.com/inward/record.url?eid=2-s2.0-0001301933&partnerID=tZOtx3y1.
[90] L. Staudenmaier, Verfahren zur Darstellung der Graphitsäure, Berichte Der Dtsch. Chem. Gesellschaft. 31 (1898) 1481–1487, https://doi.org/10.1002/cber.18980310237.
[91] W.S. Hummers, R.E. Offeman, Preparation of graphitic oxide, J. Am. Chem. Soc. 80 (1958) 1339. http://www.scopus.com/inward/record.url?eid=2-s2.0-33947461960&partnerID=tZOtx3y1.
[92] Z. Jin, J.R. Lomeda, B.K. Price, W. Lu, Y. Zhu, J.M. Tour, Mechanically assisted exfoliation and functionalization of thermally converted graphene sheets, Chem. Mater. 21 (2009) 3045–3047, https://doi.org/10.1021/cm901601g.
[93] D. Torres, J.L. Pinilla, R. Moliner, I. Suelves, On the oxidation degree of few-layer graphene oxide sheets obtained from chemically oxidized multiwall carbon nanotubes, Carbon N. Y. 81 (2015) 405–417, https://doi.org/10.1016/j.carbon.2014.09.073.
[94] D. Torres, J.L. Pinilla, I. Suelves, Unzipping of multi-wall carbon nanotubes with different diameter distributions: effect on few-layer graphene oxide obtention, Appl. Surf. Sci. 424 (2017) 101–110, https://doi.org/10.1016/J.APSUSC.2017.01.273.
[95] M. Liu, R. Zhang, W. Chen, Graphene-supported nanoelectrocatalysts for fuel cells: synthesis, properties, and applications, Chem. Rev. 114 (2014) 5117–5160, https://doi.org/10.1021/cr400523y.
[96] E. Antolini, Graphene as a new carbon support for low-temperature fuel cell catalysts, Appl. Catal. Environ. 123–124 (2012) 52–68, https://doi.org/10.1016/j.apcatb.2012.04.022.
[97] X. Tong, Q. Wei, X. Zhan, G. Zhang, S. Sun, The new graphene family materials: synthesis and applications in oxygen reduction reaction, Catalysts 7 (2016) 1, https://doi.org/10.3390/catal7010001.
[98] H.-W. Ha, I.Y. Kim, S.-J. Hwang, R.S. Ruoff, One-pot synthesis of platinum nanoparticles embedded on reduced graphene oxide for oxygen reduction in methanol fuel cells, Electrochem. Solid St. 14 (2011), https://doi.org/10.1149/1.3584092.
[99] H.J. Kim, S.M. Choi, M.H. Seo, S. Green, G.W. Huber, W.B. Kim, Efficient electrooxidation of biomass-derived glycerol over a graphene-supported PtRu electrocatalyst, Electrochem. Commun. 13 (2011) 890–893, https://doi.org/10.1016/j.elecom.2011.05.031.
[100] S. Bong, Y.-R. Kim, I. Kim, S. Woo, S. Uhm, J. Lee, H. Kim, Graphene supported electrocatalysts for methanol oxidation, Electrochem. Commun. 12 (2010) 129–131, https://doi.org/10.1016/j.elecom.2009.11.005.
[101] S.M. Choi, M.H. Seo, H.J. Kim, W.B. Kim, Synthesis of surface-functionalized graphene nanosheets with high Pt-loadings and their applications to methanol electrooxidation, Carbon N. Y. 49 (2011) 904–909, https://doi.org/10.1016/j.carbon.2010.10.055.
[102] S.-Y. Yang, K.-H. Chang, Y.-F. Lee, C.-C.M. Ma, C.-C. Hu, Constructing a hierarchical graphene-carbon nanotube architecture for enhancing exposure of graphene and electrochemical activity of Pt nanoclusters, Electrochem. Commun. 12 (2010) 1206–1209, https://doi.org/10.1016/j.elecom.2010.06.020.
[103] S. Park, Y. Shao, H. Wan, P.C. Rieke, V.V. Viswanathan, S.A. Towne, L.V. Saraf, J. Liu, Y. Lin, Y. Wang, Design of graphene sheets-supported Pt catalyst layer in PEM

fuel cells, Electrochem. Commun. 13 (2011) 258–261, https://doi.org/10.1016/j.elecom.2010.12.028.
[104] L. Zhang, Z. Xia, Mechanisms of oxygen reduction reaction on nitrogen-doped graphene for fuel cells, J. Phys. Chem. C 115 (2011) 11170–11176, https://doi.org/10.1021/jp201991j.
[105] E. Bayram, G. Yilmaz, S. Mukerjee, A solution-based procedure for synthesis of nitrogen doped graphene as an efficient electrocatalyst for oxygen reduction reactions in acidic and alkaline electrolytes, Appl. Catal. Environ. 192 (2016) 26–34, https://doi.org/10.1016/j.apcatb.2016.03.043.
[106] G. Jo, J. Sanetuntikul, S. Shanmugam, Boron and phosphorous-doped graphene as a metal-free electrocatalyst for the oxygen reduction reaction in alkaline medium, RSC Adv. 5 (2015) 53637–53643, https://doi.org/10.1039/C5RA06952A.
[107] Z. Yang, Z. Yao, G. Li, G. Fang, H. Nie, Z. Liu, X. Zhou, X. Chen, S. Huang, Sulfur-doped graphene as an efficient metal-free cathode catalyst for oxygen reduction, ACS Nano 6 (2012) 205–211, https://doi.org/10.1021/nn203393d.
[108] R. Li, Z. Wei, X. Gou, W. Xu, Phosphorus-doped graphene nanosheets as efficient metal-free oxygen reduction electrocatalysts, RSC Adv. 3 (2013) 9978, https://doi.org/10.1039/c3ra41079j.
[109] G. Lemes, D. Sebastián, E. Pastor, M.J. Lázaro, N-doped graphene catalysts with high nitrogen concentration for the oxygen reduction reaction, J. Power Sources 438 (2019) 227036, https://doi.org/10.1016/J.JPOWSOUR.2019.227036.
[110] L.T. Soo, K.S. Loh, A.B. Mohamad, W.R.W. Daud, W.Y. Wong, Effect of nitrogen precursors on the electrochemical performance of nitrogen-doped reduced graphene oxide towards oxygen reduction reaction, J. Alloys Compd. 677 (2016) 112–120, https://doi.org/10.1016/j.jallcom.2016.03.214.
[111] H.B. Yang, J. Miao, S.-F. Hung, J. Chen, H.B. Tao, X. Wang, L. Zhang, R. Chen, J. Gao, H.M. Chen, L. Dai, B. Liu, Identification of catalytic sites for oxygen reduction and oxygen evolution in N-doped graphene materials: development of highly efficient metal-free bifunctional electrocatalyst, Sci. Adv. (2016), https://doi.org/10.1126/sciadv.1501122.
[112] D. Guo, R. Shibuya, C. Akiba, S. Saji, T. Kondo, J. Nakamura, Active sites of nitrogen-doped carbon materials for oxygen reduction reaction clarified using model catalysts, Science 351 (2015) 1–24, https://doi.org/10.1126/science.aad0832.
[113] K. Mamtani, D. Jain, D. Dogu, V. Gustin, S. Gunduz, A.C. Co, U.S. Ozkan, Insights into oxygen reduction reaction (ORR) and oxygen evolution reaction (OER) active sites for nitrogen-doped carbon nanostructures (CN x) in acidic media, Appl. Catal. Environ. 220 (2017) 88–97, https://doi.org/10.1016/j.apcatb.2017.07.086.
[114] L. Lai, J.R. Potts, D. Zhan, L. Wang, C.K. Poh, C. Tang, H. Gong, Z. Shen, J. Lin, R. S. Ruoff, Exploration of the active center structure of nitrogen-doped graphene-based catalysts for oxygen reduction reaction, Energ. Environ. Sci. 5 (2012) 7936, https://doi.org/10.1039/c2ee21802j.
[115] R. Raccichini, A. Varzi, S. Passerini, B. Scrosati, The role of graphene for electrochemical energy storage, Nat. Mater. 14 (2015) 271–279, https://doi.org/10.1038/nmat4170.
[116] G. Wu, A. Santandreu, W. Kellogg, S. Gupta, O. Ogoke, H. Zhang, H.-L. Wang, L. Dai, Carbon nanocomposite catalysts for oxygen reduction and evolution reactions: from nitrogen doping to transition-metal addition, Nano Energy (2016), https://doi.org/10.1016/j.nanoen.2015.12.032.
[117] J.M. Luque-Centeno, M.V. Martínez-Huerta, D. Sebastián, G. Lemes, E. Pastor, M. J. Lázaro, Bifunctional N-doped graphene Ti and Co nanocomposites for the oxygen reduction and evolution reactions, Renew. Energy 125 (2018) 182–192, https://doi.org/10.1016/j.renene.2018.02.073.

[118] J.M. Luque-Centeno, M.V. Martínez-Huerta, D. Sebastián, J.I. Pardo, M.J. Lázaro, $CoTiO_3$/NrGO nanocomposites for oxygen evolution and oxygen reduction reactions: synthesis and electrocatalytic performance, Electrochim. Acta 331 (2020) https://doi.org/10.1016/j.electacta.2019.135396.
[119] C. Alegre, D. Sebastián, E. Baquedano, M.E. Gálvez, R. Moliner, M.J. Lázaro, Tailoring synthesis conditions of carbon xerogels towards their utilization as Pt-catalyst supports for oxygen reduction reaction (ORR), Catalysts 2 (2012) 466–489, https://doi.org/10.3390/catal2040466.
[120] S.A. Al-Muhtaseb, J.A. Ritter, Preparation and properties of resorcinol-formaldehyde organic and carbon gels, Adv. Mater. 15 (2003) 101–114, https://doi.org/10.1002/adma.200390020.
[121] M. Enterría, J.L. Figueiredo, Nanostructured mesoporous carbons: tuning texture and surface chemistry, Carbon N. Y. 108 (2016) 79–102, https://doi.org/10.1016/J.CARBON.2016.06.108.
[122] J.L. Figueiredo, M.F.R. Pereira, Synthesis and functionalization of carbon xerogels to be used as supports for fuel cell catalysts, J. Energy Chem. 22 (2013) 195–201, https://doi.org/10.1016/S2095-4956(13)60025-X.
[123] R.W. Pekala, Organic aerogels from the polycondensation of resorcinol with formaldehyde, J. Mater. Sci. 24 (1989) 3221–3227, https://doi.org/10.1007/BF01139044.
[124] C. Alegre, E. Baquedano, M.E. Galvez, R. Moliner, M.J. Lazaro, Tailoring carbon xerogels' properties to enhance catalytic activity of Pt catalysts towards methanol oxidation, Int. J. Hydrogen Energy (2015) 14736–14745, https://doi.org/10.1016/j.ijhydene.2015.05.169.
[125] P. Hu, B. Tan, M. Long, Advanced nanoarchitectures of carbon aerogels for multifunctional environmental applications, Nanotechnol. Rev. 5 (2016) 23–39, https://doi.org/10.1515/ntrev-2015-0050.
[126] A. Arenillas, J.A. Menéndez, G. Reichenauer, A. Celzard, V. Fierro, G. Reichenauer, E. Bailón-Garcia, N. Job, Organic and Carbon Gels, Springer International Publishing, 2019, https://doi.org/10.1007/978-3-030-13897-4.
[127] J.L. Figueiredo, Carbon gels with tuned properties for catalysis and energy storage, J. Sol-Gel Sci. Technol. 89 (2019) 12–20, https://doi.org/10.1007/s10971-018-4633-y.
[128] C. Alegre, M.E. Gálvez, R. Moliner, M.J. Lázaro, Comprehensive guide for mesoporous materials. vol. 4: Application and Commercialization - Nova Science Publishers, (2015).
[129] P. Serp, J.L. Figueiredo, Carbon Materials for Catalysis, John Wiley and Sons, 2008, https://doi.org/10.1002/9780470403709.
[130] E. Antolini, Structural parameters of supported fuel cell catalysts: the effect of particle size, inter-particle distance and metal loading on catalytic activity and fuel cell performance, Appl. Catal. Environ. 181 (2016) 298–313, https://doi.org/10.1016/j.apcatb.2015.08.007.
[131] E. Antolini, Carbon supports for low-temperature fuel cell catalysts, Appl. Catal. Environ. 88 (2009) 1–24, https://doi.org/10.1016/j.apcatb.2008.09.030.
[132] A. Bharti, G. Cheruvally, Application of novel carbonaceous materials as support for fuel cell electrocatalysts, in: Advanced Electrocatalysts for Low-Temperature Fuel Cells, Springer International Publishing, 2018, pp. 175–213, https://doi.org/10.1007/978-3-319-99019-4_5.
[133] H. Huang, X. Wang, S.P. Jiang, N. Deng, Y.S. Yoon, A.S. Aricò, P. Spinelli, L. Jiang, C. Wang, R. O'Hayre, F. Kang, H. Yoshikawa, S. Ueda, K. Kobayashi, T.-K. Sham, X. Sun, V. Scardaci, A.C. Ferrari, J.N. Coleman, Recent progress on carbon-based support materials for electrocatalysts of direct methanol fuel cells, J. Mater. Chem. A 2 (2014) 6266, https://doi.org/10.1039/c3ta14754a.

[134] S. Sharma, B.G. Pollet, Support materials for PEMFC and DMFC electrocatalysts—a review, J. Power Sources 208 (2012) 96–119, https://doi.org/10.1016/j.jpowsour.2012.02.011.
[135] P.Y. You, S.K. Kamarudin, Recent progress of carbonaceous materials in fuel cell applications: an overview, Chem. Eng. J. 309 (2017) 489–502, https://doi.org/10.1016/j.cej.2016.10.051.
[136] C. Moreno-Castilla, F.J. Maldonado-Hódar, Carbon aerogels for catalysis applications: an overview, Carbon N. Y. 43 (2005) 455–465, https://doi.org/10.1016/j.carbon.2004.10.022.
[137] C. Moreno-Castilla, Carbon gels in catalysis, in: Carbon Materials for Catalysis, John Wiley & Sons, Inc., 2008, pp. 373–399, https://doi.org/10.1002/9780470403709.ch10
[138] C. Alegre, M.E. Gálvez, R. Moliner, V. Baglio, A. Stassi, A.S. Aricò, M.J. Lázaro, Platinum ruthenium catalysts supported on carbon xerogel for methanol electro-oxidation: influence of the catalyst synthesis method, ChemCatChem 5 (2013) 3770–3780, https://doi.org/10.1002/cctc.201300542.
[139] C. Alegre, M.E. Gálvez, E. Baquedano, R. Moliner, E. Pastor, M.J. Lázaro, Carbon xerogel-based Pt and PtRu electrocatalysts for DMFC: influence of the carbon support's functionalization, Cienc. e Tecnol. Dos Mater. 24 (2012) 172–175. http://www.scopus.com/inward/record.url?eid=2-s2.0-84894182168&partnerID=tZOtx3y1.
[140] C. Alegre, M.E. Gálvez, R. Moliner, V. Baglio, A.S. Aricò, M.J. Lázaro, Towards an optimal synthesis route for the preparation of highly mesoporous carbon xerogel-supported Pt catalysts for the oxygen reduction reaction, Appl. Catal. Environ. 147 (2014) 947–957, https://doi.org/10.1016/j.apcatb.2013.10.031.
[141] C. Alegre, D. Sebastián, M. Gálvez, R. Moliner, A. Stassi, A. Aricò, M. Lázaro, V. Baglio, PtRu nanoparticles deposited by the sulfite complex method on highly porous carbon xerogels: effect of the thermal treatment, Catalysts 3 (2013) 744–756, https://doi.org/10.3390/catal3030744.
[142] C. Alegre, M.E. Gálvez, E. Baquedano, E. Pastor, R. Moliner, M.J. Lázaro, Influence of support's oxygen functionalization on the activity of Pt/carbon xerogels catalysts for methanol electro-oxidation, Int. J. Hydrogen Energy 37 (2012) 7180–7191, https://doi.org/10.1016/j.ijhydene.2011.11.022.
[143] C. Alegre, D. Sebastián, M.E. Gálvez, E. Baquedano, R. Moliner, A.S. Aricò, V. Baglio, M.J. Lázaro, N-doped carbon xerogels as Pt support for the electro-reduction of oxygen, Materials (Basel) 10 (2017), https://doi.org/10.3390/ma10091092.
[144] N. Job, S. Lambert, A. Zubiaur, C. Cao, J.-P. Pirard, Design of Pt/carbon xerogel catalysts for PEM fuel cells, Catalysts 5 (2015) 40–57, https://doi.org/10.3390/catal5010040.
[145] N. Job, M.F.R. Pereira, S. Lambert, A. Cabiac, G. Delahay, J.-F. Colomer, J. Marien, J.L. Figueiredo, J.-P. Pirard, Highly dispersed platinum catalysts prepared by impregnation of texture-tailored carbon xerogels, J. Catal. 240 (2006) 160–171, https://doi.org/10.1016/j.jcat.2006.03.016.
[146] A. Zubiaur, N. Job, Streamlining of the synthesis process of Pt/carbon xerogel electrocatalysts with high Pt loading for the oxygen reduction reaction in proton exchange membrane fuel cells applications, Appl. Catal. Environ. 225 (2018) 364–378, https://doi.org/10.1016/j.apcatb.2017.11.059.
[147] C. Cao, G. Yang, L. Dubau, F. Maillard, S.D. Lambert, J.P. Pirard, N. Job, Highly dispersed Pt/C catalysts prepared by the charge enhanced dry impregnation method, Appl. Catal. Environ. 150–151 (2014) 101–106, https://doi.org/10.1016/j.apcatb.2013.12.004.

[148] A. Zubiaur, M. Chatenet, F. Maillard, S.D. Lambert, J.P. Pirard, N. Job, Using the multiple SEA method to synthesize Pt/carbon xerogel electrocatalysts for PEMFC applications, in: Fuel Cells, John Wiley and Sons Ltd, 2014, pp. 343–349, https://doi.org/10.1002/fuce.201300208.

[149] C. Alegre, D. Sebastián, M.E. Gálvez, R. Moliner, M.J. Lázaro, Sulfurized carbon xerogels as Pt support with enhanced activity for fuel cell applications, Appl. Catal. Environ. 192 (2016) 260–267, https://doi.org/10.1016/j.apcatb.2016.03.070.

[150] C. Alegre, M.E. Gálvez, E. Baquedano, R. Moliner, E. Pastor, M.J. Lázaro, Oxygen-functionalized highly mesoporous carbon xerogel based catalysts for direct methanol fuel cell anodes, J. Phys. Chem. C 117 (2013) 13045–13058, https://doi.org/10.1021/jp400824n.

[151] C. Alegre, L. Calvillo, R. Moliner, J.A. González-Expósito, O. Guillén-Villafuerte, M.V.M. Huerta, E. Pastor, M.J. Lázaro, Pt and PtRu electrocatalysts supported on carbon xerogels for direct methanol fuel cells, J. Power Sources 196 (2011) 4226–4235, https://doi.org/10.1016/j.jpowsour.2010.10.049.

[152] W. Kiciński, M. Szala, M. Bystrzejewski, Sulfur-doped porous carbons: synthesis and applications, Carbon N. Y. 68 (2014) 1–32, https://doi.org/10.1016/j.carbon.2013.11.004.

[153] T.J. Bandosz, T.-Z. Ren, Porous carbon modified with sulfur in energy related applications, Carbon N. Y. 118 (2017) 561–577, https://doi.org/10.1016/J.CARBON.2017.03.095.

[154] W.Y. Wong, W.R.W. Daud, A.B. Mohamad, A.A.H. Kadhum, K.S. Loh, E. H. Majlan, Recent progress in nitrogen-doped carbon and its composites as electrocatalysts for fuel cell applications, Int. J. Hydrogen Energy 38 (2013) 9370–9386, https://doi.org/10.1016/j.ijhydene.2012.12.095.

[155] D. Sebastián, M.J. Nieto-Monge, S. Pérez-Rodríguez, E. Pastor, M.J. Lázaro, Nitrogen doped ordered mesoporous carbon as support of PtRu nanoparticles for methanol electro-oxidation, Energies 11 (2018), https://doi.org/10.3390/en11040831.

[156] A. Walcarius, Electrocatalysis, sensors and biosensors in analytical chemistry based on ordered mesoporous and macroporous carbon-modified electrodes, TrAC Trends Anal. Chem. 38 (2012) 79–97, https://doi.org/10.1016/j.trac.2012.05.003.

[157] S. Song, K. Wang, Y. Liu, C. He, Y. Liang, R. Fu, D. Wu, Y. Wang, Highly ordered mesoporous carbons as the support for Pt catalysts towards alcohol electrooxidation: the combined effect of pore size and electrical conductivity, Int. J. Hydrogen Energy 38 (2013) 1405–1412, https://doi.org/10.1016/j.ijhydene.2012.11.029.

[158] L. Calvillo, M.J. Lázaro, E. García-Bordejé, R. Moliner, P.L. Cabot, I. Esparbé, E. Pastor, J.J. Quintana, Platinum supported on functionalized ordered mesoporous carbon as electrocatalyst for direct methanol fuel cells, J. Power Sources 169 (2007) 59–64, https://doi.org/10.1016/j.jpowsour.2007.01.042.

[159] J.R.C. Salgado, F. Alcaide, G. Álvarez, L. Calvillo, M.J. Lázaro, E. Pastor, Pt–Ru electrocatalysts supported on ordered mesoporous carbon for direct methanol fuel cell, J. Power Sources 195 (2010) 4022–4029, https://doi.org/10.1016/j.jpowsour.2010.01.001.

[160] H. Nishihara, T. Kyotani, Templated nanocarbons for energy storage, Adv. Mater. 24 (2012) 4473–4498, https://doi.org/10.1002/adma.201201715.

[161] J. Lee, J. Kim, T. Hyeon, Recent progress in the synthesis of porous carbon materials, Adv. Mater. 18 (2006) 2073–2094, https://doi.org/10.1002/adma.200501576.

[162] Y. Xia, Z. Yang, R. Mokaya, Templated nanoscale porous carbons, Nanoscale 2 (2010) 639, https://doi.org/10.1039/b9nr00207c.

[163] S. Jun, S.H. Joo, R. Ryoo, M. Kruk, M. Jaroniec, Z. Liu, T. Ohsuna, O. Terasaki, Synthesis of new, nanoporous carbon with hexagonally ordered mesostructure, J. Am. Chem. Soc. 122 (2000) 10712–10713, https://doi.org/10.1021/ja002261e.

[164] S. Che, K. Lund, T. Tatsumi, S. Iijima, S.H. Joo, R. Ryoo, O. Terasaki, Direct observation of 3D mesoporous structure by scanning electron microscopy (SEM): SBA-15 silica and CMK-5 carbon, Angew. Chem. Int. Ed. 42 (2003) 2182–2185, https://doi.org/10.1002/anie.200250726.
[165] L. Calvillo, V. Celorrio, R. Moliner, P.L. Cabot, I. Esparbé, M.J. Lázaro, Control of textural properties of ordered mesoporous materials, Microporous Mesoporous Mater. 116 (2008) 292–298, https://doi.org/10.1016/j.micromeso.2008.04.015.
[166] Y. Meng, D. Gu, F. Zhang, Y. Shi, H. Yang, Z. Li, C. Yu, B. Tu, D. Zhao, Ordered mesoporous polymers and homologous carbon frameworks: amphiphilic surfactant templating and direct transformation, Angew. Chem. Int. Ed. 44 (2005) 7053–7059, https://doi.org/10.1002/anie.200501561.
[167] A. Walcarius, Mesoporous materials and electrochemistry, Chem. Soc. Rev. 42 (2013) 4098, https://doi.org/10.1039/c2cs35322a.
[168] R. Ryoo, S.H. Joo, M. Kruk, M. Jaroniec, Ordered mesoporous carbons, Adv. Mater. 13 (2001) 677–681, https://doi.org/10.1002/1521-4095(200105)13:9<677::AID-ADMA677>3.0.CO;2-C.
[169] C. Vix-Guterl, S. Saadallah, K. Jurewicz, E. Frackowiak, M. Reda, J. Parmentier, J. Patarin, F. Beguin, Supercapacitor electrodes from new ordered porous carbon materials obtained by a templating procedure, Mater. Sci. Eng. B 108 (2004) 148–155, https://doi.org/10.1016/j.mseb.2003.10.096.
[170] K. Jurewicz, C. Vix-Guterl, E. Frackowiak, S. Saadallah, M. Reda, J. Parmentier, J. Patarin, F. Béguin, Capacitance properties of ordered porous carbon materials prepared by a templating procedure, J. Phys. Chem. Solid 65 (2004) 287–293, https://doi.org/10.1016/j.jpcs.2003.10.024.
[171] C. Vix-Guterl, E. Frackowiak, K. Jurewicz, M. Friebe, J. Parmentier, F. Béguin, Electrochemical energy storage in ordered porous carbon materials, Carbon N. Y. 43 (2005) 1293–1302, https://doi.org/10.1016/j.carbon.2004.12.028.
[172] M. Hartmann, Ordered mesoporous materials for bioadsorption and biocatalysis, Chem. Mater. 17 (2005) 4577–4593, https://doi.org/10.1021/cm0485658.
[173] H. Chang, S.H. Joo, C. Pak, Synthesis and characterization of mesoporous carbon for fuel cell applications, J. Mater. Chem. 17 (2007) 3078, https://doi.org/10.1039/b700389g.
[174] G. Chai, S.B. Yoon, S. Kang, J.H. Choi, Y.E. Sung, Y.S. Ahn, H.S. Kim, J.S. Yu, Ordered uniform porous carbons as a catalyst support in a direct methanol fuel cell, Electrochim. Acta 50 (2004) 823–826, https://doi.org/10.1016/j.electacta.2003.11.041.
[175] E.P. Ambrosio, C. Francia, M. Manzoli, N. Penazzi, P. Spinelli, Platinum catalyst supported on mesoporous carbon for PEMFC, Int. J. Hydrogen Energy 33 (2008) 3142–3145, https://doi.org/10.1016/j.ijhydene.2008.03.045.
[176] L. Chuenchom, R. Kraehnert, B.M. Smarsly, Recent progress in soft-templating of porous carbon materials, Soft Matter 8 (2012) 10801, https://doi.org/10.1039/c2sm07448f.
[177] A. Eftekhari, Z. Fan, Ordered mesoporous carbon and its applications for electrochemical energy storage and conversion, Mater. Chem. Front. 1 (2017) 1001–1027, https://doi.org/10.1039/C6QM00298F.
[178] E. Lim, C. Jo, J. Lee, A mini review of designed mesoporous materials for energy-storage applications: from electric double-layer capacitors to hybrid supercapacitors, Nanoscale 8 (2016) 7827–7833, https://doi.org/10.1039/C6NR00796A.
[179] T.-Y. Ma, L. Liu, Z.-Y. Yuan, Direct synthesis of ordered mesoporous carbons, Chem. Soc. Rev. 42 (2013) 3977–4003, https://doi.org/10.1039/C2CS35301F.
[180] Y. Wan, Y. Shi, D. Zhao, Supramolecular aggregates as templates: ordered mesoporous polymers and carbons, Chem. Mater. 20 (2008) 932–945, https://doi.org/10.1021/cm7024125.

[181] H. Yang, D. Zhao, Synthesis of replica mesostructures by the nanocasting strategy, J. Mater. Chem. 15 (2005) 1217–1231, https://doi.org/10.1039/B414402C.
[182] R. Ryoo, S.H. Joo, S. Jun, Synthesis of highly ordered carbon molecular sieves via template-mediated structural transformation, J. Phys. Chem. B 103 (1999) 7743–7746.
[183] D. Zhao, J. Feng, Q. Huo, N. Melosh, G.H. Fredrickson, B.F. Chmelka, G. D. Stucky, Triblock copolymer syntheses of mesoporous silica with periodic 50 to 300 Angstrom pores, Science 279 (1998) 548–552, https://doi.org/10.1126/science.279.5350.548.
[184] Y. Deng, T. Yu, Y. Wan, Y. Shi, Y. Meng, D. Gu, L. Zhang, Y. Huang, C. Liu, X. Wu, D. Zhao, Ordered mesoporous silicas and carbons with large accessible pores templated from amphiphilic diblock copolymer poly(ethylene oxide)-b-polystyrene, J. Am. Chem. Soc. (2007), https://doi.org/10.1021/JA067379V.
[185] C. Liang, K. Hong, G.A. Guiochon, J.W. Mays, S. Dai, Synthesis of a large-scale highly ordered porous carbon film by self-assembly of block copolymers, Angew. Chem. Int. Ed. 43 (2004) 5785–5789, https://doi.org/10.1002/anie.200461051.
[186] R.T. Mayes, C. Tsouris, J.O. Kiggans Jr., S.M. Mahurin, D.W. DePaoli, S. Dai, Hierarchical ordered mesoporous carbon from phloroglucinol-glyoxal and its application in capacitive deionization of brackish water, J. Mater. Chem. 20 (2010) 8674, https://doi.org/10.1039/c0jm01911a.
[187] A. Sanchez-Sanchez, M.T. Izquierdo, G. Medjahdi, J. Ghanbaja, A. Celzard, V. Fierro, Ordered mesoporous carbons obtained by soft-templating of tannin in mild conditions, Microporous Mesoporous Mater. 270 (2018) 127–139, https://doi.org/10.1016/J.MICROMESO.2018.05.017.
[188] Y. Fang, D. Gu, Y. Zou, Z. Wu, F. Li, R. Che, Y. Deng, B. Tu, D. Zhao, A low-concentration hydrothermal synthesis of biocompatible ordered mesoporous carbon nanospheres with tunable and uniform size, Angew. Chem. Int. Ed. 49 (2010) 7987–7991, https://doi.org/10.1002/anie.201002849.
[189] M. Steinhart, C. Liang, G.W. Lynn, U. Gösele, S. Dai, Direct synthesis of mesoporous carbon microwires and nanowires, Chem. Mater. (2007), https://doi.org/10.1021/CM070455O.
[190] M.E. Gálvez, L. Calvillo, C. Alegre, D. Sebastián, I. Suelves, S. Pérez-Rodríguez, V. Celorrio, E. Pastor, J.I. Pardo, R. Moliner, M.J. Lázaro, Nanostructured carbon materials as supports in the preparation of direct methanol fuel cell electrocatalysts, Catalysts (2013), https://doi.org/10.3390/catal3030671.
[191] S.H. Joo, S.J. Choi, I. Oh, J. Kwak, Z. Liu, O. Terasaki, R. Ryoo, Ordered nanoporous arrays of carbon supporting high dispersions of platinum nanoparticles, Nature 412 (2001) 169–172, https://doi.org/10.1038/35084046.
[192] J.-H. Zhou, J.-P. He, Y.-J. Ji, W.-J. Dang, X.-L. Liu, G.-W. Zhao, C.-X. Zhang, J.-S. Zhao, Q.-B. Fu, H.-P. Hu, CTAB assisted microwave synthesis of ordered mesoporous carbon supported Pt nanoparticles for hydrogen electro-oxidation, Electrochim. Acta 52 (2007) 4691–4695, https://doi.org/10.1016/j.electacta.2007.01.007.
[193] J. Zhang, D. Liu, H. Song, Z. Liang, X. Guo, L. Du, S. Liao, Effects of tailoring and dehydrated cross-linking on morphology evolution of ordered mesoporous carbons, RSC Adv. 6 (2016) 19515–19521, https://doi.org/10.1039/C5RA26639D.
[194] L. Calvillo, M. Gangeri, S. Perathoner, G. Centi, R. Moliner, M.J. Lázaro, Synthesis and performance of platinum supported on ordered mesoporous carbons as catalyst for PEM fuel cells: effect of the surface chemistry of the support, Int. J. Hydrogen Energy 36 (2011) 9805–9814, https://doi.org/10.1016/J.IJHYDENE.2011.03.023.
[195] M.J. Lázaro, L. Calvillo, E.G. Bordejé, R. Moliner, R. Juan, C.R. Ruiz, Functionalization of ordered mesoporous carbons synthesized with SBA-15 silica as template,

Microporous Mesoporous Mater. 103 (2007) 158–165, https://doi.org/10.1016/J.MICROMESO.2007.01.047.
[196] J.R.C. Salgado, J.J. Quintana, L. Calvillo, M.J. Lázaro, P.L. Cabot, I. Esparbé, E. Pastor, Carbon monoxide and methanol oxidation at platinum catalysts supported on ordered mesoporous carbon: the influence of functionalization of the support, Phys. Chem. Chem. Phys. 10 (2008) 6796, https://doi.org/10.1039/b809227c.
[197] G. Álvarez, F. Alcaide, O. Miguel, L. Calvillo, M.J. Lázaro, J.J. Quintana, J. C. Calderón, E. Pastor, Technical electrodes catalyzed with PtRu on mesoporous ordered carbons for liquid direct methanol fuel cells, J. Solid State Electrochem. 14 (2010) 1027–1034, https://doi.org/10.1007/s10008-009-0913-3.
[198] J. Calderón Gómez, V. Celorrio, L. Calvillo, D. Sebastián, R. Moliner, M. Lázaro Elorri, J.C. Calderón Gómez, V. Celorrio, L. Calvillo, D. Sebastián, R. Moliner, M.J. Lázaro Elorri, Electrochemical behavior of Pt–Ru catalysts supported on graphitized ordered mesoporous carbons toward CO and methanol oxidation, Surfaces 2 (2019) 1–15, https://doi.org/10.3390/surfaces2010001.
[199] J. Zhou, J. He, T. Wang, D. Sun, G. Zhao, X. Chen, D. Wang, Z. Di, $NiCl_2$ assisted synthesis of ordered mesoporous carbon and a new strategy for a binary catalyst, J. Mater. Chem. 18 (2008) 5776, https://doi.org/10.1039/b811076j.
[200] D. Xiang, L. Yin, Well-dispersed and size-tuned bimetallic PtFex nanoparticle catalysts supported on ordered mesoporous carbon for enhanced electrocatalytic activity in direct methanol fuel cells, J. Mater. Chem. 22 (2012) 9584, https://doi.org/10.1039/c2jm16641k.
[201] F. Hu, C. Chen, Z. Wang, G. Wei, P.K. Shen, Mechanistic study of ethanol oxidation on Pd–NiO/C electrocatalyst, Electrochim. Acta 52 (2006) 1087–1091, https://doi.org/10.1016/J.ELECTACTA.2006.07.008.
[202] G.Z. Hu, F. Nitze, X. Jia, T. Sharifi, H.R. Barzegar, E. Gracia-Espino, T. Wågberg, Reduction free room temperature synthesis of a durable and efficient Pd/ordered mesoporous carbon composite electrocatalyst for alkaline direct alcohols fuel cell, RSC Adv. 4 (2014) 676–682, https://doi.org/10.1039/C3RA42652A.
[203] Y. Qiu, J. Huo, F. Jia, B.H. Shanks, W. Li, N- and S-doped mesoporous carbon as metal-free cathode catalysts for direct biorenewable alcohol fuel cells, J. Mater. Chem. A 4 (2016) 83–95, https://doi.org/10.1039/C5TA06039G.
[204] N. Daems, X. Sheng, I.F.J. Vankelecom, P.P. Pescarmona, Metal-free doped carbon materials as electrocatalysts for the oxygen reduction reaction, J. Mater. Chem. A 2 (2014) 4085–4110, https://doi.org/10.1039/C3TA14043A.
[205] D.-S. Yang, D. Bhattacharjya, S. Inamdar, J. Park, J.-S. Yu, Phosphorus-doped ordered mesoporous carbons with different lengths as efficient metal-free electrocatalysts for oxygen reduction reaction in alkaline media, J. Am. Chem. Soc. 134 (2012) 16127–16130, https://doi.org/10.1021/ja306376s.
[206] X. Wang, J.S. Lee, Q. Zhu, J. Liu, Y. Wang, S. Dai, Ammonia-treated ordered mesoporous carbons as catalytic materials for oxygen reduction reaction, Chem. Mater. 22 (2010) 2178–2180, https://doi.org/10.1021/cm100139d.
[207] R. Silva, D. Voiry, M. Chhowalla, T. Asefa, Efficient metal-free electrocatalysts for oxygen reduction: polyaniline-derived N- and O-doped mesoporous carbons, J. Am. Chem. Soc. 135 (2013) 7823–7826, https://doi.org/10.1021/ja402450a.
[208] J. Li, Z. Li, J. Tong, C. Xia, F. Li, Nitrogen-doped ordered mesoporous carbon sphere with short channel as an efficient metal-free catalyst for oxygen reduction reaction, RSC Adv. 5 (2015) 70010–70016, https://doi.org/10.1039/C5RA10484J.
[209] S. Jiang, Y. Sun, H. Dai, J. Hu, P. Ni, Y. Wang, Z. Li, Facile synthesis of nitrogen and sulfur dual-doped hierarchical micro/mesoporous carbon foams as efficient metal-free electrocatalysts for oxygen reduction reaction, Electrochim. Acta 174 (2015) 826–836, https://doi.org/10.1016/j.electacta.2015.06.050.

[210] Q. Shi, Y. Lei, Y. Wang, H. Wang, L. Jiang, H. Yuan, D. Fang, B. Wang, N. Wu, Y. Gou, B, N-codoped 3D micro-/mesoporous carbon nanofibers web as efficient metal-free catalysts for oxygen reduction, Curr. Appl. Phys. 15 (2015) 1606–1614, https://doi.org/10.1016/J.CAP.2015.09.012.

[211] X. Bo, L. Guo, Ordered mesoporous boron-doped carbons as metal-free electrocatalysts for the oxygen reduction reaction in alkaline solution, Phys. Chem. Chem. Phys. 15 (2013) 2459, https://doi.org/10.1039/c2cp43541a.

[212] L. Wang, W. Jia, X. Liu, J. Li, M.M. Titirici, Sulphur-doped ordered mesoporous carbon with enhanced electrocatalytic activity for the oxygen reduction reaction, J. Energy Chem. 25 (2016) 566–570, https://doi.org/10.1016/J.JECHEM.2016.02.012.

[213] X. Liu, W. Li, S. Zou, Cobalt and nitrogen-codoped ordered mesoporous carbon as highly efficient bifunctional catalysts for oxygen reduction and hydrogen evolution reactions, J. Mater. Chem. A 6 (2018) 17067–17074, https://doi.org/10.1039/C8TA06864J.

[214] H. Tang, Y. Zeng, D. Liu, D. Qu, J. Luo, K. Binnemans, D.E. De Vos, J. Fransaer, D. Qu, S.-G. Sun, Dual-doped mesoporous carbon synthesized by a novel nanocasting method with superior catalytic activity for oxygen reduction, Nano Energy 26 (2016) 131–138, https://doi.org/10.1016/J.NANOEN.2016.05.015.

[215] V. Celorrio, D. Sebastián, L. Calvillo, A.B. García, D.J. Fermin, M.J. Lázaro, Influence of thermal treatments on the stability of Pd nanoparticles supported on graphitised ordered mesoporous carbons, Int. J. Hydrogen Energy (2016)https://doi.org/10.1016/j.ijhydene.2016.05.271.

[216] S. Pérez-Rodríguez, D. Torres, M.J. Lázaro, Effect of oxygen and structural properties on the electrical conductivity of powders of nanostructured carbon materials, Powder Technol. 340 (2018) 380–388, https://doi.org/10.1016/J.POWTEC.2018.09.038.

[217] K.P. Gierszal, M. Jaroniec, T.-W. Kim, J. Kim, R. Ryoo, High temperature treatment of ordered mesoporous carbons prepared by using various carbon precursors and ordered mesoporous silica templates, New J. Chem. 32 (2008) 981, https://doi.org/10.1039/b716735k.

[218] H. Darmstadt, C. Roy, S. Kaliaguine, S.J. Choi, R. Ryoo, Surface chemistry of ordered mesoporous carbons, Carbon N. Y. 40 (2002) 2673–2683, https://doi.org/10.1016/S0008-6223(02)00187-2.

[219] H. Darmstadt, C. Roy, S. Kaliaguine, S.H. Joo, R. Ryoo, Pore structure and graphitic surface nature of ordered mesoporous carbons probed by low-pressure nitrogen adsorption, Microporous Mesoporous Mater. 60 (2003) 139–149, https://doi.org/10.1016/S1387-1811(03)00335-4.

[220] M. Sevilla, A.B. Fuertes, Catalytic graphitization of templated mesoporous carbons, Carbon N. Y. 44 (2006) 468–474, https://doi.org/10.1016/J.CARBON.2005.08.019.

[221] T.-W. Kim, I.-S. Park, R. Ryoo, A synthetic route to ordered mesoporous carbon materials with graphitic pore walls, Angew. Chem. Int. Ed. 42 (2003) 4375–4379, https://doi.org/10.1002/anie.200352224.

[222] T. Wang, C. Zhang, X. Sun, Y. Guo, H. Guo, J. Tang, H. Xue, M. Liu, X. Zhang, L. Zhu, Q. Xie, J. He, Synthesis of ordered mesoporous boron-containing carbon films and their corrosion behavior in simulated proton exchange membrane fuel cells environment, J. Power Sources 212 (2012) 1–12, https://doi.org/10.1016/J.JPOWSOUR.2012.03.068.

[223] U.B. Nasini, V.G. Bairi, S.K. Ramasahayam, S.E. Bourdo, T. Viswanathan, A. U. Shaikh, Phosphorous and nitrogen dual heteroatom doped mesoporous carbon synthesized via microwave method for supercapacitor application, J. Power Sources 250 (2014) 257–265, https://doi.org/10.1016/J.JPOWSOUR.2013.11.014.

[224] S. Pérez-Rodríguez, D. Sebastián, M.J. Lázaro, Electrochemical oxidation of ordered mesoporous carbons and the influence of graphitization, Electrochim. Acta 303 (2019) 167–175, https://doi.org/10.1016/J.ELECTACTA.2019.02.065.

[225] S. Pérez-Rodríguez, D. Sebastián, M.J. Lázaro, Insights on the electrochemical oxidation of ordered mesoporous carbons, J. Electrochem. Soc. 167 (2020) 024511. https://doi.org/10.1149/1945-7111/ab6a8f.

[226] L.M. Rivera Gavidia, G. García, V. Celorrio, M.J. Lázaro, E. Pastor, Methanol tolerant Pt2CrCo catalysts supported on ordered mesoporous carbon for the cathode of DMFC, Int. J. Hydrogen Energy 41 (2016) 19645–19655, https://doi.org/10.1016/J.IJHYDENE.2016.06.132.

[227] S. Pérez-Rodríguez, N. Rillo, M.J. Lázaro, E. Pastor, Pd catalysts supported onto nanostructured carbon materials for CO_2 valorization by electrochemical reduction, Appl. Catal. Environ. 163 (2015) 83–95, https://doi.org/10.1016/j.apcatb.2014.07.031.

[228] S. Pérez-Rodríguez, G. García, M.J. Lázaro, E. Pastor, Chapter 9: Probing CO_2 Reduction Intermediates Employing In Situ Spectroscopy and Spectrometry, (2018)https://doi.org/10.1039/9781782623809-00212.

[229] C. Comminges, N.E. Sahin, A. Le Valant, T.W. Napporn, J. Parmentier, J. Kiener, O. Ersen, G. Melinte, K.B. Kokoh, Carbon support nanostructuration for CO_2 electroreduction to formic acid on copper based catalysts, ECS Trans. (2017) 1460. Meet. Abstr. MA2017-01.

[230] N.E. Sahin, P.P. Pescarmona, Boron-, phosphorus-doped ordered mesoporous carbon electrocatalysts for the reduction of CO_2 to formic acid, J. Am. Chem. Soc. (2019) 1090. Meet. Abstr. MA2019-02.

[231] Y. Song, W. Chen, C. Zhao, S. Li, W. Wei, Y. Sun, Metal-free nitrogen-doped mesoporous carbon for electroreduction of CO_2 to ethanol, Angew. Chem. Int. Ed. 56 (2017) 10840–10844, https://doi.org/10.1002/anie.201706777.

[232] K. Fic, E. Frackowiak, F. Beguin, Unusual energy enhancement in carbon-based electrochemical capacitors, J. Mater. Chem. 22 (2012) 24213–24223, https://doi.org/10.1039/C2JM35711A.

[233] J. Castro-Gutiérrez, N. Díez, M. Sevilla, M.T. Izquierdo, J. Ghanbaja, A. Celzard, V. Fierro, High-rate capability of supercapacitors based on tannin-derived ordered mesoporous carbons, ACS Sustain. Chem. Eng. 7 (2019) 17627–17635, https://doi.org/10.1021/acssuschemeng.9b03407.

[234] D.Y. Kim, X. Jin, C.H. Lee, D.W. Kim, J. Suk, J.K. Shon, J.M. Kim, Y. Kang, Improved electrochemical performance of ordered mesoporous carbon by incorporating macropores for Li–O2 battery cathode, Carbon N. Y. 133 (2018) 118–126, https://doi.org/10.1016/J.CARBON.2018.03.016.

CHAPTER 11

Graphene materials for the electrocatalysts used for fuel cells and electrolyzers

Gonzalo García[a], Francisco Alcaide Monterrubio[b], and Elena Pastor[a]
[a]Department of Chemistry, Institute of Materials and Nanotechnology, University of La Laguna, La Laguna, Tenerife, Spain
[b]CIDETEC, Donostia-San Sebastián, Spain

1 Graphene-based electrocatalysts for fuel cells

1.1 Low-temperature fuel cells

Low-temperature fuel cells (LTFCs) have attracted great attention as a primary power source for electric devices and residential cogeneration units. LTFCs consist of two electrodes separated by an electrolyte medium (usually a membrane) that determines the type of LTFC, i.e., anion exchange membrane fuel cell (AEMFC) for LTFC working at high pH and proton-exchange fuel cell (PEMFC) for LTFC running at low pH. Similar to the majority of fuel cells, LTFCs oxidized the fuel (hydrogen at LTFCs) at the anode, while oxygen is reduced at the cathode. The ions produced by these electrochemical reactions are transported through the electrolyte from one electrode to the other, while the electrons generated at the anode flow through an external circuit to the cathode. Accordingly, LTFCs usually reveal higher efficiency than systems operating on internal combustion since no combustion occurs, and consequently, there is no limitation by the Carnot cycle [1–6].

Despite the significant advantages of LTFCs as power sources, a real LTFC has several drawbacks, which decrease their working performance and increase their cost. In this regard, one of the most important issues to solve is that related to the anode and cathode electrodes for LTFCs, which usually employ platinum-based materials that are very active catalysts, although their elevated cost, low abundance, and poor durability are main obstacles in the development of LTFCs. Hence, it is very important to reduce the amount and optimize the use of this noble metal [4–6].

Emerging Carbon Materials for Catalysis
https://doi.org/10.1016/B978-0-12-817561-3.00011-1

LTFC is strongly affected by the performance of the electrocatalysts employed to carry out the desired reaction. Since electrocatalysts mainly affect the exchange current (I_0), it is common to plot the I_0 as function of electrocatalyst surface—adsorbate bond strength and typical volcano plots are usually reported [7–9]. The last is just a translation of the Sabatier Principle in electrochemical terms, in which the catalytic activity of a specific group of catalysts toward a given reaction is qualitatively linked to the interaction between one or more reaction intermediates and the catalytic material. Therefore, volcano plots appear as a very powerful tool to search and find low-cost electrodic materials that fulfill the required catalytic activity.

In the last decade, graphene-based catalysts emerged as promising materials to overcome the reactions at both electrodes of LTFCs since they are low-cost, abundant, and electrical conductors with tuneable catalytic activity [10–12]. However, pure graphene sheet reveals low catalytic activity toward most of the electrochemical reactions, and consequently, catalysts based on reduced graphene oxide (rGO) are usually employed [13–17].

Chemical and physical approaches are the most common ways to produce rGO and heteroatom-modified rGO from the reduction of graphene oxide (GO), which is usually created from graphite by the popular Hummers' method [18]. The doping process may produce electronic and geometric distortions of the carbonaceous grid, and consequently, the activity toward several electrochemical reactions may be altered [19, 20]. Basically, two types of doping that depend on the interatomic distance between carbon-heteroatom and C—C may occur in the graphene network, e.g., B—C, N—C, and C—C have similar interatomic distance, and therefore, the heteroatom-modification will follow a similar modus operandi; on the other hand, P, Si, and S show higher interatomic distance than C—C and their modification will display an exohedral doping, i.e., the heteroatom will be located above and/or below the graphene sheet [21]. Another factor to take into account in the catalyst activity is the surface local acidity, which may be altered by introduction of functional groups in the graphene network [12, 22–24]. Fig. 1 summarizes the most common graphene modifications.

In the following sections, basic principles of the oxygen reduction reaction (ORR) and the hydrogen oxidation reaction (HOR) will be examined and graphene-containing electrocatalysts for both reactions will be reviewed.

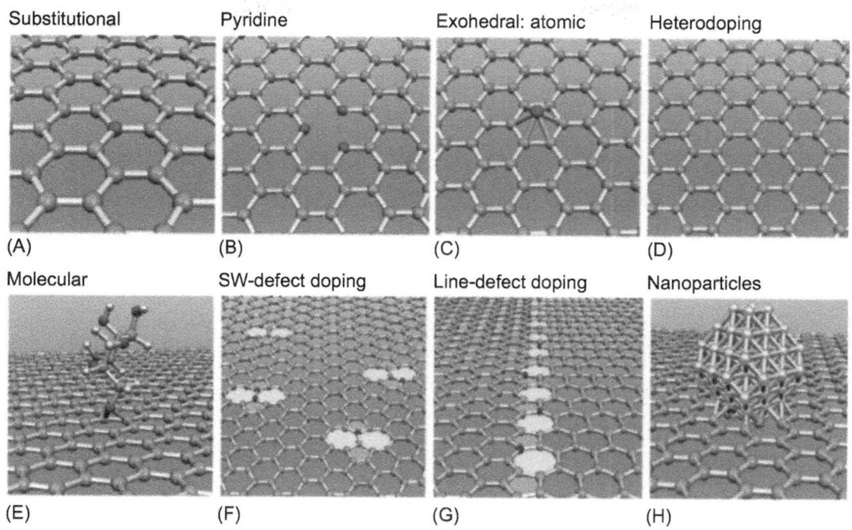

Fig. 1 Molecular models representing different types of doped graphenes. (A) Substitutional doping, in which a carbon atom is removed and replaced by the dopant; (B) pyridine-like doping, in which a single vacancy is generated and the low coordinated carbon atoms (three carbons around the vacancy) are replaced by dopants; (C) exohedral atomic doping where a dopant is hosted on the surface of graphene; (D) heterodoping which consists of doping simultaneously the graphene with two or more types of atoms different from carbon; (E) molecular doping or chemical functionalization of the graphene sheet; (F) 55–77 Thrower-Stone-Wales defects in which 5–7–5–7 rings are generated by rotating a C—C bond by 90 degrees (dopants are also included in some defects); (G) 5–7 line-defect in combination with substitutional doping; and (H) nanoparticles or clusters anchored on the graphene surface. *(Reprinted from F. López-Urías, R. Lv, H. Terrones, M. Terrones, Doped graphene: theory, synthesis, characterization, and applications, in: D.-E. Jiang, Z. Chen (Eds.), Graphene Chemistry: Theoretical Perspectives, John Wiley & Sons, Ltd, Chichester, UK, 2013, pp. 183–207, https://doi.org/10.1002/9781118691281, with permission from John Wiley & Sons.)*

1.2 Graphene materials for the oxygen reduction reaction

The sluggish kinetics of the oxygen reduction reaction (ORR) in PEMFC system is the main challenge to overcome in order to introduce this technology in the energy market [25]. Indeed, the ORR is five times slower than the hydrogen oxidation reaction (HOR) in acid media and elevated amounts of expensive materials, such as platinum and palladium, are usually employed to overcome this issue [26]. Take as an example a typical PEMFC that employs platinum as catalyst at both electrodes, i.e., cathode and anode. It usually includes about 80%–90% of the precious metal in the cathode side [26, 27].

It becomes clear that high amounts of scarce noble metals are not economically viable and a drastic reduction is mandatory as was reported by the US Department of Energy (DoE), in which a maximum loading of $0.125\,\mathrm{mg\,cm^{-2}}$ and a mass activity of $0.44\,\mathrm{mA\,g^{-1}}$ are the targets to be achieved by 2020 [26, 27]. In order to achieve these goals, the improvement and optimization of the cathode are crucial, and to accomplish these requirements, a full understanding of the ORR mechanism is necessary [6, 28–32]. The complete ORR involves four electrons:

Acidic media:

$$O_2 + 4H^+ + 4e^- \longrightarrow 2H_2O \qquad E^0 = 1.229\,\mathrm{V\ vs\ SHE} \tag{1}$$

Basic media:

$$O_2 + 2H_2O + 4e^- \longrightarrow 4OH^- \qquad E^0 = 0.401\,\mathrm{V\ vs\ SHE} \tag{2}$$

This reaction involves several elementary steps, pathways, and intermediates, which depend on the pH, dissolved ions in the electrolyte, catalyst and support materials, surface atomic structure, surface energy, and oxygen coverage and potential, among others [5, 6, 10, 33]. Briefly, three main reaction pathways for the ORR are extensively accepted, which may coexist: (i) dissociative, (ii) associative, and (iii) peroxo mechanisms. The first step during the ORR in acidic and basic media is common for all different pathways:

$$O_2 \longrightarrow O_{2,ads} \tag{3}$$

Dissociative mechanism in acidic media:

$$O_{2,ads} \longrightarrow O_{ads} + O_{ads} \tag{4}$$

$$O_{ads} + O_{ads} + 2(H^+ + e^-) \longrightarrow OH_{ads} + OH_{ads} \tag{5}$$

$$OH_{ads} + OH_{ads} + 2(H^+ + e^-) \longrightarrow 2H_2O \tag{6}$$

Associative mechanism in acidic media:

$$O_{2,ads} + (H^+ + e^-) \longrightarrow OOH_{ads} \tag{7}$$

$$OOH_{ads} \longrightarrow OH_{ads} + O_{ads} \tag{8}$$

$$O_{ads} + (H^+ + e^-) \longrightarrow OH_{ads} \tag{9}$$

$$OH_{ads} + OH_{ads} + 2(H^+ + e^-) \longrightarrow 2H_2O \tag{10}$$

Peroxo mechanism in acidic media:

$$O_{2,ads} + (H^+ + e^-) \longrightarrow OOH_{ads} \tag{11}$$

$$OOH_{ads} + (H^+ + e^-) \longrightarrow HOOH_{ads} \tag{12}$$

$$HOOH_{ads} \longrightarrow H_2O_2 \tag{13}$$

$$H_2O_2 \longrightarrow H_2O + O_2 \tag{14}$$

$$HOOH_{ads} \longrightarrow OH_{ads} + OH_{ads} \tag{15}$$

$$OH_{ads} + OH_{ads} + 2(H^+ + e^-) \longrightarrow 2H_2O \tag{16}$$

As can be observed, associative and peroxo mechanisms share the same initial reaction (Eqs. 7 and 11) and its product (OOH_{ads}) may follow two different pathways: O—O cleavage (Eq. 8) or $HOOH_{ads}$ formation (Eq. 12). The last intermediate may dissociate into two OH_{ads} (Eq. 15) or desorb in the form of hydrogen peroxide (Eq. 13) and disproportionate (Eq. 14), which drastically reduce the ORR efficiency (two instead of four electrons). Additionally, hydrogen peroxide is a very corrosive species in acidic media that may compromise the stability of fuel cell materials.

Then, it becomes clear that dissociative and associative pathways are desired, and as was described above, the prevalence of one pathway among the others depends on the properties of the electrocatalyst and the surrounding media.

On the other hand, the ORR in alkaline media comprises similar mechanisms than in acidic media:

Dissociative mechanism in basic media:

$$\frac{1}{2}O_{2,ads} \longrightarrow O_{ads} \tag{17}$$

$$O_{ads} + H_2O + e^- \longrightarrow OH_{ads} + OH^- \tag{18}$$

$$OH_{ads} + e^- \longrightarrow OH^- \tag{19}$$

$$OH_{ads} + H_2O + e^- \longrightarrow H_2O_{ads} + OH^- \tag{20}$$

Associative mechanism in basic media:

$$O_{2,ads} + H_2O + e^- \longrightarrow OOH_{ads} + OH^- \quad (21)$$

$$OOH_{ads} + e^- \longrightarrow O_{ads} + OH^- \quad (22)$$

$$O_{ads} + H_2O + e^- \longrightarrow OH_{ads} + OH^- \quad (23)$$

$$OH_{ads} + e^- \longrightarrow OH^- \quad (24)$$

$$OH_{ads} + H_2O + e^- \longrightarrow H_2O_{ads} + OH^- \quad (25)$$

Peroxo mechanism in basic media:

$$O_{2,ads} + H_2O + e^- \longrightarrow OOH_{ads} + OH^- \quad (26)$$

$$OOH_{ads} + e^- \longrightarrow OOH^- \quad (27)$$

$$2OOH^- \longrightarrow 2OH^- + O_2 \quad (28)$$

$$OOH_{ads} + e^- \longrightarrow O_{ads} + OH^- \quad (29)$$

$$O_{ads} + H_2O + e^- \longrightarrow OH_{ads} + OH^- \quad (30)$$

$$OH_{ads} + e^- \longrightarrow OH^- \quad (31)$$

$$OH_{ads} + H_2O + e^- \longrightarrow H_2O_{ads} + OH^- \quad (32)$$

Besides the associative, dissociative, and peroxo mechanisms, the water-mediated pathway was recently considered in a theoretical study for the ORR on Pt (111) in alkaline medium [34].

Water-mediated mechanism in basic media:

$$O_{2,ads} + H_2O_{ads} \longrightarrow OOH_{ads} + OH_{ads} \quad (33)$$

$$OOH_{ads} \longrightarrow O_{ads} + OH_{ads} \quad (34)$$

$$O_{ads} + H_2O_{ads} \longrightarrow 2OH_{ads} \quad (35)$$

$$OH_{ads} + e^- \longrightarrow OH^- \quad (36)$$

$$OH_{ads} + H_2O + e^- \longrightarrow H_2O_{ads} + OH^- \quad (37)$$

In this work, authors suggested that the water-mediated mechanism might be the dominant pathway in alkaline solution. Interestingly, the only contribution to the faradaic current depends on the reduction of adsorbed

OH_{ads} species (Eqs. 36 and 37) as other reactions are potential-independent surface reactions, and therefore, only the adsorption strengths of OH_{ads} and $O_{2,ads}$ will determine the catalytic activity toward the ORR.

Research works in alkaline media attract more and more attention due to [6, 10, 35–40]: (i) the ORR kinetics is faster and more efficient in it than in acidic media; (ii) it confers reduced adsorption energies of anions; (iii) it is not so corrosive; and (iv) a wide range of nonprecious materials such as graphene-based catalysts can be employed. However, several drawbacks such as the carbonation of the electrolyte by the CO_2 present in the air that severely blocks active sites and the durability of the catalyst performance need to be solved [4, 6, 33, 41].

After this general overview, this section will illustrate and summarize the recent advances of oxygen reduction on graphene-based electrocatalysts.

1.2.1 Nitrogen-doped graphene catalysts (NG)

As was described earlier, B—C and N—C reveal similar interatomic distance than C—C, and therefore, substitutional doping may occur. Main difference resides in their electronic configuration, which will introduce electrons (N) or holes (B) into the graphene sheet. In the following paragraphs, only works related to NG will be scrutinized to show this type of heteroatom effect in the graphene network.

Graphene is a noncatalytic material for the ORR due to the absence of active sites for oxygen adsorption. Thus, it is recurrent to introduce in the graphene structure active sites such as functional groups (e.g., carbonyls, alcohols, carboxylic acids, esters, and amines) and low coordinated sites (e.g., edge sites) [10]. Additionally, N can be introduced into the graphene grid with three diverse bonding configurations (Fig. 2): graphitic, pyrrolic, and pyridinic (contributes one p electron to the π system). It is remarkable that pyridinic and pyrrolic nitrogen contribute one and two p electrons to the π system, which will considerably influence the chemical and physical properties of NG, such as electron, hole, and phonon transport behaviors [42]. Thus, nitrogen insertion creates local charge sites, which will result in redistribution of the spin and charge density states in the graphene [10]. In this way, molecular oxygen adsorption and bond scission as well as adsorbate bond strength on NG may be modulated by the type of nitrogen, functional group, defect sites, and the amount of them into the carbonaceous structure [10, 19, 20, 24, 43, 44].

For example, a solvothermal method was used by Ma et al. to convert pyrrolic (G-A) to pyridinic (NG-A) species into the NG catalyst and both

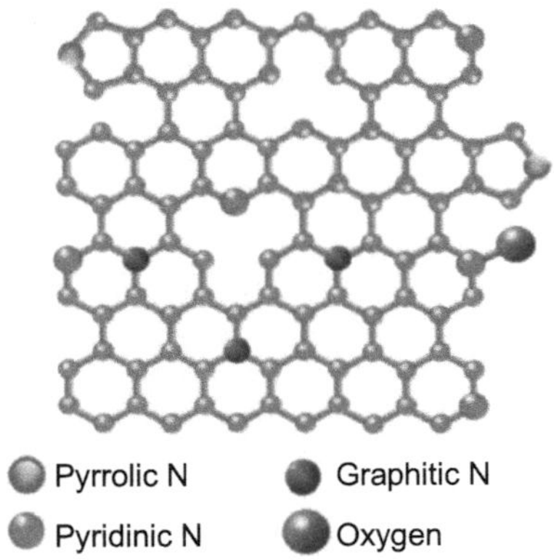

Fig. 2 Schematic illustration of the three diverse N bonding configurations into the graphene structure. *(Reprinted from R. Ma, X. Ren, B.Y. Xia, Y. Zhou, C. Sun, Q. Liu, et al., Novel synthesis of N-doped graphene as an efficient electrocatalyst towards oxygen reduction, Nano Res. 9 (2016) 808–819, https://doi.org/10.1007/s12274-015-0960-2, with permission from Springer Nature.)*

materials were studied toward the ORR in acidic and alkaline media [43]. They found a great increment of the catalytic activity toward the ORR in both media after the conversion of pyrrolic to pyridinic species, and the rise of the catalytic performance was attributed to an enhancement of O_2 adsorption strength on pyridinic species (Fig. 3). Additionally, the four-electron pathway was the dominant one at both pHs, although much higher activity was discerned in alkaline than in acidic media. The latter is the most

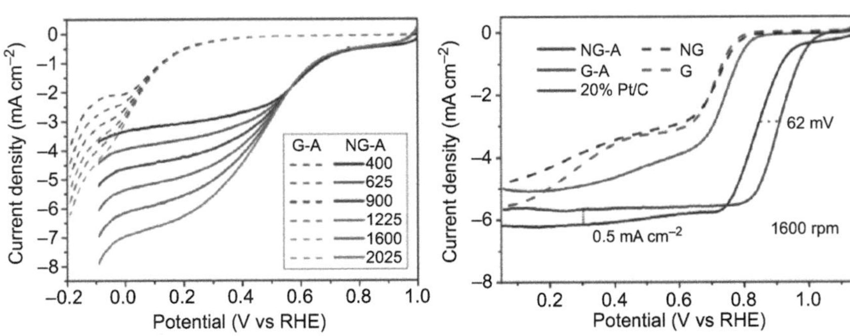

Fig. 3 Lineal sweep voltammetry curves of NG-A and G-A in 0.5 M H_2SO_4 recorded at different rotation speeds (left panel) and lineal sweep voltammetry curves of G, NG, G-A, NG-A, and Pt/C in 0.1 M KOH recorded at 1600 rpm (right panel). *(Reprinted from R. Ma, X. Ren, B.Y. Xia, Y. Zhou, C. Sun, Q. Liu, et al., Novel synthesis of N-doped graphene as an efficient electrocatalyst towards oxygen reduction, Nano Res. 9 (2016) 808–819, https://doi.org/10.1007/s12274-015-0960-2, with permission from Springer Nature.)*

common result as the ORR kinetics is enhanced at high pH (see the previous subsection).

The size effect (and therefore the density of edge sites) of nitrogen-doped graphene was analyzed by density functional theory [45]. It was corroborated that the adsorption strength of ORR intermediates, the reaction-free energy of the rate-determining step, and the overpotential increase with the rise of the size of N-doped graphene. Thus, catalyst with the smallest size revealed the best performance toward the ORR.

Several methodologies were employed for the preferential growth of nitrogen configurations into the NG. Some examples include pyrolysis, solvothermal, hydrothermal, thermal, free radical reaction, and chemical vapor deposition [10, 19, 20, 24, 42, 46–49].

1.2.2 Sulfur-doped graphene catalysts (SG)

As was described above, P, Si, and S show higher interatomic distance than C—C and their modification will mostly display an exohedral doping. In the following paragraphs, only research of SG will be explored as example for this type of heteroatom effect in the graphene structure.

S atom is larger than C atom, and hence, the C—S bond length is longer than C—C bond [21]. Additionally, the electronegativity of both atoms is very similar and negligible charge transfer is expected in the C—S bond [10, 21]. However, the graphene lattice experiences significant distortions in presence of S atom, which results in a position displacement of the surrounding carbon atoms out of the plane [21]. The last, together with the mismatch of the outermost orbitals of S and C, may induce a nonuniform spin density distribution, which could modify the SG surface activity toward the ORR [10, 21].

Fig. 4 shows polarization and power plots carried out in anion- and proton-exchange membrane fuel cells that were fed with H_2 and O_2, in which the cathode is comprised of SG [23]. In this investigation, the sulfur content was about 2.2 at.% and it was integrated in the graphene-ring matrices as thiophene motifs. Open circuit voltage of 0.37 and 0.78 V and a maximum power density of 1.19 and 2.38 mW cm^{-2} in acidic and alkaline media, respectively, were achieved at a working temperature of 95°C. Values are not so impressive in comparison to carbon-supported Pt, but they bring out the effect of doping graphene with sulfur atom.

The synthetic methodology is similar to that of NG; however, a lot of effort needs to be done to perform them without the use of toxic precursors.

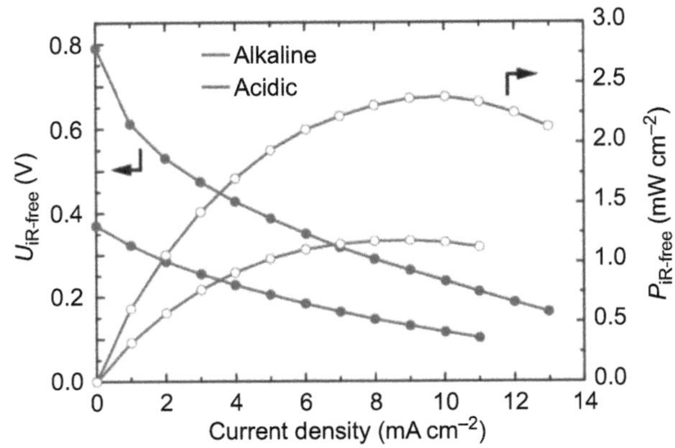

Fig. 4 Polarization and power density curves of alkaline and acidic PEMFC devices comprising S-rGO as porous cathode catalyst layer and 0.3 mg Pt/cm^2 (46% Pt/C) for the anode. Operation conditions: 0.5/0.5 slpm H_2/O_2 anode/cathode gas flow, 95°C cell temperature, 100% relative humidity, and 300 kPa absolute pressure. *(Reprinted from M. Klingele, C. Pham, K.R. Vuyyuru, B. Britton, S. Holdcroft, A. Fischer, et al., Sulfur doped reduced graphene oxide as metal-free catalyst for the oxygen reduction reaction in anion and proton exchange fuel cells, Electrochem. Commun. 77 (2017) 71–5, https://doi.org/10.1016/j.elecom.2017.02.015, with permission from Elsevier.)*

1.2.3 Nitrogen and sulfur dual-doping graphene catalysts (NSG)

Doping graphene with heteroatoms of diverse varieties enhances the overall catalyst performance toward the ORR [10, 20, 39, 45, 48, 50]. It is common to obtain NSG catalysts with high surface area, elevated conductivity, and stability. Additionally, the insertion of both heteroatoms reduces the ORR onset potential, produces high and constant limiting currents, and the four-electron pathway in basic media is usually reported.

Recently, we describe the effect of defect sites and the importance of both types of heteroatoms into the graphenic network [39]. The catalysts were synthesized using the hydrothermal reduction method and caffeine, thiourea, and ammonium thiocyanate were employed as sources of N and S. The successful insertion of heteroatoms into the graphene lattice and an increment of short basal-plane domains and edge planes into the prepared catalysts were noted. Main results indicate that short basal-plane domains, especially high density of edge planes, strongly enhances the ORR by changing the rate-determining step from the first electrochemical step on graphite (that has long basal-plane domains and low density of edge planes), to the removal of adsorbed intermediate species from the surface of graphenic nanomaterials. Additionally, it was discerned that nitrogen atoms improve the reaction by enhancing the adsorption and bond cleavage of O_2, favoring the direct four-electron pathway. However, sulfur atoms increase the catalytic activity toward the ORR by rising the desorption rate of adsorbed intermediate species [39].

1.2.4 Graphene composite catalysts (GC)

GC are highly versatile materials since the best physical and chemical properties of individual components are combined to enhance the catalytic performance of a desired reaction [51–53]. Thus, GCs with high surface area, appropriate porosity, high chemical stability, excellent conductivity, elevated alcohol tolerance, and appropriate catalytic activity toward the ORR in a wide pH range were reported [24, 51–59].

GC constituted from NG, Fe, and Co are the most popular catalysts employed for the ORR. A strong synergism between the components of the GC is usually indicated and improved catalytic performance toward the ORR is commonly reported [24, 51–59].

Remarkable is the work accomplished by Zhang et al., in which nitrogen-doped graphene quantum dots (N-GQDs) are uniformly grown in situ on thermally reduced graphene oxide (rGO) [24]. Dopamine (DA) in the presence of Fe was employed as nitrogen-rich precursor onto rGO, and after several pyrolysis and etching steps to remove Fe from the catalyst, the ORR was studied in alkaline medium (Fig. 5). The catalytic activity for the ORR was similar to that of carbon-supported Pt catalyst, although with better durability and a higher tolerance to the methanol oxidation reaction.

Metal-organic frameworks (MOFs) and covalent organic frameworks (COFs) may form one-, two-, or three-dimensional porous structures and can be employed as precursors of GC. In this regard, sandwich-like structured N-doped porous carbon@graphene composites (N-PC@G) derived from sandwich-like structured zeolitic imidazolate framework@-graphene oxide (ZIF-8@GO) were synthesized and their catalytic performance toward the ORR in alkaline solution was studied [59]. Higher catalytic activity than commercial carbon-supported Pt was observed (Fig. 6), which was attributed to the large surface area ($1094.3\,m^2\,g^{-1}$) and hierarchically porous structure that facilitate the mass transfer of the reactants (O_2) and products (OH^-).

1.3 Graphene materials for the hydrogen oxidation reaction

The electrochemical oxidation of hydrogen (HOR) takes place at the anode side of both PEMFC and anion exchange membrane fuel cell (AEMFC).

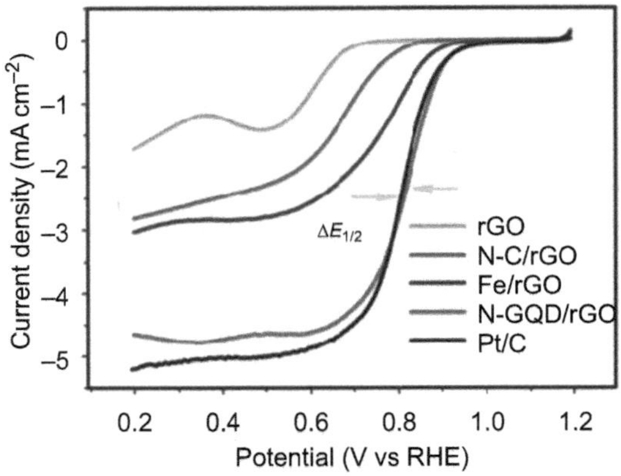

Fig. 5 Lineal sweep voltammetry curves of rGO, N-C/rGO, Fe/rGO, N-GQD/rGO, and Pt/C catalysts in O_2-saturated 0.1 M KOH solution at a rotating rate of 1600 rpm. *(Reprinted from B. Zhang, C. Xiao, Y. Xiang, B. Dong, S. Ding, Y. Tang, Nitrogen-doped graphene quantum dots anchored on thermally reduced graphene oxide as an electrocatalyst for the oxygen reduction reaction, ChemElectroChem 3 (2016) 864–870, https://doi.org/10.1002/celc.201600123, with permission from John Wiley and Sons.)*

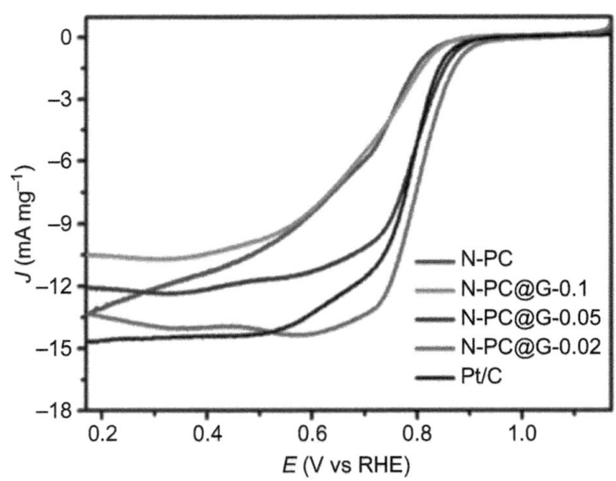

Fig. 6 Lineal sweep voltammetry curves of ZIF-8-derived carbon and commercial Pt/C catalysts in O_2-saturated 0.1 M KOH solution at a scan rate of $10\,mV\,s^{-1}$ and a rotating rate of 1600 rpm. *(Reprinted from S. Liu, H. Zhang, Q. Zhao, X. Zhang, R. Liu, X. Ge, et al., Metal-organic framework derived nitrogen-doped porous carbon@graphene sandwich-like structured composites as bifunctional electrocatalysts for oxygen reduction and evolution reactions, Carbon 106 (2016) 74–83, https://doi.org/10.1016/j.carbon.2016.05.021, with permission from Elsevier.)*

In these electrochemical devices, H_2 is used as fuel in that it is efficiently transformed into electrical energy without emission of contaminants.

The overall reaction for HOR occurs in the subsequent way:

$$H_2 \rightarrow 2H^+ + 2e^- \ (pH < 7) \tag{38}$$

$$H_2 + 2OH^- \rightarrow 2H_2O + 2e^- \ (pH > 7) \tag{39}$$

Besides its importance in fuel cell technology, HOR is a model reaction for the understanding of electrode kinetics and electrocatalysis and it has been extensively studied [33, 60–64]. The mechanism of HOR in electrolyte solutions generally includes three sequential stages [60–64]:

(1) Molecular hydrogen diffusion from the electrolyte to the electrode surface and posterior adsorption:

$$H_2 \rightarrow H_{2,solv} \rightarrow H_{2,ad} \tag{40}$$

(2) Hydrogen dissociation through either the Tafel-Volmer sequence or Heyrovsky-Volmer route:

Tafel:

$$H_{2,ad} \rightarrow 2H_{ad} \tag{41}$$

Heyrovsky:

$$H_{2,ad} \rightarrow H_{ad} + H^+ + e^- \quad (pH < 7) \tag{42}$$

$$H_{2,ad} + OH^- \rightarrow H_{ad} + H_2O + e^- \quad (pH > 7) \tag{43}$$

Volmer:

$$H_{ad} \rightarrow H^+ + e^- \ (pH < 7) \tag{44}$$

$$H_{ad} + OH^- \rightarrow H_2O + e^- \ (pH > 7) \tag{45}$$

(3) Product desorption and diffusion from electrode surface to the electrolyte.

The catalytic performance toward HOR depends strongly on the electrode material nature, surface structure, hydrogen binding energy, ions (anions at pH <7 and cations at pH >7) dissolved in the electrolyte, and the surface acidity [7, 8, 22, 33, 61].

HOR cannot occur on carbon materials, but it is extremely facile on Pt in acidic media with an exchange current density of 20–80 mA cm^{-2}, which is about 10^5–10^7 times faster than ORR [60, 65–67]. Indeed, under

PEMFC working operation, the anode overpotential is lower than 0.1 V. Furthermore, Pt is stable under acidic environments and ultralow loading (~0.03 mg cm^{-2}) of the precious metal can be employed. Carbon supported Pt-based catalysts are the compounds mostly used as anode in FC technology since they have shown superior catalytic performance than pristine Pt [2, 5, 6, 61, 68]. However, since most metals and compounds are not stable at low pH, there are not many choices of nonnoble metal catalysts to be employed in PEMFC conditions. Indeed, literature about HOR on platinum-group metals (PGM)-free catalysts in acidic media is scarce and employing graphene is uncommon, although tuning the catalyst surface acidity and/or the amount of intercalated acid into the graphene structure may help to fabricate PGM-free catalysts for the HOR in acidic media [22, 69].

On the other hand, PGM also reveal the highest catalytic activities for HOR in alkaline media, although they are at least two orders of magnitude lower than in acidic solutions [7, 33, 64]. The last is corroborated by the rise of the hydrogen binding energy with the increment of the pH, which is attributed to interactions between oxygenated species, cations, and open surface structures of PGM [7, 33, 64]. Therefore, the use of PGM-based catalysts in AEMFC requires high metal loadings at the anode, which is cost-prohibitive. In order to enhance the exchange current density of PGM, addition of a second element to weaken the hydrogen binding energy is commonly adopted [70].

Nevertheless, numerous PGM-free catalysts can survive at high pH due to the fact that alkaline solutions are less severe than acidic media [70–74]. Ni, Mo, Co, and their alloys are the most employed PGM-free catalysts for the HOR in basic solutions and their exchange current density was reasonable (0.4–28 μA cm^{-2}), although there is still room to improve it [71–74].

Remarkable is the work accomplished by Zhuang et al., in which nickel nanoparticles were supported on nitrogen-doped carbon nanotubes and the HOR in basic solution was scrutinized [73]. They demonstrate that nitrogen-doped carbon nanotubes (N-CNT) are very poor for the HOR. However, the catalytic performance of nickel nanoparticles increased by a factor of 33 (mass activity) or 21 (exchange current density) when N-CNT was employed as catalyst support (Fig. 7). Synergetic effect of the edge N atom in the CNT and Ni was proposed as the main origin of the enhanced catalytic activity, which was similar to that of Pd.

This work opens a strategy for the future application of doped graphene materials for the HOR.

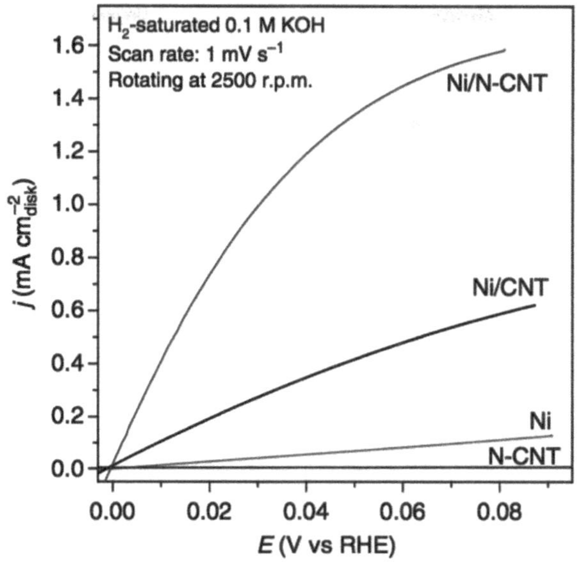

Fig. 7 Polarization curves of Ni/N-CNT, Ni/CNT, Ni (all of the three catalysts with a loading of 0.25 mg_{Ni} cm^{-2}), and N-CNT (0.1 mg_C cm^{-2}) catalysts in H_2-saturated 0.1 M KOH at a scan rate of 1 mV s^{-1} and rotating speed of 2500 rpm. *(Reprinted from Z. Zhuang, S.A. Giles, J. Zheng, G.R. Jenness, S. Caratzoulas, D.G. Vlachos, et al., Nickel supported on nitrogen-doped carbon nanotubes as hydrogen oxidation reaction catalyst in alkaline electrolyte, Nat. Commun. 7 (2016) 10141, Springer Nature.)*

2 Graphene-based electrocatalysts for electrolyzers

This section is dealing with the use of electrocatalysts containing graphene in electrochemical water splitting, intended for hydrogen production. It has been only focused on most recent developments that utilize nonnoble metals, avoiding the following noble metals: Pt, Rh, Pd, Ag, Ir, Au, Os, and Ru. In this way, two of major drawbacks of the technology, catalyst cost and scarcity of such metals, could be overcome.

2.1 Electrolysis

The increasing global warming caused by the combustion of fossil fuels is leading to a search for alternatives that are environmentally friendly, accessible, and economically attractive. Hydrogen is considered as a clean fuel of future because it acts as a green energy carrier and provides a method for the storage and transport of energy. A variety of processes are available for H_2 production, based on conventional or renewable technologies. The latter includes those technologies that utilize renewable resources. Water electrolysis offers a practical route for sustainable hydrogen production by utilizing a renewable electrical energy source for water splitting.

A typical electrolysis unit consists of a cathode and an anode in contact with an electrolyte. When a potential difference or electrical current is applied, water splits and H_2 is evolved at the cathode and O_2 is evolved on the anode side, according to the following reaction:

$$H_2O_{(l/g)} \longrightarrow H_{2(g)} + \frac{1}{2}O_{2(g)} \quad (46)$$

being the backward reaction occurring in the H_2 fuel cell.

Depending on the electrolyte used in the electrolysis cell, three different electrolyzers are available: alkaline water electrolyzer (AWE), solid polymer electrolyte (SPE) or polymer electrolyte membrane (PEM) water electrolyzer (WE), and solid oxide electrolyte cell (SOEC).

As a mature technology, alkaline water electrolysis is the dominating technology due to the lower cost of production. Typically, alkaline units operate at temperature around 60–80°C. PEM water electrolysis, on the other hand, may offer advantages like improved energy efficiency, higher production rate, purer hydrogen, and a more compact design. PEM technology usually works at around 50–80°C. SOEC technology operates at high temperatures, such as 800°C, and offers significant reduction in consumption of electrical energy, but usually requires a source of high-temperature heat.

Nowadays, PEM water electrolysis technology is receiving growing attention, because it offers several advantages over other electrolysis technologies and benefits from PEM fuel cell technology. However, high cost of the components and the acidic corrosive environment are the main challenges toward better durability and commercialization of the PEMWE technology. The development of new electrode materials is contributing to overcome these drawbacks to guarantee the future deployment of the technology. In this line, new nonnoble metal-supported electrocatalysts are playing a capital role in cost reduction.

Graphene is a two-dimensional carbon nanostructure consisting of single-layer of sp^2 hybridized carbon atoms. Recently, it has attracted great technological and scientific attention as a carbonaceous support, because of its outstanding properties such as unique structure, high surface area, good electrical conductivity, high chemical and thermal stability, and potentially low cost [75]. Thus, it can be said that its most prominent role in electrocatalysis is that of the catalyst support, with high electronic conductivity that enables anchoring of different active nanostructures, acting at the same time as a current collector.

2.2 Graphene materials for the hydrogen evolution reaction

The electrochemical hydrogen evolution reaction (HER):

$$2H^+ + 2e^- \rightarrow H_2 \qquad (pH < 7) \tag{47}$$

$$2H_2O + 2e^- \rightarrow H_2 + 2OH^- \quad (pH > 7) \tag{48}$$

requires advanced catalysts with a high current density at low overpotential. The most effective HER electrocatalysts in acidic media are those based on Pt. However, the large-scale application of these catalysts is limited due to their high cost and scarcity. Therefore, the development of new advanced HER electrocatalysts, especially those composed of low-cost and abundant materials, is of capital importance.

It is generally accepted that graphene has poor catalytic HER activities, which renders it the role of the catalyst support in the form of aerogel, hydrogel, or foam, among others, or doped with nonmetals, like S or N, which provides high electronic conductivity. However, some experimental studies indicate that the role of graphene-based materials is more complex than a simple supportive role, although a detailed explanation of the mechanism on the molecular level is still missing [76].

Several investigations about transition metal compounds for HER, such as carbides, oxides, and chalcogenides, have been reported, because of their outstanding performance for HER in acidic conditions [77].

In recent years, molybdenum sulfide nanosheets [78] and other morphologies, capable of accepting electrons and protons, have been considered as promising H_2 evolution catalyst. Recently, MoS_2 supported on graphene nanocomposite has been successfully synthesized avoiding restacking. This material shows improved electrocatalytic activity toward HER in acidic media due to the intrinsic properties of graphene and the interaction between Mo and the support as can be seen in the Fig. 8 [79]. Tafel slopes are in the range of 40–55 $mV\,dec^{-1}$, and the overpotential to achieve $10\,mA\,cm^{-2}$ is around 85–100 mV.

MoS_2 supported on reduced graphene oxide was evaluated as a cathode catalyst in PEM electrolyzer, yielding reasonable performance (see Fig. 9) [80].

Other low-cost and active HER graphene-supported catalysts have been explored, like Ni, Ni-based alloy, sulfide, and selenide; CoS_2, FeS, NiS,

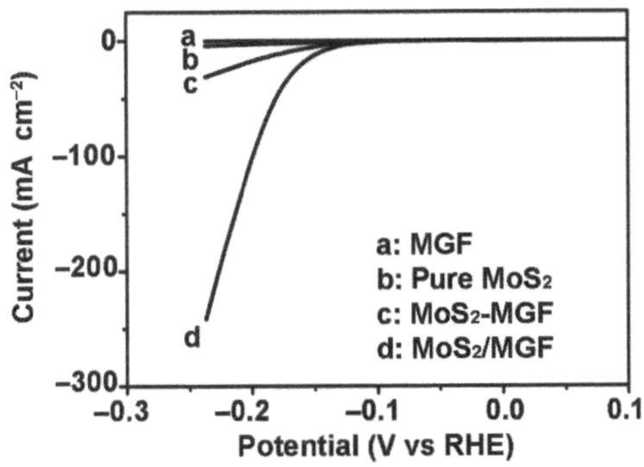

Fig. 8 Linear sweep voltammetry polarization curves for glassy carbon electrodes modified with (a) MGF, (b) pure MoS_2, (c) MoS_2 physically mixed with MGFs (MoS_2-MGF), and (d) MoS_2 formed on MGFs (MoS_2/MGF) in 0.5 M H_2SO_4 at room temperature with a hybrid catalyst loading of 0.21 mg cm^{-2} (MFG, mesoporous graphene foam). *(Reprinted from L. Liao, J. Zhu, X. Bian, L. Zhu, M.D. Scanlon, H.H. Girault, et al., MoS_2 formed on mesoporous graphene as a highly active catalyst for hydrogen evolution, Adv. Funct. Mater. 23 (2013) 5326–5333, https://doi.org/10.1002/adfm.201300318, with permission from John Wiley and Sons.)*

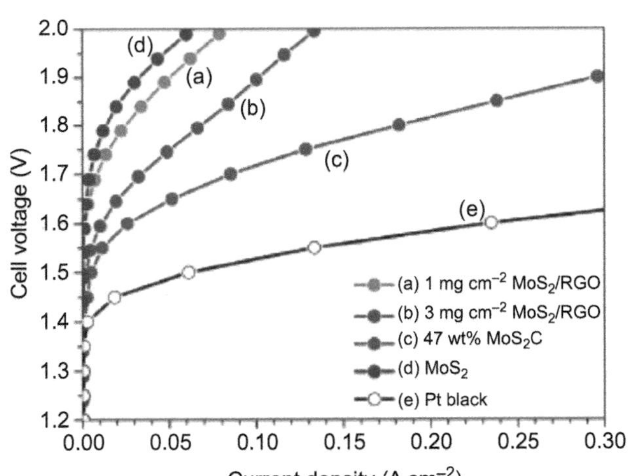

Fig. 9 Cell voltage vs current density curves (potential sweep from 1.2 to 2.0 V, $\Delta V = 0.05$ V) for PEM electrolysis using two different catalyst loadings of MoS_2/rGO over the membrane (a) 1 mg cm^{-2} and (b) 3 mg cm^{-2}. For the sake of better comparison, the performances of (c) 47 wt% and (d) 100 wt% MoS_2/Vulcan as well as (e) Pt black are shown. Anode: IrO_2 2 mg cm^{-2}. Temperature: 80°C. *(Reprinted from T. Corrales-Sánchez, J. Ampurdanés, A. Urakawa, MoS_2-based materials as alternative cathode catalyst for PEM electrolysis, Int. J. Hydrogen Energy 39 (2014) 20837–20843, https://doi.org/10.1016/j.ijhydene.2014.08.078, with permission from Elsevier.)*

$CoSe_2$, $MoSe_2$, and MoN. Transition metal phosphides, such as Ni_2P, CoP, and FeP, have been demonstrated to be active and robust HER catalysts [81], as well as, phosphorus-modified tungsten nitride/reduced graphene oxide [82]. More recently, using first-principle calculations, the potential of nonprecious Co and N-codoped graphene system as catalyst for the HER was demonstrated [83].

On the other hand, regarding alkaline media, CeO_2 supported on graphene hydrogel [84] has been reported as electrocatalyst for HER in alkaline media. Ni-graphene composites prepared by electrodeposition [85] and, more recently, Ni and NiMo nanostructures supported on reduced graphene oxide by reduction with ethylene glycol showed enhanced activity toward the HER [86]. In this work, the use of differential electrochemical mass spectrometry (DEMS) technique allowed to determine the exact onset potential for the hydrogen evolution reaction.

2.3 Graphene materials for oxygen evolution reaction

It is well-known that metal oxides are the most active and durable electrocatalysts for the oxygen evolution reaction (OER):

$$2H_2O \longrightarrow O_2 + 4H^+ + 4e^- \qquad (pH < 7) \qquad (49)$$

$$4OH^- \longrightarrow O_2 + 2H_2O + 4e^- \qquad (pH > 7) \qquad (50)$$

among which RuO_2 and IrO_2 are thought to be best OER catalysts, respectively, in both acidic and alkaline solutions. However, the high cost and low-abundance hinder the widespread use of these noble metal oxide electrocatalysts. Hence, it is desirable to develop alternative electrocatalysts based on inexpensive and earth-abundant elements, with good activity and durability for OER. The nonprecious catalysts of choice for OER are 3d transition metal materials, especially Co and Ni. According to theoretical calculations, these metal and metal oxides are near to the peak of the electrocatalytic activity volcano for OER. It is worth to mention that the same occurs for HER, so it is not surprising to find Co- and Ni-based materials that act now as electrocatalyst for OER [87].

Recent investigations indicate that a completely nonnoble metal-based electrocatalyst could be developed for OER in acidic media. Thus, at Argonne National Laboratory (USA), new developed Co-MOF catalyst supported on graphene and activated by a controlled thermal treatment showed excellent activity and durability in 0.5 M H_2SO_4 aqueous electrolyte toward OER, namely, a potential of 1.644 V vs RHE at 10 mA cm^{-2} [88].

In recent times, extensive efforts have been made to use perovskites and first-row transition-metal-based materials as low-cost catalysts for OER in alkaline media. Thus, $NiCo_2O_4$ and reduced graphene oxide hybrid nanostructures have been reported as electrocatalyst for OER in alkaline media [89]. The nanonet and microsphere-like $NiCo_2O_4$/rGO catalysts require overpotentials of 0.450 and 0.530 V at 10 mA cm^{-2}, and their corresponding Tafel slopes are between 53 and 62 mV dec^{-1} in 0.1 KOH, respectively.

On the other hand, despite graphene-based electrocatalysts have enough activity for OER application, their utility is hindered because of their chemical inertness. In this line, the combination with layered double hydroxides complements each other, increasing performance. A strongly coupled graphene and FeNi double hydroxide hybrid has been reported for oxygen evolution reaction [90]. The overpotential of catalytic OER is among the lowest of nonnoble metal catalysts (as low as 0.195 V) and the Tafel slope is close to 40 mV dec^{-1}. Furthermore, to tune their catalytic activity toward OER in alkaline media, FeCoNi ternary alloys have been encapsulated by direct annealing of different MOFs between N-doped graphene sheets [91]. Fig. 10 shows the activity toward the OER.

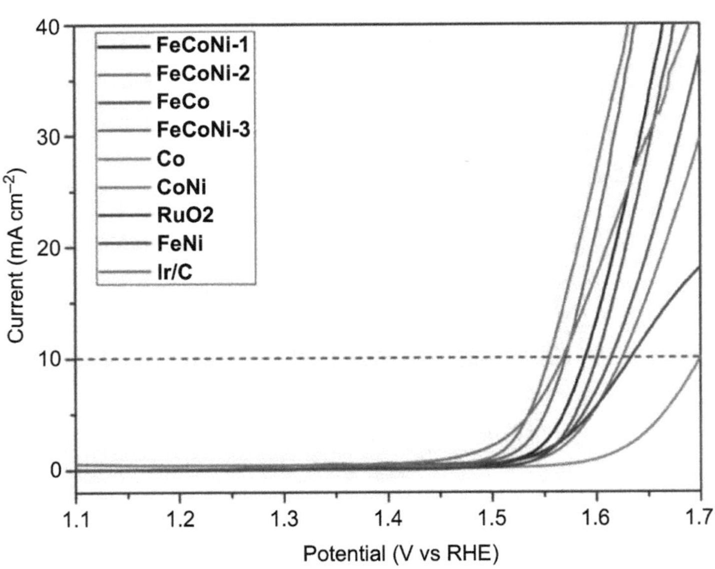

Fig. 10 Electrocatalytic OER performance tests of different metal, binary and ternary alloys in 1 M KOH solution compared with Ir/C and RuO_2 with same mass loading at room temperature. $s_r = 5\,mV\,s^{-1}$. *(Reprinted from Y. Yang, Z. Lin, S. Gao, J. Su, Z. Lun, G. Xia, et al., Tuning electronic structures of nonprecious ternary alloys encapsulated in graphene layers for optimizing overall water splitting activity, ACS Catal. 7 (2017) 469–479, https://doi.org/10.1021/acscatal.6b02573, with permission from American Chemical Society.)*

Ni_2P nanocrystals supported on graphene have been also reported as an efficient catalyst for oxygen evolution reaction in alkaline media, showing 0.285 mV at 10 mA cm^{-2} and Tafel slope of 56 mV dec^{-1} in 1 M KOH. In this material, graphene plays a role of protective layer and enhancing matrix for the Ni_2P nanoparticles [92]. More recently, Co_3O_4/graphene nanocomposites have been synthesized following rational procedures for OER in alkaline media, in an attempt to overcome low intrinsic conductivity and thermal stability of Co_3O_4 [93].

3 Conclusions

In this chapter, the general concept of fuel cells and electrolyzers was discussed, with special attention to graphene-based catalysts. With the aim to solve the principal catalytic, durability, and cost problems of the electrodes of fuel cells and electrolyzers, a fundamental and applied study of all involved electrochemical reactions on graphene-based materials in a wide pH range was considered. All information reviewed and scrutinized here may help to improve the fabrication of novel catalysts in order to decrease the cost and enhance the performance of both technologies.

Acknowledgment

This work has been supported by the Spanish Ministry of Science (MICINN) under project ENE2017-83976-C2-2-R (cofounded FEDER). G. G. acknowledges the "Viera y Clavijo" program (ACIISI & ULL).

References

[1] B. Sørensen, G. Spazzafumo, 4—Fuel cell systems BT, in: Hydrogen and Fuel Cells, third ed., Academic Press, 2018, pp. 221–272, https://doi.org/10.1016/B978-0-08-100708-2.00004-7.

[2] J. Divisek, Low temperature fuel cells, in: W. Vielstich, A. Lamm, H. Gasteiger (Eds.), Handbook of Fuel Cells: Fundamentals, Technology and Applications, vol. 1, John Wiley & Sons, Ltd, Chichester, UK, 2003, pp. 99–114, https://doi.org/10.1002/9780470974001.f103008.

[3] J.R. Ladebeck, J.P. Wagner, Catalyst development for water-gas shift, in: W. Vielstich, A. Lamm, H. Gasteiger (Eds.), Handbook of Fuel Cells: Fundamentals, Technology and Applications, vol. 3, John Wiley & Sons, Ltd, Chichester, UK, 2003, pp. 190–202, https://doi.org/10.1002/9780470974001.f302017.

[4] G. García, M.T.M. Koper, Carbon monoxide oxidation on pt single crystal electrodes: understanding the catalysis for low temperature fuel cells, ChemPhysChem 12 (2011) 2064–2072, https://doi.org/10.1002/cphc.201100247.

[5] M.V. Martínez Huerta, G. García, D.Y.C. Leung, Fabrication of electrocatalytic nanoparticles and applications to proton exchange membrane fuel cells, in: J. Xuan (Ed.),

Micro Nano-Engineering Fuel Cells, CRC Press, 2015, pp. 95–129, https://doi.org/10.1201/b18394-5.
[6] M. Roca-Ayats, S. Pérez-Rodríguez, G. García, E. Pastor, Recent advances on electrocatalysts for PEM and AEM fuel cells, in: F.J. Rodríguez-Varela, T.W. Napporn (Eds.), Advanced Electrocatalysts for Low-Temperature Fuel Cells, Springer, Cham, 2018, pp. 51–89, https://doi.org/10.1007/978-3-319-99019-4_2.
[7] W. Sheng, M. Myint, J.G. Chen, Y. Yan, Correlating the hydrogen evolution reaction activity in alkaline electrolytes with the hydrogen binding energy on monometallic surfaces, Energy Environ. Sci. 6 (2013) 1509–1512, https://doi.org/10.1039/c3ee00045a.
[8] S. Trasatti, Work function, electronegativity, and electrochemical behaviour of metals: III. Electrolytic hydrogen evolution in acid solutions, J. Electroanal. Chem. Interfacial Electrochem. 39 (1972) 163–184, https://doi.org/10.1016/S0022-0728(72)80485-6.
[9] J.K. Nørskov, J. Rossmeisl, A. Logadottir, L. Lindqvist, J.R. Kitchin, T. Bligaard, et al., Origin of the overpotential for oxygen reduction at a fuel-cell cathode, J. Phys. Chem. B 108 (2004) 17886–17892, https://doi.org/10.1021/jp047349j.
[10] L.M. Rivera, G. García, E. Pastor, Novel graphene materials for the oxygen reduction reaction, Curr. Opin. Electrochem. 9 (2018) 233–239, https://doi.org/10.1016/j.coelec.2018.05.009.
[11] C.R. Raj, A. Samanta, S.H. Noh, S. Mondal, T. Okajima, T. Ohsaka, Emerging new generation electrocatalysts for the oxygen reduction reaction, J. Mater. Chem. A 4 (2016) 11156–11178, https://doi.org/10.1039/c6ta03300h.
[12] X. Liu, L. Dai, Carbon-based metal-free catalysts, Nat. Rev. Mater. 1 (2018) 16064, https://doi.org/10.1002/9783527811458.
[13] S. Pei, H.-M. Cheng, The reduction of graphene oxide, Carbon 50 (2012) 3210–3228, https://doi.org/10.1016/j.carbon.2011.11.010.
[14] L. Tao, Q. Wang, S. Dou, Z. Ma, J. Huo, S. Wang, et al., Edge-rich and dopant-free graphene as a highly efficient metal-free electrocatalyst for the oxygen reduction reaction, Chem. Commun. 52 (2016) 2764–2767, https://doi.org/10.1039/C5CC09173J.
[15] M. Borghei, N. Laocharoen, E. Kibena-Põldsepp, L.-S. Johansson, J. Campbell, E. Kauppinen, et al., Porous N,P-doped carbon from coconut shells with high electrocatalytic activity for oxygen reduction: alternative to Pt-C for akaline fuel cells, Appl. Catal. B Environ. 204 (2017) 394–402, https://doi.org/10.1016/j.apcatb.2016.11.029.
[16] J. Lilloja, E. Kibena-Põldsepp, M. Merisalu, P. Rauwel, L. Matisen, A. Niilisk, et al., An oxygen reduction study of graphene-based nanomaterials of different origin, Catalysts 6 (2016) 108, https://doi.org/10.3390/catal6070108.
[17] A. Sarapuu, E. Kibena-Poldsepp, M. Borghei, K. Tammeveski, Electrocatalysis of oxygen reduction on heteroatom-doped nanocarbons and transition metal-nitrogen-carbon catalysts for alkaline membrane fuel cells, J. Mater. Chem. A 6 (2018) 776–804, https://doi.org/10.1039/C7TA08690C.
[18] W.S. Hummers, R.E. Offeman, Preparation of graphitic oxide, J. Am. Chem. Soc. 80 (1958) 1339, https://doi.org/10.1021/ja01539a017.
[19] A. Yadegari, L. Samiee, S. Tasharrofi, S. Tajik, A. Rashidi, F. Shoghi, et al., Nitrogen doped nanoporous graphene: an efficient metal-free electrocatalyst for the oxygen reduction reaction, RSC Adv. 7 (2017) 55555–55566, https://doi.org/10.1039/C7RA10626B.
[20] F. Pan, Y. Duan, X. Zhang, J. Zhang, A facile synthesis of nitrogen/sulfur co-doped graphene for the oxygen reduction reaction, ChemCatChem 8 (2016) 163–170, https://doi.org/10.1002/cctc.201500893.
[21] F. López-Urías, R. Lv, H. Terrones, M. Terrones, Doped graphene: theory, synthesis, characterization, and applications, in: D.-E. Jiang, Z. Chen (Eds.), Graphene Chemistry: Theoretical Perspectives, John Wiley & Sons, Ltd, Chichester, UK, 2013, pp. 183–207, https://doi.org/10.1002/9781118691281.

[22] S. Pérez-Rodríguez, G. García, M.J. Lázaro, E. Pastor, DEMS strategy for the determination of the difference in surface acidity of carbon materials, Electrochem. Commun. 90 (2018) 87–90, https://doi.org/10.1016/j.elecom.2018.04.014.
[23] M. Klingele, C. Pham, K.R. Vuyyuru, B. Britton, S. Holdcroft, A. Fischer, et al., Sulfur doped reduced graphene oxide as metal-free catalyst for the oxygen reduction reaction in anion and proton exchange fuel cells, Electrochem. Commun. 77 (2017) 71–75, https://doi.org/10.1016/j.elecom.2017.02.015.
[24] B. Zhang, C. Xiao, Y. Xiang, B. Dong, S. Ding, Y. Tang, Nitrogen-doped graphene quantum dots anchored on thermally reduced graphene oxide as an lectrocatalyst for the oxygen reduction reaction, ChemElectroChem 3 (2016) 864–870, https://doi.org/10.1002/celc.201600123.
[25] H.A. Gasteiger, S.S. Kocha, B. Sompalli, F.T. Wagner, Activity benchmarks and requirements for Pt, Pt-alloy, and non-Pt oxygen reduction catalysts for PEMFCs, Appl. Catal. B Environ. 56 (2005) 9–35, https://doi.org/10.1016/j.apcatb.2004.06.021.
[26] S. Sui, X. Wang, X. Zhou, Y. Su, S. Riffat, C.-j. Liu, A caomprehensive review of Pt electrocatalysts for the oxygen reduction reaction: nanostructure, activity, mechanism and carbon support in PEM fuel cells, J. Mater. Chem. A 5 (2017) 1808–1825, https://doi.org/10.1039/C6TA08580F.
[27] Office of Energy Efficiency & Renewable Energy. https://www.energy.gov/eere/fuelcells/doe-technical-targets-polymer-electrolyte-membrane-fuel-cell-components.
[28] N.A. Anastasijević, V. Vesović, R.R. Adžić, Determination of the kinetic parameters of the oxygen reduction reaction using the rotating ring-disk electrode: Part I. Theory, J. Electroanal. Chem. Interfacial Electrochem. 229 (1987) 305–316, https://doi.org/10.1016/0022-0728(87)85148-3.
[29] N.A. Anastasijević, V. Vesović, R.R. Adžić, Determination of the kinetic parameters of the oxygen reduction reaction using the rotating ring-disk electrode: Part II. Applications, J. Electroanal. Chem. Interfacial Electrochem. 229 (1987) 317–325, https://doi.org/10.1016/0022-0728(87)85149-5.
[30] J. Greeley, N.M. Markovic, The road from animal electricity to green energy: combining experiment and theory in electrocatalysis, Energy Environ. Sci. 5 (2012) 9246–9256, https://doi.org/10.1039/C2EE21754F.
[31] Y.J. Tong, Unconventional promoters of catalytic activity in electrocatalysis, Chem. Soc. Rev. 41 (2012) 8195–8209, https://doi.org/10.1039/C2CS35381D.
[32] C.F. Zinola, A.J. Arvia, G.L. Estiu, E.A. Castro, A quantum chemical approach to the influence of platinum surface structure on the oxygen electroreduction reaction, J. Phys. Chem. 98 (1994) 7566–7576, https://doi.org/10.1021/j100082a030.
[33] G. García, Correlation between CO oxidation and H adsorption/desorption on Pt surfaces in a wide pH range: the role of alkali cations, ChemElectroChem 4 (2017) 459–462, https://doi.org/10.1002/celc.201600731.
[34] S. Liu, M.G. White, P. Liu, Mechanism of oxygen reduction reaction on Pt(111) in alkaline solution: importance of chemisorbed water on surface, J. Phys. Chem. C 120 (2016) 15288–15298, https://doi.org/10.1021/acs.jpcc.6b05126.
[35] L.M. Rivera Gavidia, G. García, D. Anaya, A. Querejeta, F. Alcaide, E. Pastor, Carbon-supported Pt-free catalysts with high specificity and activity toward the oxygen reduction reaction in acidic medium, Appl. Catal. B Environ. 184 (2016) 12–19, https://doi.org/10.1016/j.apcatb.2015.11.021.
[36] J. Mora-Hernández, Y. Luo, N. Alonso-Vante, What can we learn in electrocatalysis, from nanoparticulated precious and/or non-precious catalytic centers interacting with their support? Catalysts 6 (2016) 145, https://doi.org/10.3390/catal6090145.
[37] E. Antolini, E.R. Gonzalez, Alkaline direct alcohol fuel cells, J. Power Sources 195 (2010) 3431–3450, https://doi.org/10.1016/j.jpowsour.2009.11.145.

[38] Y. Holade, N.E. Sahin, K. Servat, T.W. Napporn, K.B. Kokoh, Recent advances in carbon supported metal nanoparticles preparation for oxygen reduction reaction in low temperature fuel cells, Catalysts 5 (2015) 310–348, https://doi.org/10.3390/catal5010310.
[39] L.M. Rivera, S. Fajardo, M.C. Arévalo, G. García, E. Pastor, S- and N-doped graphene nanomaterials for the oxygen reduction reaction, Catalysts 7 (2017) 278, https://doi.org/10.3390/catal7090278.
[40] X. Ge, A. Sumboja, D. Wuu, T. An, B. Li, F.W.T. Goh, et al., Oxygen reduction in alkaline media: from mechanisms to recent advances of catalysts, ACS Catal. 5 (2015) 4643–4667, https://doi.org/10.1021/acscatal.5b00524.
[41] G. García, P. Rodríguez, V. Rosca, M.T.M. Koper, Fourier transform infrared spectroscopy study of CO electro-oxidation on Pt(111) in alkaline media, Langmuir 25 (2009) 13661–13666, https://doi.org/10.1021/la902251z.
[42] J. Zhang, C. Zhao, N. Liu, H. Zhang, J. Liu, Y.Q. Fu, et al., Tunable electronic properties of graphene through controlling bonding configurations of doped nitrogen atoms, Sci. Rep. 6 (2016) 28330.
[43] R. Ma, X. Ren, B.Y. Xia, Y. Zhou, C. Sun, Q. Liu, et al., Novel synthesis of N-doped graphene as an efficient electrocatalyst towards oxygen reduction, Nano Res. 9 (2016) 808–819, https://doi.org/10.1007/s12274-015-0960-2.
[44] X. Bai, Y. Shi, J. Guo, L. Gao, K. Wang, Y. Du, et al., Catalytic activities enhanced by abundant structural defects and balanced N distribution of N-doped graphene in oxygen reduction reaction, J. Power Sources 306 (2016) 85–91, https://doi.org/10.1016/j.jpowsour.2015.10.081.
[45] P. Zhang, Q. Hu, X. Yang, X. Hou, J. Mi, L. Liu, et al., Size effect of oxygen reduction reaction on nitrogen-doped graphene quantum dots, RSC Adv. 8 (2018) 531–536, https://doi.org/10.1039/c7ra10104j.
[46] V.K. Das, Z.B. Shifrina, L.M. Bronstein, Graphene and graphene-like materials in biomass conversion: paving the way to the future, J. Mater. Chem. A 5 (2017) 25131–25143, https://doi.org/10.1039/C7TA09418C.
[47] J. Molina, Graphene-based fabrics and their applications: a review, RSC Adv. 6 (2016) 68261–68291, https://doi.org/10.1039/C6RA12365A.
[48] D.K. Perivoliotis, N. Tagmatarchis, Recent advancements in metal-based hybrid electrocatalysts supported on graphene and related 2D materials for the oxygen reduction reaction, Carbon 118 (2017) 493–510, https://doi.org/10.1016/j.carbon.2017.03.073.
[49] H. Wang, T. Maiyalagan, X. Wang, Review on recent progress in nitrogen-doped graphene: synthesis, characterization, and its potential applications, ACS Catal. 2 (2012) 781–794, https://doi.org/10.1021/cs200652y.
[50] M. Wu, Z. Dou, J. Chang, L. Cui, Nitrogen and sulfur co-doped graphene aerogels as an efficient metal-free catalyst for oxygen reduction reaction in an alkaline solution, RSC Adv. 6 (2016) 22781–22790, https://doi.org/10.1039/c5ra22136f.
[51] M. Sun, H. Liu, Y. Liu, J. Qu, J. Li, Graphene-based transition metal oxide nanocomposites for the oxygen reduction reaction, Nanoscale 7 (2015) 1250–1269, https://doi.org/10.1039/c4nr05838k.
[52] S. Ratso, I. Kruusenberg, U. Joost, R. Saar, K. Tammeveski, Enhanced oxygen reduction reaction activity of nitrogen-doped graphene/multi-walled carbon nanotube catalysts in alkaline media, Int. J. Hydrogen Energy 41 (2016) 22510–22519, https://doi.org/10.1016/j.ijhydene.2016.02.021.
[53] S. Ratso, I. Kruusenberg, M. Vikkisk, U. Joost, E. Shulga, I. Kink, et al., Highly active nitrogen-doped few-layer graphene/carbon nanotube composite electrocatalyst for oxygen reduction reaction in alkaline media, Carbon 73 (2014) 361–370, https://doi.org/10.1016/j.carbon.2014.02.076.

[54] Y. Niu, X. Huang, W. Hu, Fe_3C nanoparticle decorated Fe/N doped graphene for efficient oxygen reduction reaction electrocatalysis, J. Power Sources 332 (2016) 305–311, https://doi.org/10.1016/j.jpowsour.2016.09.130.
[55] G. Zhang, W. Lu, F. Cao, Z. Xiao, X. Zheng, N-doped graphene coupled with co nanoparticles as an efficient electrocatalyst for oxygen reduction in alkaline media, J. Power Sources 302 (2016) 114–125, https://doi.org/10.1016/j.jpowsour.2015.10.055.
[56] X. Fu, J.Y. Choi, P. Zamani, G. Jiang, M.A. Hoque, F.M. Hassan, et al., Co-N decorated hierarchically porous graphene aerogel for efficient oxygen reduction reaction in acid, ACS Appl. Mater. Interfaces 8 (2016) 6488–6495, https://doi.org/10.1021/acsami.5b12746.
[57] T. Palaniselvam, V. Kashyap, S.N. Bhange, J.-B. Baek, S. Kurungot, Nanoporous graphene enriched with Fe/co-N active sites as a promising oxygen reduction electrocatalyst for anion exchange membrane fuel cells, Adv. Funct. Mater. 26 (2016) 2150–2162, https://doi.org/10.1002/adfm.201504765.
[58] C. Chen, X. Zhang, Z.-Y. Zhou, X.-D. Yang, X.-S. Zhang, S.-G. Sun, Highly active Fe, N co-doped graphene nanoribbon/carbon nanotube composite catalyst for oxygen reduction reaction, Electrochim. Acta 222 (2016) 1922–1930, https://doi.org/10.1016/j.electacta.2016.12.005.
[59] S. Liu, H. Zhang, Q. Zhao, X. Zhang, R. Liu, X. Ge, et al., Metal-organic framework derived nitrogen-doped porous carbon@graphene sandwich-like structured composites as bifunctional electrocatalysts for oxygen reduction and evolution reactions, Carbon 106 (2016) 74–83, https://doi.org/10.1016/j.carbon.2016.05.021.
[60] W. Sheng, H.A. Gasteiger, Y. Shao-Horn, Hydrogen oxidation and evolution reaction kinetics on platinum: acid vs alkaline electrolytes, J. Electrochem. Soc. 157 (2010) B1529, https://doi.org/10.1149/1.3483106.
[61] N. Marković, The hydrogen electrode reaction and the electrooxidation of CO and H_2/CO mixtures on well-characterized Pt and Pt-bimetallic surfaces, in: Handbook of Fuel Cells, vol. 2, John Wiley & Sons, Ltd, Chichester, UK, 2010, pp. 1–26, https://doi.org/10.1002/9780470974001.f204027m.
[62] M.R. Tarasevich, O.V. Korchagin, Electrocatalysis and pH (a review), Russ. J. Electrochem. 49 (2013) 600–618, https://doi.org/10.1134/S102319351307015X.
[63] W. Sheng, Z. Zhuang, M. Gao, J. Zheng, J.G. Chen, Y. Yan, Correlating hydrogen oxidation and evolution activity on platinum at different pH with measured hydrogen binding energy, Nat. Commun. 6 (2015) 5848, https://doi.org/10.1038/ncomms6848.
[64] J. Zheng, W. Sheng, Z. Zhuang, B. Xu, Y. Yan, Universal dependence of hydrogen oxidation and evolution reaction activity of platinum-group metals on pH and hydrogen binding energy, Sci. Adv. 2 (2016), https://doi.org/10.1126/sciadv.1501602.
[65] J. Durst, C. Simon, F. Hasche, H.A. Gasteiger, Hydrogen oxidation and evolution reaction kinetics on carbon supported Pt, Ir, Rh, and Pd electrocatalysts in acidic media, J. Electrochem. Soc. 162 (2014) F190–F203, https://doi.org/10.1149/2.0981501jes.
[66] J. Maruyama, M. Inaba, K. Katakura, Z. Ogumi, Z. Takehara, Influence of Nafion® film on the kinetics of anodic hydrogen oxidation, J. Electroanal. Chem. 447 (1998) 201–209, https://doi.org/10.1016/S0022-0728(98)00004-7.
[67] A. Parthasarathy, C.R. Martin, S. Srinivasan, Investigations of the O_2 reduction reaction at the platinum/Nafion® interface using a solid-state electrochemical cell, J. Electrochem. Soc. 138 (1991) 916, https://doi.org/10.1149/1.2085747.
[68] M. Watanabe, H. Uchida, Catalysts for the electro-oxidation of small molecules, in: W. Vielstich, H. Yokokawa, H.A. Gastaiger (Eds.), Handbook of Fuel Cells: Fundamentals, Technology, and Applications, John Wiley & Sons Ltd, Chichester, 2010, pp. 1–10.
[69] R.K. Das, Y. Wang, S.V. Vasilyeva, E. Donoghue, I. Pucher, G. Kamenov, et al., Extraordinary hydrogen evolution and oxidation reaction activity from carbon nanotubes and graphitic carbons, ACS Nano 8 (2014) 8447–8456, https://doi.org/10.1021/nn5030225.

[70] S. Lu, Z. Zhuang, Electrocatalysts for hydrogen oxidation and evolution reactions, Sci. China Mater. 59 (2016) 217–238, https://doi.org/10.1007/s40843-016-0127-9.
[71] K. Mund, G. Richter, F. von Sturm, Titanium-containing raney nickel catalyst for hydrogen electrodes in alkaline fuel cell systems, J. Electrochem. Soc. 124 (1977) 1–6, https://doi.org/10.1149/1.2133234.
[72] W. Sheng, A.P. Bivens, M. Myint, Z. Zhuang, R.V. Forest, Q. Fang, et al., Non-precious metal electrocatalysts with high activity for hydrogen oxidation reaction in alkaline electrolytes, Energy Environ. Sci. 7 (2014) 1719–1724, https://doi.org/10.1039/C3EE43899F.
[73] Z. Zhuang, S.A. Giles, J. Zheng, G.R. Jenness, S. Caratzoulas, D.G. Vlachos, et al., Nickel supported on nitrogen-doped carbon nanotubes as hydrogen oxidation reaction catalyst in alkaline electrolyte, Nat. Commun. 7 (2016) 10141.
[74] S. Lu, J. Pan, A. Huang, L. Zhuang, J. Lu, Alkaline polymer electrolyte fuel cells completely free from noble metal catalysts, Proc. Natl. Acad. Sci. 105 (2008) 20611–20614, https://doi.org/10.1073/pnas.0810041106.
[75] S. Shiva Kumar, S.U.B. Ramakrishna, B. Rama Devi, V. Himabindu, Phosphorus-doped graphene supported palladium (Pd/PG) electrocatalyst for the hydrogen evolution reaction in PEM water electrolysis, Int. J. Green Energy 15 (2018) 558–567, https://doi.org/10.1080/15435075.2018.1508468.
[76] S.J. Gutić, A.S. Dobrota, M. Leetmaa, N.V. Skorodumova, S.V. Mentus, I.A. Pašti, Improved catalysts for hydrogen evolution reaction in alkaline solutions through the electrochemical formation of nickel-reduced graphene oxide interface, Phys. Chem. Chem. Phys. 19 (2017) 13281–13293, https://doi.org/10.1039/C7CP01237C.
[77] G. Hu, J. Li, P. Liu, X. Zhu, X. Li, R.N. Ali, et al., Enhanced electrocatalytic activity of WO_3@NPRGO composite in a hydrogen evolution reaction, Appl. Surf. Sci. 463 (2019) 275–282, https://doi.org/10.1016/j.apsusc.2018.08.227.
[78] X. Zhang, M. Zhang, Y. Tian, J. You, C. Yang, J. Su, et al., In situ synthesis of MoS_2/graphene nanosheets as free-standing and flexible electrode paper for high-efficiency hydrogen evolution reaction, RSC Adv. 8 (2018) 10698–10705, https://doi.org/10.1039/C8RA01226A.
[79] L. Liao, J. Zhu, X. Bian, L. Zhu, M.D. Scanlon, H.H. Girault, et al., MoS_2 formed on mesoporous graphene as a highly active catalyst for hydrogen evolution, Adv. Funct. Mater. 23 (2013) 5326–5333, https://doi.org/10.1002/adfm.201300318.
[80] T. Corrales-Sánchez, J. Ampurdanés, A. Urakawa, MoS_2-based materials as alternative cathode catalyst for PEM electrolysis, Int. J. Hydrogen Energy 39 (2014) 20837–20843, https://doi.org/10.1016/j.ijhydene.2014.08.078.
[81] X. Wang, W. Yuan, Y. Yu, C.M. Li, Synthesis of cobalt phosphide nanoparticles supported on pristine graphene by dynamically self-assembled graphene quantum dots for hydrogen evolution, ChemSusChem 10 (2017) 1014–1021, https://doi.org/10.1002/cssc.201601761.
[82] H. Yan, C. Tian, L. Wang, A. Wu, M. Meng, L. Zhao, et al., Phosphorus-modified tungsten nitride/reduced graphene oxide as a high-performance, non-noble-metal electrocatalyst for the hydrogen evolution reaction, Angew. Chem. Int. Ed. 54 (2015) 6325–6329, https://doi.org/10.1002/anie.201501419.
[83] X. Zhang, Z. Yang, X. Yang, Y. Wang, Z. Lu, Engineering the activity of CoN_x-graphene for hydrogen evolution, Int. J. Hydrogen Energy 43 (2018) 20573–20579, https://doi.org/10.1016/j.ijhydene.2018.09.043.
[84] M. Liu, Z. Ji, X. Shen, H. Zhou, J. Zhu, X. Xie, et al., An electrocatalyst for a hydrogen evolution reaction in an alkaline medium: three-dimensional graphene supported CeO_2 hollow microspheres, Eur. J. Inorg. Chem. 2018 (2018) 3952–3959, https://doi.org/10.1002/ejic.201800757.

[85] Y. Huang, H. Fan, Z. Chen, C. Gu, M. Sun, H. Wang, et al., The effect of graphene for the hydrogen evolution reaction in alkaline medium, Int. J. Hydrogen Energy 41 (2016) 3786–3793, https://doi.org/10.1016/j.ijhydene.2015.12.113.
[86] Flórez-Montaño J, Rodríguez JL, Pastor E. Ni and NiMo nanostructures supported on reduced graphene oxide with enhanced activity towards the hydrogen evolution reaction (In preparation).
[87] M.A. Khan, H. Zhao, W. Zou, Z. Chen, W. Cao, J. Fang, et al., Recent progresses in electrocatalysts for water electrolysis, Electrochem. Energy Rev. 1 (2018) 483–530, https://doi.org/10.1007/s41918-018-0014-z.
[88] D.-J. Liu, L. Chong, H. Wang, G. Wu, H. Xu, PGM-free OER catalysts for PEM electrolyzer, in: DOE Hydrogen and Fuel Cells Program Annual Merit Review and Peer Evaluation Meeting, June 2017.
[89] E. Umeshbabu, G. Rajeshkhanna, P. Justin, G.R. Rao, $NiCo_2O_4$/rGO hybrid nanostructures for efficient electrocatalytic oxygen evolution, J. Solid State Electrochem. 20 (2016) 2725–2736, https://doi.org/10.1007/s10008-016-3278-4.
[90] X. Long, J. Li, S. Xiao, K. Yan, Z. Wang, H. Chen, et al., A strongly coupled graphene and FeNi double hydroxide hybrid as an excellent electrocatalyst for the oxygen evolution reaction, Angew. Chem. Int. Ed. 53 (2014) 7584–7588, https://doi.org/10.1002/anie.201402822.
[91] Y. Yang, Z. Lin, S. Gao, J. Su, Z. Lun, G. Xia, et al., Tuning electronic structures of nonprecious ternary alloys encapsulated in graphene layers for optimizing overall water splitting activity, ACS Catal. 7 (2017) 469–479, https://doi.org/10.1021/acscatal.6b02573.
[92] M. Wang, M. Lin, J. Li, L. Huang, Z. Zhuang, C. Lin, et al., Metal–organic framework derived carbon-confined Ni_2P nanocrystals supported on graphene for an efficient oxygen evolution reaction, Chem. Commun. 53 (2017) 8372–8375, https://doi.org/10.1039/C7CC03558F.
[93] M. Leng, X. Huang, W. Xiao, J. Ding, B. Liu, Y. Du, et al., Enhanced oxygen evolution reaction by Co-O-C bonds in rationally designed Co_3O_4/graphene nanocomposites, Nano Energy 33 (2017) 445–452, https://doi.org/10.1016/j.nanoen.2017.01.061.

Index

Note: Page numbers followed by *f* indicate figures, *t* indicate tables, and *s* indicate schemes.

D

FSC
www.fsc.org
MIX
Papier | Fördert
gute Waldnutzung
FSC® C083411

Kolibri 360 GmbH
Ferdinand-Jühlke-Straße 7
99095 Erfurt, Deutschland
produktsicherheit@kolibri360.de